Lecture Notes in Energy

Volume 22

For further volumes:
http://www.springer.com/series/8874

Nazim Muradov

Liberating Energy from Carbon: Introduction to Decarbonization

Springer

Nazim Muradov
Florida Solar Energy Center
University of Central Florida
Cocoa, FL, USA

ISSN 2195-1284 ISSN 2195-1292 (electronic)
ISBN 978-1-4939-0544-7 ISBN 978-1-4939-0545-4 (eBook)
DOI 10.1007/978-1-4939-0545-4
Springer New York Heidelberg Dordrecht London

Library of Congress Control Number: 2014934293

Printed on acid-free paper

Springer is part of Springer Science+Business Media (www.springer.com)

To my wife Pervin and children Esther and Orhan whose love and support made this book possible.

Preface

Carbon is the basis of life on our planet (and, possibly, in the Universe). We owe our good fortune to carbon-bearing fossil fuels (carbon fuels) that powered the Industrial Revolution and brought about the unprecedented standard of living we currently enjoy. Global economy runs on energy, and energy runs on carbon fuels: virtually all goods and services require their input, and, as the demand for these goods and services keeps growing, so does the amount of carbon fuels consumed. Because of the most critical role carbon fuels played and continue playing in the making and sustaining of our industrial civilization, the latter is often called *Carbon Civilization*.

Just a few years ago, a prevailing opinion among experts was that the concerns about "oil peak" and looming depletion of oil and gas reserves would drive their prices so high that switching from fossil fuels to alternative energy sources would become inevitable in the near future. That judgment has proved illusive. Thanks to technological innovations, crude oil and gas production is now growing in many countries; the world seems to be leaving behind the worst fears about resource scarcity and moving toward the new opportunities presented by the potential resource abundance.

Because of the incredible convenience of carbon fuels, our entire way of life is physically constructed around them; we became heavily addicted to carbon fuels and invested enormous resources in their infrastructure that proved extremely profitable. It is clear that neither developed nor developing countries are willing or will be able to break the fossil fuel addiction anytime soon, because for a significant part of our planet's population burning more and more fossil fuels is the only way of getting out of energy poverty and improving their standard of living.

The Secretary General of the Organization of Economic Cooperation and Development (OECD), Angel Gurria, recently warned about the powerful "carbon entanglement" factor, which will make the introduction of alternative energy sources to the market extremely difficult as they will be "swimming against very strong tides." The carbon entanglement paradigm is the primary reason for the very slow and modest progress of carbon mitigation and climate change policies over the last decades. All the ongoing trends and energy scenarios (even "optimistic" ones) point to the world economy moving along the carbon-intensive path until at least the middle of the century and it could be much longer.

Recently, carbon fuels received a lot of negative publicity; it is impossible to pick up a newspaper or watch TV news without being reminded that they are responsible for many troubles on our planet: air pollution and health problems, oil spills and catastrophic explosions, acid rains and the disappearance of biospecies, changes in climate patterns, and other ecological cataclysms. At no point in history, have we come more close to the realization of potential risks of the high-carbon economic model. At this junction, society has to make important choices with regard to the present and future role of carbon fuels. Success will depend on a significant decoupling of energy use from economic activity, which would require changes in economic structure, technology development, and individual behavior. If we learned how to extract energy from carbon fuels without harming our environment, that would have solved many problems and opened the path to a cleaner and brighter energy future. But, is it technically feasible, and, if so, could it be done within a reasonable time frame and cost?

This book attempts to answer these and many other questions with regard to the future role of carbon fuels in the carbon-constrained world. The major tenor of this book is about decoupling energy from carbon through an approach called "decarbonization," which aims at eliminating or drastically reducing the amount of carbon dioxide (CO_2) emitted from the use of primary fossil fuel resources. Many experts believe that the inclusion of fossil fuel decarbonization in the portfolio of carbon mitigation options would greatly facilitate achieving "safe" atmospheric CO_2 stabilization goals, and it may potentially extend the fossil fuel era by perhaps 100 years (purportedly) without an adverse impact on our planet's ecosystems and inhabitants. However, opponents of this approach are concerned that it could provide only a temporary relief, and would make humankind even more dependent on fossil fuels, thus making the necessary changes later even more difficult.

In a broader context of decarbonizing the fossil fuel-based economy, this book examines three main decarbonization strategies: (1) *carbon reduction* (through energy efficiency improvements and energy conservation), (2) *carbon rejection* (through carbon capture and storage (CCS)), and (3) *carbon abandonment* (through switching to zero-carbon energy sources and fuels, such as nuclear, renewables, hydrogen, biofuels). It highlights the current status of science and technology as well as economic, environmental, societal, and commercial development aspects of the decarbonization concept.

The second major motif of this book is CO_2. Many people look at CO_2 only in negative light and consider it an unfortunate by-product of our techno-civilization and a noxious gas creating lots of problems: from asphyxiation to climate change. In May 2013, the researchers monitoring atmospheric CO_2 concentration at Mauna Loa Observatory in Hawaii reported that for the first time since humans became humans, the CO_2 concentration in the air reached 400 parts per million (ppm) (or 0.04 vol.%). Some people see this as an alarming and ominous milestone with grave implications for humankind, but for others it is just a number. Many are puzzled: how it is even possible that extra few ppm of CO_2 in air could do such an enormous harm to our habitat.

This volume is a brief handbook of CO_2—from its origins on our planet to its role in making our planet inhabitable, to its function in providing energy and fuels to humans, to its utility as a valuable industrial resource. The state-of-the-art technologies and commercial processes for CO_2 capture, transport and storage, as well as its conversion to value-added products and clean fuels are highlighted in this book. It attempts to prove that CO_2 is not only a cause of problems, but it could be part of the solution by reducing our dependence on petroleum-based fuels and feedstocks.

The book is organized in 11 chapters starting with the introductory Chap. 1 describing a brief history of carbon fuels, their origin, diversity, abundance, and crucial role in sustaining our well-being in the past, present, and future. The chapter examines the main grounds of our addiction to carbon fuels, controversies around "Peak Oil" theory, and a new paradigm of dealing with the "tide" of carbon fuels and coping with their environmental impact. Chapter 2 explores what is so unique about the CO_2 molecule that makes it so essential for humans' survival. Without CO_2 our planet would be too cold and not livable, but it also could become too hot if too much of CO_2 is in the atmosphere. Where is the "sweet spot"? This chapter seeks to address this and other questions by explaining such phenomena as greenhouse effect, radiative forcing, global warming potential, global carbon cycle, and other factors that control the livability of our planet and are linked to unique physicochemical properties of CO_2.

Carbon fuels as the main source of anthropogenic CO_2 emissions is the topic of Chap. 3. It classifies major CO_2 sources by industrial sector, scale of emissions, CO_2 content in flue gases, and geographical distribution; current and future trends in CO_2 emission sources are analyzed. Chapter 4 examines the issue of "acceptable risk" limits of atmospheric CO_2 concentrations in terms of the global mean temperature rise, and analyzes proposed CO_2 stabilization scenarios and roadmaps. The historical trends in carbon intensity of energy and the current status of decarbonization of global economy using the Kaya Identity (KI) modeling tool are examined in Chap. 5. The KI analysis shows that dramatic reductions in the energy and carbon intensities of world economy would be necessary to stop and reverse current recarbonizing trends, and it points to prospective carbon mitigation options helping reach that goal. Chapter 6 seeks to address the question, are there alternative carbon-neutral solutions ready to replace incumbent fossil fuel technologies without major perturbations affecting the health of the world economy? The chapter puts nuclear and renewable energy technologies in spotlight as the major decarbonizing techniques.

The role of CCS as a critical component of the portfolio of carbon mitigation options is addressed in detail in Chap. 7. The state-of-the-art technologies encompassing all three major stages of the CCS chain—CO_2 capture, transport and storage, as well as economic, environmental, and societal aspects of the large-scale CCS deployment—are examined in this chapter. Chapter 8 focuses on the range of strategies and pathways to transitioning from high-carbon to low- and zero-carbon energy carriers and fuels. The increasing role of the integrated electricity, methane, and hydrogen grids in the decarbonization of the global energy system is emphasized.

Carbon capture and utilization (CCU) as an important carbon abatement option is highlighted in Chap. 9. Existing and emerging CO_2 utilization technologies are analyzed in terms of their technological maturity, environmental impact, potential revenue generation, and carbon mitigation potential. Chapter 10 identifies the opportunities for carbon-negative technologies such as bioenergy coupled with CCS (Bio-CCS), biochar production, and removal of CO_2 from atmosphere (air capture). Chapter 11 is concerned with the range of radical geoengineering strategies aiming at reducing CO_2 levels in the atmosphere. The current status of major geoengineering projects, their economic feasibility, technical challenges, and risks associated with the global deployment of the technology are analyzed in this chapter.

The uniqueness of this book is that it takes a holistic approach to carbon fuels by tracking a complete transformation chain from fossil fuel sources to the fuel's end-use efficiency, to CCS, and, finally, to CO_2 industrial utilization. This approach allows comparison of different technological options from a "cradle-to-grave" viewpoint, thus providing better understanding of the challenges of transition from carbon-intensive to low-to-zero-carbon technologies. Being aware of the complexity and still-unknown factors behind climate change science, and taking into consideration the divergence of opinions and viewpoints on the role of nuclear energy, carbon storage, and geoengineering, the author tried to present a balanced view of the subject providing a podium to both sides of the debate.

This book is intended for a broad readership. Newcomers and nonexperts may find it a thorough introduction to the field of decarbonization of fossil fuels and CO_2 technologies (to help them, excessive technical details and jargon are mostly avoided in this book). At the same time, the book presents a large amount of up-to-date technical information and analysis that experts may find useful in their work. In general, the book will be handy to all scientists, engineers, and students working and studying in practically all areas of energy technology and alternative energy sources and fuels, and it will be a good supplement to textbooks on environmental technology, CCS, renewable energy sources, and alternative fuels.

This sourcebook provides a comprehensive overview of decarbonization and CO_2 utilization technologies that will play an increasingly important role in the near-to-mid term future in response to the ecological challenges of the carbon-intensive economy. It tries to answer a simple but vital question: will we be able to continue to rely on carbon fuels and live in harmony with the environment against a backdrop of an ever-growing demand for energy? The author hopes this book will contribute to an improved understanding and appreciation of the unique role carbon fuels and CO_2 play in today's life and will do so in the decarbonized energy future. The author also hopes this book will help the readers recognize the scope of problems and available options in order to make educated choices and set priorities with regard to adjusting to new realities of the carbon-constrained world.

Orlando, FL, USA Nazim Muradov

About the Author

Nazim Muradov is a research professor at the University of Central Florida (UCF)—Florida Solar Energy Center. He holds a Doctor of Science degree in Physical Chemistry, Ph.D. in Kinetics and Catalysis and M.S./B.S. in Petrochemical Engineering. Dr. Muradov's main areas of research include thermocatalytic and photocatalytic hydrogen production systems, fossil fuel decarbonization, solar-powered water-splitting cycles, advanced biofuels, radiant detoxification of hazardous wastes and nanostructured carbon materials.

Dr. Muradov is a member of the Board of Directors of the International Association for Hydrogen Energy (IAHE), and a member of the Board of Trustees and the Scientific Council of the Madrid Institute for Advanced Studies, IMDEA Energia (Spain). He is an Associate Editor of the *International Journal of Hydrogen Energy* since 2006.

Dr. Muradov has authored and co-authored two books, several book-chapters, close to 100 peer-reviewed papers and over 50 patents. Dr. Muradov is a recipient of the UCF awards for excellence in research (1996, 2003, 2012). In 2010, he was granted the honorary title of the *IAHE Fellow*.

Contents

Abbreviations

μm	Micrometer (10^{-6} m)
2DS	2 °C Scenario
450S	450 ppm Scenario
4DS	4 °C Scenario
6DS	6 °C Scenario
AC	Air conditioning
AF	Airborne fraction
AFC	Alkaline fuel cell
AGT	Aeroderivative gas turbines
ALM	Asset lifecycle model
ANG	Adsorbed natural gas
ASU	Air separation unit
AZEP	Advanced zero-emissions power plant
B&W	Babcock & Wilcox
BAT	Best available technology
BAUS	Business-as-usual scenario
BB	Billion barrels
BC	Black carbon
BCM	Billion (10^9) cubic meters
BEV	Battery electric vehicle
Bio-CCS	Bioenergy coupled with carbon capture and storage
BrC	Brown carbon
BTL	Biomass-to-liquid
CAC	Capture of atmospheric CO_2
CAES	Compressed air energy storage
CBM	Coal bed methane
CCA	Cost of CO_2 avoided
CCC	Cost of CO_2 captured
CCGT	Combined cycle gas turbine
CCS	Carbon capture and storage
CCU	Carbon capture and utilization

CDM	Clean development mechanism
CE	Cellulosic ethanol
CEM	Clean Energy Ministerial
CEMS	Cluster energy management system
CFB	Circulating fluidized bed
CFC	Chlorofluorocarbons
CFL	Compact fluorescent lamp
CHP	Combined heat and power
CLC	Chemical looping combustion
CNG	Compressed natural gas
CO_2-equiv.	CO_2 equivalent
CPG	Carbon dioxide plume geothermal technology
CrCC	Cryogenic carbon capture
CSP	Concentrating solar power
CTL	Coal-to-liquid
DCFC	Direct carbon fuel cell
DIC	Dissolved inorganic carbon
DOGR	Depleted oil and gas reservoirs
DOPB	Dynamic operation of packed beds
DSF	Deep saline formations
DT	Decarbonization Triangle concept
EBS	Environmentally "benign" sequestration
EBTP	European Biofuels Technology Platform
E_{cap}	Energy consumption per capita
ECBM	Enhanced coal bed methane
EGS	Enhanced geothermal systems
EIA	Energy Information Administration (USA)
EJ	Exajoule (10^{18} J)
ENEA	Agency for Energy and New Technologies and Environment (Italy)
EOR	Enhanced oil recovery
EPA	Environmental Protection Agency (USA)
ESAS	East Siberian Arctic Shelf
ESCII	Energy Sector Carbon Intensity Index
ETP	Energy Technology Perspectives report
ETS	Emissions Trading System
EU	European Union
FAME	Fatty acid methyl ester
FBG	Fluidized bed gasification
FC	Fuel cell
FCCC	Fuel cell combined cycle
FCEV	Fuel cell electric vehicle
FIREX	Fast ignition realization experiment
FOAK	First-of-a-kind
FT	Fischer–Tropsch
FTE	Fuel-to-electricity

G20	Group of Twenty
GAC	Gas-powered air conditioning
GAO	General Accountability Office (USA)
GDP	Gross domestic product
gge	Gallon gasoline equivalent
GGM	Greenhouse Gas Management
GHG	Greenhouse gas
GT	Gas turbine
Gt	Gigaton (10^9 t)
GtC	Gigaton carbon
$GtCO_2$	Gigaton carbon dioxide
GTL	Gas-to-liquid
GW	Gigawatt (10^9 W)
GWP	Global warming potential
H/C	Hydrogen to carbon atomic ratio in fossil fuels
HC	Halocarbons
HDI	Human Development Index
HDV	Heavy-duty vehicles
HEV	Hybrid electric vehicle
HF	Hydraulic fracturing
HHV	Higher heating value
HICE	Hydrogen internal combustion engines
HT	Hydrogen turbine
HTFC	High temperature fuel cell
HTL	Hydrothermal liquefaction
ICE	Internal combustion engine
IEA	International Energy Agency
IFCGT	Integrated fuel cell and gas turbine
IGCC	Integrated gasification combined cycle
IIASA	International Institute for Applied Systems Analysis
IL	Ionic liquids
IPCC	Intergovernmental Panel on Climate Change
ITER	International Thermonuclear Experimental Reactor
ITM	Ion-transport membrane
KI	Kaya Identity
kW	Kilowatt (10^3 W)
kWh	Kilowatt-hour
LANL	Los Alamos National Laboratory (USA)
L-CO_2	Liquefied CO_2
LCOE	Levelized cost of electricity
LDV	Light-duty vehicles
LEED	Leadership in Energy and Environmental Design
Lge	Liters gasoline equivalent
LHF	Liquid hydrocarbon fuel
LHV	Lower heating value

L-NG	Liquid natural gas
LSIP	Large-scale integrated CCS projects
LUCF	Land-use change and forestry
MBD	Million barrels per day
MCFC	Molten carbonate fuel cell
MEA	Monoethanol amine
MHI	Mitsubishi Heavy Industries, Ltd. (Japan)
MIECM	Mixed ionic-electronic conducting membrane
MIT	Massachusetts Institute of Technology
MJ	Megajoule (10^6 J)
MOF	Metal organic frameworks
MOSES	Model of Short-Term Energy Security
MPa	Megapascal (10^6 Pa)
MSW	Municipal solid waste
Mt	Megaton (10^6 t)
MTG	Microturbine generators
Mtoe	Million ton oil equivalent
MW	Megawatt (10^6 W)
MWh	Megawatt-hour
NETL	National Energy Technology Laboratory (USA)
NG	Natural gas
NGCC	Natural gas combined cycle
NGL	Natural gas liquids
NGV	Natural gas vehicle
NIF	National Ignition Facility (USA)
NOAA	National Oceanic and Atmospheric Administration
NOAK	*n*th of a kind
NOM	Natural organic matter
NO_x	Nitrogen oxides (mixture of NO and NO_2)
NPP	Net primary productivity
NRC	National Research Council (USA)
NYMBY	Not-in-my-backyard
OECD	Organization for Economic Cooperation and Development
OFC	Oxyfuel combustion
OFS	Oxyfuel system
OPEC	Organization of Petroleum Exporting Countries
OTEC	Ocean thermal energy conversion
PAFC	Phosphoric acid fuel cell
PAN	Peroxyacetylnitrate
PBTE	Peak brake thermal efficiency
PDU	Process development unit
PEC	Polyethylene carbonate
PEF	Petroleum-equivalent fuel
PEM	Polymer electrolyte membrane
PETM	Paleocene–Eocene Thermal Maximum

PFC	Perfluorocompounds (include CHF_3, NF_3, and SF_6)
Pg	Petagram (10^{15} g)
PHEV	Plug-in hybrid electric vehicle
PM	Particulate matter
PNNL	Pacific Northwest National Laboratory (USA)
Post-CCC	Post-combustion carbon capture
ppb	Parts per billion
PPC	Polypropylene carbonate
ppm	Parts per million
Pre-CCC	Pre-combustion carbon capture
PSA	Pressure swing adsorption
PV	Photovoltaic
R&D	Research and development
R/P	Reserve-to-production ratio
RE	Renewable electricity
RF	Radiative forcing
RFS	Renewable fuel standard
RS	Reference scenario
SC	Supercritical
SMR	Steam methane reforming
SNG	Substitute (synthetic) natural gas
SOFC	Solid oxide fuel cell
SO_x	Sulfur oxides (mixture of SO_2 and SO_3)
SRM	Solar Radiation Management
ST	Steam turbine
SW	"Stabilization Wedges" concept
TAG	Triacylglycerides
TCC	Triple combined cycle
TCD	Thermocatalytic decomposition
TCM	Trillion (10^{12}) cubic meters
TE	Transesterification
TEG	Triethylene glycol
Tg	Teragram (10^{12} g)
TMI	Three Mile Island (USA)
toe	Ton oil equivalent (toe equals to 42 GJ)
TPES	Total primary energy supply
TRL	Technology readiness level
TW	Terawatt (10^{12} W)
TWC	Three-way catalyst
TWh	Terawatt-hour
UBH	Unburned hydrocarbons
UN	United Nations
UNCLOS	UN Convention on the Law of the Sea
UNDP	United Nations Development Programme
UNEP	United Nations Environment Programme
UNFCCC	United Nations Framework Convention on Climate Change

USC	Ultra-supercritical
USGS	United States Geological Survey
US-NRC	US Nuclear Regulatory Commission
UV	Ultraviolet
UYB	Urea yield boosting
VHTR	Very high temperature reactor
VOC	Volatile organic compound
WEO	World Energy Outlook
WGS	Water gas shift
WTO	World Trade Organization

Chapter 1
Introduction to Carbon Civilization

Abstract Carbon is the basis of life on our planet: starting from the discovery of fire, our civilization vitally depends on carbon for its energy and livelihood; for this reason, it is often called "*Carbon Civilization.*" Our entire way of life is physically constructed around carbon fuels, and this "carbon entanglement" factor is the primary reason for the very slow and modest progress of carbon mitigation policies over the last couple of decades. There are clear indications, however, that the high-carbon economical model may face serious challenges; the continued heavy reliance on a narrow set of conventional fossil fuel-based technologies is a significant threat to energy security, stable economic growth, and the environment. A brief history of carbon fuels, their origin, diversity, abundance, and crucial role in supporting and sustaining humans' well-being in the past, present, and future is discussed in this chapter. It examines the main grounds of our addiction to carbon fuels, controversies around "peak oil" concept, and new paradigms of dealing with the "tide" of carbon fuels and coping with their environmental impact.

1.1 Earth's Carbon Inventory: Its Origin and Abundance

1.1.1 Earth's Major Carbon Reservoirs

Carbon is the backbone of life on the Earth and, possibly, in the Universe. (According to the *Carbon Chauvinism* hypothesis, due to the unique chemical properties of carbon, life can only exist on the planets where it could be evolved from carbon-based structural units.) Carbon's unique capacity for forming multiple bonds and long-chain molecules (biopolymers) makes life possible; carbon comprises about half the dry weight of most living organisms. Starting from the discovery of fire, our civilization vitally depends on carbon for its livelihood. Carbon-based fossil fuels powered the Industrial Revolution and brought about the rise in the standard of living we currently enjoy. Almost everything we get energy from, whether through

N. Muradov, *Liberating Energy from Carbon: Introduction to Decarbonization*,
Lecture Notes in Energy 22, DOI 10.1007/978-1-4939-0545-4_1,

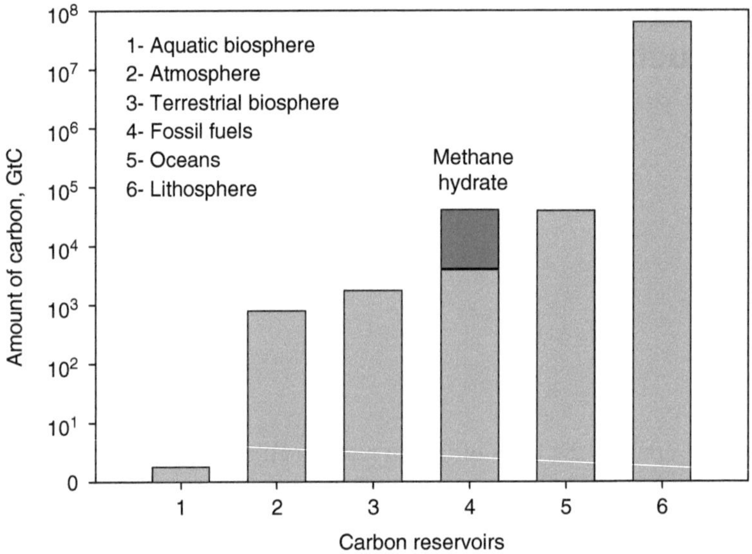

Fig. 1.1 The abundance of major carbon reservoirs on Earth. *Dark gray bar* corresponds to methane hydrate reserves. *Source* [1]

food (carbohydrates) or through fuels at power stations (gas, coal) and transport (gasoline, jet, and diesel fuels), is based on one form of carbon-based compounds or another. For this very reason, our civilization is rightfully called "*Carbon Civilization.*"

Carbon's abundance on our planet is surprisingly low: the lithosphere has only 0.032 wt.% of carbon in all its forms (for comparison, iron's abundance is 5 wt.%). Over geologic timescale, most of the carbon on the Earth became locked up in sedimentary rocks as carbonates and fossil fuels, and significant part of it got dissolved into the oceans as CO_2, carbonate (CO_3^{2-}), and bicarbonate (HCO_3^-) ions. Available data show that the atmospheric CO_2 concentration gradually reached the level of about 0.02–0.03 vol.% and fluctuated within this range for about half a million years.

Carbon is stored on our planet in the following major carbon reservoirs:

- Carbonates and other sedimentary rock deposits in the lithosphere
- Dissolved CO_2 and carbonates in the ocean
- Soil organic matter
- Fossil fuel deposits
- Living and dead organisms in the biosphere
- CO_2 in the atmosphere

Figure 1.1 depicts the relative abundance of the major carbon reservoirs on the Earth.

Inorganic deposits of carbon in the lithosphere in the form of limestone, dolomite, chalk, and other carbonates (representing the most thermodynamically stable form of carbon) constitute the largest reservoir of carbon on our planet. Organic forms of carbon, e.g., carbon in biosphere (plants, living organisms) and soil organic matter (e.g., humus), represent significantly lesser share of the total carbon inventory compared to inorganic forms of carbon. The amount of carbon in the form of carbon-bearing fossil fuels,[1] i.e., coal, oil, natural gas (NG), peat, tar, and bitumen, is estimated at about 5,000 Gt, however, if the potentially recoverable resources of methane hydrates would be factored in this figure would increase by almost one order of magnitude [1] (Gt is gigaton or 10^9 ton).

1.1.2 *Origin of Carbon Fuels: Biotic vs. Abiotic*

It is widely recognized that the occurrence of CO_2 in the early atmosphere and near surface environment was the result of degassing of the Earth's interior: as its surface cooled, the volcanoes released massive amounts of CO_2, steam (H_2O), ammonia (NH_3), and methane (CH_4) [2]. The early primitive life forms started photosynthesizing food, energy, and oxygen (O_2) using sunlight, CO_2, and water:

$$CO_2 + H_2O + \text{sunlight} \rightarrow (CH_2O) + O_2 \quad (1.1)$$

where (CH_2O) refers to a photosynthesis product.

During this early evolution process, CO_2 concentration in the atmosphere was gradually reduced and the concentration of O_2—increased (according to reaction 1.1). Green plants further facilitated the conversion of CO_2 to O_2. Nitrogen (N_2) was built up in the atmosphere partly through the oxidation of NH_3 with O_2, but predominantly from denitrifying bacteria. Atmospheric methane concentration decreased via oxidative pathways (the reaction with O_2). As O_2 levels increased in the atmosphere, the ozone layer was formed, which started to filter out harmful ultraviolet (UV) radiation. This facilitated the evolution of living organisms and species first in the shallow seas and later throughout the Earth. Buried under thick layers of rocks, the remains of marine organisms, swamp plants, and incompletely decayed plant matter exposed to high pressures and temperatures were transformed to fossil fuels: coal and hydrocarbons (liquid and gaseous), through an anaerobic decomposition process over the geological time scale of hundreds of millions of years (according to some estimates, about 650 million years). Because of a nonuniformity of the "feedstock" and different conditions of the transformation process, globally, no two coals or oils or gases have the same chemical composition.

According to this theory, from the historical perspectives, all types of carbon-bearing fuels, including biomass and fossil fuels, have been originated from

[1] In this book, fossil fuels are interchangeably called "carbon fuels" to emphasize the significance of carbon as a basic element of these fuels.

solar-powered photosynthesis of a biological matter (that was eventually converted into different types of fuels as we know them today). These carbonaceous fuels are differentiated based on the timescale required for their formation: from million to hundred million years for coal and hydrocarbons (oil and gas), and from hundreds to thousands of years for peat, and from days to hundreds of years for biomass. On the scale of human lifespan, coal, oil, and gas are defined as *fossil* or *nonrenewable* fuels, whereas biomass and associated biofuels as *renewable* fuels [3]. It should be noted, however, that this classification is arbitrary: for example, peat is considered fossil fuel, although in terms of its formation timescale it overlaps with biomass.

There are, however, competing theories of the carbon fuels origin on the Earth; for example, one of them infers that carbon first arrived on our planet in a reduced form, as found in almost all meteorites, and it was *abiotic* (or *abiogenic*) in origin [4]. The supporters of this theory hypothesize that an early ocean contained a high concentration of photochemically produced complex organic compounds formed under reducing conditions, which led to the formation of a reduced carbon reservoir near Earth's surface. The oxidation of subducted organic rich sediments during upper-mantle magma genesis slowly released CO_2 to the surface environment on a timescale consistent with the rate of oxygenation of the surface environment by photosynthetic cyanobacteria, with the record of carbon isotopes in sedimentary rocks and with the record of carbonate sedimentation. One of the strengths of this hypothesis is that the proposed "reduced carbon reservoir" is a more favorable environment for the emergence of life (compared to an oxidized carbon route via CO_2). This model also provides a suitable explanation of the early methane-enhanced greenhouse effect.

According to other (older) abiotic hypothesis, fossil fuels (e.g., oil) were formed from deep carbon deposits, most likely, during the formation of the Earth. This hypothesis suggests that petroleum originated from carbon-bearing fluids that migrated upward from the mantle, which implies that more oil could exist on our planet than previously estimated. The presence of methane on other planets, e.g., Jupiter, Saturn, Uranus, and Neptune, supports this theory, since this fact is cited as an evidence of the formation of hydrocarbons without the involvement of biological processes. Lately, this theory fell out of favor because it failed to make any useful prediction for the discovery of large oil deposits. However, the abiotic theory still has many supporters, and it cannot be dismissed because the mainstream theory of fossil fuels origin still has to be established conclusively.

1.2 Carbon Fuels: The Backbone of Industrial Civilization

1.2.1 From Biomass to … Biomass

For many millennia, wood was the only source of energy upon which humans relied for their very survival and basic needs, such as the preparation of food and the provision of heat. Advantageously, wood provided the easiest way to acquire and store energy by simply finding, cutting, and collecting it. Due to its abundance, storage of wood in warmer parts of the world was not such a necessity as in colder regions where wood was collected during summer and stored for use during winter months. Our ancestors discovered that wood could be thermally converted into charcoal, which has much higher heating value than wood (30.4 MJ/kg vs. 14.7 MJ/kg), thus, providing more efficient way of storing and producing heat. Fuelwood along with other traditional energy sources (such as draft animals, water, and windmills) dominated primary energy until about the 1870s; at this time the average energy consumption typically did not exceed 0.5 toe per capita per year [5] (*toe* is ton oil equivalent). Although the role of wood as a primary energy source gradually eroded with the emergence of more efficient fossil fuels, due to the availability and conveniences of its storage and utilization, fuelwood is still widely being used as an energy source in many parts of the world. (Currently, wood energy accounts for 3.4 % of the global primary energy supply and 38.9 % of the renewable energy supply [6].)

Recently, there has been a surge in renewed interest in biomass energy for the production of electricity and heat (bioenergy) and transportation fuels (biofuels). The main driving force for this renewed interest can be linked to environmental concerns associated with the excessive use of fossil fuels and energy security matters. The energy aspects of biomass use will be discussed in detail in Chap. 6 of this book.

1.2.2 Veteran Carbon Fuel: Coal

With the beginning of the Industrial Revolution in the mid-eighteenth century, wood started yielding its dominance to more energy-dense fuel—coal (carbon content of coal is in the range of 90–98 %, and its energy content varies in the range of 24–33 MJ/kg). Coals used in industry differ by their rank, i.e., its degree of maturity, which is determined by the stage coal reached during the so-called coalification process: the sequence of transformation processes leading to the formation of coals with increasingly higher energy content [7] is as follows:

$$\text{Peat} \rightarrow \text{lignite}\left(\text{brown coal}\right) \rightarrow \text{sub-bituminous coal} \rightarrow \text{bituminous coal} \rightarrow \text{anthracite}$$

Coal catalyzed the industrialization process through a radical technological innovation: a steam engine. The importance of the coal-powered steam engine was that for the first time fossil energy was converted into work with relatively high efficiency. Second, the steam engine allowed energy supply to be flexible and site independent, because coal could be transported, stored, and used on demand. Third, the steam engines enabled reaching rather high power densities, which spurred their widespread use at power plants, steel-making and other factories, steam ships, and locomotives. The latter enabled the first transport revolution, as railway networks were rabidly expanding and reaching remote locations. At the peak of the "steam age" (mid-nineteenth century), in industrial countries such as England, the average energy consumption levels were about 2 toe per capita per year [5, 8].

The second radical technological innovation after the steam engine was the introduction of electricity. Electricity was the first energy carrier that could be easily converted into work, heat, or light. Coal-derived electricity further revolutionized the industrialization process and led to other innovations such as an electric motor, an electric light bulb, power grid, and others. Most importantly, coal-fueled technological revolution facilitated far reaching societal and economic structural changes: an increased employment, the division of labor, specialization, urbanization, monetization of the economy, local and international trade, etc. By the end of the nineteenth century, coal supplied practically all the primary energy needs of industrialized countries.

The global coal-proven reserves and recoverable resources are estimated at 1,004 and 21,208 billion tons [9]. Currently, coal represents the largest and (in many countries) the least expensive fossil fuel resource currently accounting for around 30 % of the world primary energy demand [10]. Eighty percent of coal is used in power generation and industrial sectors, with a small percentage used in transport (0.5 %) and other sectors. Globally, coal-fired power generation rose by about 6 % from 2010 to 2012, and it continues to grow faster than non-fossil energy sources on an absolute basis [11]. Approximately half of all coal-fired power plants built in 2011 use inefficient technologies, which offsets the measures to close older, inefficient plants. In 2011, China closed 85 GW, and the USA closed 9 GW of capacity in 2012 [11]. China's and India's coal consumption represented 46 and 11 % of global coal demand in 2011 [11].

Although coal is still the global backbone fuel for electricity, its global share of total generation, according to International Energy Agency's (IEA) 2012 World Energy Outlook (WEO) report, is projected to decline from 41 % in 2010 to 33 % in 2035 [12]. In the USA, coal's share of electricity generation will drop from nearly 50 % today to 39 % in 2035 [13]. The decline in coal's use will be due to a combination of several factors, such as slower growth in electricity demand, competition from NG and renewables, and stricter environmental regulations. Although the decline will continue in the USA and European Union (EU), coal will remain the dominant fuel in many developing countries. Close to 1,200 new coal-fired facilities with a total installed capacity of 1,401 GW spread across 59 countries were proposed as of July 2012 [14]. China and India account for more than three-fourths (76 %) of the globally proposed coal-fired power generation capacities. Below is the list of the ten countries—global leaders in the coal-fired power generation (the total proposed capacity is in MW) [14]:

China	557,938
India	519,396
Russia	48,000
Turkey	36,719
Vietnam	34,725
S. Africa	22,633
USA	20,236
Ukraine	14,000
Poland	12,086
Germany	12,060

It has been recommended that, in order to reduce the environmental impact of the increasing coal usage, especially, in developing countries, ultra-supercritical units at coal-fired plants should be installed [15].

1.2.3 Oil: The Greatest Gift of Nature

Oil is less carbon-intensive fuel than coal: its carbon content is in average 85 wt.%. As fuel, oil was introduced to the world at the turn of the twentieth century (though it was known to ancient people in many parts of the world, where crude oil naturally seeped out of the ground). Although the initial discoveries of oil at a commercial scale occurred as early as in the 1850s in Baku (Azerbaijan), Bend (Romania), Oil Springs (Canada), and Titusville (Pennsylvania, USA), oil started entering the world energy market after major oil field discoveries in early 1900s in Texas, Oklahoma, California (USA), and in the 1920s–1940s in the Middle East (Bahrain, Iraq, Iran, Saudi Arabia). The 1960s–1970s witnessed the discovery of major oil fields in Alaska (USA) and North Sea areas (UK, Norway).

Unlike coal, oil is liquid, which makes it more versatile, convenient, easily transportable, and valuable primary fuel in a great variety of applications. It also has the highest gravimetric and volumetric energy content among all fossil fuels: 46 MJ/kg and 37 MJ/L, respectively (on average). Only two of these features make oil an ultimate carbon fuel, surpassing in value all other types of fossil fuels. Oil is the greatest gift given by Nature to humankind. Just to imagine what would have happened if oil never existed on our planet and all that was available to our predecessors were only coal and gas, in all likelihood, people would have spent enormous resources to convert them into more convenient, versatile, and energy-rich liquid hydrocarbon fuels (as some countries were compelled to do that in a response to necessity).

The introduction of oil and oil-derived products to the world market led to another radical technological innovation: an internal combustion engine (ICE), which revolutionized individual, commercial, and public transport through the use of cars, buses, trucks, and first-generation aircrafts. Since mid-twentieth century, oil assumed a dominant role at the energy market as the automotive, petrochemical, and other oil-reliant industries have matured. As more and more oil was discovered and gasoline and diesel fuel driven transportation was rapidly expanding all over the world, oil's share in the world's total final energy consumption steadily grew and reached 33 % in 2011 [10]. Because of its unique properties and value, oil has become the world's strategic commodity: it is produced in a few oil-rich regions, but is shipped all over the world via pipelines, railroads, and marine tankers. No wonder, oil is sometimes called "*blood of industry*."

Crude oil is rarely used as is; so, the first step in its utilization by consumers is its preprocessing and refining at large refineries, which transform crude oil into a variety of products including motor fuels (gasoline, diesel fuel), aviation fuels (jet fuel, kerosene), and heating oil, coke, and feedstocks for petrochemical and chemical industries. Transportation sector has been and remains the major consumer of oil-derived products (consuming more than half of all petroleum products). In the USA, gasoline is the primary transportation fuel (318.5 million gallons per day), followed by diesel fuel (153.1 million gallons per day) and jet fuel (61.8 million gallons per day) [16] (1 US gallon is equal to 3.8 l).

According to IEA, oil demand is projected to grow and reach 105.2 million barrels per day (MBD) in 2030 [17] (1 barrel is equal to ~159 l). The transport sector will remain the main driver of the oil demand increase worldwide accounting for 97 % of the increase in the world oil use until 2030 [17]. Most of the projected increase will be covered by OPEC: its share of the world oil production will increase from the current 44 to 52 % in 2030 (OPEC stands for Organization of Petroleum Exporting Countries). Non-OPEC conventional oil production has already peaked (or is expected to peak in the near future); however, this decline will be offset by the increase in unconventional oil production.

The report from the HIS Cambridge Energy Research Associates states that oil demand in developed countries has probably already peaked and will not exceed the prerecession (2008) levels, mostly, due to the combination of several factors, such as [18, 19]:

- Demographics and socioeconomical changes (vehicle ownership rates in developed countries have already reached a "saturation" level).
- Introduction of more fuel efficient vehicles (by 2016, mileage of cars and light trucks is projected to increase by 42 and 30 %, respectively).
- Introduction of new more energy efficient technologies.

At the same time, the global demand for oil from 2010 to 2020 is projected to increase by almost 14 %, mostly due to developing countries, predominantly, China and India. The report notes that China's fleet will grow from 12 million vehicles in 2005 to 110 million by 2030. The share of developing countries in the global oil demand will increase from 39 % in 1990 to 51 % by 2020, whereas the share of developed countries will drop from 61 % in 1990 to 49 % in 2020.

The current trends show that the new oil reserves that are being exploited are not only more expensive to develop and recover, but the time span between times when the well is drilled and when oil is produced becomes much longer. Currently, it takes longer for oil supply to respond to changes in oil price, which implies that the oil supply is becoming less *elastic* (*Elasticity* is the term used by economists to describe how much supply or demand would respond to changes in price.) [20]. Worldwide, the oil supply is becoming less elastic as new oil supplies come increasingly from hard-to-reach reserves and unconventional oil. For example, Brazil's giant pre-salt fields and deep-water discoveries on the Gulf of Mexico and elsewhere are much more difficult, expensive, and slow to develop compared to past discoveries. Likewise, Canada's tar sands are expensive and slow to develop.

The USA produced 221 million barrels of crude in April 2013, with more than half coming from Texas and the Gulf of Mexico [21]. In March 2013, Texas oil production reached its highest level since 1984. That month, the State pumped more than 74 million barrels of crude from the ground. (If Texas were a country, it would be one of the 15 largest oil producers in the world.)

Unconventional oil plays increasingly important role in the overall supply of liquid fuels to the energy market. Unconventional oil resources include extra-heavy oil, oil sands (tar sands, bituminous sands), shale oil, gas-to-liquids (GTL), and coal-to-liquids (CTL). Canadian (Alberta) oil sands and Venezuela's Orinoco Belt bituminous sands are typical representatives of unconventional oil resources. Oil sands represent a thick mixture of heavy organic matter (bitumen), sand, clay, and water. The estimates of Canadian oil sands reserves vary between 178 billion barrels [17] and one trillion barrels [22]. In the USA, tar sand resources are mostly concentrated in the state of Utah, and their recoverable reserves are somewhat less than that of Canadian tar sands: 12–20 billion barrels of oil [23].

Advantageously, tar sands in Alberta can be recovered by open pit mining technique, which substantially reduces their cost. Nevertheless, the oil recovery from tar sands is an extremely laborious and energy-intensive process: 2 tons of the sand yields only one barrel of oil, and it requires large amounts of steam and water (2–4.5 volumes of water per one volume of oil although most of the water is recycled). Roughly, 75 % of oil (bitumen) can be recovered from the sand. Although the development of tar sands is net energy positive: providing 7–10 units of energy for every unit consumed, this index is substantially lesser than that of conventional oil [24]. After recovery, tar sands require chemical manipulation with heat, pressure, and chemicals to become crude oil that could be further processed to diesel, jet fuels, and other petroleum products. In order to transport the tar sands through a pipeline (e.g., from Canada to the USA), it has to be diluted with light liquid hydrocarbons to become "dilbit" (which stands for "diluted bitumen").

Oil sands have the potential to contribute to global energy security via diversification of oil supply (e.g., it makes the USA less dependent on OPEC's oil). Increasingly higher crude oil prices would stimulate the increase in the output of the Canadian oil sands and other unconventional oil sources (For the Canadian oil sands, the profitability relies on oil prices with the threshold around $75–80 per barrel.) [17]. The global unconventional oil production is projected to increase from 1.8 MBD in 2008 to 7.4 MBD in 2030 [17].

Recently, there have been concerns that oil sands could exact a heavy toll on the environment, and their increased production could move our planet to a disastrous tipping point for climate change [24]. The adverse environmental impact of the oil sands industry is linked to the immense amount of water and fuel resulting in 20 % more CO_2 emissions than conventional oil on a "well-to-wheel" basis [17]. Many experts consider oil sands as one of the dirtiest, most carbon-intensive fuels in par with coal. According to reports, CO_2 pollution from oil sands has risen 36 % since 2007 [24]. In an attempt to limit CO_2 emissions from oil sands, IEA suggested that their production should not exceed 3.3 MBD, and yet approved oil sand production is projected to surpass 5 MBD (NASA's climatologist James Hansen called this move "*game over for climate change*") [25]. Estimates indicate that just from burning Alberta's tar sands alone there will be additional temperature rise of nearly 0.4 °C.

As it stands now, tar sands are part of fossil fuel addiction. In 2011, the industry produced 1.8 million barrels per day of oil resulting in the emission of 47.1 million metric tons of CO_2 (equivalent) into the atmosphere, about 2 % more than the year before, and the production is still growing [25]. There are, however, positive developments in the industry aiming at reducing its carbon footprint. For example, at its facilities, Shell has introduced alternative less carbon-intensive approaches to thermal cracking of bitumen that involve adding hydrogen to the process [25]. Additionally, Shell has recently begun deploying carbon capture and storage (CCS) technology to some of its bitumen upgraders. When completed (in 2015), the project (called "Quest") will capture and store underground about one million metric tons of CO_2 per year [25]. In another recent development, tar sand producers could now face carbon tax; in particular, Alberta province imposes carbon price of $15 per metric ton for any emission above a target of reducing by 12 % the total amount of GHG emitted per total number of barrels produced. Although tangible, this carbon price would unlikely compel tar sand developers to pursue CCS, because to implement the technology it would be necessary to impose a carbon price of about $100 per metric ton or even more. Therefore, any future carbon regulations may adversely affect the competitiveness of the unconventional oil industry.

According to the US Energy Information Administration (EIA), oil and NG production in the USA has jumped 14 % and 10 %, respectively, since 2008 [26]. An oil boom the USA is experiencing now is largely due at least three main reasons:

- Breakthroughs in hydraulic fracturing and horizontal drilling techniques facilitated new oil production in rich oil shale formations in North Dakota (Bakken Shale) and Texas (Eagle Ford)
- High oil prices spurred the record investments by oil companies for new production
- Higher oil prices rendered the production of marginal oil economically viable (made possible by the first two factors)

The US oil production is projected to further increase in the near future. EIA estimates that the country's oil production will grow another 20 % by 2020, and as a result of that and higher fuel efficiency standards, the USA will reduce its share of petroleum imports from 49 % in 2010 to 38 % by 2020 to 36 % in 2035 [13]. A new oil shale formation has been discovered in California: Monterey Shale is estimated to hold 400 billion barrels of oil, according to the US Geological Survey, which is more oil than in North Dakota's Bakken Shale and nearly half the conventional oil in Saudi Arabia [27]. But getting it out will be a challenge considering California's specific geological structure (due to the San Andres fault), which may render the horizontal drilling combined with hydraulic fracturing unpractical here. California-based company Occidental will try to utilize a new technology known as "deep acid injection" to recover shale oil. The technique involves injecting hydrofluoric or other acids deep underground, where they dissolve shale rock and allow the oil to flow. The developers claim that this method is cheaper and less controversial than fracking, since much lesser volumes and pressures are involved.

In the USA, growing oil production coupled with shrinking consumption (due to energy efficiency gains in transportation and industry) has created positive trends that may have not only economic but also geopolitical repercussions.

According to the IEA's WEO-2012 report, North America's oil supply will grow by nearly 4 million barrels per day between 2012 and 2018, amounting to nearly 50 % of the global output growth over that period [28]. By 2020, the USA is projected to be the world's largest oil and gas producer, overtaking Saudi Arabia, and by 2030, the USA will become a net exporter of oil (most of the growth will come from drilling in the Gulf of Mexico and hydraulic fracturing of shale formations) [12, 29]. By that time, the USA will also become self-sufficient in terms of net energy use (today, the US energy imports provide 20 % of the US energy needs) [9].

A rapidly rising demand for oil from developing economies is another important factor in the oil equation: for the first time, developing countries are set to consume more oil that developed countries (IEA estimates that developing countries will hit 54 % of the global total by 2018, up from 49 % in 2012.) [28]. China will lead this move: its demand for oil will rise by 25 % between 2012 and 2018. Developing economies and emerging markets are also heavily investing in oil refining and infrastructure, and they will be responsible for virtually all net crude distillation capacity growth.

The IEA projects that under the right conditions the world could produce increasing amounts of oil through 2035 and potentially meet the world's growing demand for oil [12, 30]. The key here is "under the right conditions," because the main challenge here is that the world will unlikely produce familiar (conventional) crude oil at the rate it did at the peak of crude production (ca. 2005) [30]. In the IEA's 2012 Outlook scenario, increasing world's population accompanied with the rising standards of living in the developing countries would push the oil demand from 87.4 MBD in 2011 to 99.7 MBD in 2035. The scenario projects that to meet this demand oil-producing countries would have to double their production of unconventional oil. In particular, the USA would have to triple its production of tight oil (trapped in nearly impermeable rocks, which would require tens of thousands of new hydro-fractured wells) and bring it to 3.2 MBD in 2020. This will become economically feasible only with continued high oil prices.

Besides tight oil, NG liquids (NGL) will be another major player in the IEA's future liquid fuel supply scenario. Although NGL are mostly a by-product of NG production, typically, they are lumped with crude oil and generalized as "liquids" (NGL include the range of hydrocarbons from the lightest ones that could be liquefied only when pressurized, e.g., C_3H_8, C_4H_{10}, to larger molecules that could exist as liquid at atmospheric pressure and are present in crude oil, e.g., C5–C7 hydrocarbons). Currently, almost half of all NGL are converted into petrochemicals and plastics, with the remainder almost equally split between transportation and fuel usage applications. IEA projects 50 % increase in NGL production by 2035 with significant part of it going to the transportation sector. The experts are concerned that money will be the most uncertain factor in the IEA scenario [30]. IEA predicts the price of oil rising to $125 per barrel in real terms by 2035, which will help fund the maintenance of crude oil production and drive up unconventional oil production and facilitate transportation's shift to NGL-based fuel. For this to happen, OPEC would have to restrain its production, while non-OPEC production would surge to allow prices to rise. It is too early to say if this scenario will be even plausible given geopolitical circumstances on the ground.

1.2.4 *Entering "Golden Age" of Gas*

1.2.4.1 Conventional and Unconventional Gas

As oil was still solidifying its leading role, in the mid-twentieth century, NG emerged as a new player and a major competitor in the energy and industry markets, and it is still gaining momentum. NG is the lightest and cleanest (from carbon emissions viewpoint) form of fossil fuels: its carbon content averages at about 76 wt.%. According to IEA, humankind is now entering the "*Golden Age of Gas*"—a tribute to an ever-increasing role of gas on the global energy arena combined with its environmental advantages over other fossil fuels.

In most cases, NG occurs near crude oil reservoirs forming a gas cap between oil and a capping (impervious) rock. At high pressure, NG is dissolved in oil and is released when oil is pumped out to the surface (more often than not, this gas is combusted at the site forming the so-called gas flares). Being gas, it can also migrate through the porous layers of the Earth's crust and accumulate in locations with favorable temperature and pressure conditions. Drilling gas wells is generally less expensive and faster than drilling for oil, and there are more gas wells than oil wells in some countries, e.g., in the USA [4].

Methane (CH_4) is the main component of NG with its content typically varying in the range of 70–90 vol.%. Other light hydrocarbons (ethane, propane, butane) and (in many cases) CO_2 are also present in NG along with small amounts of N_2, He, H_2S, and water vapor. H_2S (due to its toxicity and chemical aggressiveness) and CO_2 (due to its capacity to lower NG heating value) are the most undesirable components in NG, and they are usually removed from NG before its transport or liquefaction. Liquefied petroleum gas (LPG, mostly consisting of propane and butane) can be recovered during NG processing. NG can be conveniently and economically transported by pipelines, or it can be liquefied, stored, and transported in refrigerated vessels by railroads or designated marine tankers.

NG is widely used in a number of important industrial and residential applications, such as power generation, transportation, industrial and residential heating, and chemical feedstock for production of fertilizers, rubber, plastics. Over the last half century, the world's NG production steadily increased, and in 2011, it accounted for 23.6 % of total global energy consumption [10]. The demand for NG is projected to continue to grow at about 1.5 % per year with new gas-fired power stations using combined-cycle technology accounting for the most of the increase [17]. Today, nearly all projections through the middle of twenty-first century and beyond show the role of gas (especially, unconventional gas) in the global energy supply increasing and that of coal and oil decreasing.

Unconventional gas sources include shale gas, tight sand gas, coal bed methane (CBM), and methane hydrates. In contrast to conventional gas, which is extracted from porous sandstones and carbonate formations, where it has been trapped under impermeable caprocks (seal), unconventional gas is typically recovered from low-permeability reservoirs such as tight sand formations, coal seams, and fine-grained

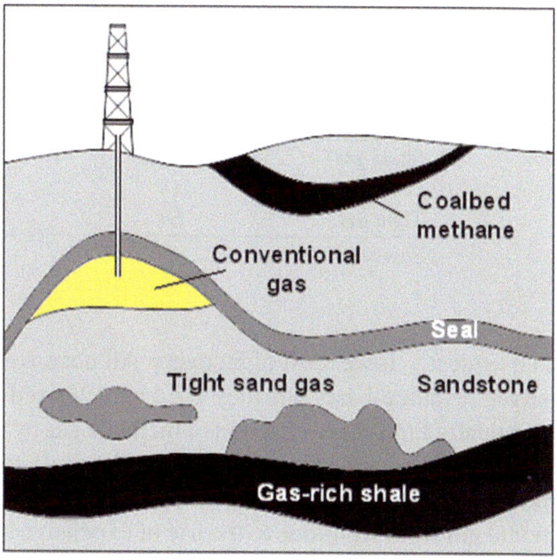

Fig. 1.2 Geologic nature of major sources of NG. *Source* [31]

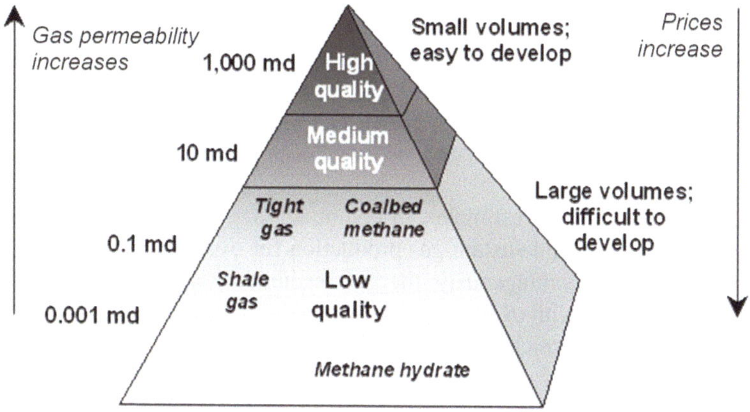

Fig. 1.3 NG resource pyramid. Both conventional and unconventional resources are included. Gas permeability is in millidarcy (md). *Source* [32]

gas-rich shale rocks. The difference between conventional and various types of unconventional gas (except methane hydrates) is depicted in Fig. 1.2.

Typically, shale gas occurs in rocks of Paleozoic and Mesozoic age, whereas tight sand gas could be found in sandstone formations where it gets trapped due to inability to further migrate upward. CBM is formed during the transformation of organic matter to coal over a geological time span. Figure 1.3 depicts NG resource pyramid, which elucidates the interrelation between gas resource, gas permeability, and the cost of gas recovering from sources.

It can be seen that high-quality gas (typically, of conventional type) has the lowest resource base, and it is characterized by high gas permeability and,

Table 1.1 The estimates of ultimately recoverable resources of unconventional gas and production costs

Unconventional gas source	Ultimately recoverable resources (in TCM)	Production cost (in US$/GJ)
Shale gas	204	2.9–6.7
Coal bed methane	118	2.9–7.6
Tight gas	84	2.9–7.6

Source [33]

subsequently, lower cost of recovery. All unconventional gas formations have much larger resource base, but they are characterized by very low permeability which drastically limits the extraction of methane gas by off-the-shelf methods and requires additional (usually, expensive nontraditional) techniques to achieve economical flow rates of gas. As a result, shale gas is much more costly to produce than conventional gas from wells due to the use of expensive hydraulic fracturing and horizontal drilling equipment.

The IEA's 2012 report provides the estimates of NG-proven reserves and recoverable resources as follows (in trillion cubic meters, TCM): 232 (with 28 in OECD and 205 in non-OECD countries) and 790 (with 193 in OECD and remaining in non-OECD countries), respectively [9, 12] (OECD stands for the Organization for Economic Cooperation and Development). The estimates of ultimately recoverable resources of unconventional gas and the range of production costs are presented in Table 1.1.

According to the IEA's estimates, in combination, conventional and unconventional gas resources could sustain gas production for over 250 years at the current consumption rate. Advantageously, the gas resource base is geographically well spread over the globe with every region having at least 75 years worth of gas at current consumption rate, which has very important geopolitical implications. Countries with the largest share of unconventional gas produced by 2035 will be the USA (shale gas), China (CBM and shale), Canada (shale), Australia (CBM), India (CBM and shale), and Russia (tight gas).

The worldwide production of unconventional gas (shale, tight sand, and coal bed) is rapidly picking up the pace. In the USA, the share of shale gas of total gas supply increased from about 1 % in 2000 to 20 % in 2010 and to 30 % in 2011 [34–36]. According to the US EIA 2011 report, technically recoverable shale gas resources of onshore lower 48 states amount to a total of 21.2 TCM of gas with the largest shares in the Northeast (63 %), Gulf Coast (13 %), and Southwest (10 %), respectively [37]. The largest shale gas plays are in Marcellus (11.6 TCM, or 55 % of the total), followed by Haynesville (2.1 TCM, or 10 % of the total), and Barnett (1.2 TCM, 6 % of the total). Since 2005, NG prices in the USA are lower than that of crude oil. Besides the USA, shale gas development is rapidly increasing across several regions of the world such as the UK, China, Poland, Ukraine, Australia, and Brazil [38]. The US EIA reported that China has the world's largest shale-gas reserves estimated at 36 TCM [39].

According to the IEA's GAS Scenario projections, global primary NG demand will increase from 3.3 TCM in 2010 to 5.1 TCM in 2035 (about 50 % increase) and will account for 25 % of the world's energy mix (overtaking coal between 2025 and 2030) [33]. The estimates of the increase in gas production in the USA vary from 20 % (according to EIA) to 100 % (according to the US Geological Survey) [40]. Some optimistic analysts project that shale gas will fuel the USA for the next century, but more cautious estimates predict that the US and Canadian gas production will likely peak sometime between 2020 and 2040 [34]. Nevertheless, it is evident that the gas reserves are much greater than previously thought, and taking into account cleanness of gas compared to coal and oil, many experts believe that it has enormous potential to provide economic and environmental benefits for the society. But, on the other hand, there are concerns that the large-scale development of shale gas could cause different types of environmental problems (see detailed discussion in Chap. 8).

1.2.4.2 "Shale Revolution" and Its Implications

Recent surge in the supply of unconventional gas, mostly shale gas, is reshaping the energy landscapes of gas-rich countries, revitalizing their economies, and impacting long-term geopolitical interests around the world. Called "*Shale Revolution*," it is substantially altering the US energy mix (increasing gas share to 32 % at the expense of coal that dropped to 34 %, in 2012), reducing dependence on oil imports (from 60 to 42 %) and reducing GHG emissions (1.7 % in 2011) [41]. It is also spurring manufacturing in downstream industries: petrochemical, chemical, metallurgical, and other energy-intensive industries. At the same time, low gas prices are impacting the economics of renewables (wind and solar) and, especially, nuclear. But there are also cautious attitudes, e.g., are the gas resources substantial enough to warrant serious investments in converting transport from gasoline to gas for fleets and private vehicles? How availability of relatively cheap gas will affect the prospects of carbon sequestration technology, and, in the long term, carbon mitigation policies? These questions will need to be answered as part of the energy policy adjustments to the new energy landscape.

Shale gas boom in the USA is having impact not only on renewables and nuclear sectors, but, quite surprisingly (and unexpectedly), it is promoting coal usage by European utilities, despite the EU's environmental policies dedicated to curbing the share of coal in their energy mix (the EU environmental policy calls for a 20 % reduction in carbon emissions[2] from 1990 levels by 2020 via growing role of renewables in electricity generation) [42]. While North America's surge in shale gas production pushed down NG prices to decade's lows, prompting power plants to switch from coal to gas, unwanted at home American coal has increasingly found its way to European markets, where it displaced more expensive gas. This trend shows how disruptive could shale gas become for traditional industries such as power

[2] The terms "carbon emissions" and 'CO_2 emissions" are interchangeably used in this book.

generation, leading to unforeseen (and often even perverse) outcomes across the global energy system. The US coal exports to Europe increased by 29 % in 2012 (against the backdrop of a reduced Chinese demand), dropping European coal prices from $130 per ton in March 2011 down to $86 per ton in February 2013 [42].

This trend was exacerbated by a sharp fall in the price of carbon allowances under the EU's flagship Emission Trading System (ETS) and rising European prices for NG. Prices for permits in the EU's ETS were at about €4 ($5.35) per metric ton in January 2013 (Note that in 2005, UN study estimated that penalties for emitting CO_2 would have to be at least $25–$30 per metric ton to make it work.) [43]. As a result, gas-fired electricity output in Germany fell 16 % in 2012, while coal-fired plants added output by the same amount. In the EU, in 2011, gas-fired generation fell by 17 % and coal-fired generation increased by 11 % [14]. Despite the recent increase in coal usage in Europe, many experts believe that this trend is short-lived and temporary, and is just a matter of the economics of the current energy market. IEA projects that the trend of European demand for coal is close to peaking, and by 2017 it will drop to levels close to those in 2011.

Besides economical, there are quite serious geopolitical ramifications of the Shale Revolution. For example, because the USA can now sell gas at 75 % below what Russian Gazprom charges East European customers, Gazprom has been forced to lower gas prices sold to Europe, and it is being investigated by the European Commission for price fixing [41]. Because of cheaper gas prices, the Gazprom's market value in 2012 dropped threefold compared to 2008, and some gas projects in the Arctic have been canceled. Analysts now question the Gazprom's future as Russia's veritable cash cow [41].

1.2.4.3 Methane Hydrates: Ocean of Energy Under the Ocean

Naturally occurring methane hydrates (or gas hydrates) are another form of the unconventional gas resource. Much less developed than shale gas or coal bed methane, methane hydrates represent an additional enormous source of methane fuel. Vast resources of methane hydrates exist in subsurface sediments in permafrost and in deep oceans widely scattered around the world, and it is believed that upon a successful development, they might become a major source of energy for the foreseeable future. The estimates of methane hydrate resources vary widely, from thousands to millions of TCM, which are several orders of magnitude greater than estimated conventional gas resources [44, 45]. For example, the estimate of the National Resources Canada put the global amount of methane in gas hydrates at about 10^4 Gt of carbon, which is about twice the amount of carbon held in all fossil fuels on the Earth [46]. The US Geological Survey estimates that there is about 2.4 TCM of technically recoverable gas hydrates in Northern Alaska alone [47]. In Canada, Mackenzie River and Beaufort Sea regions contain some of the most concentrated deposits of gas hydrates in the world with estimated resources of 8.8–10.2 TCM [48]. Other countries, Russia, India, Japan, and China, also have substantial marine gas hydrate deposits.

Most of the methane hydrate resources is considered not commercially recoverable with present-day technologies. The USA and Canada are leading the worldwide efforts on the commercial development of methane hydrates. The Mallik Gas Hydrate Research Well Project tested a major gas hydrate accumulation and conducted the first modern production test of methane hydrates [49]. The initial success of the project led to the formation of another international consortium that started in 2001–2002 and involved the drilling of a 1,200-m deep main production research well and two nearby observation wells. Another methane hydrate project, Integrated Ocean Drilling Program Expedition 311, started drilling gas hydrate cores offshore Vancouver Island during the fall of 2005 [46]. In its 2010 report, the US National Research Council (NRC) underscored an increasing confidence that, although a number of technical challenges still remain, none of them are insurmountable, and commercial production of methane fuel from hydrates could start in the USA by 2025 [50].

1.3 Why Are We So Addicted to Carbon Fuels?

Even after recent price increases for oil and other primary energy sources, fossil fuels remain extremely cheap. How do we know they are cheap? Here is one way of estimating the relative value of fuel. A physically strong man, working at the peak of his efficiency, would be able to sustain a power output of about 0.8 kW for several hours a day doing a low-skill physical job [51]. Assuming the wage of $10/h for this muscular worker, the cost of his labor would be $8/kWh. This is about 40–80 times more expensive than the cost of electricity (at $0.10–0.20/kWh[3]) and about 125–250 times more expensive than the cost of gasoline (at about US$1.00–2.00/l). The price of crude oil could (hypothetically) increase one order of magnitude, and it still would be astonishingly good value for money.

One of the main reasons for our addiction to oil and other forms of fossil energy is that they are so convenient and cheap that humans physically constructed their entire way of living around them and unlikely to be willing to replace them (whatever alternatives could be around). It is also clear that as the standard of living in the developing world is increasing, more and more people would tend to replace their labor with rather affordable electricity and gasoline, resulting in ever-increasing consumption of fossil energy. In the agriculture and industrial sectors, labor has been (and is being) replaced by sophisticated machinery and cheap fossil-derived energy. In air-travel, the jet-fuel cost remains a small portion of an airline ticket; for example, EasyJet (a large budget airline) had a fuel bill of only £7.50 per passenger at the average price of a ticket of £48 (about 15 %), in 2005 [51]. Cheap aviation, on the other hand, makes overseas holidays and a fresh produce all-year-round affordable to an increasing share of the world's population, thus, delivering higher

[3] In this book, kWh corresponds to electric kWh (or kWh_{el}), unless otherwise indicated.

standard of leaving, of course, at the expense of the increased use of fossil fuels. Keeping carbon fuels cheap allows industry to grow and humankind to prosper.

Due to availability and affordability of fossil fuels, the underlying demand on them is astonishingly unresponsive to price changes; from 2001 to predepression year of 2008, the price of crude oil quadrupled from $23.00 to $91.48 per barrel, and still world oil demand increased from 77.5 to 85.8 MBD, a rise of 10.7 % [52, 53]. The recent oil price trends show that despite the substantial and sustained price increases, there is an incredible resilience in demand for oil, and the world economy shows no signs of easing its insatiable appetite for oil and other carbon fuels. Increased oil consumption spurs economic growth, and the improved economic prosperity motivates people to travel more and buy more and bigger products: larger TV sets, larger appliances, larger homes that consume ever greater amount of fossil-derived energy, and so on. Breaking this vicious cycle becomes increasingly difficult with every passing year.

Paradoxically, the great improvements in the efficiency of fossil fuel usage brought about by decades of technological progress (e.g., in power generation, transport) did not result in the diminished demand for fuels, as could be expected, just the opposite. The main reason for that is that the fuel savings resulting from the improved energy efficiency of a product manufacturing process were in almost every instance heavily outweighed by the increased usage of fossil fuels to cover the greater demand for the product (or the service). For example, in the airline industry, the fuel efficiency is growing in average at the rate of 1–2 % a year, but this advantage is completely wiped out by the increasing levels of the passenger miles rising at about 5 % a year [51, 54]. The increased fuel efficiency of modern cars is outweighed by the dramatically increased number of personal vehicles in the world and the increased level of comfort for a driver; most of cars now have fuel-gobbling air conditioning and power-activated systems, e.g., power steering, power window (this is a vivid illustration of human labor being replaced by relatively cheap fossil energy). Another example: the fuel savings due to the increased energy efficiency of modern TV sets (resulting from the switch from cathode ray tubes to liquid crystal displays, LCD) are completely negated by the astonishing increase in the number of TV sets and game consoles (typically, several units per household, continuously working 12–18 h a day). The examples of increased fuel usage in parallel with the technological advancements are countless.

The above examples show that, in most cases, the gains we make in fuel and energy efficiency are taken back in increased comfort, reduced manual work, easier and more pleasant lifestyle, etc. [51]. The main driving forces and trends in the modern society seem to point to ever-increasing personal usage of carbon fuels, which manifests itself in:

- Greater prosperity through replacing manual work with fossil fuels.
- Social changes (e.g., employment, health services, entertainment) stimulate people to have more independent mobility and inevitable push toward car ownership.
- The maintenance of a personal status and prestige in the modern society motivates the purchase of bigger and better-quality products, services, and other symbols of affluence.

- People in developing world (rightfully) strive for reaching the same standard of living as people in the developed world.

Are these ongoing trends in ever-increasing usage of fossil fuels by individuals and the society as a whole sustainable? For how long humankind can satisfy its insatiable appetite for carbon fuels before the "plate" is empty? The next section will shed some light on these questions.

1.4 Is the Depletion of Carbon Fuels a Real Problem?

It is a common knowledge that fossil fuels will not be infinitely available; as nonrenewable resources, they are getting depleted as more and more of their content is removed from the underground. At the peak of oil field discoveries in the 1920s through the 1950s, it was thought that oil depletion would not become a problem, because new oil fields would be found to replace depleted ones. American geophysicist M. King Hubbert was the first who disagreed with this perception; in 1956, he predicted that annual oil (or any other fossil fuel) production in different countries and the world, in general, would follow a bell-shaped curve: initially, production nearly exponentially increases reaching a peak output, which is followed by an exponential decline [55]. Mathematically, the Hubbert's curve follows the logistic function:

$$Q(t) = \frac{Q_{max}}{1 + ae^{-bt}} \tag{1.2}$$

where Q_{max} and $Q(t)$ are the total resource available (i.e., ultimately recoverable) and the cumulative production, respectively, t is time, a and b are constants.

The year when the maximum annual production (peak output) will be reached can be found from

$$t_{max} = \frac{1}{b}\ln(a) \tag{1.3}$$

Although the Hubbert's theory initially attracted strong criticism and skepticism, especially from the oil industry, his prediction turned out to be true—the US (conventional) oil production peaked in 1970/1971. With regard to the world oil production, Hubbert predicted that it would peak in 2000 [56]. Other noteworthy predictions put the oil peak at somewhat later time; Deffeyes predicted the oil peak to occur in 2005 [57], and Campbell in 2005–2010 [58]. However, these predictions did not prove correct; the global oil production kept rising (with ups and downs) after 2000 and 2005, but the year 2011 witnessed a new global oil production record of 83.6 MBD (Fig. 1.4).

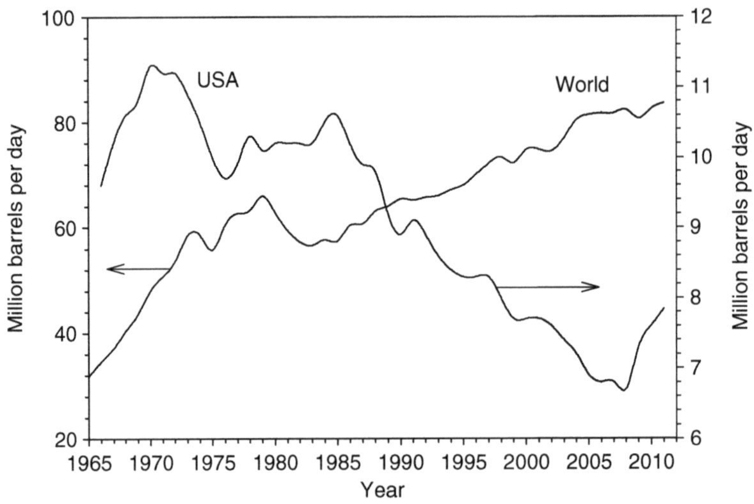

Fig. 1.4 Global and the US oil (includes NGL) production from 1965 to 2011. *Source*: BP 2012 statistical data [59]

The data in Fig. 1.4 show that the global oil production grew between 1965 and 2011 at the average annual growth rate of 2.1 %. Moreover, the production in 2011 was about 2.7 % higher than the 2005 production level (projected by many as the peak oil level) [60]. However, the average annual growth rate from 2005 to 2011 was only 0.4 %, i.e., much lower than the historical average of 2.1 % (since 1965). One might argue that the Hubbert's prognosis and other predictions took into account conventional oil only, but the share of unconventional oil production during the 2000–2011 period was rather insignificant (e.g., in 2008, only about 2 % of total production).

It is interesting to compare the trends in global and the US oil productions. As seen from Fig. 1.4, the US oil production peaked in 1970 (as predicted by Hubbert) at the production level of 11.3 MBD (including NGL) [59]. This was followed by rather a steep decline until 1977, when Alaskan oil fields began delivering oil to the market. After several years of rising production (until 1985), the decline followed and continued until 2008, when oil production slowed down to 6.6 MBD (i.e., 40 % below the 1970 peak). Since 2008, the US oil production is on the rise reaching 7.84 MBD in 2011. It should be noted, however, that it was still 31 % below the peak level of 1970, and because the global oil production has risen, the US share of global crude oil production has declined from 24 % in 1970 to just 9 % in 2011; in that year, the USA was the third largest oil producer after Saudi Arabia (11.1 MBD) and Russia (10.3 MBD) [60]. The IEA projects the US oil production in 2015 at 10 MBD, and above 11 MBD in 2020, followed by slow gradual decline, but still above 10 MBD in 2030 [61].

As a general trend, the rise in oil production in most cases is accompanied by the worsening of its quality: crude is getting heavier and its sulfur and minerals contents are on the rise. Although this issue gets a little attention in the press (compared to the quantitative aspects of oil production), this factor is getting increasingly important because it requires more energy and resources to produce and refine oil. As a result, refineries have become larger and more complex (e.g., requiring additional hydrogen production and hydrotreatment capacities), and the cost of oil processing is getting higher (which passes on to the higher prices for gasoline, diesel fuel, and other petroleum products). Most importantly, the net energy obtained from a barrel of oil is getting smaller and smaller, which effectively translates into the reduced level of oil production. This issue is usually hidden behind cheerful reports on increased oil production, and is rarely discussed, although this worrisome trend will continue as the remaining supplies of sweet crude oil are rapidly disappearing.

According to the IEA's estimates, oil global-proven reserves and recoverable resources are (in billion tons): 1,694 (with 244 in OECD and 1450 in non-OECD countries) and 5,871 (with 2,345 in OECD and 3,526 in non-OECD countries), respectively [9, 12]. The USA possess the largest reserves of shale oil (estimated at 1.5–2.6 trillion barrels); other countries having large shale oil inventories are China, Brazil, and Estonia (they are currently producing oil from shale formations). Global technically recoverable oil shale reserves have been estimated at about 2.8–3.3 trillion barrels [62]. It is likely that many more large shale oil formations will be discovered in different locations throughout the world in the future. Combined with tar sands and heavy oil, all these newly developed immense reserves of hydrocarbon fuels will most likely rewrite international oil trade routes and have a heavy footprint on a geopolitical situation in the world. But how will that impact the "peak oil" problem?

The timeframe for which global oil production is expected to reach the peak output and eventually decrease to minimum is one of the most burning topics and a subject to much debate nowadays. The seriousness of this issue is underscored by the fact that more accurately we can predict how much of technically recoverable oil still remains underground, better we can plan and be prepared for inevitability of depletion of this precious resource. But if you ask a geologist, an oil company executive, or an economist: for how long we can rely on oil to power our economy, you will get completely different estimates from a few years to hundred years. The reason is that the answer to this question is associated with consideration of a myriad of different factors, criteria, and assumptions of technical, economical, and political nature.

In general, the majority of experts agree that the issue is not as much about running out of oil, as about the rates: the rates of oil discovery and production *vs* the rates of oil consumption/demand growth all correlated with the rate of technological changes in the oil industry [63]. There is an argument that despite the finiteness of the global fossil resources, they tend to become less scarce over time, because the declining rates of new oil discovery could be offset by increasing technical capacities

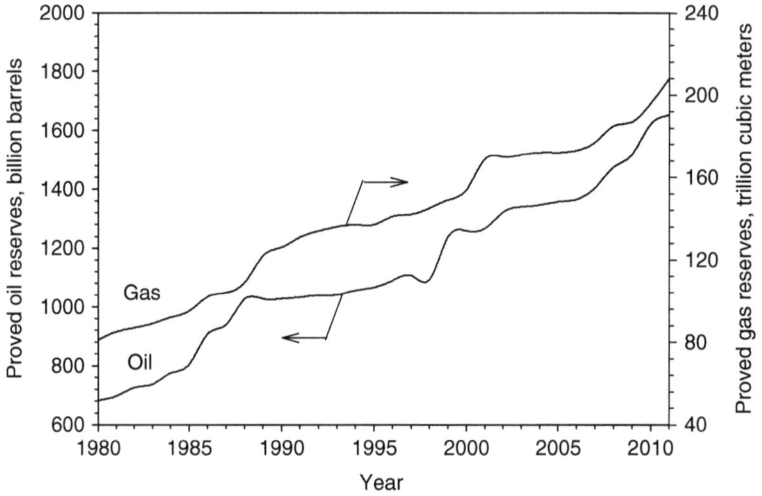

Fig. 1.5 History of global oil- and gas-proven reserves. *Source*: BP 2012 statistical data [65]

to recover a greater fraction of oil from the existing reservoirs, and also through the development of revolutionary technological approaches to extract oil from the formations that only a decade ago deemed to be impractical (e.g., shale oil). Recent developments in the energy arena indicate that many worries about hydrocarbon resource scarcity proved unjustified [64], and Fig. 1.5 showing history of the global oil- and gas-proven reserves[4] seems to prove that, during the last three decades, the estimates of oil reserves increased 2.4-fold [65].

The decrease in the value of oil reserves estimates in 1998 was mostly due to the significant reduction in the estimates for North America: from 127.1 billion barrels (BB) in 1997 to 100.0 BB in 1998; this, however, was followed by a surge to 232.8 BB in 1999. The world-proven oil reserves at the end of 2011 reached 1652.6 BB. Most notably, in 2011, Iraq added 28 BB to its reserves, and Russia, Brazil, and Saudi Arabia all increased reserves by 1 BB.

Interestingly, the historical trend of proved gas reserves almost repeat that of oil, with the estimates increasing 2.6-fold over the same period (Fig. 1.5). Note that there were almost synchronous ups and downs in the estimates of proved oil and gas reserves. Three prominent surges in oil/gas reserves estimates can be seen on the historical data curves: in 1985–1989, 1998–2002, and the last one starting in 2006 and still ongoing (although there are signs of leveling up for the oil reserve estimate in 2011). The last two surges could be traced to technological advancements in gas/ oil industry (e.g., hydraulic fracking, horizontal drilling). Overall, the long-term

[4]According to British Petroleum (BP) definition, *proved reserves* of oil are generally those quantities that geological and engineering information indicates with reasonable certainty can be recovered in the future from known reservoirs under existing economic and geological conditions.

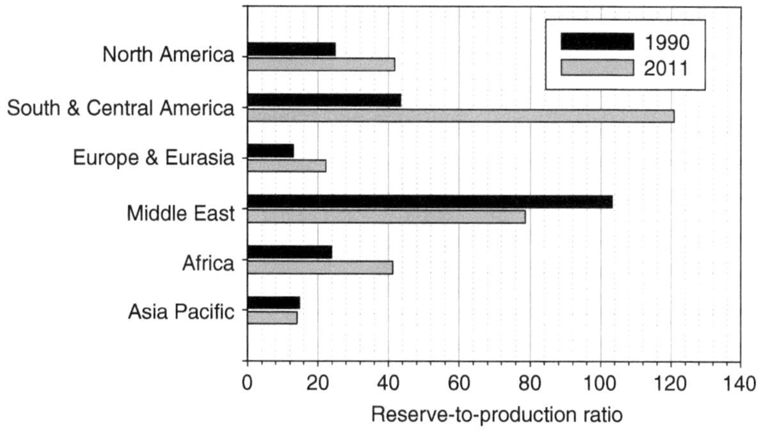

Fig. 1.6 Oil reserve-to-production (R/P) ratio for different regions. *Source*: BP 2012 statistical data [65]

trends in oil and gas estimates indeed show that (up until now) the world continues to add more reserves than it uses. Proved reserves of oil remain concentrated in OPEC, which controls 72 % of the world's oil reserves, the highest proportion since 1998. Most of the world's largest gas reserves are concentrated in Middle East and Eurasia (predominantly, Russia and former Soviet Republics) [65].

The timing of the depletion of fossil fuel reserves could be roughly estimated from the Reserve-to-Production (R/P) ratio. R/P ratio at any given year represents the length of time that the remaining reserves would last if production were to continue at the previous year's rate. R/P is calculated by dividing remaining reserves at the end of the year by the production in that year [65]. Figure 1.6 shows the changes in R/P ratios for different regions in the years of 1990 and 2011.

The continuing increase in Venezuelan unconventional oil reserves pushed the South and Central American R/P ratio above 100 (the largest increase among all regions), while the large increase in Middle Eastern production reduced the region's R/P ratio despite an overall increase in reserves (the region holds 48.1 % of global-proven reserves). North America saw substantial increase in R/P ratio from 24.9 in 1990 to 41.7 in 2011 (mostly due to unconventional oil contribution). R/P ratios for other regions stayed relatively flat. The data imply that South/Central American oil reserves will last longest (more than100 years at current consumption rate) among all regions, followed by Middle East (close to 80 years) and North America and Africa (both, about 40 years). According to BP 2012 data [65], the global R/P ratio stands at 54.2 at the end of 2011, which implies that the world-proven oil reserves are sufficient to meet 54.2 years of global production.

Returning to the "peak oil" debate: are the worries of those who believe in "running-out-of-oil" justified, or the doomsday predictions will prove wrong again? The best answer to this dilemma could be attributed to an oil economist, M. Adelman, who wrote after the "oil crisis" of the 1970s that oil reserves

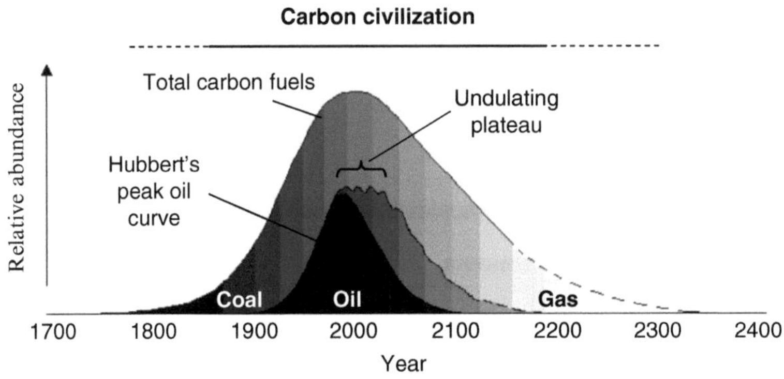

Fig. 1.7 The relationship between the Hubbert's oil peak and undulating plateau concepts

"are no gift of nature. They (are) a growth of knowledge, paid for by heavy investment" [66]. One way of interpreting this statement is that the world's oil reserves are still finite, but the coming of the "peak oil" could be postponed far into the future, timing of which will be determined by the combination of gained knowledge and investments.

The IEA's position on the peak oil argument is that while there are ample oil and liquid fuel resources for the foreseeable future, the rate at which new supplies can be developed and the break-even prices for those new supplies are constantly changing [67]. Declining oil production in any given year can occur for a variety of reasons unrelated to physical peak oil production, e.g., the OPEC's production policy and decisions, unplanned field stoppages, the impact of earlier investment decisions by the oil industry. Political motives by the governments of major oil-producing countries could also greatly affect the oil output. A combination of sustained high oil prices and energy policies promoting diversification in energy supplies and greater end-use efficiency might actually reduce oil consumption to the point that peak oil demand would occur before the resource base is even nearly exhausted.

The US Department of Energy (DOE) dismisses the "peak oil" theory; instead, it supports the alternative hypothesis of an "*undulating plateau.*" The US DOE official Lauren Mayne explains this term as follows: "*Once maximum world oil production is reached, that level will be approximately maintained for several years thereafter, creating an undulating plateau. After this plateau period, production will experience a decline*" [68]. According to this source, there is a chance of a decline in liquid fuels production between 2011 and 2015 as the first stage of the undulating plateau pattern (if adequate investment is not there), which will start "*once maximum world oil production is reached.*" Figure 1.7 depicts the relationship between the Hubbert's oil peak and undulating plateau concepts.

Many energy analysts support the notion of the undulating plateau, and argue that oil production appears to be on an undulating plateau, the shape of which will be determined by oil prices [69]. Higher prices would spur oil production, while lower prices would quickly crush the producer's cash flows considering today's high cost of marginal production. The catalyst for the shale oil revolution was a new much higher price deck.

1.5 Dealing with the "Tide" of Carbon Fuels

It is widely recognized that ending our reliance on fossil fuels is not going to be an easy task, especially, considering that now they generate two-thirds of electricity and 95 % of transportation fuels. Our industrial civilization so heavily relies on them that any alternatives that will try competing with carbon fuels will be "*swimming against very strong tides*" (as OECD's Angel Gurria put it) [64].

The first strong tide comes from the fossil fuel resource abundance. Just a few years ago, a prevailing opinion among experts was that the impending depletion of oil and gas resources would drive their prices so high that a switch to alternatives would become inevitable in not so distant future. That has proved illusive. Thanks to technological innovations, crude oil and gas production is growing in many countries around the world (e.g., the USA, Canada, Brazil, Russia, Kazakhstan, Algeria, Argentina, to name few). In 2012, US$674 billion was spent on finding and developing new sources of oil and gas [64]; so, it will not be a big surprise to hear more about the discoveries of new large oil and gas fields.

The second tide comes from the fact that the incumbent fossil-based technologies have a huge advantage over alternative low-carbon technologies as a result of enormous investments to the fossil infrastructure over the last century. Those investments proved very profitable and still continue to attract new capital. For example, in 2012, more than half of newly installed electricity-generating capacity was still based on fossil fuels; about 1,200 new coal-fired power plants are now at the planning stage (and most of them will likely have a very long life) [64]. At the current rate of capital expenditure, more that US$6 trillion will be spent on the development of fossil fuel infrastructure.

The third strong tide deals with the so-called *carbon entanglement* factor. This term implies that governments have major stakes in bringing fossil fuels to the market and taking their share of the rents. The size of the royalty payments, taxes, and other revenue streams associated with upstream oil and gas rents varies in a very wide range from country to country: from 1 to 4 % of the total government revenues for OECD countries to 28 % for Russia (or US$150 billion) and close to a third of total revenues for Mexico (OPEC countries extract revenues of US$600–700 billion a year) [64]. The reliance of governments on fossil fuel revenues cannot be overestimated: they have heavily invested in them, and they will do everything to keep these flows of income undisturbed. (And, if necessary, they will go to the deep ocean or places like Arctic to exploit new reserves.)

Another aspect of the carbon entanglement has to do with the potential impact of carbon policies on the valuation of many companies and, consequently, common investors. According to published data, over 55 % of pension funds' portfolios are invested in "high carbon assets," or, simply, fossil fuel infrastructure; thus, for policymakers it may come to the choice of "*either stranding those assets, or stranding the planet*" [64]. The impact of these tides is extremely strong and will not be easy to deal with. The very slow and modest progress of carbon mitigation and climate change-related policies and regulations over the last couple of decades is in part a testament to that.

1.6 Environmental Impact of Carbon Fuels

1.6.1 Carbon Fuels and Climate: Facts and Uncertainties

Ironically, the same carbon-bearing fuels that drove technological progress and brought us the prosperity and high standard of living are now being blamed for a host of problems and miseries of the present-day life: unbreathable air and premature deaths, oil spills and catastrophic explosions, unbearably hot summers and unusually strong hurricanes, acid rains and disappearance of biospecies, and so on. The adverse impact of fossil fuels on our planet's environment, climate and, in the final analysis, our way of life has drawn much attention during the last decades. Among main concerns are:

- Tens of billion tons of anthropogenic[5] CO_2 and other GHG emissions that are released straight to the atmosphere, where they will stay for hundreds of years and impact our planet's ecosystems
- Release into the atmosphere of significant amounts of neurotoxic metals—mercury (Hg) and arsenic (As) by coal-fired power plants
- Emission of sulfur and nitrogen oxides and particulate matter from burning low-quality coals and petroleum products
- Millions gallons of oil spilled into the environment (During the accident at The Deepwater Horizon rig 185 million gallons of oil was released into the Gulf of Mexico in 2010, the total ecological impact of which we may never know.)
- Practicing controversial mining techniques (e.g., coal industry is now practicing a "mountaintop removal" method of extraction in many coal deposit sites, where the top of a mountain is blasted off)
- The full extent of the ecological impact of hydraulic fracturing technology that extracts NG and oil from shale beds is still unclear.

Some disquieting signs of the adverse impact of the increased usage of fossil fuels on our planet's ecosystems and climate have already started manifesting themselves in the form of retreating glaciers, rising sea levels, shifting rainfall patterns, stronger and more frequent hurricanes, increasing floods, etc. In the years between 1951 and 1980, extremely hot temperatures covered less than 0.2 % of the

[5] The term "*anthropogenic*" refers to greenhouse gas emissions that are the direct result of human activities or are the result of natural processes that have been affected by human activities [70].

planet; now this expanded to about 10 % of the land area [71]. The 2003 heat wave in Europe, 2010 heat wave in Russia, 2011 drought in Texas, 2012 wildfires across Australia, 2012 superstorm Sandy in the USA: too many extreme weather events in relatively short time period to consider them a natural weather variations.

In its 2013 Assessment Report, IPCC states that "*Warming of the climate system is unequivocal, and since the 1950s, many of the observed changes are unprecedented over decades to millennia.*" [72]. According to the report, human fingerprints have been detected in

- Warming of the atmosphere and the ocean
- Changes in global water cycle
- Reductions in snow and ice
- Global mean sea level rise
- Changes in some climate extremes

The globally averaged combined land and ocean surface temperature increased by 0.85 °C (the range of 0.65–1.06 °C) [72]. The report provides data showing that there is 90 % certainty that 1981–2010 was the warmest span in the last eight centuries and 66 % chance that it was the warmest 30-year period in the last 1,400 years (although last 15 years have not warmed as quickly).

The authors of a recent study reported in the *Science* magazine reconstructed regional and global temperature anomalies for the past 11,300 years (to distinguish anthropogenic influences on climate from natural variability) (this rebuts the arguments of critics of climate change research, who try to make a case that all the current studies cover short periods of time, typically, 1,500–2,000 years, and they do not take into account warming the Earth experienced many thousands of years ago due to natural causes.) [73]. The study showed that the 1900–1909 decade was colder than 95 % of the last 11,300 years, whereas the decade of 2000–2009 was hotter than 75 % of the last 11,300 years. Thus, the Earth's climate was propelled from one of its coldest decades since the last ice age to one of its hottest—in just one century: a very short period of time for such a spike.

Variations due to the Earth's tilt and orbit and other natural factors cannot explain this sudden anomalous increase in global temperature, which "incidentally" coincided with the surge in the consumption of fossil fuels; just in contrary, based on the historical trends related to the Earth's tilt and position relative to the sun, our planet is supposed to be cooling. The majority of climate scientists hold that if carbon emissions continue to rise, as currently projected, by 2100, global temperatures will rise well above anything seen in the last eleven millennia. If that scenario will prove true, this would bring a misery to hundreds of millions people around the world over the span of several generations. Does all this imply that we will be paying an increasingly heavy environmental toll for the economic prosperity brought about by fossil fuels?

Despite an access to capable scientific instrumentation and extremely sophisticated computer models, climate scientists are far from understanding all the observations and sometimes unexpected trends; for example, they are struggling to explain a slowdown in climate warming in the last decade [74]. Predominantly focused on century-long climate trends, most climate models failed to predict the slowdown trend in the average temperature rise starting at the turn of the current century and millennium (i.e., around 2000). This exposed some gaps in the understanding of many climatic phenomena and, as could be expected, provided more ammunition to those who question the link between the growth in GHG emissions and rise in planet's mean temperature and climate change. Although many scientists expect a revival of the warming trend in the coming years, they have hard time to determine whether the current break will be a brief or more lasting phenomenon.

According to IPCC, the temperature records since 1850 indicate fluctuations of 10–15 years in duration, but the overall warming trend is unmistakable. Among explanations of the warming slowdown are theories that the deep oceans have taken up more heat leading to cooler surface than expected, or that industrial pollution produced by booming Asian economies is blocking the sun, or that GHG trap less heat that previously thought, or it could be a result of an observed decline in heat-trapping water vapor in the atmosphere at high altitude, or a combination of different factors and unknown or poorly understood natural variations. Regardless of the scientific basis behind the current counterintuitive observations, these uncertainties do not reinforce trust in climate science among general population, although many appreciate the complexity of the climate system.

There is also a lack of agreement among the climate scientists with regard to long-term impact of past CO_2 emissions on present and future global mean temperatures. One school of thought holds that even if humankind moves quickly and starts cutting CO_2 emissions at unprecedented rates, global temperatures would still continue to rise for many years due to two types of inertia:

- Thermal inertia of the oceans (the estimated value of the temperature rise is about 0.6 °C, in addition to 0.76 °C rise that has already occurred [75])
- Institutional or infrastructural inertia (fossil fuel infrastructure that currently powers 80 % of the world economy represents an extremely large investment; it will ensure that emissions will continue for decades to come [76])

Some climate experts go even further and claim that even if the atmospheric CO_2 concentrations would remain fixed at the current level, there still be additional future warming due to past emissions; this implies that the increase in the Earth's global temperature is inevitable regardless of the scope of carbon emission reduction. For example, IPCC in its Fourth Assessment Report (2007) states that "*Adaptation will be necessary to address impacts resulting from the warming which is already unavoidable due to past emissions. Past emissions are estimated to involve some unavoidable warming (about a further 0.6 °C by the end of the century relative to 1980-1999) even if atmospheric greenhouse gas concentrations remain at 2000 levels*" [77]. In its 2013 Assessment Report, IPCC held to the same viewpoint "*Surface temperatures will remain approximately constant at elevated levels for many centuries after a complete cessation of net anthropogenic CO_2 emissions. Due to the long time scales of heat transfer from the ocean surface to depth, ocean warming will continue for centuries*" [72].

In an article in *Science* magazine, Matthews and Solomon debate this viewpoint [78]. According to the authors, because of the equal and opposing effects of physical climate inertia and carbon cycle inertia, there is practically no delayed warming due to the past CO_2 emissions. This implies that if CO_2 emissions were to cease immediately, global average temperatures would not increase and remain nearly constant for many centuries. In other words, any further increase in CO_2-induced warming would entirely result from the current CO_2 emissions, and warming at the end of this century will be caused by the cumulative CO_2 emissions humankind produces between now and then. But the main (optimistic) conclusion of this study is that future warming is not unavoidable: any tangible reductions in man-made CO_2 emissions would lead to an immediate drop in the rate of global climate change. All these conflicting views on the future climate impact of anthropogenic CO_2 show that despite advances in computer modeling and accumulated knowledge in the field, climate science is far from understanding of all the climate-related phenomena that could affect humans in not so distant future.

Public opinion on the climate change issue and the human link is as conflicting as some theories and concepts in climate change science. Although the world has seen a steady trend of increasingly hot years, public's belief in climate change has remained relatively stagnant over the past decade. Ironically, recent polls show that people are more likely to believe in climate change during hot years, when they are starting to see global warming as an important issue. According to one survey study, Americans' opinions on climate change literally "blow with the wind"—with more concern shown in the years that are much warmer or much colder than normal [79]. The study also shows that most Americans (and, probably, the majority of the world's population) aren't steadfast in their opinions on climate change, whether they are believers or skeptics; their opinion is malleable depending on the weather. A report published by the UK Energy Research Centre shows that 19 % of people do not believe climate change is real—up from just 4 % in 2005—while 9 % did not know [80].Climate change skeptics do not constitute a homogeneous crowd; among them, it is easy to distinguish three main types:

- "Die-hard" deniers who refute the very possibility of global warming; they claim the data presented by IPCC and other scientific bodies are misleading, their models cannot be trusted, "the Earth is actually cooling not warming," "weather is not climate," etc. (one example is [81]).
- Skeptics admitting that the climate change may be real, but the available data do not provide a solid proof that human activities are to blame; the real cause of the change is still unknown.
- Skeptics believing that climate change is solely due to natural cycles and variations, and humans have nothing to do with that.

Summarizing, the science of climate change has become politically controversial, and there are diametrically opposite interpretations of the same climatic events. Multiple lines of evidence point to changes in climate over the last 150 years. Debate continues, however, on what is causing changes in the temperature and precipitation patterns since the late nineteenth century. As will be shown in Chap. 2, the changes in atmospheric chemistry due to human activities could lead to warming (due to GHG) or cooling (due to aerosols), which seems to explain a large part of the surface temperature oscillations at a short-term scale. The IPCC 2013 report in line with its earlier deductions emphasized that "*It is extremely likely that more than half of the observed increase in global average surface temperature from 1951 to 2010 was caused by the anthropogenic increase in greenhouse gas concentrations and other anthropogenic forcings together*" (in the IPCC report the term "extremely likely" corresponds to 95 % probability) [72]. Note that the above probability estimate marks a sharp increase in the IPCC's confidence level compared to its 2007 and 2001 reports, where it was 90 % and 66 %, respectively, confident of the similar conclusion. More than 850 experts and 50 editors from 85 countries have contributed to the 2013 report, which underscores the significance of the main conclusions of the report.

Although the ever-increasing body of evidence points to the human imprint in the current climate change trend, it is not the intent of this book to revisit the range of views with regard to the ongoing heated "climate debate," or some purported controversies surrounding the climate change science and the IPCC reports. There are many excellent books and reviews dedicated to this topic.

1.6.2 Economy–Environment Trade-Off Dilemma

There is a deep-rooted belief among many economists and policymakers that a trade-off between economy and environment is inevitable: economic growth and cleaner environment are incompatible. Those who believe that the trade-off is unavoidable reason as follows: either we increase, for example, the number of coal-fired power plants or passenger cars to facilitate the economic growth and prosperity or we decrease their number to reduce GHG emissions and clean up the environment, but we cannot do both at the same time. Some advocates of this idea even suggest that a "planned recession" would be necessary in order to reduce fossil-based GHG emissions to prevent climate change [82]. Pielke critically analyzed the notion that taking any action on climate change will necessarily lead to tangible economic sacrifices [83]. He referred to a recent poll conducted in the USA that asked respondents about their willingness to support a climate bill in the US Congress if that would entail three different annual costs per household: $80, $175, and $770. In the first case, more than 50 % of respondents said that they would support the bill; however, in the second case, the support dropped by almost half, with the majority opposing such a bill. In the third case, the opposition exceeded support by ten-to-one ratio. The results of the poll imply that when the environment and economy are presented as a trade-off, the economic considerations prevail. This conclusion seems to agree with the Pielke's "iron law" of climate policy, which states that "*when policies focused on economic growth confront policies focused on emissions reductions, it is economic growth that will win every time*" [83]. In more simple terms, the implications are that even if people are willing to bear some costs to reduce GHG emissions, they are willing to go only so far, and these limits exist at different thresholds around the world.

Although the economy–environment trade-offs are inevitable, the successful policies dealing with decarbonization of global economy and the climate change problem must be designed such that economic growth and environmental progress do not confront each other but rather go hand in hand. What this would ultimately mean is that the actions to achieve environmental goals will have to be fully compatible with the desire of people in rich as well as poor countries around the world to meet their economic goals [83].

1.6.3 Local and Global Impacts of Carbon Fuels

In general, negative environmental and societal impacts of fossil fuels on humans' well-being can be categorized at the local and global levels. Among local factors, the impact of air pollution is the most recognized and well studied. Local air pollution could be a result of at least three factors:

- The direct effect of the products of fossil fuel combustion such as CO and CO_2 (summarily, CO_x), sulfur oxides: SO_2 and SO_3 (summarily, SO_x), nitrogen oxides: NO and NO_2 (summarily, NO_x), unburned hydrocarbons (UBH), ammonia (NH_3), mercury vapor (Hg), soot, and a variety of organic and inorganic aerosols.
- Photochemical reactions in the atmosphere involving some of the above products of fossil fuel combustion (e.g., SO_x, NO_x, UBH).
- Formation of ground-level ozone and its reactions with the products of fossil fuel combustion to form smog.

Indoor and outdoor air pollution is the sixth-leading cause of death on our planet, resulting in over 2.4 million pollution-related fatalities worldwide [84]. The effect is especially prevalent in the heavily populated areas of large cities and industrial regions. In China, for example, approximately 1.2 million people die prematurely from exposure to polluted air [64]. It was estimated that emissions due to the use of battery-electric vehicles powered by coal-derived electricity would kill prematurely (as a result of pollution-related health problems) between 2,880 and 6,900 people per year in 2020 [85]. Carbon monoxide is a very toxic gas formed when combustors and engines operate with an insufficient supply of air. SO_x are formed during combustion of sulfur-containing fuels, e.g., high-sulfur coals or residual fuels. (Note that most of sulfur is typically removed from petroleum products by installing hydrodesulfurization units at refineries; sulfur content of gasoline and diesel fuel is strictly regulated.) When SO_2 is released to the atmosphere, it is further photochemically oxidized to SO_3 which through the reaction with water vapor forms sulfuric acid (H_2SO_4), the major component of "acid rain." Acid rain could cause health-related problems, the damage to vegetation, buildings, metallic structures, decrease in pH in water reservoirs, and other negative effects on ecosystems. Although most of modern power plants and industrial installations are equipped with sulfur removal units, still significant quantities of SO_x are released to the atmosphere, especially, in developing countries.

NO_x are formed predominantly by oxidation of nitrogen of air in the high-temperature environment existing in the combustion chambers of boilers, furnaces, engines, turbines, etc. (to a lesser extent, NO_x could also originate from N-containing

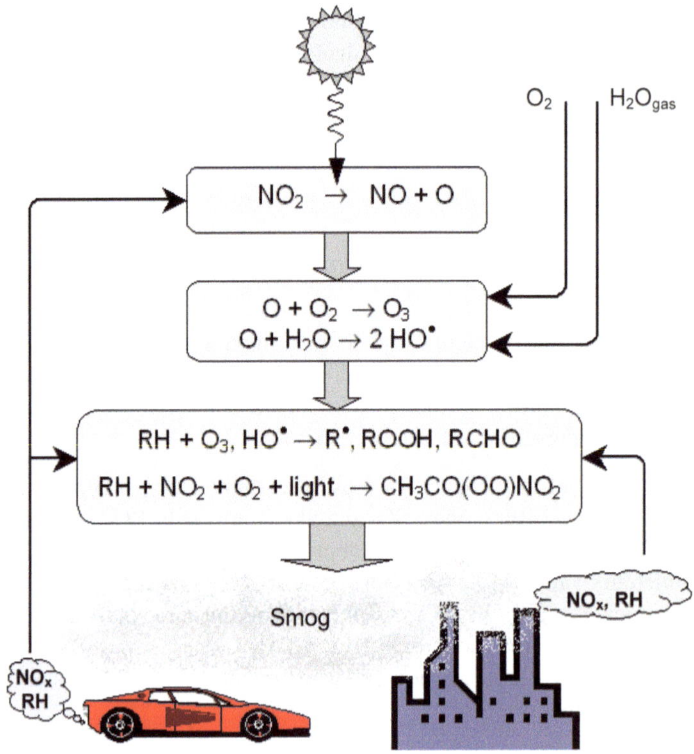

Fig. 1.8 Major chemical pathways to photochemical smog formation

compounds in fossil fuels). A prolonged exposure to NO_x can cause lung damage and aggravate such conditions as asthma and bronchitis. (It could also affect an immune system and increase susceptibility to flu and cold.) Exposed to sunlight NO_2 is photochemically split into NO and atomic oxygen, which reacts with oxygen causing the formation of relatively high concentration of ground-level ozone (O_3). This explains the fact that the ozone concentration is considerably higher in urban areas compared to rural areas. Ozone could cause breathing difficulties, irritation to the lungs and eyes, fatigue, and headaches, and can aggravate respiratory problems. As a very reactive oxidant, ozone can destroy vegetation and synthetic materials such as rubber and plastics [86].

Ozone, atomic oxygen, and hydroxyl radicals react with hydrocarbons or UBH forming highly reactive radicals, aldehydes, peroxides, and peroxyacids. In the presence of NO_2 these compounds are further (photochemically) oxidized to N-containing peroxycompounds, predominantly peroxyacetylnitrate: $CH_3CO(O_2)NO_2$ (PAN). These final products of chemical transformations of NO_x and UBH (i.e., aldehydes, aldehyde peroxides, PAN) form photochemical smog, which typically occurs in large urban centers where there are plenty of sources and precursors for its formation. Figure 1.8 schematically represents major chemical pathways to photochemical smog formation. Smog contains compounds irritating sensitive biological tissues, and it could cause severe health problems especially for people with asthma and other ailments.

In response to stringent regulations on NO_x emissions in some countries, efficient methods for reducing NO_x emissions in industrial burners and engines are being developed. The main approaches to the problem include (1) the reduction in combustion temperature and (2) catalytic reduction of NO_x to N_2. In transportation, the introduction of three-way catalytic converters has been particularly effective in reducing air pollution from vehicles [87]. In the catalytic converters (typically, containing Pt or Pt–Rh catalysts), CO and UBH are nearly stoichiometrically oxidized to CO_2 and H_2O, with NO being reduced to N_2:

$$2CO + 2NO \xrightarrow{TWC} 2CO_2 + N_2 \tag{1.4}$$

where TWC is three-way catalyst.

In the converter, the rate of reduction of NO to N_2 must exceed that of CO oxidation to CO_2 so that there is an excess of CO in the reacting medium to complete reaction 1.4. Catalytic converters in gasoline vehicles reduce NO_x emission by as much as 90 % [86].

Among other negative impacts of fossil fuel use is the formation of *particulate matter* (*PM*), especially in diesel engine exhaust and flue gases of coal and heavy oil combustion. The term PM covers a wide range of particles of different chemical composition and size that are formed via multistage complex processes involving unburned hydrocarbons, SO_x, NO_x, and other compounds. (The fuel-rich flames that could be found in diesel engines and pulverized coal combustion systems provide particularly favorable conditions for PM generation.) Soot from diesel cars is a big problem as they are widely used in urban centers with a high population density. A particular health-related concern is the formation and release of particles with the dimension less than 10 micron (micron, μ, is equal to 10^{-6} m), that are referred to as PM_{10} (the emission of PM_{10} is regulated in many countries). These particles can enter deep into the lungs and could potentially cause serious heart and lung problems, such as asthma, bronchitis, lung cancer, and premature death [86]. The developers and manufacturers of diesel engines are trying different techniques to reduce the amount of PM emitted from the engines (e.g., by increasing fuel injection pressure, installing different traps and filters); however, these devices still need more development work to improve their reliability and reduce cost. Although the release of relatively large-size PM at coal-fired power plants can be controlled through the use of electrostatic precipitators and baghouses, the removal of very small PM presents a challenge. Recent studies indicated that the negative impact of soot (or black carbon) on climate has been greatly underestimated [88]. Eight million tons of soot is being generated each year from different sources such as coal-fired power plants, oil-fired ship boilers, and diesel engines. Soot particles are excellent absorbers of solar radiation that they pass to the atmosphere.

Estimates have been reported on the monetized cost of CO_2 emissions and other air pollutants (i.e., the relative cost impacts of increased emissions) on human health and agricultural productivity as well as ecosystem damages and losses.

According to 2013 review by a working group of 11 federal agencies, the monetized cost of CO_2 emissions to the US economy in 2020 is projected to be from US$12 to US$129 per metric ton of CO_2 depending on a variety of scenarios [89]. An earlier estimate reported in 2010 put the cost of CO_2 emissions in 2020 at the range of US$7 to US$81 per metric ton of CO_2. These assessments serve as a guide to regulatory agencies in determining the costs and benefits of reducing CO_2 emissions.

The US National Research Council (NRC) estimated that fossil-based energy production, mostly from coal and oil, causes US$120 billion worth of health and other non-climate-related damages each year in the USA only that are not included in the price of fossil energy [90]. These cost numbers are primarily based on health impacts and premature deaths of nearly 20,000 people annually that are caused by air pollution from coal-fired power plants (about $63 billion) and ground transportation (about $56 billion), over their full life cycles, with two-thirds of it from extraction and production of fuels (the remaining $1 billion is due to heating). The NRC report underscores that these estimates are rather conservative, because they do not account for the adverse effect of toxic air pollutants, such as mercury and lead, the impact of climate change, and negative impacts to ecosystems. The report also shows that most of these damages come from coal-fired plants, with the share of NG-fired power plants of about $1 billion in health and non-climate-related damages.

1.6.4 Coping with the Environmental Impact of Carbon Fuels

In general, one can envision three options in dealing with the negative environmental impacts of the increased usage of fossil fuels:

- "Do nothing" approach (with a hope that nothing bad will happen)
- Adaptation to new living conditions (hopefully, with minimal sacrifices)
- Mitigation efforts to prevent, avoid, or allay as much as possible the negative impacts

The supporters of the "do nothing" approach, in general, question the necessity of any action with regard to possible climate change, with some of them even arguing that it could be beneficial at certain conditions. For example, there were reports that several establishment Russian scientists came up with a notion that global warming could be good for their country, because it might pep up vast cold regions (e.g., Siberia) and allow more grain and potatoes to be grown, making the country wealthier [83]. They argued that from the Russia's perspective nothing needs to be done to stop climate change; quite opposite, it will result in what might be called "*beneficial interference*." The absurdity of such an approach is obvious and does not need any further comments.

In its Fourth Assessment Report (2007) on climate change, IPCC outlined a portfolio of strategies for dealing with the effects of potential climate change that included *adaptation* and *mitigation* as main policies [77].

1.6.4.1 Adaptation

Adaptation policies focus on taking necessary steps to counter or better prepare for the anticipated negative impacts of global climate change, and/or to make social, biological, and ecological systems more resilient (or less vulnerable) to the effects of the change. Adaptation is considered a necessary and very important strategy that may complement other strategies (e.g., mitigation) dealing with the impacts of the climate change, because there is no 100 % certainty that all climate effects can be successfully mitigated.

Many experts believe that even in the best-case scenario, e.g., if global anthropogenic GHG emissions are dramatically reduced and stabilized, the climate change and its effects will most likely last decades; thus, the adaptation will be necessary in any case. IPCC considers the adaptation one of the cornerstones of climate policy; in its 2007 Report, IPCC states that due to extensive man-made CO_2 emissions "*There are some impacts for which adaptation is the only available and appropriate response.*" [77]. The range of available adaptive responses and measures is very broad and includes the following strategies [77]:

- Technological (e.g., sea defenses, infrastructure design)
- Behavioral (e.g., altered food and recreational choices)
- Managerial (e.g., altered farm practices, land-use planning, measures to reduce vulnerability in existing disaster risk reduction strategies)
- Policy (e.g., planning regulations)

While most technologies and adaptation strategies are already known, it is still not clear how effective various options will be, especially, at the higher levels of temperature rise and associated impacts, and, particularly, for vulnerable groups. In addition, there are daunting environmental, economic, social, cultural, and behavioral barriers to the implementation of the adaptation policies in many countries. According to IPCC, "*Adaptation alone is not expected to cope with all the projected effects of climate change, and especially not over the long term as most impacts increase in magnitude*" [77].

Adaptation is not a new concept: humankind and natural systems have been adopting to changing living conditions and surroundings for millennia; thus, it is believed that they will most likely adopt autonomously to climate change. Planned and organized adaptation can to some extent supplement autonomous adaptation, although it is recognized that there will be more options and incentives with regard to humans' adaptation compared to the adaptation to protect the Earth's ecosystems. It is also widely acknowledged that the ability of humans to adopt to climate change depends on such factors as technological advancements, an access to resources, an existing infrastructure, education and organizational capabilities, and overall wealth. From this viewpoint, developing countries, especially poorest of them located in tropical regions, will find themselves in a great disadvantage in their capacity to adopt to the climate change compared to rich developed countries. One of the approaches to deal with this inequality is to implement an international adaptation policy according to which the climate "winners" of the world would bear some responsibility for the climate "losers" of the world [83]. Although some relief and assistance programs have been practiced in the international community for many years (e.g., a disaster relief, debt forgiveness, development assistance), the new approach to adaptation policy would distinguish such issues from those associated with GHG emissions (i.e., the impact of human activities).

At the United Nations Framework Convention on Climate Change (UNFCCC), country signatories have made commitments to financially assist the countries most vulnerable with regard to adopting to climate change. UNFCCC Adaptation Fund has been established under the Kyoto Protocol in 2007 with the Clean Development Mechanism (CDM) set up as a main source of income for the Fund (the CDM is subject to a 2 % levy) [91].

In principle, the adaptation could be implemented in two modes, with some measures being introduced in response to the changes (*reactive adaptation*) and some in the anticipation of the changes (*anticipatory adaptation*) [92]. Most of the present adaptation measures are in response to current climate trends and variability, e.g., increased irrigation in response to reduced rainfall, the increase in the use of air-conditioning systems in response to elevated humidity and temperatures, the increased use of artificial snow making in the European Alps in response to reduced amount of snowfall, and the adaptation of crops to local climate conditions.

Adaptation to climate change has already started in some countries, for example, Australia, USA, Canada, and others are planning adaptation strategies and are in the process of implementing some adaptation measures. In 2009, the state of California has issued "California Climate Adaptation Strategy" discussion draft that analyzes the best-known science, technology, and practices on climate change impacts in several specific sectors and provides recommendations on how to deal with those threats [93]. Some cities started planning for adapting to the possible changes in climate patterns. For example, Chicago and New York have begun adaptation initiatives such as planting heat-tolerant tree varieties, changing to water permeable pavements to absorb higher rainfalls, adding air conditioners in public schools, and careful planning of water storage [94, 95]. Another example of the anticipatory adaptation measure is the construction of the Confederation Bridge in Canada at a higher elevation, which takes into account the possible impact of future sea-level rise on ship clearance under the bridge [96]. Since climate change is expected to modify rainfall, evaporation, and soil moisture storage, there are plans to develop crop varieties with improved drought tolerance.

Some suggested adaptation measures could raise a few eyebrows: Dr. Derocher, a biologist at the University of Alberta (Canada) recently proposed several emergency actions that will have to be taken soon to save Arctic bears [97]. Polar bears live and hunt from sea ice; if it is gone, they cannot survive. Among proposed scenarios is delivering food to polar bears using a helicopter as their icy habitat continues to melt (at a rate of 13 % per decade). This operation will cost about $32,000 per day for the "most accessible" bears, and it is hoped that such interventions would last days per year, not months. Other ideas include moving the bears northward, where the ice is less likely (or more slowly) to melt. However, none of the countries having polar bears, the USA, Russia, Norway, Canada, have a plan for responding to polar bear emergencies, and for many people, the very fact that such an idea is even up for the consideration is a wake-up call, a reminder that climate change is real and happening now [97].

Despite overwhelming support of adaptation as a policy response to climate change, many climate scientists point to its limitations. They reason that the excessive reliance on adaptation as the main or even the only strategy would carry enormous risks since many of the predicted climate change impacts are likely to exceed the capacity for humans and ecosystems to adapt [98]. Moriarty and Honnery point out that: "*adaptation is a slippery slope, since if we commit ourselves to it as our main strategy, we may not be able to mitigate the change in climate if our ability to adapt to climate change diminishes*" [76].

1.6.4.2 Mitigation

In a broad definition of the term, climate change "mitigation" is a set of actions to alter the Earth's radiative energy balance in order to prevent or greatly reduce the effect of global climate change. In a more specific and simplistic terms, the objective of climate change mitigation is to either reduce the number of GHG sources and the volume of GHG emissions, and/or enhance the sinks of GHG. The examples of the first approach include using fossil fuels more efficiently, better insulating buildings, switching to low-carbon energy sources, and using renewable and nuclear energy, whereas the examples of the second approach deal with capturing and sequestering GHG emissions, expanding forests, etc. All these technological approaches will be analyzed in detail in the following chapters of this book.

1.6.4.3 Role of UN Conventions

The world's 165 countries are signatories to the UNFCCC, the treaty that calls for the stabilization of atmospheric GHG. The 1997 Kyoto Protocol required the participating industrialized countries to cut their GHG emissions by 5 % from the 1990 levels during the period 2008–2012. The UN climate change conference in Doha, Qatar (2012), agreed on the extension of the Kyoto Protocol to reduce GHG and set the stage for negotiations on a new global climate change treaty (the negotiations are supposed to start in 2013 to have a new treaty in place by 2015) [99, 100]. It is realized, however, that the extension is largely symbolic, and it will have a little impact on global GHG emissions. Many people are frustrated that the pace of the progress on climate change is still very slow and the political will for greater changes and stronger actions remains very weak; and some even question the usefulness of the UNFCCC or other climate-related meetings, and whether they could have any impact on climate change.

Despite the lack of "breaking-news" outcome, the UN climate convention is the only venue that brings all countries together and provides a common arena for each participant to have a voice; it also enhances transparency and accountability among countries. However, the international body cannot solve the problem on its own; there is a need for more national leadership. Unfortunately, when it comes to climate change issues, there is a lack of adequate commitment, or strong political will, or ambition from national leaders. Although leaders agreed to cut emissions to limit average global temperature increase, the business-as-usual scenarios project a steady increase in the man-made GHG emissions throughout the century. Such projections are based on a growing global economy and abundant fossil fuel resources.

On the other hand, it is widely recognized that the time we have to cut GHG emissions to avoid the worst consequences of climate change is running short, and the stakes are high [100]. It is time for business leaders, government officials, scientists and engineers, and the general public to step up and develop more specific strategies, climate mitigation policies, and technical solutions to GHG emission reduction. Mario Molina, the winner of the 1995 Nobel Prize in Chemistry, said at the American Chemical Society 2012 convention: "*Climate change is a grand challenge facing humanity, but it can be solved. We have solutions in hand.*" Solutions suggested by Molina included putting a price on carbon emissions, dramatically increasing investments in energy research, and expanding international collaboration. Although these solutions are likely to cost on the order of 1–2 % of the world's gross domestic product (GDP), it is still much less than the cost of unchecked climate change, which, according to estimates, could easily reach 4–5 % of the world's GDP [101].

References

1. Lackner K (2003) A guide to CO_2 sequestration. Science 300:1677–1678
2. Moorland School (2012) The origin of the Earth, Earth Institute. http://www.moorlandschool.co.uk/earth/earthorigin.htm. Accessed 10 Jan 2013
3. UN Intergovernmental Panel on Climate Change (2005) Special report on carbon dioxide capture and storage. In: Metz B, Davidson O, de Coninck H et al (eds) Prepared by working group III of the intergovernmental panel on climate change. Cambridge University Press, Cambridge, UK
4. Shaw G (2006) A (not so) brief history of carbon on Earth. NASA Astrobiology Institute seminar. https://astrobiology.nasa.gov/nai/seminars/other-seminar-series/university-of-washington-seminars/2006/10/24/a-not-so-brief-history-of-carbon-on-earth/. Accessed 10 Jan 2013
5. Grubler A, Nakicenovic N (1996) Decarbonizing the global energy system. Technol Forecast Soc Change 53:97–110
6. United Nations Economic Commission for Europe (2013) UNECE/FAO survey confirms wood as leading source of renewable energy in the UNECE region. Press release. http://www.unece.org/index.php?id=32790. Accessed 12 May 2013
7. Carpenter A (1988) Coal classification. IEA coal research report, IEACR/12, London, UK. 104 pp
8. Nakićenović N (1996) Freeing energy from carbon. Daedalus 125:95–112
9. Johnson J (2012) The new Saudi Arabia. Chem Eng News 90:30–31
10. British Petroleum (2012) Statistical review of world energy. http://www.bp.com/content/dam/bp/pdf/Statistical-Review-2012/statistical_review_of_world_energy_2012.pdf. Accessed 1 Sep 2013

11. International Energy Agency (2013) Tracking clean energy progress. www.iea.org/publications/tcep_web.pdf. Accessed 10 Aug 2013
12. International Energy Agency (2012) World Energy Outlook. OECD/IEA. Paris, France. http://www.iea.org/publications/freepublications/publication/English.pdf. Accessed 25 Jan 2013
13. Johnson J (2012) US energy supply and demand. Chem Eng News 90:10
14. Patel S (2013) World energy outlook foresees distinct generation shift. Power January:8–12
15. Johnson J (2013) CO_2 capture effort progresses. Chem Eng News 91:33
16. US Energy Information Agency (2013) Petroleum Marketing Monthly. July 2013. http://www.eia.gov/petroleum/marketing/monthly/pdf/hilites.pdf. Accessed 15 Sep 2013
17. International Energy Agency (2009) World Energy Outlook. OECD/IEA. Paris, France. ISBN: 978 92 64 06130 9
18. Jackson P (2009) The future of global oil supply: Understanding the building blocks. Cambridge Energy Research Associates. Special Report. www.cera.com Accessed 10 May 2010
19. Schmit J (2009) Fuel-efficient cars help steady oil demand. USA Today, October 14. p. 3B
20. Konrad T (2012) The end of elastic oil. Forbes, January 26. http://www.forbes.com/sites/tomkonrad/2012/01/26/the-end-of-elastic-oil/. Accessed 12 Oct 2012
21. Sebastian S (2013) FuelFix (Statoil). Texas pumping more oil than some OPEC countries. July 10, 2013. http://fuelfix.com/blog/2013/07/10/texas-oil-surges-to-highest-level-since-1984/?cmpid=hpts/. Accessed 20 Aug 2013
22. Koerner B (2007) The trillion barrel tar pit. WIRED Magazine. http://wired.com/wired/archive/12.07/oil.html. Accessed 19 Sep 2007
23. Argonne National Laboratory (2007) Tar sands. http://ostseis.anl.gov/guide/tarsands/index.cfm. Accessed 19 Sep 2007
24. Biello D (2013) A dirty business. Sci Am. February 14
25. Biello D (2013) How much will tar sands oil add to global warming? Sci Am. http://www.scientificamerican.com/article.cfm?id=tar-sands-and-keystone-xl-pipeline-impact-on-global-warming&print=true. Accessed 26 Jul 2013
26. Hargreaves S (2012) Obama's energy plan: The winners and winners, CNN Money. http://money.cnn.com/2012/01/30/news/companies/obama_energy/index.htm?hpt=hp_t3. Accessed 4 Feb 2012
27. Hargreeves S (2013) California could be next oil boom state, CNN Money. http://money.cnn.com/2013/01/14/news/economy/california-oil-boom/index.html?iid=HP_LN&htp=hp_t2. Accessed 10 Feb 2013
28. Petroff A (2013) US oil boom causing energy upheaval. CNN Money. http://money.cnn.com/2013/05/14/news/world/oil-iea-demand/index.html?hpt=p_t2. Accessed 1 Jun 2013
29. Reuters (2012) UPDATE 5-U.S. to overtake Saudi as top oil producer—IEA. http://www.reuters.com/article/2012/11/12/iea-oil-report- idUSL5E8MC7GA20121112. Accessed 15 Jan 2013
30. Kerr R (2012) An oil gusher in the offing, but will it be enough? Science 338:1139
31. US Energy Information Agency (2010) Schematic geology of natural gas resources. http://www.eia.doe.gov/oil_gas/natural_gas/special/ngresources/ngresources.html, Accessed 29 Mar 2010.
32. Holditch S (2006) Tight gas sands. J Petrol Technol June:84–90
33. International Energy Agency (2011) World Energy Outlook. Are we entering a Golden Age of Gas? IEA. Paris, France.
34. Kerr R (2010) Natural gas from shale bursts onto the scene. Science 328:1624–1626
35. US Energy Information Agency (2013) Annual Energy Outlook 2013. Early release. www.eia.gov/forecasts/aeo/erindex.cfm. Accessed 15 Sep 2013.
36. Hess G (2011) Energy panel sees risks of gas drilling. Chem Eng News 89:25
37. US Energy Information Agency (2011) Review of emerging resources: US shale gas and shale oil plays produced by INTEK. July 2011. www.eia.gov/analysis/studies/usshalegas/pdf/usshaleplays.pdf. Accessed 15 Aug 2012

38. Vidic R, Brantley S, Vandenbossche J et al (2013) Impact of shale gas development on regional water quality. Science 340:826. doi:10.1126/science.1235009
39. US Energy Information Agency (2011) World shale gas resources: An initial assessment of 14 regions outside the Unites States. www.eia.gov/analysis/studies/worldshalegas/pdf/fullreport.pdf. Accessed 20 Jul 2012
40. Hargreaves S (2010) Natural gas: fuel of the future. CNN. http://money.cnn.com/2010/03/29/news/economy/natural_gas/. Accessed 29 Mar 2010
41. Manning R (2012) Shale revolution shakes the world. Ideas Laboratory. Atlantic Council. 22 Oct 2012. http://www.ideaslaboratory.com/2012/10/22/robert-manning-shale-revolution-shakes-the-world/?utm_source=outbrain&utm_medium=relatedcontent&utm_campaign=ilo#. Accessed 19 Dec 2012
42. Chazan G, Wiesmann G (2013) Shale gas boom sparks EU coal revival. CNN edition. Business. http://edition.cnn.com/2013/02/03/business/europe-shale-gas-revival/index.html?hpt=hp_t3. Accessed 28 Aug 2013
43. Doyle A (2013) Norwegian sea can hold 100 years of Norway's CO_2. Sci Am. http://www.scientificAmercian.com/article.cfm?id=Norwegian-sea-can-hold100-years-of-norway-s-co2-emissions/. Accessed 20 Sep 2013
44. US National Energy Technology Laboratory (2011) Energy resource potential of methane hydrate. www.netl.doe.gov/technologies/oil-gas/.../Hydrates/.../MH_Primer2011.pdf. Accessed 20 Jul 2013
45. Milkov A (2004) Global estimates of hydrate-bound gas in marine sediments: how much is really there. Earth-Sci Rev 66:183–197
46. National Resources Canada (2006) Earth sciences sector. http://ess.ntcan.gc.ca/2002)2006/ghff/index_e.php. Accessed 28 Jun 2010
47. US Geological Survey (2008) Assessment of gas hydrate resources on the north slope Alaska, fact sheet 2008-3073. USGS, Washington, DC
48. Counsel of Canadian Academies (2008) Energy from gas hydrates: assessing the opportunities and challenges for Canada. www.scienceadvice.ca/documents/(2008_07_07)_GH_Report_in_Focus.pdf. Accessed 22 Sept 2008
49. US National Energy Technology Laboratory (2011) Methane hydrates. http://www.netl.doe.gov/technologies/oil-gas/future supply/methanehydrates/projects/. Accessed 28 Jun 2010
50. The National Academies Press (2010) Realizing the energy potential of methane hydrate for the United States. Committee chair: C. Paull. The National Academies Press, Washington, DC. ISBN-10: 0-309-14889-8
51. Goodall C (2007) How to live a low-carbon life. Earthscan, London, UK
52. InflationData.Com (2012) Historical crude oil prices. Updated 14 June 2012. http://inflationdata.com/inflation/inflation_rate/historical_oil_prices_table.asp. Accessed 10 Jan 2013
53. International Energy Agency (2012) World energy demand and supply. 15 March 2012. www.iaea.org/.../schools/.../04_Rogner_World_Energy_D%26S.pdf. Accessed 10 Feb 2013
54. International Air Transport Association (IATA) (2012) Fuel conservation. http://www.iata.org/whatwedo/ops-infra/fuel/Pages/fuel-conservation.aspx. Accessed 12 Dec 2012
55. Hubbert MK (1956) Nuclear energy and the fossil fuels. Amer. Petr. Inst. drilling and production practice proceedings, 1956, p. 5
56. Hubbert MK (1969) Energy resources. Nat. research council, committee on resources and man, resources and man. W.H. Freeman and Co., San Francisco, CA, p 196
57. Deffeyes KS (2005) Beyond oil: the view from Hubbert's peak. Hill and Wang, New York
58. Campbell C (2000) Peak oil, presentation at Technical Univ. of Clausthal, Germany. http://www.oilcrisis.com/de/lecture.html. Accessed 24 Jul 2012
59. British Petroleum (2012) Statistical data http://www.bp.com/sectiongenericarticle800.do?categoryID=9037169&contentId=7068608. Accessed 15 Mar 2013
60. Papier R (2012) How much oil does the world produce? Consumer energy report. http://www.consumerenergyreport.com/2012/06/25/how-much-oil-does-the-world-produce/. Accessed 10 Jan 2013

61. Thompson M (2012) US to become biggest oil producer—IEA, CNN Money. http://money.cnn.com/2012/11/12/news/economy/us-oil-production-energy/index.html?iid=EL. Accessed 10 Dec 2012
62. World Energy Council (WEC) (2007) Survey of energy resources (21 ed.) ISBN 0-946121-26-5
63. Greene D (2010) Oil peak or panic? Science 328:828
64. Gurria A (2013) Carbon price vital to address climate change. Responding to climate change (RTCC). 9 Oct 2013. http://www.rtcc.org/2013/10/09/oecd-chief-carbon-price-vital-to-address-climate-change/. Accessed 14 Oct 2013
65. British Petroleum (2012) Statistical review of world energy. http://www.bp.com/liveassets/bp_internet/globalbp/globalbp_uk_english/reports_and_publications/statistical_energy_review_2011/STAGING/local_assets/pdf/statistical_review_of_world_energy_full_report_2012.pdf. Accessed 16 Jan 2013
66. Frum D (2013) "Peak oil" doomsayers proved wrong, CNN. http://www.cnn.com/2013/03/04/opinion/frum-peak-oil/index.html?hpt=hp_t3. Accessed 10 Apr 2013
67. International Energy Agency (2012) Oil. http://www.iea.org/aboutus/faqs/oil/. Accessed 17 Jan 2013
68. Auzanneau M (2010) Washington considers a decline of world oil production as of 2011. Le Monde. http://petrole.blog.lemonde.fr/2010/03/25/washington-considers-a -decline-of-world-oil-production-as-of-2011/. Accessed 12 Jan 2013
69. Butler J (2012) Peak oil: undulating plateau shaped by price, Seeking alpha. http://seekingalpha.com/article/846571-peak-oil-undulating-plateau-shaped-by-price. Accessed 15 Jan 2013
70. US Environmental Protection Agency (2011) Inventory of U.S. greenhouse gas emissions and sinks: 1990–2011. US-GHG-Inventory-2011-Chapter-1-Introduction.pdf. Accessed 12 Dec 2012
71. Cable Network News (2012) NASA scientist links climate change, extreme weather. http://www.cnn.com/2012/08/05/us/climate-change/index.html. Accessed 10 Sept 2012
72. U.N. Intergovernmental Panel on Climate Change (2013) Working group I contribution to the ipcc fifth assessment report climate change 2013: the physical science basis. Summary for policymakers. http://www.climatechange2013.org./images/uploads/WGIAR5-SPM_Approved 27Sep2013.pdf. Accessed 27 Sep 2013
73. Marcott S, Shakun J, Clark P et al (2013) A reconstruction of regional and global temperature for the past 11,300 years. Science 339:1198–1202
74. Doyle A (2013) Climate scientists struggle to explain warming slowdown. Reuters. http://www.reuters.com/assets/print?aid=USBRE93F0AJ20130416. Accessed 10 May 2013
75. Pittock B (2009) Climate change: the science, impacts and solutions, 2nd edn. CSIRO, Melbourne
76. Morriarty P, Honnery D (2011) Rise and fall of the carbon civilization. Springer, London, p 218
77. U.N. Intergovernmental Panel on Climate Change (2007) Fourth assessment report: climate change 2007 climate change 2007: working group ii: impacts, adaptation and vulnerability. http://www.ipcc.ch/publications_and_data/ar4/wg2/en/spmsspm-d.html. Accessed 10 Mar 2010
78. Matthews D, Solomon S (2013) Irreversible does not mean unavoidable. Science 340:438–439
79. Koebler J (2013) Study: Opinions on climate change rise and fall with the temperature. US News.http://www.usnews.com/news/articles/2013/02/06/study-opinions-on-climate-change-rise-and-fall-with-the-temperatures. Accessed 10 Mar 2013
80. Bennett O (2013) What climate change? Fewer people than EVER believe the world is really warming up. Express. http://www.express.co.uk/news/uk/430649/What-climate-change-Fewer-people-than-EVER-believe-the-world-is-really-warming-up. Accessed 20 Oct 2013
81. Peltier R (2013) Where's the warming? Power February:6
82. Gray L (2009) "Planned recession" could avoid catastrophic climate change. The Telegraph. http://www.telegraph.co.uk/earth/earthnews/6248257/planned-recession-could-avoid-catastrophic-climate-change.html. Accessed 20 Oct 2013

83. Pielke R (2010) The climate fix. Basic Books, NY, pp 61–80
84. World health Organization (2002) The world health report, annex table 9, 2002, http://www.who.int/whr/2002/en/whr2002_annex9_10.pdf. Accessed 12 Oct 2010
85. Jacobson M (2009) Review of solutions to global warming, air pollution and energy security. Energy Environ Sci 2:148–173
86. Evans R (2007) Fueling our future. An introduction to sustainable energy. Cambridge Univ. Press, Cambridge, UK
87. Votsmeier M, Kreuzer T, Gieshoff J et al (2002) Automobile exhaust control. Wiley-VCH, Ullmann's encyclopedia of industrial chemistry. doi:10.1002/14356007.a03_189.pub2
88. Kerr R (2013) Soot is warming the world even more than thought. Science 339:382
89. Johnson J (2013) Projected CO_2 cost impacts rise. Chem Eng News 91:21
90. Johnson J (2009) Fossil-fuel costs. Chem Eng News 87:6
91. World Bank (2010) World development report 2010: development and climate change. The Int. Bank for Reconstruction and Development. doi:10.1596/978-0-8213-7987-5. ISBN 978-0-8213-7987-5. http://go.worldbank.org/BKLQ9DSDDU0. Accessed 2 Feb 2011
92. Adger W, Arnell N, Tompkins E (2005) Successful adaptation to climate change across scales. Environ Change 15:77–86. doi:10.1016/j.gloenvcha.2004.12.005
93. California Climate Change Adaptation (2011) http://www.climatechange.ca.gov/adaptation/index.html. Accessed 15 Mar 2012
94. Kaufman L (2011) City prepares for a warm long-term forecast. New York Times. Accessed 22 May 2011
95. Revkin A (2011) Cities embrace the adaptation imperative. The New York Times, May 23, 2011. http://dotearth.blogs.nytimes.com/2011/05/23/Cities-embrace-the-adaptation-imperative/?ref=earth. Accessed 22 May 2012
96. U.N. Intergovernmental Panel on Climate Change (2007) Assessment of adaptation practices, options, constraints and capacity. http://www.ipcc.ch/pdf/assessment-report/ar4/wg2/ar4-wg2-chapter17.pdf. Accessed 29 Aug 2010
97. Sutter J (2013) Don't feed the (polar) bears. CNN. http://www.cnn.com/2013/02/20/opinion/sutter-polar-bears-/index.html?hpt=hp_c2. Accessed 1 Mar 2013
98. Mann M (2009) Do global warming and climate change represent a serious threat to our welfare and environment? Soc Philos Policy Foundation 26:193–230. doi:10.1017/S0265052509090220
99. Hogue C, Scott A (2012) Steps toward new climate treaty. Chem Eng News 90:8
100. Morgan J (2012) More voices needed in climate debate. CNN. Special report. http://edition.cnn.com/2012/12/09/opinion/morgan-climate-change-talks/index.html?hpt=hp_bn8. Accessed 1 Mar 2013
101. Baum R (2012) Once more on climate change. Chem Eng News 90:3

Chapter 2
What Is So Unique About CO_2?

Abstract Considering that the concentration of CO_2 in the atmosphere is extremely low: only 400 ppm or 0.04 vol.%, it is surprising how much impact this gas exerts on life on our planet. What is so unique about CO_2? In this chapter, Greenhouse effect, radiative forcing, global warming potential, global carbon cycle, and other phenomena that control the livability of our planet are linked to unique optical and physicochemical properties of CO_2. An increasing body of scientific evidence suggests that humans are affecting the Earth's radiative and carbon balances mainly through increased emissions of greenhouse gases originating from industrial activities, land-use change, deforestation, and other practices that became prevalent during the rapid industrial development of the last two and half centuries.

2.1 Carbon and Greenhouse Effect

Considering that CO_2 concentration in the atmosphere is extremely low: only 400 ppm, and the overall amount of fossil carbon on our planet is miniscule, 3×10^{-21} wt.%, one can only wonder why there is so much anxiety and agitation in media recently about carbon— "carbon footprint," "carbon price," "carbon entanglement," "carbon tax," "carbon trading," "carbon allowances," "carbon credit," etc.—and what it has to do with our planet's environment and climate. Since not carbon itself, but the product of its combustion, CO_2, is at the center of the world's attention, then the question becomes: what is so special about CO_2, and how much of it in the air is too much?

Based on the total amount of carbon in fossil fuels, it is easy to estimate how high the atmospheric CO_2 concentration could rise if all the global resources of fossil fuels are burned to CO_2. Figure 2.1 depicts the equivalent amount of CO_2 to be released upon combusting the global resources of major types of fossil fuels: coal, oil, and natural gas (light bars), and the corresponding equivalent atmospheric CO_2 concentration (dark bars).

N. Muradov, *Liberating Energy from Carbon: Introduction to Decarbonization*, Lecture Notes in Energy 22, DOI 10.1007/978-1-4939-0545-4_2,

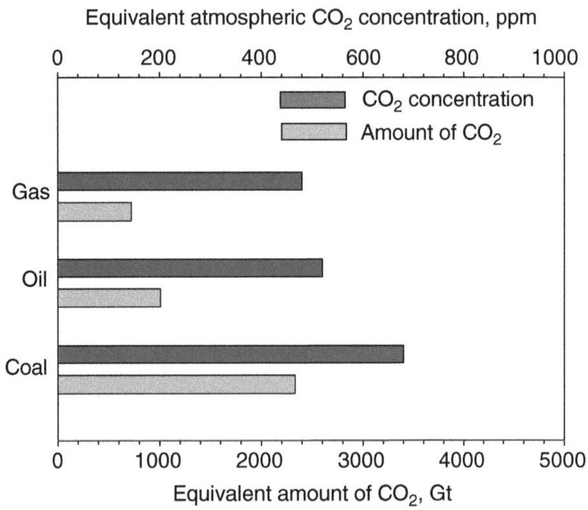

Fig. 2.1 Equivalent amount of CO_2 corresponding to the global resources of coal, oil, and natural gas (*light bars*), and the equivalent atmospheric CO_2 concentration (*dark bars*). Fossil fuel resources include recoverable reserves and do not include methane hydrates and nonconventional fossil fuels. *Source* [1]

Calculations indicate that if all fossil fuel resources are completely combusted the atmospheric CO_2 concentration would rise to 917 ppm[1] (this value is a net of any carbon absorption by natural sinks). Clearly, this is a hypothetical scenario, and it may never materialize; however, one burning question might arise: can humans live with 917 ppm CO_2 in the atmosphere? Note that this is a miniscule increase—"only" 0.052 vol.%—in the atmospheric CO_2 concentration against the current level, and, besides, CO_2 is a benign gas produced by human body that can tolerate CO_2 concentrations in air many times higher than that. In order to answer this question, let's first find out what makes CO_2 so unique.

We all know that life on the Earth originated and is sustained due to upcoming solar radiation. Quantitatively, the incident solar radiation flux is equal to about 1,370 W/m^2, which represents the amount of solar energy hitting 1 m^2 of the top of the Earth's atmosphere facing the Sun in 1 s during daytime (If averaged over the entire planet, the amount of received radiant energy would be 341 W/m.^{2}) (W is watt, or 1 J/s) [2]. This amount of radiant energy is balanced by a number of energy reflection and emission processes resulting in an equilibrium or zero-energy balance. (Otherwise, the Earth's surface temperature would have continuously increased.) Roughly a third (102 W/m^2) of the incoming solar radiation is reflected back to space by the Earth's atmosphere, clouds, and aerosols, and the main portion of solar radiant energy (239 W/m^2) is absorbed by the Earth's atmosphere and surface. In order to maintain the energy balance, our planet has to release the same amount of energy back to the space, which it does by emitting radiation. According to the Stefan–Boltzmann law, the amount of radiated energy is proportional to the object's surface temperature to the fourth power:

---

[1] ppm stands for "part per million"; hereafter ppm relates to volume units (unless otherwise indicated), e.g., 1 ppm=0.0001 vol%.

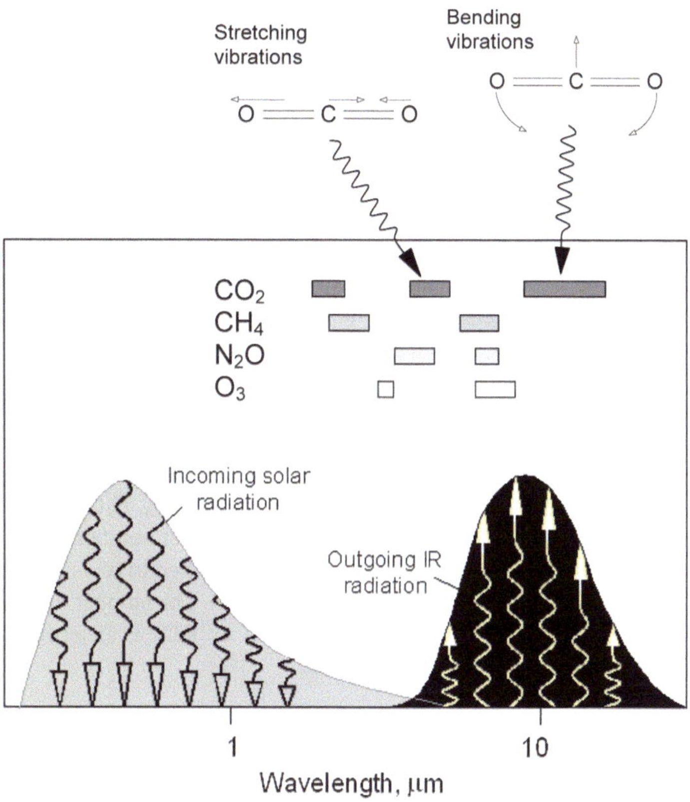

Fig. 2.2 Simplified schematic diagram of greenhouse effect. *Source* [4, 5]

$$E = \varepsilon\sigma T^4 \tag{2.1}$$

where E is the total energy radiated by a body (e.g., in W/m^2), ε is emissivity coefficient, σ is Stefan-Boltzmann coefficient, and T is absolute temperature.

If we estimate the temperature of a body that emits 239 W/m^2 of radiant energy based on this equation, we would arrive to temperature of about −19 °C [3], which is far below the average temperature on the Earth's surface: +14 °C. That temperature gap of about 33 °C is attributed to the presence of heat-trapping agents in the atmosphere called greenhouse gases (GHG): predominantly, CO_2 and water vapor, and, to a smaller extent, methane, ozone (O_3), N_2O and other GHG (note that nitrogen and oxygen—two major constituents of the atmosphere—do not exert such effect for the reasons explained below). GHG absorb and reflect radiant energy within the atmosphere, which in turn emits most of this long wavelength radiation energy back to the Earth's surface and a smaller fraction out to space. Trenberth et al. estimated that an imbalance of 0.9 W/m^2 in the energy fluxes could be attributed to the enhanced greenhouse effect [2]. Figure 2.2 depicts the schematic diagram of the greenhouse effect mechanism involving major heat-trapping gases.

The left curve shows the simplified representation of the spectrum of incoming solar radiation, which includes the wavelengths varying from about 0.2 μm to more than 4 μm, which could be broken down to ultraviolet (UV) light (0.2–0.4 μm), visible light region (0.4–0.8 μm), and infrared (IR) region (longer than 0.8 μm) (μm is micrometer, equal to 10^{-6} m). This curve closely follows the spectrum of a black body heated to about 5,500 K with the peak of the spectral curve in the visible area at about 0.5–0.6 μm. The right curve corresponds to the Earth's radiation spectrum which is associated with black body radiation extending from wavelengths of 1 to 3 μm to about 70–80 μm with the peak at about 10 μm. Due to structural and electronic properties of H_2O_{gas}, CO_2, CH_4, N_2O, and O_3 molecules, they are almost transparent to the upcoming sunlight, but very efficiently (up to 80 %) absorb outgoing IR radiation directed from the Earth surface to the space. The horizontal bars on the diagram correspond to radiation-absorbing capacities of these gases with the length of the bars being approximately proportional to the absorption bandwidth of the corresponding molecules in the IR area of the spectrum (only most important bands are shown).

The common feature of all GHG (e.g., H_2O_{gas}, CO_2, CH_4, N_2O, O_3) is that they contain at least three atoms which allow for a much greater number of fundamental molecular vibrations in response to IR excitation, compared to two-atom molecules (e.g., O_2 and N_2 that do not exert the greenhouse effect). The number of fundamental vibrations for linear three-atom molecules, such as CO_2 is determined by the following formula:

$$V = 3N - 5 \tag{2.2}$$

where V is the number of possible fundamental vibrations and N is the number of atoms in the molecule. (Note that non-linear molecules have $V = 3N - 6$ vibrations.)

Based on the above formula, CO_2 molecule has four vibrations: two stretching (symmetric and asymmetric) and two bending (in-plane and out-of-plane) vibrations (some of them are shown in Fig. 2.2). Thus, CO_2 and other GHG efficiently uptake IR radiation in their respective absorption areas and convert it into thermal vibrational energy (Note that the contribution of H_2O and CO_2 to the greenhouse effect is much greater than that of other GHG due to their relative abundance in the atmosphere.) Collectively, all GHG absorb most of the radiation emitted by the Earth surface leaving a relatively narrow gap permitting some of the thermal radiation to escape into the space and, thus, preventing thermal runaway. In order to further elucidate and quantify the impact of CO_2 and other GHG on the Earth's mean temperature and climate system, it would be useful to introduce the concepts of radiative forcing and global warming potential.

2.1.1 *Radiative Forcing Concept*

The concept of radiative forcing (RF) has been introduced by IPCC in its early assessment reports for the quantitative comparison of the impact of different natural and man-made drivers on the climate system. RF (expressed in W/m^2) is quantified as the rate of radiative energy change per unit area of the globe measured at the top

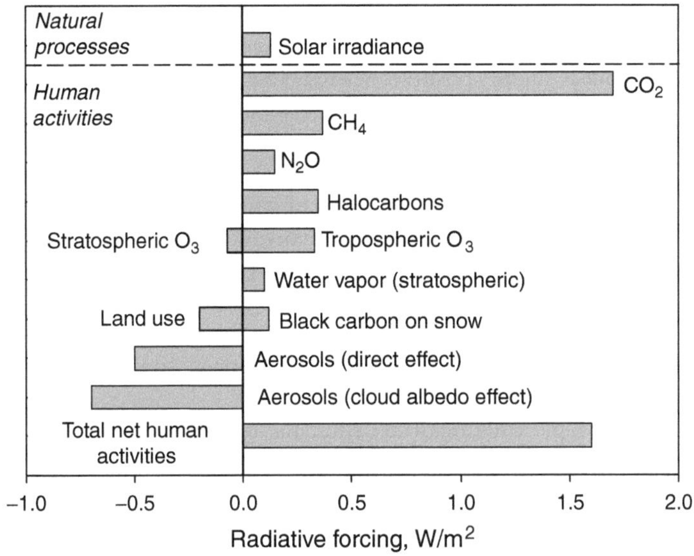

Fig. 2.3 Radiative forcings of the major natural and anthropogenic factors. RF changes since the beginning of the industrial era (ca. 1750) until 2005 are presented. *Source* [6]

of the atmosphere [6]. RF is correlated with the global mean equilibrium temperature change (ΔT_S) at the Earth's surface as follows:

$$\Delta T_S = \lambda \text{RF} \tag{2.3}$$

where λ is the climate sensitivity parameter.

For a greenhouse gas (e.g., CO_2), the change in RF can be calculated as a function of changing CO_2 concentration. For example, in a simplified first-order approximation form, the algebraic expression for ΔRF for CO_2 is [7]

$$\Delta RF = 5.35 \ln \frac{C_{CO_2}}{C^0_{\ CO_2}} \tag{2.4}$$

where ΔRF is radiative forcing (in W/m^2), C_{CO_2} and $C^0_{\ CO_2}$ are the CO_2 variable and reference concentrations, respectively, in ppm.

Since the relationship between CO_2 concentration and radiative forcing (ΔRF) is logarithmic, the increase in atmospheric CO_2 concentrations would have a progressively smaller warming effect. At a typical λ value of 0.8 K per W/m^2, doubling of CO_2 concentration would result in a mean temperature increase of 3 K.

According to the definition, the RF value is positive when the energy of the Earth–atmosphere system increases (i.e., warming effect), and, correspondingly, the RF is negative if the energy of the Earth–atmosphere system decreases (cooling effect) in response to affecting factors. The RF values for some major natural and human-induced factors are shown in Fig. 2.3.

Natural radiative forcings mainly result from changes in solar irradiance and major volcanic eruptions. Solar source forcing arises from several direct and indirect factors and its value is slightly positive, whereas volcanic eruptions spewing immense amounts of aerosols into the atmosphere create short-lived negative forcing [6]. Human activities have greatly contributed to the changes in radiative forcings since the Industrial Revolution. As can be seen in Fig. 2.3, CO_2 has the greatest increase in the RF values compared to other GHG (all of which have positive RF). The forcing effect of aerosols is rather complex and involves a number of direct (e.g., reflection and absorption of solar and long wavelength radiation in the atmosphere) and indirect (e.g., the changes aerosol particles exert on the optical properties of clouds) effects. The net effect of all aerosol types results in negative RF values. All anthropogenic climate agents, both cooling and warming ones, summarily add up to the forcing value of 1.6 W/m^2.

Over the last couple centuries (i.e., since the beginning of the industrial era), human activities not only altered the abundance of atmospheric GHG (via burning of fossil fuels) but also changed the land cover over the vast areas on the Earth's surface mainly through agriculture and deforestation. These activities either directly (e.g., via altering the reflectivity of the land surface) or indirectly (e.g., via increasing the CO_2 and CH_4 concentrations in the atmosphere) resulted in appreciable negative changes in the RF values. When assessing the impact of natural factors and human activities on climate system, one should also take into consideration timescales during which a given RF term would persist in the atmosphere after associated emissions or changes are ceased. The available data indicate that the lifetime of various RF factors could last from days for aerosols to 100 years for long-lived GHG and surface albedo (including land use changes) [6].

IPCC 2007 report underscored the interconnection between radiative forcing factors and climate. Figure 2.4 illustrates how RF factor is linked to other aspects of climate change.

Natural processes and influences (e.g., processes on the Sun, volcanoes, changes in the Earth's orbit) as well as human activities (e.g., burning of fossil fuels, industrial GHG emissions, land use) cause direct and indirect changes in the climate change drivers (e.g., release of GHG, aerosols, changes in clouds and solar irradiation, etc.). These changes could lead to specific RF changes (either positive or negative) and noninitial radiative effects (e.g., changes in evaporation). RF and noninitial radiative effects cause climate perturbations and responses. A variety of biogeochemical processes could generate feedback from climate change to its drivers (e.g., increase in methane emissions from wetland during warmer climate). Altering human activities could be one of the potential approaches to mitigating climate change (dashed line).

Recently, the contribution of soot (or black carbon, BC, one of the products of fossil fuel combustion) to overall RF has been revisited [8]. Soot could affect the Earth radiative balance and climate via many routes: absorbing solar radiation, darkening ice and snow, shrinking cloud droplets, etc. The new estimate for the soot RF value −1.1 W/m^2 is roughly twice as large as the RF estimate reported by IPCC in its 2007 report. This puts soot second behind the major RF agent—CO_2—which

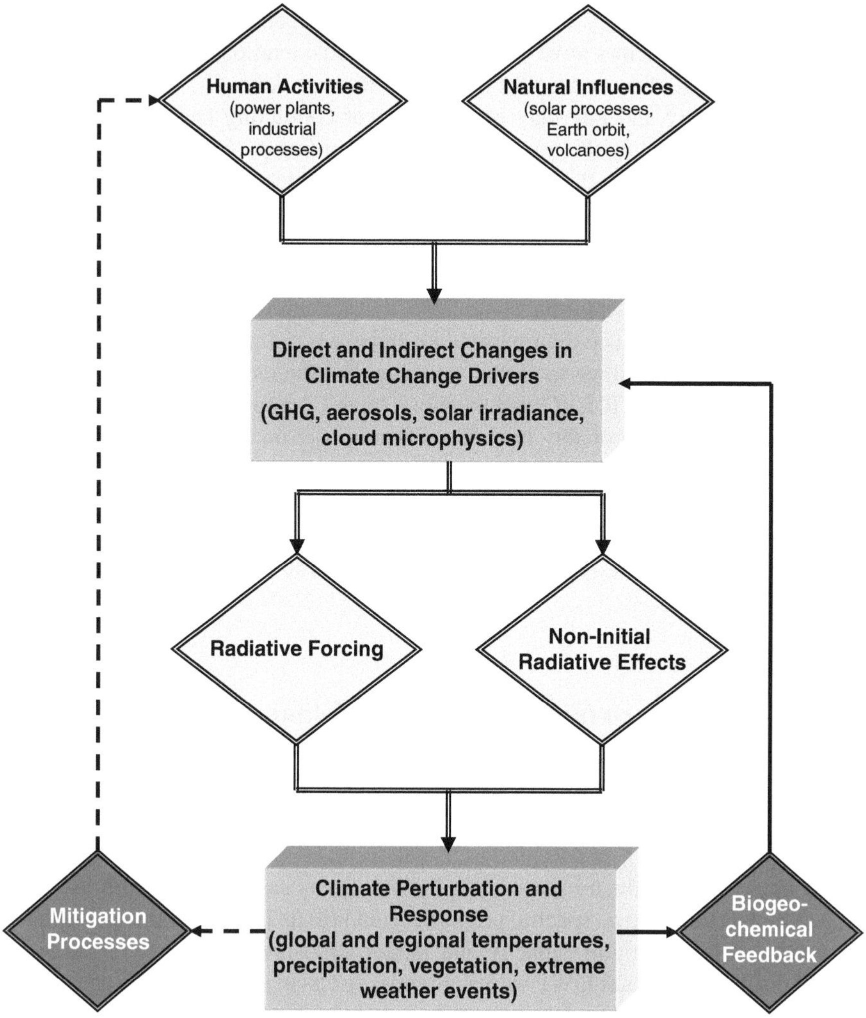

Fig. 2.4 Interconnection between radiative forcing and climate change. *Source* [6]

accounts for 1.66 W/m^2. On the other hand, many processes that produce BC (e.g., burning of coal or heavy oil) also produce compounds (e.g., sulfur aerosols) that might exert a cooling effect by reflecting sunlight back into space. Forest or brush fires produce soot, but they also produce microscopic particles of unburned organic carbon that can brighten clouds and reflect more radiation, thus, providing some cooling effect. To add to the complexity, besides BC, the atmosphere contains light-absorbing organic "brown" carbon (BrC). According to some estimates, BrC accounts for 15–50 % of light absorption in the atmosphere and in snow and ice [9]. In many climate models, BrC is combined with BC and included in the same inventories, although they have different optical properties and source and sink patterns. All these factors have to be taken into consideration when estimating the overall RF impact of carbonaceous nano- and microparticles.

The complexity and interplay of the different conflicting factors affecting RF could be underscored by the impact of sulfur-containing particulates. Burning coal is the main way of adding the vast amounts of tiny sulfate particles (sulfate aerosols) in the atmosphere. These particles can either directly (i.e., by reflecting sunlight) or indirectly (i.e., acting as condensation nuclei for cloud formation that reflect solar radiation) cause a cooling effect on climate. Taking this effect into consideration, there has been a counterintuitive conclusion, that if leading world emitters (e.g., the USA, China, India) reduce their particulate emissions in the near future as planned, it would actually contribute to global warming. But the recent findings of the joint project conducted by CICERO and the Norwegian Computing Center indicate that particulate emissions probably have less of an impact on climate through indirect cooling effects than previously thought [10].

Summarizing, the above data clearly indicate that changes in RF resulting from human activities far exceed that from natural sources, which implies that human activities can potentially impact the climate system more profoundly compared to natural sources.

2.1.2 Global Warming Potential of Carbonaceous Gases

Global warming potential (GWP) is a widely used metric tool that provides a means of comparing the capacity of different greenhouse agents to contribute to global warming. GWP of CO_2 is accepted as a unity (i.e., $GWP_{CO_2} = 1$). GWP of a greenhouse agent is determined by two main factors: (a) its capacity to absorb IR radiation (which is linked to its spectral properties), and (b) its lifetime in the atmosphere. The greater IR absorption capacity and longer atmospheric lifetime of the agent would result in its greater GWP values. In order to calculate GWP of the agent *i*, the time-integrated global mean RF value of the agent (*i*) 1 kg of which was pulse emitted to the atmosphere is to be divided by the corresponding RF value of the reference gas CO_2 as follows [11]:

$$GWP_i = \frac{\int_0^{TH} RF_i(t)\,dt}{\int_0^{TH} RF_{CO_2}(t)\,dt} = \frac{\int_0^{TH} a_i\left[C_i(t)\right]\mathrm{d}t}{\int_0^{TH} a_{CO_2}\left[C_{CO_2}(t)\right]\mathrm{d}t} \tag{2.5}$$

where TH is a time horizon, RF_i is the global mean RF of the agent *i*, a_i is the RF per unit mass increase in atmospheric abundance of the agent *i*, and $[C_i(t)]$ is the time-dependent abundance of the agent *i*. The denominator includes the corresponding values for the reference gas CO_2.

Table 2.1 Global warming potentials of selected greenhouse gases for the time horizons of 20 and 100 years

	Time horizon (years)	
Greenhouse gas	20	100
Methane (CH_4)	72	25
Nitrous oxide (N_2O)	289	298
Chlorofluorocarbon (CCl_2F_2)	11,000	10,900

Source [12]

GWP values for major GHG for the time horizons of 20 and 100 years (average uncertainty of the GWP values are ±35 %) are shown in Table 2.1.

GWP values for the different GHG vary in the wide range: from a few units to tens of thousands. GWP of methane is 72 and 25 times greater than that of CO_2 over the time horizon of 20 and 100 years, respectively. GWP of N_2O and chlorofluorocarbons are about 300 times and 4 orders of magnitude greater than that of CO_2, respectively (Note that they do not change markedly over the shown time horizon.) The above data indicate that compared to other GHG, CO_2 is a rather mild greenhouse agent, but, nevertheless, it can exert a profound impact on the climate system for the reasons explained below.

2.2 Trends in Atmospheric Greenhouse Gases

2.2.1 *Increase in Atmospheric CO_2: Natural vs. Human-Induced Factors*

The atmospheric CO_2 concentration has been measured with a great accuracy since the 1950s by C. Keeling at Mauna Loa, Hawaii (USA) using a high-precision IR gas analyzer. Since then, continuous CO_2 measurements are being conducted at other sites in both hemispheres: Baring Head (New Zealand), Cape Grim (Australia) and South Pole (the selection of these particular locations is dictated by the lack of significant CO_2 sources or sinks nearby). The results of direct and indirect measurements of atmospheric CO_2 level imply that it has increased from preindustrial concentration of 275–285 ppm to the benchmark 400 ppm measured in May 2013 by Mauna Loa Observatory (i.e., the growth by 45 % over two and half centuries) [6, 13]. The measurements also indicated that the rate of CO_2 growth is accelerating: the average annual rate of increase was 0.7 ppm per year during the 1950s, 1.4 ppm per year during 1955–1995, and 1.9 ppm per year during 1995–2005 [13, 14]. In its 2007 report, IPCC projected the rate of CO_2 growth at 1.9 ppm per year [6], but these projections have proved rather conservative, because during the decade of 2002–2011, the CO_2 growth rate increased to 2.07 ppm per year [14].

In order to ascertain whether this increase in atmospheric CO_2 concentrations is due to natural causes or the result of human activity, let's consider several factors:

- There is a clear correlation between the amount of anthropogenic CO_2 released to the atmosphere and the increase in atmospheric CO_2 concentration during last decades.
- Atmospheric oxygen is declining proportionately to CO_2 increase and fossil fuel combustion.
- For the last half century, the CO_2 airborne fraction (AF) parameter remained consistent and averaged at 0.55 (the AF parameter is the ratio of the increase in atmospheric CO_2 concentration to fossil fuel-derived CO_2 emissions). AF has been introduced to assess short- and long-term changes in the atmospheric carbon content; in particular, AF of 0.55 indicates that the oceans and terrestrial ecosystems have cumulatively removed about 45 % of anthropogenic CO_2 from the atmosphere over the last half century [6].
- The isotopic signature of fossil fuels (e.g., the lack of ^{14}C and the depleted level of ^{13}C carbon isotopes) is detected in atmospheric CO_2.
- There exists an interhemispheric gradient in the atmospheric CO_2 concentrations in the Northern and Southern Hemispheres. In particular, the predominance of fossil-derived CO_2 emissions in more industrially developed Northern Hemisphere (compared to the Southern Hemisphere) causes the occurrence of the atmospheric CO_2 gradient in the amount of about 0.5 ppm per GtC per year [6].
- There have been dramatic changes in RF_{CO2} values over the last decades. For example, during 1995–2005, the RF_{CO2} increased by about 0.28 W/m^2 (or about 20 % increase), which represents the largest increase in RF_{CO2} for any decade since the beginning of the industrial era. RF_{CO2} in 2005 was estimated at $RF_{CO2} = 1.66 \pm 0.17$ W/m^2 (corresponding to the atmospheric CO_2 concentration of 379 ± 0.65 ppm), which is the largest RF among all major forcing factors shown in Fig. 2.3.
- The data show that the changes in the land use greatly contributed to the RF_{CO2} value in the amount of about 0.4 W/m^2 (since the beginning of the industrial era). This implies that the remaining three quarters of RF_{CO2} can be attributed to burning fossil fuels, cement manufacturing, and other industrial CO_2 emitters [6].

The above facts suggest it is *extremely likely* that there is a link between the increase in the atmospheric CO_2 concentration and the greater than before levels of CO_2 emissions from human activities. The IPCC 2013 Assessment Report underscored that "*The atmospheric concentrations of greenhouse gases carbon dioxide (CO_2), methane (CH_4), and nitrous oxide (N_2O) have all increased since 1750 due to human activity*" [15]. Atmospheric concentrations of these gases substantially exceed the highest concentrations recorded in ice cores during the past 800,000 years, and the mean rates of their increase over the past century are (with very high confidence) unprecedented in the last 22,000 years [15].

In the next sections, the potential impact of two other major GHG, water vapor and methane, on our planet's ecosphere and climate will be discussed. Other GHG (N_2O, ozone, halocarbons) are out of scope of this book; the information on these GHG could be found in [6] and other publications.

2.2.2 Role of Water Vapor

Water vapor is the most abundant and dominant GHG in the atmosphere and one of the key variables of the climate system. Water vapor accounts for the lion's share of IR absorption in the atmosphere (according to some estimates, about 60 % of the natural greenhouse effect, for clear skies) [16]. The atmospheric concentration of water vapor depends largely on temperature and varies in a very wide range, from less than 0.01 % in extremely cold regions up to 3 % by mass in saturated air at about 32 °C [17]; compared to other GHG, it is neither long-lived nor well mixed in the atmosphere. As an additional distinction from other GHG, atmospheric water can exist in several physical states: gaseous, liquid, and solid (these forms of water can coexist in the atmosphere for a short period of time).

Despite the abundance of water vapor in the atmosphere, it is believed that human activities *do not directly* affect the average global concentration of water vapor in the atmosphere (stratospheric water vapor has relatively low RF value). However, the radiative forcing produced by the increased concentrations of other greenhouse gases may *indirectly* affect the hydrologic cycle. As a result, human activities could profoundly *indirectly* affect the atmospheric water vapor concentration and, ultimately, the climate system, via a number of mechanisms. For example, the accumulation of CO_2 (or CH_4) could lead to warming of the atmosphere and increasing its water vapor content by enhancing the evaporation process. Studies show that for every 1 °C increase in the global temperature, the specific humidity rises in average by 5.7 % and 4.3 % over the ocean and land surfaces, respectively (global average is about 4.9 % increase per 1 °C) [6]. Warmer atmosphere via an increased water holding capacity (i.e., increased concentrations of water vapor) could potentially affect the formation of clouds, which can either absorb or reflect solar radiation.

Aircraft contrails, which consist of water vapor and other aircraft exhausts, have a radiative forcing effect similar to that of clouds [18]. Another mechanism dealing with the indirect effect of water vapor formation in the stratosphere involves photochemical oxidation of organic and inorganic compounds originating from human activities, such as anthropogenic methane, ammonia, uncombusted hydrocarbons, H_2S, and volatile organic compounds. Due to the sensitivity of water vapor content to temperature, a variety of models predict that this factor could provide the largest positive feedback with regard to the climate system [19] (see discussion of the water vapor feedback mechanism in Sect. 2.5.3).

A direct link between the changes in the concentration of stratospheric water vapor and mean surface temperature has been reported by a group of scientists from the US National Oceanic and Atmospheric Administration (NOAA) and other institutions [20]. It was shown that stratospheric water vapor concentrations gradually increased between 1980 and 2000 causing a near decade-long (1990–2000) warming by about 30 % (compared to the scenario neglecting this change). However, since 2000, the stratospheric water vapor concentration dropped by about 10 %, which is consistent with the observation of global surface temperature remaining almost flat since the late 1990s, despite ongoing increases in GHG emissions. These trends indicate that stratospheric water vapor is a very important factor in global surface temperature variations, and the terms related to water vapor impact should be fully represented in advanced climate models.

2.3 Methane: Just Another Greenhouse Gas or a Sleeping Giant?

About 55 million years ago (during the Paleocene–Eocene Thermal Maximum, PETM), our planet experienced "a hot flash," i.e., sudden surge in surface temperature, which lasted for over hundred millennia [21]. Many researchers linked that extreme warming to a dramatic perturbation of the Earth's carbon cycle caused by massive methane emissions. What event or phenomenon triggered these massive methane emissions is still open to debates. But even more important question remains: if this giant outbreak of methane so dramatically impacted the global radiative energy balance and climate, could this happen again? Indeed, some scientists raised alarm that one of the dire consequences of the climate change could be the widespread liberation of (currently dormant) immense quantities of methane gas trapped in the marine and permafrost sediments [22–25].

2.3.1 Methane as a Potent Greenhouse Gas

As GHG, methane (CH_4) is significantly more potent than CO_2 (based on its GWP values), and it has second (next to CO_2) largest radiative forcing ($RF_{CH4} = 0.48 \pm 0.05$ W/m^2) among all GHG (see Fig. 2.3); as such, it could potentially exert a substantial impact on the global radiative energy balance and, consequently, climate. While the IPCC 2007 report [6] recommended $GWP_{methane}$ values of 25 and 72 over integrated 100 and 20 years time horizons, respectively, more recent data reported by Shindell et al. in *Science* magazine indicate that better accounting for the interaction of methane with other radiatively active matter in the atmosphere would put mean $GWP_{methane}$ values at 105 and 33 for the same 20 and 100 years time horizon, respectively [26]. Using this value of methane GWP, Howarth et al. estimated that methane would be responsible for 44 % of the warming impact of the entire US GHG inventory (including CO_2 and other man-made emissions) over 20 years time frame [27]. An important implication of these data was that, in the short term, the climate system is more responsive to changes in methane emissions than CO_2 emissions [28]. This underscores the urgency of taking immediate and strong measures to reduce methane emissions worldwide.

Although methane is a very potent GHG, in contrast to CO_2, it does not remain in the atmosphere for too long: its residence time in the atmosphere is estimated at

8.4 years [6] (compared to hundreds years for CO_2). Methane is primarily removed (via oxidation to CO_2) from the atmosphere via a chain of chemical and photochemical reactions involving very reactive hydroxyl radicals ($OH^{\bullet}$), the primary source of which is UV-assisted ozone photodissociation in the presence of water vapor [29]:

$$O_3 + h\nu \rightarrow O_2 + O(^1D) \quad (radiation\ wavelengths\ \nu\ below\ 350\ nm) \tag{2.6}$$

where $O(^1D)$ is an electronically excited state of oxygen atom, and $h\nu$ is a light photon

$$O(^1D) + H_2O \rightarrow 2OH^{\bullet} \tag{2.7}$$

$OH^{\bullet}$-radicals attack methane molecules converting them first into oxygenated compounds (e.g., formaldehyde, CH_2O) and then to CO_2 and water as follows:

$$OH^{\bullet} + CH_4 \rightarrow CH_3^{\bullet} + H_2O \tag{2.8}$$

$$CH_3^{\bullet} + O_2 \rightarrow \ldots \rightarrow CH_2O \xrightarrow{O_2 + h\nu} \ldots \rightarrow CO_2 + H_2O \tag{2.9}$$

Thus, methane released to the atmosphere is gradually converted into CO_2 and ultimately increases CO_2 levels in the atmosphere. Other important methane sinks include biological oxidation in soil, the loss to the stratosphere, and reactions with halogen (e.g., $Cl^{\bullet}$) radicals. The amount of methane removed by the sinks is estimated at 581 Mt (CH_4) per year [6].

2.3.2 *Historical Trends in Atmospheric Methane Concentration*

Over the last millennia, the atmospheric methane concentration varied in the range of 400–700 parts per billion (ppb) [30]. Similar to CO_2, the atmospheric concentration of methane has dramatically increased since the beginning of the Industrial Revolution. The atmospheric CH_4 concentration measurements conducted in the Northern (Mace Head, Ireland) and Southern (Cape Grim, Tasmania) hemispheres gave the values of 1,865 and 1,741 ppb, respectively [31]. In 2011, the atmospheric concentration of methane was 1,803 ppb [15]. Although this value is about 200 times less than that of CO_2, one should take into consideration that its concentration has risen by about 150 % since preindustrial times, compared to about 45 % for CO_2 [15, 32].

Atmospheric methane concentration is controlled by the balance between its sources (both natural and anthropogenic) and sinks. The major natural sources of methane are wetlands, permafrost, vegetation, termites, oceans, methane hydrates, and geothermal sources (mud volcanoes, marine and land seepage, etc.). The amount of methane (in carbon equivalent) in wetlands and permafrost (subsea and soils)

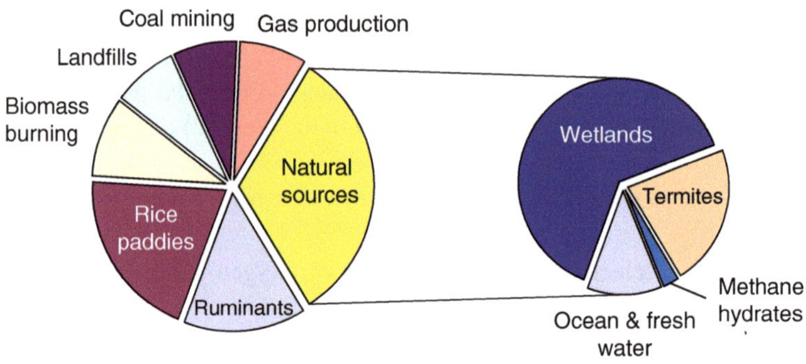

Fig. 2.5 Relative distribution of natural and man-made methane emissions from different sources. *Source* [35]

is immense: it is about twice the amount of CO_2 in the atmosphere [33]. The anthropogenic sources of methane include natural gas processing facilities, fossil fuel production and use, coal mining, landfills, ruminant animals (cattle, sheep, etc.), rice agriculture, biomass processing, and combustion. Methane emissions from living vegetation accounts for 10–30 % of the total methane emissions [34]. Figure 2.5 shows relative distribution of methane emissions from different sources. The overall amount of atmospheric methane is estimated at 4,932 Mt (CH_4) [6].

The available data indicate that although the atmospheric methane concentration increased by about 30 % over the last quarter of a century, its growth rate substantially slowed down in the late 1990s. There are several explanations to this phenomenon. Based on the comparison of isotopic signatures of methane from fossil fuels and microbial sources, researchers at the University of California have come to a conclusion that this decline is partly due to an increase in fertilizer use combined with decreased water use in Asian rice agriculture [36]. An alternative explanation of the slowdown in methane emission rate links it to the decrease in the rate of fossil fuels combustion beginning in the 1980s. This conclusion was based on measurements of the concentration of trace gas, ethane, in air bubbles in Greenland and Antarctica (Note that ethane and methane are both produced during fossil fuels combustion.) and a relevant atmospheric model [36]. These conflicting conclusions illustrate the complexity of the physico-biochemical processes behind this phenomenon.

2.3.3 Natural Sources of Methane

A potential impact of methane on the Earth's climate system is linked to the high sensitivity of methane biochemistry to temperature and water level changes. Several authors reported a significant increase in CH_4 emissions from northern peatlands due to permafrost melting [37]. The results of modeling studies indicate that methane emissions from Scotland wetlands could increase by 17, 30, and 60 % if climate warms up by 1.5,

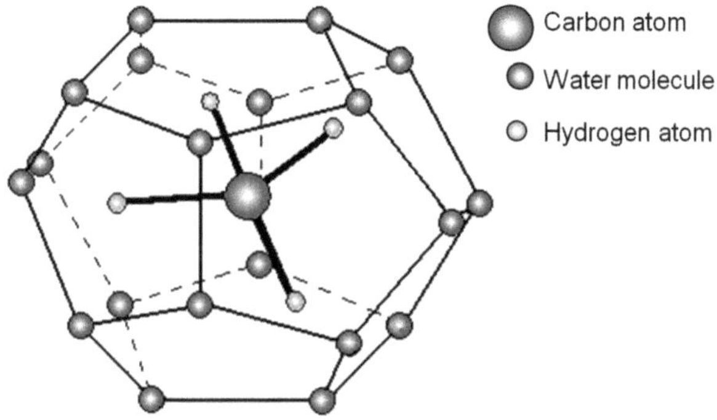

Fig. 2.6 One of possible structures of methane hydrates: 12-hedron (Dodecahedron)

2.5, and 4.5 °C, respectively [38]. Based on simulation studies, Shindel et.al. [39] projected that doubling in the atmospheric CO_2 concentration would result in warming by 3.4 °C, and, consequently, in 78 % increase in CH_4 emissions from wetlands.

2.3.3.1 Methane Hydrates

No other source of methane emissions instigates so much concern and fear with regard to its potentially destructive impact on climate and life on our planet as methane hydrate (it is often referred to in the literature as "*ticking time bomb*," "*the harbinger of impending catastrophe*," "*a sleeping giant*" [24]. (Methane hydrate as a potential source of unconventional gas is discussed in Chap. 1.) Methane hydrate is an ice-like solid substance formed from methane and water molecules under high pressure (greater than 3–5 MPa, which corresponds to water/sediment depths of 300–500 m) and relatively low temperature (up to about 25 °C) [40], and can be represented by a general formula $CH_4 \cdot nH_2O$ (the nominal methane clathrate composition is $CH_4 \cdot 5.75H_2O$). Figure 2.6 depicts the diagram of one of the possible methane hydrate structures.

Methane hydrate can be easily transformed back to gas if one or both parameters (i.e., pressure or temperature) are altered such that hydrate molecules move out of the thermodynamic stability zone. When destabilized, one cubic meter of methane hydrate releases 164 m^3 of methane (at near ambient conditions) [41]. One of the main causes of methane hydrate destabilization that worries the majority of climate scientists relates to the increase in global mean temperature. In principle, the susceptibility of gas hydrates to warming climate depends on a number of factors, such as the duration of the warming event, the depth at which methane hydrates lie beneath the ocean floor or tundra surface, and the amount of heat required to warm up sediments to the point of hydrates dissociation.

Methane hydrates typically occur in shallow sediments in cold regions (e.g., Arctic area) or in deep-water (depths greater than 500–600 m) marine sediments where sufficiently low temperature and high-pressure conditions favor forming and sustaining of the hydrates. Methane that forms hydrate can be of *biogenic* nature (created by biological activity in sediments, e.g., the microbial decomposition or

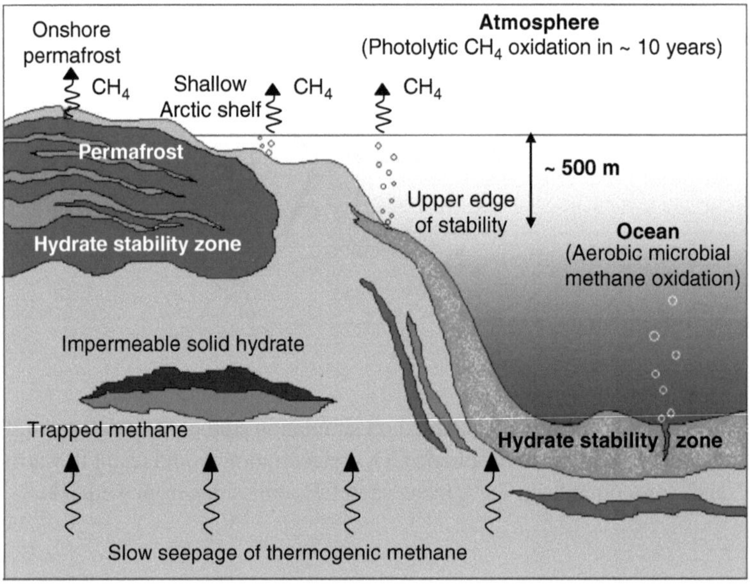

Fig. 2.7 Diagram of different methane hydrate deposits. *Source* [40]

deep burial and/or heating of organic matter) or *thermogenic* nature (created by geological processes deep within the Earth crust). Most of the Earth's methane hydrates (estimated at about 95 %) occur in the ocean depths greater than 1,000 m (where an estimated amount of ~4×10^6 million ton of CH_4 are stored [42]), and their stability is another unknown in climate change models. Figure 2.7 provides a schematic diagram of different methane hydrate deposits.

It was estimated that, currently, about 2 % of atmospheric methane might have originated from dissociation of global deposits of methane hydrates (this is a rough estimate, because there are no tools available to scientists that can distinguish between methane originated from methane hydrates or other sources) [35].

The fate of methane hydrates during warming climate is a highly debated subject. In recent years, a number of research and popular articles have put methane hydrates in focus of the climate change dispute and explored consequences of a catastrophic methane outburst for our civilization [22–24, 43, 44]. It has been emphasized that even if only a fraction of liberated methane was to reach the atmosphere, its strong heat-trapping properties combined with the persistence of its oxidation product (CO_2) could potentially represent a tipping point for the Earth's carbon cycle and contemporary period of climate change [25, 43]. Modeling studies showed that the anthropogenic CO_2 could cause the release of about 2,000 Gt of methane from hydrates, and the increase in deep-water temperature of 3 °C would result in the release of about 85 % of methane from methane hydrates [42].

What fuels the fears most is that, according to available data, such a historical large-scale climate-driven destabilization of methane hydrates on a global scale has already occurred in the past: extreme warming during PETM is attributed to a massive release of methane from global methane hydrates [40]. Some climate scientists advanced the "*clathrate gun*" hypothesis, which postulates that repeated warming of intermediate ocean waters triggered periodic catastrophic dissociation of methane hydrates during the Late Quaternary taking place 400,000–10,000 years ago [35, 45]. Of particular worry is the fact that climate changes in the past were incredibly rapid. The research on Red Sea sediments shows that during the last warm period between ice ages (about 125,000 years ago), sea levels rose and fell by as much as 2 m within a century [46]. Also surprising is how little forcing was required to trigger past climate swings, e.g., research shows that PETM was apparently sparked by a preceding increase of about 2 °C in the Earth's temperature, which was already warmer than today. That warming may have been amplified by positive feedback mechanisms [47].

Less alarming voices could also be heard in the dispute over potential climate perturbations due to methane hydrates destabilization. Carolyn Ruppel, the head of the US Geological Survey's Gas Hydrates Project, suggests that if our planet's warming continues at rates documented by IPCC for the twentieth century (0.2 °C per decade), this should not result in catastrophic breakdown of methane hydrates and major leakage of methane to the ocean–atmosphere system [40]. She holds that most of the methane hydrates would have to experience sustained warming over thousands of years before their massive destabilization could be triggered, although, in some places, methane hydrates could dissociate now in response to short- and long-term climatic processes.

Different types of methane hydrates could experience different rates of dissociation. For example, methane hydrates beneath thick onshore permafrost (lying deeper than about 190 m below the Earth's surface) will remain largely stable even if climate warming lasts hundreds of years. On the other hand, subsea permafrost beneath shallow Arctic shelf is thawing, and associated methane hydrates are likely dissociating now. Luckily, only 1 % of the world's methane hydrates probably occur in this setting (but this estimate could be revised as more data become available). Methane hydrates occurring in upper continental slopes, beneath 300–500 m of water, lie at the borderline of hydrates thermodynamic stability. Methane hydrates exposed to warming ocean waters could completely dissociate in less than hundred years; however, due to dissolution and aerobic oxidation in the water column, only a fraction of released methane will likely reach the atmosphere. About 3.5 % of the global methane hydrates occur in this climate-sensitive setting [35, 40].

Geophysical calculations conducted by the Ruppel's group indicate that over 10, 100, and 1,000 years, the methane hydrates lying at the depths of 18 m, 56 m, and 178 m, respectively, will be affected by warming climate [40]. Even over 1,000 years time horizon, only methane hydrates located close to the seafloor and approaching the thermodynamic stability boundary (about 5 % of the total methane hydrate inventory) might experience dissociation in response to reasonable rates of warming. There are several factors that can alleviate the impact of liberated methane on the ocean-atmosphere system and, ultimately, on climate. First, in marine sediments, the released methane may remain trapped as gas, and up to 90 % of methane that reaches the near-seafloor sediments (especially, in sulfate reduction zone) may be consumed by anaerobic microbial oxidation [48, 49]. Second, methane bubbles released at the seafloor rarely survive the trip to the surface. At the depths greater than about 100 m, dissolved oxygen and nitrogen almost completely replace methane within the rising bubbles; methane dissolved in the water column becomes an easy target of aerobic microbes that serve as an important sink for methane over a wide depth range [50].

Although microbial oxidation of methane in water column mitigates its direct impact as GHG, it also depletes dissolved oxygen and generates oxidation product CO_2, hence, leading to the acidification of ocean waters and eventual release of CO_2 to the atmosphere after residence times of less than 50 years to several hundred years (from water depths of up to 500 m to more profound depths, respectively) [40, 51]. According to Ruppel's calculations, even in the unlikely event that 0.1 % (or 1.8 GtC) of global methane inventory of ~1.8×10^3 GtC in hydrates (according to [52]) was instantaneously released to the atmosphere, CH_4 atmospheric concentrations would increase to ~2,900 ppb from the 2005 value of ~1,774 ppb (according to [6]), i.e., by 63 % [40] (or by about 1 ppm). Considering that this would be an isolated incident, the warming is supposed to be relatively short-lived given the relatively short atmospheric residence time for methane (about a decade). According to the study, CO_2 produced by the oxidation of CH_4 released from dissociating methane hydrates will likely have a greater impact on the Earth ecosystems (e.g., on ocean chemistry and atmospheric CO_2 concentrations) and climate than will methane that remains after passing through various sinks.

Summarizing, although there appears to be a consensus against the looming methane-induced catastrophe (at least in the short-to-mid term), the available scientific information highlights gaps in our understanding of the methane release impact on climate, and points to the need for monitoring of changes to the methane cycle and the development of better models to predict future changes.

2.3.3.2 Permafrost

Methane hydrate is not the only natural source of methane emissions to the atmosphere. As can be seen from Fig. 2.6, tropical wetlands, agriculture, landfills, and fossil fuel production are much bigger players (cumulative annual global methane emissions are estimated at 440–500 million tons CH_4 of which anthropogenic emissions make up about 60 %.) [21, 33]. It has been reported that methane is leaking out of thawing permafrost and regions of glacial retreat across Alaska (estimated at about 2 million tons of methane per year) [53]. The researchers emphasized that the most active sites emitting "old" biogenic methane occur in the areas that have only recently lost their capping ice due to warming. If warming continues, this could pop-up the "cork" and lead to a relatively rapid release (pulse) of methane into the atmosphere.

A recent study conducted by Russian researchers reported that a large amount of methane is seeping into the atmosphere from East Siberian Arctic Shelf (ESAS) sediments [54]. The authors of the study attributed the sustained release of methane from thawing Arctic permafrost to perforations in the seal due to ongoing warming and a possible positive feedback effect. The annual methane outgassing from the shallow ESAS areas was estimated at about 8 million ton CH_4 per year, which is of the same magnitude as the total methane emissions from all the oceans. The authors of the study contend that although the amount of methane emissions seems to be insignificant compared to cumulative global emissions of methane (only about 2 % of the total 440–500 million ton), in a worst-case scenario, it could signal the triggering of worrisome positive feedback and the beginning of massive methane release with unpredictable consequences for climate.

The importance of permafrost factor is that it could completely throw off global climate change forecasts, because existing models and scientific assessments (including those of IPCC) don't factor in the emissions from thawing permafrost. The United Nations Environment Programme (UNEP) 2012 report points out that human-induced climate change is expected to cause significant amounts of permafrost to thaw [55]. As a result, organic material in the soil frozen for millennia (which is found beneath 24 % of exposed land in Northern Hemisphere, i.e., under tundra, boreal forests, and alpine regions) will decompose and irreversibly release both CO_2 and CH_4 [56].

Most of the current permafrost was formed during or since the last Ice Age and extends to depths of more than 700 meters in parts of northern Siberia and Canada and contains 1,700 Gt of carbon (twice the amount currently in the atmosphere). Once the thawing process begins, it will trigger a feedback loop known as the *permafrost carbon feedback* (see Sect. 2.5.3.4 for details), which has the effect of accelerating the further warming of permafrost—a process that would be irreversible on a human life timescale. Warming permafrost could emit 43–135 Gt of CO_2 equivalent by 2100 and 246–415 Gt by 2200 [55]. The emissions could start within the next few decades and continue for several centuries, ultimately accounting for up to 39 % of total emissions. According to the UNEP report [55], Arctic and alpine air temperatures are expected to increase at roughly twice the global rate: a global temperature increase of 3 °C would mean a 6 °C increase in the Arctic, resulting in an irreversible loss of anywhere between 30 and 85 % of near-surface permafrost.

Warming permafrost could bring negative consequences in terms of both ecosystems and infrastructure damage. Since thawing permafrost is structurally weak, it may result in foundational settling that can damage or even destroy buildings, roads, pipelines, railways and power lines (For example, in the 1994, the pipeline to the Vozei oilfield in northern Russia, was broken down, resulting in a spill of 160,000 t of oil, the world's largest terrestrial oil spill.) [55]. Economic impact of permafrost thawing could be huge: it could add up to US$6.1 billion to future costs for public infrastructure in the US state of Alaska between now and 2030. UNEP report recommends that permafrost emissions must be factored into the treaty addressing global climate change expected to replace the Kyoto Protocol. In particular, the

IPCC may consider preparing a special assessment report on how CO_2 and methane emissions from warming permafrost would influence global climate change policies. The report also recommends the countries with the most permafrost: Russia, Canada, China, and the USA, to initiate adaptation policies, and evaluate the potential risks, damage and costs of permafrost degradation to critical infrastructures.

2.3.3.3 Wetlands

Wetlands are another potential source of methane emissions. Estimated 39 % of global methane emissions originate from natural and agricultural wetlands [57]. Wetlands are significant carbon sinks, storing about 20 % of the world's soil carbon in only 5 % of land [58]. However, if more lands become submerged due to rising temperatures or sea-level rise, methane and nitrous oxide (N_2O) could be released, which will change wetlands from being a net sink to a net source of GHG emissions.

A wide variety of wetlands with a broad range of GHG fluxes could be found around the world from high latitudes to tropical zones. In wetlands located in temperate and tropical zones, water depth and temperature changes determine whether these wetlands are net sources or sinks of methane emissions. It is very difficult to quantify the amount of carbon sequestration and methane emissions in wetlands using existing computer models, since many factors need to be considered including variability in landscape, salinity, and plant species.

2.3.4 Anthropogenic Sources of Methane

Methane emissions linked to human activities make up almost two-thirds of the overall methane emissions to the atmosphere. In the USA, in 2011, total methane emissions amounted to 551 Tg CO_2-equiv., which is about 8.2 % of all GHG (i.e., CO_2, CH_4, N_2O, and halocarbons) emissions from all sources amounting to 6,708 Tg CO_2-equiv. [59] (Tg is teragram, or 10^{12} g). Figure 2.8 depicts the contributions of different sources (energy sector, agriculture, waste, etc.) to overall methane emissions in the USA in 2011.

NG systems and agriculture are two major sources of man-made methane emissions. Agricultural processes such as wetland rice cultivation, enteric fermentation in animals, and the anaerobic decomposition of animal wastes and municipal solid waste (MSW) emit large amounts of CH_4. Methane is also emitted during the production, transportation, and distribution of NG and petroleum, and it is released as a by-product of coal mining and incomplete fossil fuel combustion (methane emissions from NG systems are discussed in more details in Sect. 8.2.4).

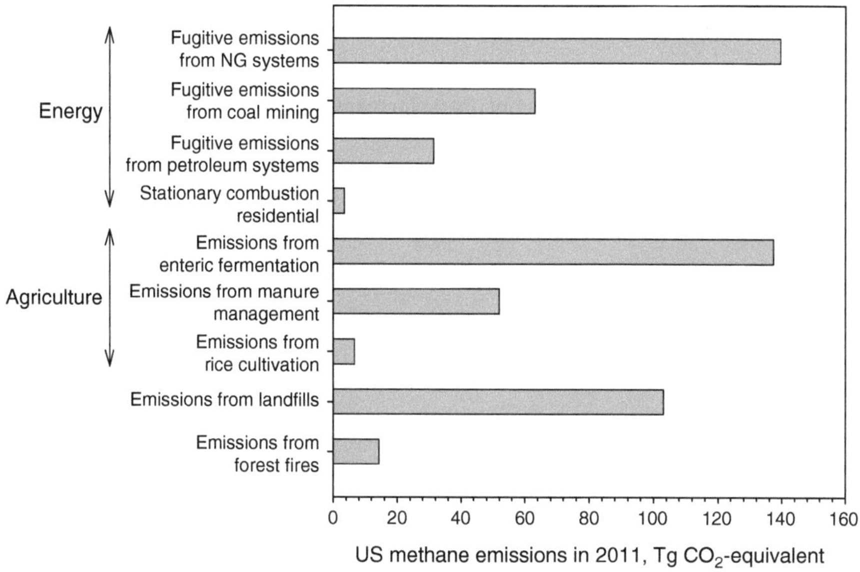

Fig. 2.8 Inventory of US methane emissions from human activities. *Source* [59]

2.4 Global Carbon Cycle

Carbon cycle is a series of cyclic processes by which carbon is exchanged between four major carbon reservoirs: atmosphere, geosphere, hydrosphere, and biosphere. Over geological time, photosynthetic CO_2 fixation exceeded respiratory oxidation of organic carbon to CO_2, which resulted in the reduction of CO_2 to organic carbon followed by its burial in marine sediments [60, 61]. The amount of carbon exchanged between the major carbon reservoirs depends on a variety of factors that are yet to be fully understood. There are several versions of the carbon cycles reported in the literature; in all of them, CO_2 plays a critical role as an exchange "currency" between the reservoirs (in some versions of the carbon cycles, methane is also involved but it plays a minor role). Figure 2.9 depicts a simplified schematic diagram of one version of the natural carbon cycle. It shows unperturbed carbon exchanges between the ocean, the atmosphere, and land in the form of arrows with the values of the carbon fluxes between the reservoirs shown by numbers between the arrows (in GtC/year).

In this diagram, the positive value of the carbon flux indicates carbon input to the atmosphere (e.g., fossil-derived emissions plus emissions from cement manufacturing plants), and negative flux values correspond to carbon losses from the atmosphere to sinks.

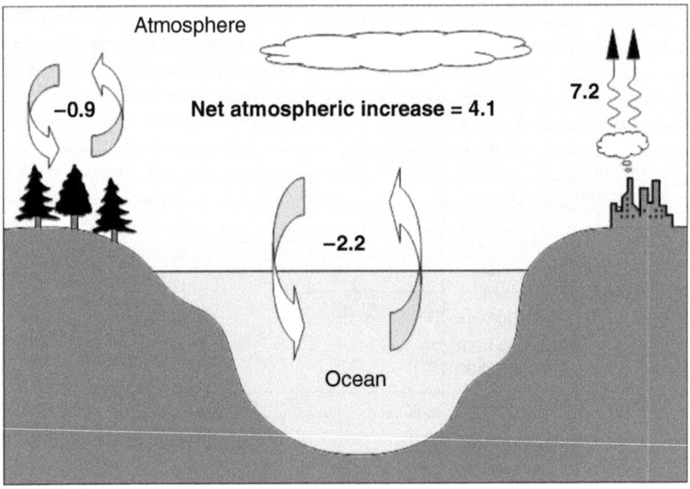

Fig. 2.9 Simplified schematic diagram of the natural carbon cycle. *Source* [6]

2.4.1 *Ocean and Terrestrial Carbon Cycles*

2.4.1.1 Ocean Carbon Cycle

CO_2 exchange between the atmosphere and the ocean is a relatively slow process: the timescale for reaching equilibrium between the atmosphere and the ocean surface depends on many factors such as wind speed, temperature, precipitation, and heat flux, and the majority of estimates agree on a timeframe of about 1 year. It was determined that the rate-limiting step of the overall atmosphere–ocean exchange process is the rate of the mixing of surface waters with the intermediate and deep ocean, which is much slower process than air-sea gas exchange. In principle, the ocean can theoretically absorb up to 70–80 % of the anthropogenic CO_2 emissions, but it would take several centuries to complete that due to the slow surface-deep ocean exchange rate [62].

CO_2 enters the surface ocean by diffusion and dissolution processes followed by a series of reactions leading to the formation of bicarbonate (HCO_3^-) and carbonate (CO_3^{2-}) ions (collectively, dissolved CO_2, bicarbonate and carbonate ions are designated as dissolved inorganic carbon, DIC):

$$CO_{2(gas)} + \mathrm{H_2O} \leftrightarrow \left(HCO_3^-\right)_{aq} + \mathrm{H}^+ \leftrightarrow \left(CO_3^{2-}\right)_{aq} + 2\mathrm{H}^+ \qquad (2.10)$$

The approximate ratio between the dissolved CO_2, bicarbonate, and carbonate ions in the ocean is $(CO_2)_{aq}:(HCO_3^-)_{aq}:(CO_3^{2-})_{aq} \approx 1:100:10$ [6]. The addition of CO_2 to seawater results in an increase in the $(HCO_3^-)_{aq}$ concentration and the ocean acidity. The ocean surface waters have slight alkalinity (pH 7.9–8.25), due to slow dissolution of minerals, which greatly facilitates the CO_2 uptake. Since the beginning of industrial era, the pH of ocean surface water has decreased by 0.1 unit, which corresponds to a 26 % increase in hydrogen ion (H^+) concentration [15]. The lifetime of DIC in the surface ocean, relative to the exchange with the atmosphere and the deep and intermediate ocean layers, is less than a decade [6].

2.4.1.2 Terrestrial Carbon Cycle

In terrestrial biosphere, carbon is released (as CO_2) via two main routes: plant and animal respiration and detritus food chain (i.e., decomposition of organic matter). The carbon flux between the biosphere (which includes vegetation, soil, and detritus) and the atmosphere is estimated at 120 GtC/year (averaged over long periods of time). About 1 GtC/year is transported from land to the ocean by rivers in the form of DIC or suspended particles [6]. Other natural carbon fluxes include rock weathering, sediment accumulation, volcanic activity, and conversion of terrestrial organic matter into inert forms of carbon in soils. These carbon fluxes when averaged over decades do not exceed about 0.1 GtC/year [6]. According to Smith et al., the feedback between the terrestrial carbon cycle and climate will be one of the key determinants of the dynamics of the Earth system over the coming decades and centuries [63].

2.4.2 Interaction Between Carbon Cycle and Climate System

Available data indicate that there are multiple interactions between the Earth's natural carbon cycle and its climate system via a variety of mechanisms involving complex physical, chemical, photochemical, biological, and biogeochemical processes that, in many cases, are not adequately quantified or even understood. In particular, such components of the carbon cycle as the ocean, biosphere, and human activities can affect the GHG concentration in the atmosphere and, thus, indirectly influence the climate system. For example, biomass (vegetation) takes up CO_2 from the atmosphere during its growth and stores it in the form of carbohydrates, thus, diminishing the heat-trapping effect of CO_2 and its impact on climate. On the other hand, human activities cause CO_2 levels in the atmosphere to increase (e.g., via burning fossil fuels, cement manufacturing, land-use changes), thus, amplifying the greenhouse effect and warming of the climate.

2.5 Impact of Human Activities on Carbon Cycle

Throughout history, humans have been modifying their surroundings, and the magnitude of the modifications has risen with the growth of population and the improvement in the standard of living. Rapid growth of human population, especially during last two centuries, was fueled by the increasingly aggressive extraction and consumption of natural resources, which are energy and land intensive activities.

2.5.1 Human Activities and Carbon Cycle

An increasing body of scientific evidence suggests that human activities are affecting radiative forcing factors and carbon cycle [6, 64]. Impacts of human activities on carbon cycle are mainly attributed to the increased emissions of fossil-derived GHG (predominantly, CO_2) through industrial activity (e.g., power generation, cement manufacturing), land-use change, deforestation: all these practices became prevalent during the rapid industrial development of the past two and half centuries.

Although CO_2 as a main GHG of concern could come from different sources, there are many lines of the evidence that recent drastic increases in CO_2 emissions have humans' fingerprints [64, 65] (see also Sect. 2.2.1). Besides CO_2, humans are responsible for the recent growth in emissions of other potent GHG such as methane, nitrous oxide, and halocarbons (HC). For example, man-made methane emissions to the atmosphere (mostly, from agricultural practices: livestock farming, rice cultivation) account for nearly 70 % of annual methane emissions [64]. The concentration of nitrous oxide in the atmosphere has increased from preindustrial levels of about 270–319 ppb in 2005, primarily due to fuel burning at high temperatures [6].

Human activities impact the global carbon cycle by causing changes in the following two main carbon fluxes to the atmosphere [6]:

- CO_2 originating from combustion of fossil fuels and cement manufacture (about 80 % of total).
- CO_2 flux related to the land use changes (e.g., deforestation, agricultural development) (remaining 20 % of the total)

Cumulatively, CO_2 emissions due to global fossil fuel burning and cement manufacturing have increased by 70 % over the last three decades and reached 9.5 GtC[2] in 2011 [15]. In the land-use changes category, CO_2 emissions are estimated at 0.5–2.7 GtC per year range, summarily contributing to 6–39 % of the carbon emission growth rate, which is an equivalent of increasing total atmospheric CO_2 concentration by 12–35 ppm from the preindustrial period to the year 2000 [66]. Tropical deforestation is a main contributor to the increase in CO_2 flux to the atmosphere due to the land-use change.

Although the carbon fluxes caused by human activities constitute only a small fraction of the gross natural carbon fluxes (which constitute hundreds billion tons) within the atmosphere–ocean–land system, they still are responsible for the appreciable changes in the major carbon reservoirs compared to the preindustrial period (because the land and ocean cannot absorb all of the extra CO_2, but only part of it). From 1750 to 2011, CO_2 emissions to the atmosphere from fossil fuel combustion and cement manufacturing were 365 GtC, while the land-use change (including deforestation) was responsible for 180 GtC emissions (cumulatively from all sources, 545 GtC) [15]. Of these 545 GtC anthropogenic emissions, 240 GtC

[2] In this book, the amount of CO_2 may be presented in the units of carbon (C) or CO_2. For example, 1 GtC is equivalent of 3.66 $GtCO_2$.

(44.0 %) have accumulated in the atmosphere, 155 GtC (28.4 %) have been absorbed by the ocean, and 150 GtC (27.6 %) have been taken up by natural terrestrial ecosystems [15].

Thus, depending on the specifics of the carbon reservoir, human-induced carbon fluxes might cause certain perturbations to the natural carbon cycle. For example, in the ocean carbon reservoir, the biological pump might not directly absorb and store anthropogenic carbon, but rather do it through marine biological cycling of carbon facilitated by the increased atmospheric CO_2 concentrations [6]. Once becoming part of the global carbon cycle, anthropogenic CO_2 will be absorbed by ocean with the efficiency and the speed controlled by the rate of the movement of surface waters and their mixing with deeper ocean layers. It was estimated that about half of the amount of CO_2 added to the atmosphere will be removed via the carbon cycle within 30 years, and 20 % may stay in the atmosphere for thousands of years (and slowly neutralize by dissolved $CaCO_3$ from sediments) [43].

2.5.2 *Natural vs. Human-Induced Climate Drivers*

Climate is usually defined in terms of average temperature, precipitation, and wind over period of time, typically, 30 years [6]. It has long been recognized that the Earth's climate is controlled by the Sun's radiation and a variety of physical phenomena involving the Earth's atmosphere and surface such as absorption, reflection, dissipation, and emission of radiant energy. The climate system itself is a complex multifaceted system that is constantly changing governed by its own internal dynamics and a number of external factors such as variations in solar irradiance, and myriad interactions between atmosphere, land surface, oceans and seas, snow and glaciers, deserts, and terrestrial biosphere.

As discussed in Sect. 2.1, the radiation balance on the Earth's surface is fundamentally governed by three main factors [6]:

- Incoming solar radiation
- The fraction of reflected solar radiation (i.e., its albedo)
- The fraction of long wavelength radiation from the Earth back to space

Some of these factors (e.g., incoming solar radiation) are controlled exclusively by natural drivers, whereas the second and third factors could be influenced by both natural and anthropogenic drivers.

2.5.2.1 Natural Drivers

It is widely recognized that the variations in incoming solar radiation are due to the so-called *planetary* or *orbital forcing*, which is caused by the eccentricity, axial tilt, and precession of the Earth's orbit in relation to the Sun (the theory has been developed by James Croll and Milutin Milankovitch in the early twentieth century) [67]. The eccentricity (or ellipticity) of the Earth's orbit varies from 0 to 5 % on a cycle of roughly 100,000 years, while its axial tilt varies from about 21.4° to 24.5° on a roughly 41,000-year cycle (currently, the axial tilt is 23.5°) [64]. Slight changes in these parameters could directly impact the amount of solar radiation reaching the Earth and, it is widely acknowledged that they drive historic glacial–interglacial climate variations (the so-called, *Milankovitch cycles*). Evidence from deep-sea sediments and ice cores suggests that considerable climate variability is associated with orbital forcing [68].

The changes in sunspots could also cause the variations in solar flux intensity and, hence, changes in climate [69]. Long-term observations indicate that the number of sunspots varies on a roughly 11-year cycle, which could potentially alter solar radiation intensity output by about 0.01 %. The available data suggest that during last two and half centuries, increased solar irradiance has contributed to an increase in positive RF of 0.06–0.30 W/m^2 [6]. Although this change is sufficient to contribute to moderate increase in temperature in the upper atmosphere, it cannot account for most of the observed increases in the Earth's surface temperature [64].

The second factor (changes in albedo) is linked to the changes in terrestrial ecosystems (e.g., vegetation), or ice/snow cover, or a cloud cover, or presence of atmospheric aerosols, etc. Natural drivers include volcanic eruptions that emit immense quantities of aerosols (i.e., suspensions of microscopic and submicroscopic solid particles in air), sulfur gases (mainly, SO_x), and CO_2 into the atmosphere. Ash aerosols and sulfur gases reach stratosphere and can contribute to global cooling by reflecting and scattering incoming solar radiation back to space. Although volcanic ash is rather quickly removed from the atmosphere (typically, within a month after the eruption by sedimentation), sulfur gases stay much longer and are largely responsible for the climatic effects associated with the volcanic eruptions (sulfur gases from volcanoes make up about 36 % of the annual tropospheric sulfur emissions) [64, 70]. Natural variations in the Earth's albedo could also result from changes in land and cloud cover, since it could directly affect the amount of solar radiation reflected or absorbed by the Earth's surface and, thus, impact climate patterns. Increased cloud or snow cover can increase reflectance and provide cooling effect, whereas increased vegetation results in increased absorption of radiation, thus, providing warming effect (via the so-called *vegetative forcing*).

The third factor deals with the fraction of long wavelength radiation from the Earth back to space, which is affected by the atmospheric concentration of GHG. The examples of natural GHG sources include volcanoes releasing immense volumes of CO_2 during eruption, or various natural CO_2 vents, plant respiration, methane from permafrost, biogas (mixture of CH_4 and CO_2) from anaerobic digestion, plant decay, and weatherization of carbonate rocks. The magnitude of these emissions varies in a wide range (e.g., the amount of volcanic CO_2 is less than 1 % of annual total CO_2 emissions [71]).

2.5.2.2 Anthropogenic Drivers

Human activities contribute to the changes in the climate system by perturbing the fine balance between incoming/reflected solar radiation and outgoing infrared radiation due to alterations in the amount of GHG, aerosols, and cloudiness in the Earth's atmosphere. The changes in the Earth's albedo could be caused by such anthropogenic drivers as the release of ash aerosols and sulfur gases from coal-fired power plants, burning of vegetation, land surface change, urbanization, and release of GHG, etc. Each of these drivers has a complex and sometimes unpredictable impact on the albedo. Man-made aerosols have a different chemical composition (depending on the source), causing them differently interact with the atmosphere and affect the Earth's albedo (exerting cooling or warming effect). Aerosols (including black carbon particles) and sulfur gases produced from fossil fuel combustion and the burning of vegetation are the primary sources of man-made aerosols. Although anthropogenic aerosol emissions have declined in North America and Europe due to more stringent regulations, their level has increased in Asia (predominantly, China and India) with the dramatic rise in urbanization [64].

The increase in man-made GHG emissions and their accumulation in the atmosphere directly impacts the balance of long-wavelength radiation between the Earth and space. Although the increase in GHG levels in the atmosphere could be attributed to both natural and anthropogenic drivers, one should take into consideration that most of the natural sources existed for millions of years and, to a large extent, are responsible for the levels of GHG currently existing in the atmosphere. With the beginning of the Industrial Revolution in the mid-eighteenth century, a human factor started playing an increasingly bigger role in the growth of GHG emissions to the atmosphere and its impact on global carbon cycle and climate system.

Humans have also been altering the Earth's albedo via widespread transformation of land surface. It was estimated that humans have transformed or degraded between 39 and 50 % of Earth's surface (via population growth, development of necessary resources, etc.) [72]. Land surface change was mostly carried out through deforestation, reforestation, and urbanization, which substantially affected the Earth's albedo [64]. The reports show that the impact of land surface transformation on the Earth's albedo accounts for a loss of RF = 0.4 W/m^2, thus, affecting the energy balance of the Earth's surface [6].

CO_2 flows between ocean and biosphere in the natural "breathing" of our planet, but the uptake of added man-made emissions depends on the net change between these flows that occur over decades to centuries to millennia [73]. According to many climate scientists, this implies that the climate changes caused by CO_2 will most likely persist for many centuries even if emissions were to be stopped at any point in time. Such an extreme "persistence" is unique to CO_2 among major GHG and warming agents. Emissions of such agents as black carbon, aerosols, methane, and ozone can potentially affect climate change over a period of years and decades, but they exert relatively little impact on the Earth's climate over centuries. CO_2, on the other hand, is a much more persistent agent, and, as such, it primarily controls long-term impacts on climate.

2.5.3 *Role of Feedback Mechanisms*

The Earth's climate system is very sensitive not only to radiative energy balance variations but also to the changes in GHG in the atmosphere, and it could react via a variety of direct and indirect *feedback mechanisms*. The feedback mechanism could be of positive or negative nature, depending on whether it amplifies or negates, respectively, the effect of the change. The positive feedback (or "*feedback loops*") is of a particular concern, because it could easily lead to "runaway" situations. There are several types of feedback mechanisms, most important ones are discussed below.

2.5.3.1 CO_2–Water Vapor Feedback

One of the most significant feedback effects relate to a CO_2–water vapor feedback loop. According to this feedback mechanism, the water vapor level in the atmosphere is increased in response to rising concentration of atmospheric CO_2 and resulting GHG effect-induced warming. This, in turn, will cause additional CO_2 flux from the ocean to the atmosphere, thus, further intensifying the warming and further increasing water vapor concentration, and so on. It was reported that the effect of the CO_2–water vapor feedback loop could double the intensity of greenhouse effect compared to CO_2 acting alone [6]. Indeed, the reported data show that global precipitation has increased by about 2 % in response to the higher evaporation rates of the ocean waters during the twentieth century [64]. The magnitude of rainfall events has noticeably increased in many areas of the Northern Hemisphere and Australia. This trend, however, is contrasted with the decreased precipitation and increased aridity at low latitudes, e.g., northern Africa and Asia (which indicates that climate shifts will not be uniform). To some extent, the observed variability in precipitation patterns could be attributed to the El Niño Southern Oscillation (warm ocean currents and associated atmosphere influencing continental climate in many regions of the world).

2.5.3.2 Ice-Albedo Feedback

Another important example of a potentially powerful positive feedback mechanism relates to a so-called *ice-albedo feedback* [6]. The mechanism functions as follows: the rising levels of GHG in the atmosphere trap increasingly more heat and make the Earth's atmosphere warmer causing snow, glaciers, and polar ice caps (that reflect sunlight back to space) to melt. This would result in exposing increasingly broader areas with the "dark" land and ocean surfaces that much better absorb solar heat than snow-covered surfaces. The resulting increase in the atmosphere temperature would melt more snow and ice, and so on. The consequence of this positive feedback could be an uncontrolled increase in temperature on the Earth's surface in response to relatively low (in an absolute value) increase in GHG atmospheric concentration.

The dramatic shrinking of sea ice in the Arctic in recent summers was quite unexpected, because it was not predicted by many climate models [46]. In Antarctica and Greenland, large ice shelves are collapsing: warmer ocean waters are melting away the ice from below, while warmer air is opening cracks from above. Ice loss is feared not just because of ice-albedo feedback but also because of sea-level rise: although the loss of floating ice does not rise sea levels, the submerging glaciers do. Recent reports by glaciologists confirmed that ice losses from Greenland and West Antarctica have been accelerating, showing that some ice sheets are disconcertingly sensitive to climate change [74]. A current annual loss of 344 billion tons of glacial ice accounts for 20 % of current seal level rise (Greenland and West Antarctica have lost, respectively, 263 and 81 billion tons of ice per year from 2005 to 2010). Glaciologists are particularly concerned about the acceleration of losses, which shows that the glaciers are very sensitive to the changes in temperature.

2.5.3.3 Ocean Current Feedback

The most rapid of the feedback mechanisms involves the ocean currents that carry heat around our planet. In warmer climate, collapsing glaciers and/or increased precipitation could dump a massive amount of freshwater into the northern seas causing warm currents to slow down or completely stop, and, as a result, disrupting the engine that drives global ocean currents and, thus, causing more warming [46]. If fully realized, the impact of this feedback might appear very quickly, e.g., Greenland could turn from cool to warm within a decade or so. The rate of adding freshwater is a critical parameter here, and some studies suggest that, fortunately, at the present rate freshwater is not added fast enough to fundamentally alter climate.

Another possible source of the ocean current feedback is located in much warmer areas of our planet. It has been reported that warming waters in the Gulf Stream (that transports warm waters from the Gulf of Mexico into the North Atlantic ocean) can potentially thaw and destabilize hundreds of gigatons of methane hydrates trapped below the seafloor, increasing the risk of the slope failure and methane release into the atmosphere [75]. Additionally, slight changes in the Gulf Stream flow direction can also destabilize methane hydrate by redirecting warm waters to regions previously exposed only to cold bottom currents. The Gulf Stream consists of anomalously warm water at the depths as great as 1,000 m below sea level; at the intermediate water depths of 300–1,000 m, the Gulf Stream temperature is 8 ± 1.1 °C warmer than ocean temperatures outside the current [75]. The destabilization of methane hydrates could extend along hundreds of kilometers of the western North Atlantic margin, it may continue for centuries and release approximately 2.5 Gt of methane (about 0.2 % of the amount of methane hydrates required to cause such

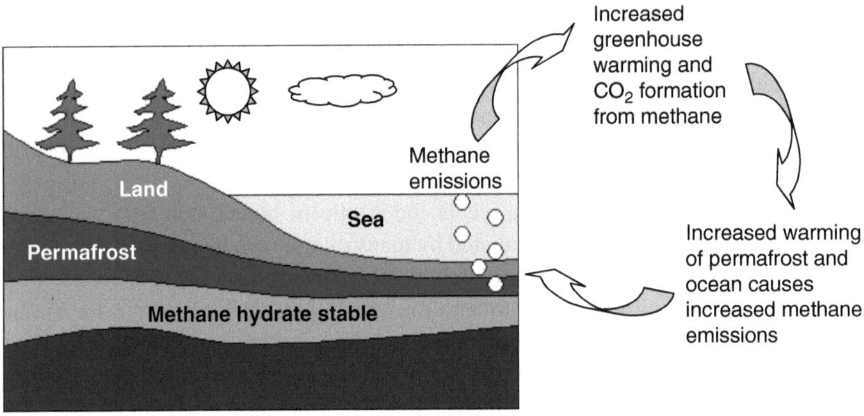

Fig. 2.10 The schematic of a positive methane-induced feedback mechanism. *Source* [35]

catastrophic event as PETM). (The authors of the study note that there could be other areas experiencing changing ocean current; thus, the quantity of destabilized methane hydrates could be much greater.) If most of the released methane will find its way to the atmosphere, this could potentially trigger the feedback loop, however, the prospect of this event and, consequently, its impact on climate still remains uncertain [76].

2.5.3.4 Permafrost Feedback

A more immediate feedback that is already manifesting itself in several locations worldwide involves permafrost (see also Sect. 2.3.3.2). It was not until recently that scientists realized the scale of the potentially devastating impact of the permafrost feedback. Permafrost stores hundreds of billions of tons of methane, roughly double the amount of carbon in the atmosphere [46]. Siberia (Russia) is dotted with giant hills of organic-rich permafrost. There are two types of feedback mechanisms that could result in accelerating release of methane from permafrost in a response to climate change. According to the first (biological) route, in warming climate, more thawing stimulates microbial anaerobic digestion of organic carbon with production of two potent GHG: methane and CO_2. In another feedback mechanism, methane hydrates that occur within or beneath thick terrestrial permafrost may begin to dissociate with the release of methane as surface temperature increases. Pronounced regional warming increases methane emissions from permafrost and destabilized methane hydrates, which strengthens the greenhouse effect (directly by methane and indirectly by CO_2 generated by methane oxidation in the atmosphere) and further warms the surface [21]. This warming, in turn, triggers the additional methane emissions via an auto-accelerating cyclic mechanism. Figure 2.10 depicts the scheme of a positive methane-induced feedback mechanism.

There are multiple observations of methane bubbling up from the shallow lake bottoms formed by melting permafrost. It was found that permafrost could crack open into the so-called thermokarsts (mini canyons), which facilitates melting and the release of GHG (see also Fig. 2.7). Recent expeditions in Arctic regions, e.g., off Spitsbergen, Norway, and Siberia, have detected plumes of methane rising from the shallow waters [46].

There is also another potentially significant cause of methane feedback: tropical wetlands, which present the predominant natural source of methane. As the atmosphere warms, rainfall increases in the tropics, and, consequently, the wetlands expand and become more productive, which creates more anaerobic digestion with the release of methane and CO_2. According to estimates, expanded wetlands could release as much additional methane as that from Arctic warming [46].

2.5.3.5 Ecological Feedback

Ecological feedback mechanism involves changes in the Earth's ecosystem in response to warming climate. For example, warmer temperatures in the northwestern USA and western Canada have triggered an epidemic of mountain pine beetles. The insects destroyed hundreds of thousands of hectares of trees, threatening to turn forests from carbon sinks into carbon sources (due to decomposition of dead trees) [46]. Another example of the ecological feedback: warming in Siberia is starting to transform vast forests of larches into spruce and fir woodlands. This change could markedly affect radiative balance as follows: larches drop their leaves in winter, thereby allowing solar radiation to reflect off the snow cover, whereas spruces and firs keep their needles, thus, absorbing solar heat and minimizing light reflection from snow. Feedback from vegetation changes alone could warm the planet by about 1.5 °C [46].

2.5.3.6 Cloud Feedback

Cloud feedback is an example of the feedback mechanism that could be both negative and positive, and it relates to clouds, which play an important role in regulating the Earth's climate. On the one hand, clouds are effective absorbers of long wavelength radiation, and, as such, amplify the greenhouse effect (i.e., via the positive feedback mechanism since more clouds would result in the more pronounced greenhouse effect and the Earth surface warming). On the other hand, clouds also very effectively reflect solar radiation, thus, potentially causing a negative feedback effect, since more clouds result in more reflection and cooling the Earth. Even minor changes in clouds' consistency and location could cause either a positive or negative feedback effect.

Summarizing, the climate feedback phenomenon and its implications are of vital importance to the very survival of humankind; therefore, this issue is at the center of intensive worldwide research efforts. Although climate experts may not agree on the pace of climate change, there is a realization that specific feedback mechanisms might already be working and amplifying the change, and this causes a profound concern about our planet's future. Climate models, although good at explaining the past and present climatic events, are still unsatisfactory in predicting potential future impacts of various feedback loops. To put it simply: "*scientists know the direction, but not the rate*" [46]. Yet the scientists agree that uncertainties do not justify inaction; on the contrary, they call for the immediate worldwide efforts to reduce GHG emissions, which are one of the main factors triggering feedback loops.

References

1. National Energy Technology Laboratory (2006) Carbon sequestration technology roadmap and program plan. US DOE Office of Fossil Energy, Washington, DC
2. Trenberth K, Fasullo J, Kiehl J (2008) Earth's global energy budget. Bull Am Meteorol Soc. doi:10.1175/2008BAMS2634.1
3. Kreith F (2000) The CRC handbook of thermal engineering. CRC, New York. ISBN 3540663495
4. Goody R (1964) Atmospheric radiation: I. Theoretical basis. Clarendon, Oxford, UK, p 436
5. Okabe H (1978) Photochemistry of small molecules. Wiley, New York
6. U.N. Intergovernmental Panel on Climate Change (2007) 4th assessment report climate change 2007. The physical science basis. Cambridge University Press, Cambridge
7. Myhre G, Highwood E, Keith P et al (1998) New estimates of radiative forcing due to well mixed greenhouse gases. Geophys Res Lett 25:2715–2718
8. Kerr R (2013) Soot is warming the world even more than thought. Science 339:382
9. Andreae M, Ramanathan V (2013) Climate's dark forcings. Science 340:280–281
10. Amunden B, Lie E (2013) The Research Council of Norway. Global warming less extreme than feared? http://www.forskningsradet.no/en/Newsarticle/Global_warming_less_extreme_than_feared/125398344535/p1177315753918?WT.ac=forside_nyhet. Accessed 10 May 2013
11. Intergovernmental Panel on Climate Change (1990) Scientific assessment. In: Houghton J et al (eds) Climate change. Cambridge Univ. Press, Cambridge, UK, p 364
12. Intergovernmental Panel on Climate Change (2005) In: Metz B et al (eds.) Special report on safeguarding the ozone layer and the global climate system: issues related to fluorohydrocarbons and perfluorocarbons. Cambridge University Press, Cambridge. p. 488
13. Landau E (2013) CO_2 levels hit new peak at key observatory. CNN http://edition.cnn.com/2013/05/10/us/climate-change/index.html?hpt=hp_t4. Accessed 1 June 2013
14. CO2 Now (2013) Atmospheric CO_2 for May 2013. http://co2now.org/. Accessed 2 Jun 2013
15. U.N. Intergovernmental Panel on Climate Change (2013) Working group I contribution to the IPCC fifth assessment report climate change 2013: the physical science basis. Summary for policymakers. http://www.climatechange2013.org./images/uploads/WGIAR5-SPM_Approved27Sep2013.pdf. Accessed 27 Sep 2013
16. Kiehl J, Trenberth K (1997) Earth's annual global mean energy budget. Bull Am Meteorol Soc 78:197–208
17. Evans K (2005) The greenhouse effect and climate change. The environment: a revolution in attitudes. Thomson Gale, Detroit. ISBN 0-7876-9082-1
18. Sausen R, Isaksen I, Grewe V et al (2005) Aviation radiative forcing in 2000: an update on IPCC (1999). Meteorol Z 114:555–561
19. Held I, Soden B (2000) Water vapor feedback and global warming. Annu Rev Energy Environ 25:441–475
20. Solomon S, Rosenlof K, Portmann R et al (2010) Contributions of stratospheric water vapor to decadal changes in the rate of global warming. Science 327:1219–1223

21. Bondre N (2012) Methane: not a damp squib, not yet a time bomb. Global Change 79:12–15
22. Bohannon H (2008) Weighing the climate risks of an untapped fossil fuel. Science 319:1753
23. Krey V, Canadell J, Nakicenovic N et al (2009) Gas hydrates: entrance to a methane age or climate threat? Environ Res Lett 4:034007. doi:10.1088/1748-9326/4/3/034007
24. Mascarelli A (2009) A sleeping giant? Nature reports. Clim Change 3:46–49
25. Archer D, Buffett B, Brovkin V et al (2008) Ocean methane hydrates as a slow tipping point in the global carbon cycle. Proc Natl Acad Sci U S A 106:20956–20601. doi:10.1073/pnas.0800885105
26. Shindell D, Faluvegi G, Koch D et al (2009) Improved attribution of climate forcing to emissions. Science 326:716–718
27. Howarth R, Santoro R, Ingraffea A (2011) Methane and greenhouse gas footprint of natural gas from shale formations. Clim Change. doi:10.1007/s10584-011-0061-5
28. Shindell D, Kuylenstierna J, Vignati E et al (2012) Simultaneously mitigating near-term climate change and improving human health and food security. Science 335:183–189
29. Hofzumahaus A, Lefer B, Monks P et al (2004) Photolysis frequency of O_3 to O1D: measurements and modeling the IPMMI. J Geophys Res 109:D08S90. doi:1029/2003JD004333
30. Spahni R, Chapellaz J, Stocker T et al (2005) Atmospheric methane and nitrous oxide of the late Pleistocene from Antarctic ice cores. Science 310:1317–1321
31. Blasing T (2009) Recent greenhouse gas concentrations. Carbon dioxide information analysis center. Oak Ridge National Laboratory. updated December 2009. http://cdiac.ornl.gov/pns/current_ghg.html. Accessed 9 July 2010
32. Intergovernmental Panel on Climate Change (2001) Climate change 2001: the scientific basis. Cambridge Univ. Press, New York
33. Heimann M (2010) How stable is the methane cycle? Science 327:1211–1212
34. Keppler F, Hamilton J, Brass M et al (2006) Methane emissions from terrestrial plants under aerobic conditions. Nature 439:187–191
35. Ruppel C, Noserale D (2012) Gas hydrates and climate warming—why a methane catastrophe is unlikely. USGS, Sound Waves, Volume FY 2012, Issue No. 140, May/June 2012. http://soundwaves.usgs.gov/2012/06/SW201206.pdf. Accessed 23 Aug 2012
36. Wilson E (2011) Examining methane's slowing increase. Chem Eng News 89:31
37. Wickland K, Striegl R, Neff J et al (2006) Effects of permafrost melting on CO_2 and CH_4 exchange of a poorly drained black spruce lowland. J Geophys Res 111, G02011. doi:10.1029/2005JG000099
38. Chapman S, Thurlow M (1996) The influence of climate on CO_2 and CH_4 emissions from organic soils. J Agric Forest Meteorol 79:205–217
39. Shindel D, Walter B, Faluvegi G (2004) Impacts of climate change on methane emissions from wetlands. Geophys Res Lett 31, L21202. doi:10.1029/2004GL021009
40. Ruppel C (2011) Methane hydrates and contemporary climate change. Nat Educ Knowledge 3:29 http://www.nature.com/scitable/knowledge/library/methane-hydrates-and-contemporary-climate-change-24314790. Accessed 10 Jan 2012
41. U.S. Department of Energy (2012) Methane hydrates R&D program. http://fossil.energy.gov/programs/oilgas/hydrates. Accessed 15 Sep 2012
42. Buffet B, Archer D (2004) Global inventory of methane clathrate: sensitivity to changes in the deep ocean. Earth Planet Sci Lett 227:185–199
43. Archer D (2005) The fate of fossil fuel CO_2 in geologic time. J Geophys Res 110(C9):C09S05. doi:10.1029/2004JC002625
44. Behl R (2000) Carbon isotopic evidence for methane hydrate instability during quaternary interstadials. Science 288:128–133
45. Kennet J, Cannariato K, Hendy I et al (2003) Methane hydrates in quarternary climate change—the Clathrate gun hypothesis. American Geophysical Union, Washington, DC
46. Carey J (2012) Global warming: faster than expected? Scient Am November:51–55
47. Kump L (2011) The last great global warming. Scient Am. http://www.scientificamerican.com/article.cfm?id=the-last-great-global-warming. Accessed 20 Jul 2012

48. Reeburgh W (2007) Oceanic methane biogeochemistry. Chem Rev 107:486–513
49. Hinrichs K, Boetius A (2002) The anerobic oxidation of methane: new insights in microbial ecology and biochemistry. In: Wefer F et al (eds) Ocean margin systems. Springer, Berlin, pp 457–477
50. Mau S, David V, Clark J et al (2007) Dissolved methane distributions and air-sea flux in the plum of a massive seep field. Coal Oil Point, California. Geophys Res Lett 34, L22603. doi:10.1029/2005JC003183
51. Liro C, Adams E, Herzog H (1992) Modeling the release of CO_2 in the deep ocean. Energy Conver Manag 33:667–674
52. Boswell R, Collett T (2011) Current perspectives on gas hydrate resources. Energy Environ Sci 4:1206–1215
53. Walter A, Anthony P, Grosse G et al (2012) Geologic methane seeps along boundaries of Arctic permafrost thaw and melting glaciers. Nat Geosci 5:419–426. doi:10.1038/ngeo1480
54. Shakhova N, Semiletov I, Salyuk A et al (2010) Extensive methane venting to the atmosphere from sediments of the East Siberian Arctic Shelf (ESAS). Science 327:1246–1250
55. United Nations Environment Programme (2012). Policy implications of warming permafrost. UNEP press release: thawing of permafrost expected to cause significant additional global warming, not yet accounted for in climate predictions. November 2012. http://www.unep.org/Documents.Multilingual/Default.Print.asp?DocumentID=2698&ArticleID=9338&l=en. Accessed 10 Jan 2013
56. Hogue C (2012) Thawing permafrost throws off global warming forecasts, report says. Chem Eng News 90:9
57. Laanbroek H (2010) Methane emission from natural wetlands: interplay between emergent macrophytes and soil microbial processes. A mini-review. Ann Bot 105:141–153
58. Milne J, Field C (2012) Assessment report from the GCEP workshop on energy supply with negative carbon emissions. Stanford University. Global Climate and Energy Project. http://gcep.stanford.edu/pdfs/rfpp/Report%20from%20GCEP%20Workshop%20on%20Energy%20Supply%20with%20Negative%20Emissions.pdf. Accessed 12 Feb 2013
59. US Environmental Protection Agency (2011) Inventory of U.S. greenhouse gas emissions and sinks: 1990–2011. US-GHG-Inventory-2011-Chapter-1-Introduction.pdf. Accessed 20 Jul 2012
60. Falkowski P, Scholes R, Boyle E et al (2000) The global carbon cycle: a test of our knowledge of Earth as a system. Science 290:291–296
61. Canfield D, Kamp L (2013) Carbon cycle makeover. Science 339:533–534
62. Turley C, Blackford J, Widdicombe S et al (2006) Reviewing the impact of increased atmospheric CO_2 on oceanic pH and the marine ecosystem. In: Schellnhuber H et al (eds) Avoiding dangerous climate change. Cambridge Univ. Press, Cambridge, UK, pp 347–353
63. Smith M, Vanderwel M, Lyutsarev V et al (2012) The climate dependence of the terrestrial carbon cycle; including parameter and structural uncertainties. Biogeosci Discuss 9:13439–13496. doi:10.5194/bgd-9-13439-2012
64. Flower C, Lynch D, Gonzalez-Meler M (2012) Global climate change. Berkshire encyclopedia of sustainability, v. 5, Ecosystem, management and sustainability. pp. 162–167
65. Prentice I, Farquhar M, Fasham R et al (2001) The carbon cycle and atmospheric carbon dioxide. In: Houghton J et al (eds) Climate change 2001: the scientific basis. Contributions of working group I to the third assessment report of the intergovernmental panel on climate change. UK, Cambridge Univ. Press, Cambridge, pp 185–237
66. Brovkin V, Sitch S, Von Bloh W et al (2004) Role of land cover changes for atmospheric CO_2 increase and climate change during the last 150 years. Global Change Biol 10:1253–1266
67. Imbrie J, Imbrie K (1979) Ice ages: solving the mystery. Harvard University Press, Cambridge, MA
68. Imbrie J, Boyle E, Clemens S et al (1992) On the structure and origin of major glaciation cycles. Linear responses to Milankovitch forcing. Paleoceanography 7:701–738
69. Lean J (2010) Cycle and trends in solar irradiance and climate. Wiley Interdiscip Rev Clim Change 1:111–122
70. Graf H, Feichter J, Langmann B (1997) Volcanic sulfur emissions: estimates of source strength and its contribution to the global sulfate distribution. J Geophys Res 102:10727–10738

71. Gerlach T (1991) Present-day CO_2 emissions from volcanoes. Eos Trans Am Geophys Union 72:254–255
72. Vitousek P, Mooney H, Lubchenko J et al (1997) Human domination of Earth's ecosystems. Science 277:494–499
73. National Academy of Sciences (2011) Climate stabilization targets: emissions, concentrations and impacts over decades to millennia. The National Academic Press, Washington, DC
74. Kerr R (2012) Experts agree global warming is melting the world rapidly. Science 338:1138
75. Phrampus B, Hornbach W (2012) Recent changes to the Gulf Stream causing widespread gas hydrate destabilization. Nature 490:527–530
76. Lewis T (2013) Greenhouse gas may be flowing into ocean waters off US east coast. Science News. http://www.sciencenews.org/view/generic/id/346009/description/Gulf_stream_might_be_releasing_seafloor_methane. Accessed 12 Mar 2013

Chapter 3
Anthropogenic CO_2 Emissions: Sources and Trends

Abstract CO_2 and other greenhouse gases are emitted to the atmosphere as a result of both natural processes (e.g., volcanoes, natural vents, and respiration) and human activities. Although the carbon fluxes caused by human activities constitute only a small fraction of the gross natural carbon fluxes between land, the ocean, and the atmosphere, they are responsible for the appreciable changes in the global carbon balance compared to the preindustrial period. Fossil fuels are the main contributors to overall anthropogenic CO_2 emissions with most of them coming from energy-related sources. The classification of major CO_2 sources by fuel type, industrial sector, CO_2 content, and the scale of emissions, as well as current and future trends in CO_2 emission sources is analyzed in this chapter. Geographically, the significant redistribution of CO_2 emission sources throughout the world between now and 2030 could be expected with developing countries getting most of the gain, and the share of the developed countries being continuously reduced.

3.1 Greenhouse Gas Sources: Natural vs Anthropogenic

GHG are emitted as a result of natural processes (e.g., volcanoes, natural CO_2 vents, respiration, and weatherization) and human activities. Figure 3.1 shows the share of major anthropogenic GHG emission sources classified by their origin [1]. CO_2 makes up the lion's share of the total man-made GHG emissions, with most of the CO_2 coming from energy-related sources. Methane sources such as coal mines, agriculture, gas leakages, and fugitive emissions make a sizable contribution (15 %) to the overall anthropogenic GHG emissions [2]. Combined, carbonaceous gases (CO_2, CH_4, and halocarbons) account for 92 % of the total anthropogenic GHG emissions (the balance is made up by N_2O originating mostly from industry and waste) [2].

Most of the following discussion will be focused on CO_2 as the predominant GHG potentially having the greatest impact on our planet's ecosystems. The objective of this chapter is to classify the diverse sources of anthropogenic CO_2 emissions

N. Muradov, *Liberating Energy from Carbon: Introduction to Decarbonization*, Lecture Notes in Energy 22, DOI 10.1007/978-1-4939-0545-4_3,

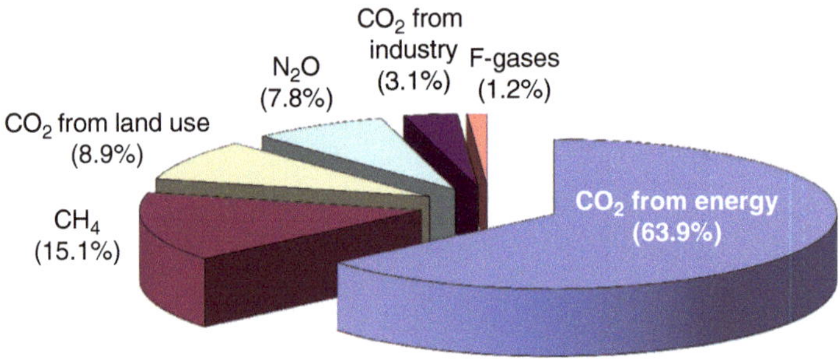

Fig. 3.1 Global anthropogenic GHG emissions by origin. *Source*: International Energy Agency [1]

in order to better appreciate the technical challenges in dealing with these CO_2 sources and evaluate possible technological solutions to mitigating carbon emissions.

3.2 Fossil Fuels as a Main Source of Anthropogenic CO_2 Emissions

Fossil fuels are important part of the global carbon cycle: carbon locked in these fuels for millions of years is released in the form of CO_2 as a result of humans' activities ending up predominantly in the atmosphere. Since the beginning of the Industrial Revolution, the ever-increasing amounts of CO_2 emissions are being released to the atmosphere as a "by-product" of the growing standard of living. From the mid-eighteenth century to 2011, CO_2 emissions to the atmosphere from fossil fuel combustion and cement production amounted to 335–395 GtC, 44 % of which have accumulated in the atmosphere [3]. The historical statistical data on the global and regional CO_2 emissions could be found in a number of databases, e.g., [4]. In recent history, after a 1.5 % decline in 2009 due to the worldwide economic downturn, global CO_2 emissions increased by 5.5 % in 2010 and 3.3 % in 2011 reaching 31.6 Gt CO_2/year [5].

The amount of CO_2 emitted from fossil fuel combustion is proportional to its carbon content: high-carbon content fuels tend to produce larger amounts of CO_2 emissions per unit of product produced (e.g., electricity, or fertilizers). The relative share of coal, oil and NG in the world's energy mix varied widely throughout history (see Chap. 1). The share of coal, oil, and gas in global CO_2 emissions in 2011 is shown below (in % of total) [5]:

Gas	20.5
Oil	36.0
Coal	43.5

Coal, due to its high carbon content relative to its energy content, produces significantly higher levels of CO_2 emissions per unit of generated power compared to other fossil fuels (e.g., nearly twice as much as NG). Besides power generation, coal is heavily used in the industrial sector: iron and steel manufacturing sector predominantly utilizes coal and petroleum coke in a blast furnace operation causing significant CO_2 emissions. Cement industry also relies on coal in many countries, e.g., China and India [6]. IEA projects that, absent additional abatement measures, emissions from coal will grow to 15.3 $GtCO_2$ in 2035 [7]. The adoption of carbon mitigation pathways (e.g., increased use of carbon-free energy sources, carbon capture, and storage) could reduce CO_2 emissions from coal to 5.6 $GtCO_2$ by 2035 [7].

CO_2 emissions from oil amounted to 10.9 and 11.1 $GtCO_2$ in 2010 and 2011, respectively [4, 5]. Oil refining and petrochemical sectors are extensively using oil and gas as primary fuel (the use of petroleum coke as supplemental fuel is also practiced). According to the latest IEA projections, emissions from oil will grow to 12.6 $GtCO_2$ in 2035, mostly, fueled by rapidly increasing demand in transportation, especially, in developing countries [4]. In 2011, CO_2 emissions from gas were equal to 6.3 $GtCO_2$, which represented an increase from the previous year, principally, due to switching to NG in power generation sector [4]. IEA projects that this trend will continue and the emissions from gas will further grow reaching 9.2 $GtCO_2$ in 2035 [7].

3.3 Classification of Anthropogenic CO_2 Emissions Sources

3.3.1 Classification by CO_2 Source Type

Four major types of CO_2 sources are summarized in Fig. 3.2.

Fuel combustion-related sources are by far the largest CO_2 emitters; in these sources, carbonaceous fuels directly react with air (or O_2) producing CO_2 and water. The examples of these sources are coal- and gas-fired power plants, or gasoline- and diesel-fueled vehicles. CO_2 can also be emitted from industrial processes, where carbonaceous feedstocks participate in chemical reactions releasing CO_2 as a by-product. Examples of such industrial processes are [8]:

- Cement manufacturing (dry process/suspension preheater rotary kiln with or without precalciner/grate cooler)
- Limestone calcination
- Iron and steel manufacturing:
  - Blast furnace (pig iron)
  - Direct reduced iron
  - FINEX and Hlsarna steelmaking processes

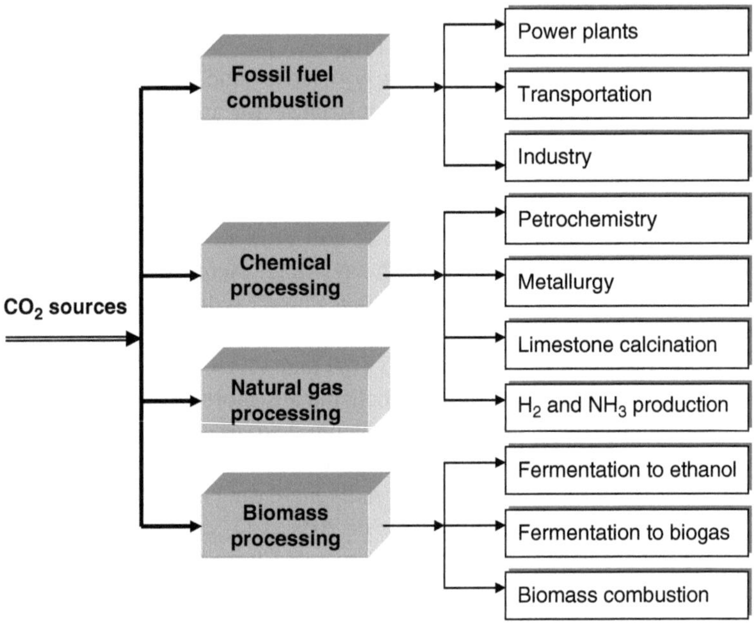

Fig. 3.2 Classification of CO_2 sources by type

- Refineries and petrochemical plants:
  - Steam methane reforming
  - Steam-oxygen gasification of oil residue
  - Fluidized bed catalytic cracking
  - Ethylene and ethylene oxide production

In most cases, CO_2 is produced not only as a reaction by-product, but as a result of fuel combustion in order to supply heat for the processes (e.g., steam methane reforming).

NG commonly contains certain percentage of naturally occurring CO_2: from several tenths of percent to concentrations as high as 71 vol.% (as in Natuna gas field, Indonesia). When CO_2 levels in NG are above a certain threshold value dictated by the economics of NG pipeline transport or NG liquefaction process, CO_2 has to be removed. (If there are no end users in proximity that could use this CO_2, it is most likely to be vented into the atmosphere.)

Although considered a carbon-neutral source, biomass could be a significant source of CO_2 emissions via a variety of fermentations, gasification, and combustion processes. For example, the following biomass conversion processes produce large amounts of CO_2 [8]:

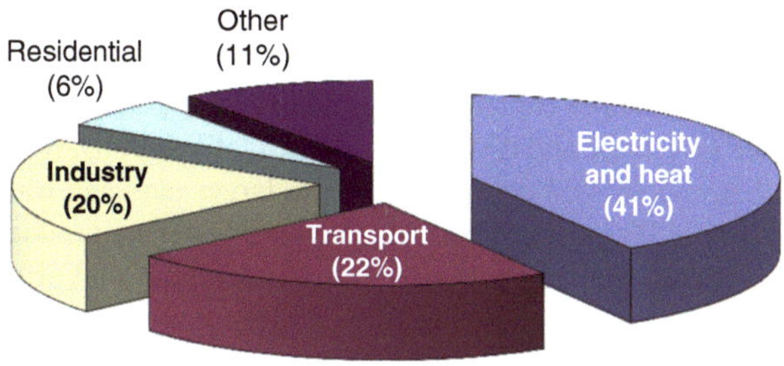

Fig. 3.3 Global CO_2 emissions from fossil fuel combustion and their distribution by industry sector (2010 data). "Other" sources include agriculture, forestry, commercial/public services, and other emissions. *Source*: International Energy Agency [4]

- Ethanol production
- Biosynthetic gas production
- Biomass-to-liquids
- Black liquor processing in pulp and paper manufacturing
- Sugar production

Fermentative ethanol plants in the USA and Brazil emit CO_2 emissions in the order of 0.1–0.14 $MtCO_2$ per year (each) [9].

3.3.2 Classification by Industrial Sector

Figure 3.3 shows the relative distribution of CO_2 sources by industry sector.

Power and heat generation sector is by far the largest contributor to the global CO_2 emissions (41 % in 2010) [4]. The emissions from electricity generation increased at faster rates compared to other sectors. As an example, in order to satisfy the exploding global electricity demand caused by the introduction of flat-screen TVs, personal computers, and other electronic gadgets, the equivalent of 560 coal-fired power plants will have to be built over the next two decades. Over 70 % of world electricity and heat generation is now provided by fossil fuels. Future developments of carbon emission intensity of this sector will strongly depend on the composition of the fuel mix for electricity generation, and on the share of non-carbon sources (nuclear, renewables) and progress on the deployment of CCS technology in this sector.

Transportation is the second largest producer of CO_2 emissions (22 % of total in 2010), with almost three-quarters of emissions coming from road travel [4]. In a contrast to the fuel-flexible electricity/heat generation sector, the transportation sector currently almost exclusively relies on oil (about 94 %) [10]. The recent increase in CO_2 emissions from the transport sector can be attributed to a rapid economic growth in populous developing countries, especially in China and India (since car ownership grows with the per capita income). Fuel prices could also be factor determining the choice of vehicle and distance traveled. Until recently, relatively low fuel prices in the USA encouraged consumers to use larger cars and travel longer distances (in average, 25,000 km per person per year [4]), thus, producing enormous amounts of carbon emissions. In contrast, in Europe, higher fuel prices contributed to the trend of using smaller cars with improved fuel economy and, thus, lesser CO_2 emissions.

IEA projects that the transportation sector will grow by 45 % by 2030 [10] and the transport fuel demand will grow by nearly 40 % by 2035 [4, 10]. In order to reduce CO_2 emissions from this sector, policymakers are trying to implement a number of measures to encourage an improvement in fuel economy, a shift from individual to public transportation, and a switch to low-to-zero-carbon fuels (including, biofuels, methane, and hydrogen) and advanced electric, plug-in hybrids and, in perspective, fuel cell vehicles. Most OECD countries now have adopted fleet average fuel economy standards that have led to significant improvements in an average fuel economy. These measures will reduce the adverse environmental impact of transport and ease the demand for oil.

Among industrial sources of CO_2, cement manufacturing is by far the most significant emitter. World cement production in 2011 reached 3.6 billion tons, which resulted in over 2 billion tons of CO_2 being produced from both the calcination of limestone and fuels used (mainly coal and gas) to drive this process [11]. Three levers are considered by global cement manufacturing sector to reduce its emissions:

- Improvements in energy efficiency
- Use of alternative fuels (including biofuels)
- Clinker substitution

Through some of these measures, the CO_2 emissions per ton of cement have been reduced by 16 % from the 1990 levels of 750 kg CO_2 per ton of cement [11]. Although the rate of CO_2 emissions could be further reduced through continued implementation and expansion of these three levers, it is recognized that there is a practical limit on the carbon reductions that can be achieved (due to technological limitations). Hence, deep cuts in CO_2 emissions from cement production can only be realistically achieved through a combination with CCS technology.

3.3.3 *Classification by Scale of Emissions*

CO_2 is emitted by a great variety of sources: large stationary sources such as power plants and industrial facilities, small-to-medium sources, such as large ships, industrial and commercial buildings, as well as a myriad of very small mobile sources such as personal transportation. IEA has developed a database for CO_2 stationary sources that includes power plants, refineries, cement manufacturing plants, and

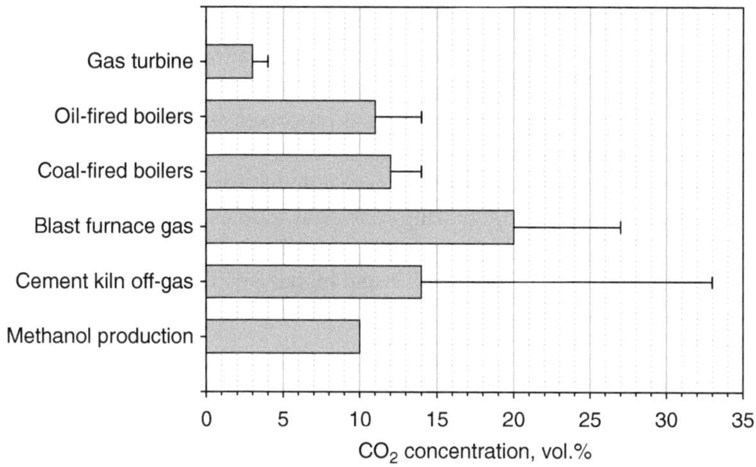

Fig. 3.4 Typical CO_2 concentrations (on dry basis) in exhaust gases from fuel combustion and noncombustion type CO_2 sources. *Source*: Intergovernmental Panel on Climate Change [6]

other major industrial sources of CO_2 emissions [12]. Of the total number of CO_2 emitting sources, more than half (56 %) are stationary sources with the emission levels greater than 0.1 $MtCO_2$/year, with 85 % of the overall CO_2 emissions emitted by large stationary sources emitting more than 1 $MtCO_2$/year [12]. At the lower end, 44 % of sources are the emitters with the individual capacity of less than 0.1 $MtCO_2$/year; the sources emitting between 0.1 and 0.5 $MtCO_2$ per year account for less than 10 % of overall emissions. In the USA, the power generation sector is the largest CO_2 emitter (each plant producing about 1 $GtCO_2$) [13]. Among other large emitters are refineries and chemical plants, NG processing facilities, cement manufacturing plants, ethanol plants, and others.

3.3.4 Classification by CO_2 Content

CO_2 concentration in the gas streams of industrial CO_2 emitters varies in a wide range: from 3 to 4 vol.% in gas turbines exhaust to about 65 vol.% in the vent gases of NG processing facilities to almost 100 vol.% in the off-gases of fermentation plants (all concentrations are on a dry basis). Figure 3.4 summarizes the data on typical CO_2 concentrations in exhaust gases from a variety of industrial sources that include both fuel combustion (power/heat generation) and noncombustion (iron/steel and cement manufacturing, methanol production) type CO_2 sources. As could be expected, CO_2 content in flue gases from fuel combustion depends on the type of fuel and oxidizer used: flue gases from NG combustion in air have the lowest CO_2 concentrations, and from coal, coke, or residual oil combustion in O_2 or O_2-enriched air the highest CO_2 concentrations.

In general, off-gases from non-combustion type industrial CO_2 sources (e.g., cement plants) tend to contain CO_2 in higher concentrations compared to flue gases from the combustion type sources (as can be seen in Fig. 3.4). Advantageously, CO_2 partial pressure in the off-gases of chemical processes (typically, 0.1–0.5 MPa) is much greater than that in the flue gases of combustion-type sources (typically, 0.005–0.014 MPa), which makes them more suitable for a subsequent recovery of CO_2 [6]. The majority of CO_2-emitting industrial sources produce exhaust streams with CO_2 content below 15 vol.%, and only a small fraction (less than 2 %) produce by-product CO_2 with a purity of 95 vol.% and higher [6]. The off-gases from ethanol fermentation plants contain almost pure CO_2. Other sources of high-purity CO_2 include [8]:

- NG processing (onshore, offshore)
- H_2 production from NG, coal, or biomass
- Ethylene oxide production
- Coal-to-liquids
- Ammonia

3.3.5 Geographical Distribution of CO_2 Sources

Geographically, the largest CO_2 emitting sources are clustered in four regions: Asia (30 %), North America (24 %), transitional economies (Central and Eastern Europe and former USSR republics) (13 %), and OECD-West (12 %) [14]. Figure 3.5 depicts the historical trends in CO_2 emissions from 1971 to 2010 for USA, EU, China, and the world.

In the USA, CO_2 emissions were gradually increasing until the turn of the century, when they practically leveled off. Since 2006, CO_2 emissions in the USA have fallen by 430 million ton (or 7.7 %) (this is the largest reduction of all countries), and in 2011, they fell by 92 million ton (or 1.7 %) [6]. In 2011, the US GHG emissions were 6.7 $GtCO_2$-equiv. [15]. The ongoing drastic reductions in the US CO_2 emissions can be attributed to a number of factors, including wide-scale switching from coal- to NG-fired power plants, significant investments in energy efficiency and conservation, lower petroleum use in transportation (due to efficiency improvements and high oil prices), reduced demand for heating oil (due to relatively mild winters), and the recent economic downturn. In the EU, CO_2 emissions in 2011 dropped by 69 million ton (or 1.9 %) mostly due to a sluggish economic growth [5]. The industrialized countries that have ratified the Kyoto Protocol plus the nonratifying USA have emitted about 7.5 % less CO_2 in 2010 than in 1990, and, thus, collectively remain on the target to meet the Protocol's 5.2 % carbon reduction objective [16].

In 2007, China overtook the USA as the world's largest CO_2 emitter, and its CO_2 emissions are currently growing at an alarming rate. In 2011, China was the largest contributor to the global carbon emissions increase by adding 720 million tons

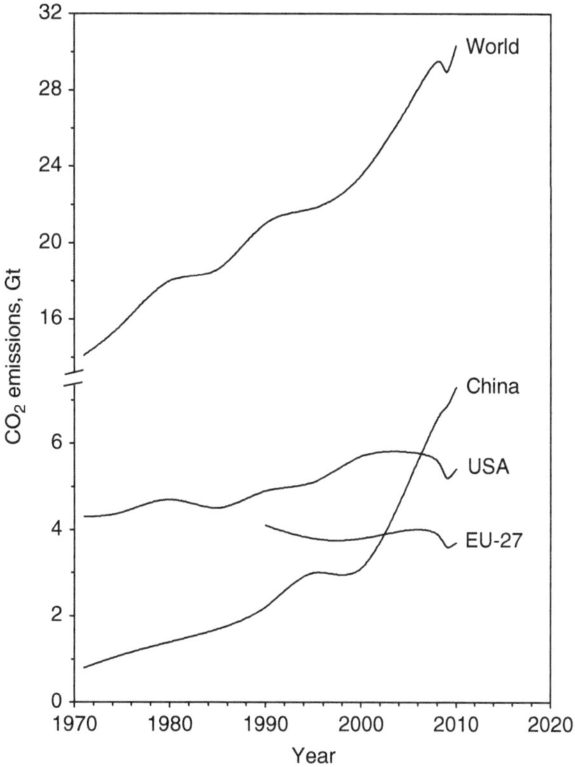

Fig. 3.5 Historical trends in CO_2 emissions from 1971 to 2010 for USA, EU, China, and the world. *Source*: International Energy Agency [4]

(or 9.3 %) of CO_2 emissions, mostly due to coal consumption in power generation sector [5]. The same year, India's emissions rose by 140 million tons (or 8.7 %), which put it ahead of Russia as the third largest emitter [5]. Almost two-thirds of the world CO_2 emissions in 2010 originated from the following ten countries (in a descending order): China, USA, India, Russian Federation, Japan, Germany, South Korea, Canada, Iran, and UK [4]. As a result, in 2011, a 0.6 % reduction in carbon emissions in the OECD countries, were overwhelmingly offset by 6.1 % increase in non-OECD countries [5]. Figure 3.6 shows the change (in %) in CO_2 emissions produced in 1990 and 2010 in different countries and regions of the world.

Figure 3.6 shows that among all regions and countries, the EU was the only one that reduced its CO_2 emissions (by about 10 %) from 1990 to 2010.

IEA in its New Policies Scenario projects that in the medium term, the global CO_2 emissions from fossil fuel combustion will continue to grow unabated, albeit at a somewhat lower rate reaching 37.0 $GtCO_2$ by 2035 [4, 7]. Developed countries will continue moving on the path of reducing carbon intensity of their economy. In the USA, the glut of cheap gas and tightening regulations on air pollutants have prompted the planned closure of 175 coal-fired plants by 2016, which amounts to

Fig. 3.6 Percentage change in CO_2 emissions between 1990 and 2010 in the world and different countries and regions. *Source*: International Energy Agency [4]

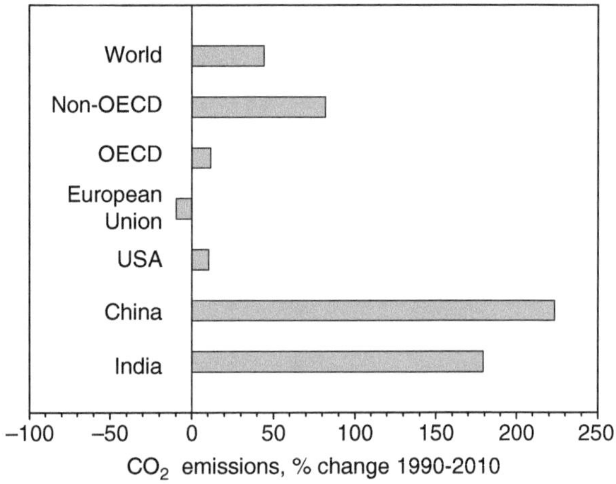

8.5 % of all coal-powered electric capacity in the country [17]. Because of the shift to higher efficiency NG power plants and the greater use of renewable energy sources, the US Energy Information Agency predicts that the US CO_2 emissions will stay flat 5 % below the 2005 level through 2040 [18].

Reported analytical studies predict that the dynamics of future carbon emissions produced by developed and developing world is subject to a change. In particular, geographically, a significant redistribution of CO_2 emission sources throughout the world between 2000 and 2030 could be expected with developing countries getting most of the gain, and the share of the OECD countries being continuously reduced. Currently, overall CO_2 emissions from the developed countries exceed those from developing countries. However, according to reported analyses, rapid economic growth in emerging economies in Asia, South America, and Africa is expected to reverse this trend within a few decades [19].

3.4 Concluding Remarks

If technological advancements and investments in low-carbon energy sources in developing countries would make these sources cost-competitive and increase their availability and affordability to people, this would greatly improve their lives without driving carbon emissions to increasingly dangerous levels. If, on the other hand, low-carbon energy sources and associated technologies will not advance fast enough, a great deal of conventional carbon-intensive infrastructure will be put in place in the emerging economies, potentially resulting in a vast and persisting societal commitment to further increase in global CO_2 emissions and associated climate change.

References

1. International Energy Agency (2012) Energy technology perspectives. Pathways to a clean energy system. IEA/OECD, Paris
2. International Energy Agency (2009) World energy outlook 2009 edition. ISBN: 978-92-64-06130-9. Paris
3. U.N. Intergovernmental Panel on Climate Change (2013) Working group I contribution to the IPCC fifth assessment report climate change 2013: The physical science basis. Summary for policymakers http://www.climatechange2013.org./images/uploads/WGIAR5-SPM_Approved27Sep2013.pdf. Accessed 27 Sep 2013
4. International Energy Agency (2012) CO_2 emissions from fuel combustion. 2012 edition. OECD/IEA. 2012. Paris. http://www.iea.org/co2highlights/co2highlights.pdf. Accessed 20 Jan 2013
5. International Energy Agency (2012) Global carbon-dioxide emissions increase by 1.0 Gt in 2011 to record high. News room and events. http://iea.org/newsroomandevents/news/2012/may/name.27216.en.html. Accessed 30 Jun 2012
6. Intergovernmental Panel on Climate Change (2005) IPCC special report on carbon dioxide capture and storage. In: Metz B, Davidson O, de Coninck H et al. (eds) Working group III of the IPCC. Cambridge University Press, Cambridge
7. International Energy Agency (2012) World energy outlook. OECD/IEA. Paris. http://www.iea.org/publications/freepublications/publication/English.pdf. Accessed 25 Jan 2013
8. International Energy Agency (2012) Carbon capture and storage in industrial applications. ISBN: 9789264130661, doi: 10.1787/9789264130661-en. http://www.oecd-ilibrary.org/energy/carbon-capture-and-storage-in-industrial-applications_9789264130661-en. Accessed 20 Feb 2013
9. Kheshgi H, Prince R (2005) Sequestration of fermentation CO_2 from ethanol production. Energy 30:1865–1871
10. International Energy Agency (2009) IEA statistics. CO_2 emissions from fossil fuel combustion 2009 edition, IEA, Paris
11. Van Puyvelde D (2013) An update on CO_2 capture from cement production. Global CCS institute. http://www.globalccsinstitute.com/insights/authors/dennisvanpuyvelde/2013/02/20/update-co2-capture-cement-production. Accessed 23 Aug 2013
12. International Energy Agency (2002) GHG, building the cost curves for CO_2 storage, part 1: sources of CO_2, report PH4/9. Paris
13. Global CCS Institute (2011) Accelerating the uptake of CCS: Industrial use of captured carbon dioxide. http://www.globalccsinstitute.com/resources/publications/accelerating-uptake-ccs-industrial-use-captured-carbon-dioxide. Accessed 25 May 2011
14. International Energy Agency (2009) Key World energy statistics 2009 edition. Paris. www.iea.org. Accessed 5 May 2010
15. US Environmental Protection Agency (2011) Inventory of U.S. greenhouse gas emissions and sinks: 1990–2011. US-GHG-Inventory-2011-Chapter-1-Introduction.pdf. Accessed 20 Jul 2012
16. Olivier J, Janssens-Maenhaut G, Peters J et al (2011) Long-term trend in global CO_2 emissions. European Commission's Joint Research Center, PBL Netherlands Environmental Assessment Agency, The Hague
17. Jenkins J (2013) Natural gas boom rewrites the energy rules. Discover. Jan-Feb 2013 Issue. http://discovermagazine.com/2013/jan-feb/10-natural-gass-boom-rewrites-the-energy-rules#.UyH6Uc66DdM. Accessed 10 Apr 2013
18. Johnson J (2012) More gas means more growth. Chem Eng News 90:36
19. Matthews D, Solomon S (2013) Irreversible does not mean unavoidable. Science 340:438–439

Chapter 4
Stabilization of Atmospheric CO_2: Prospects and Implications

Abstract In the face of ever-increasing amounts of anthropogenic CO_2 emissions, there have been attempts to estimate the "safe" limits of atmospheric CO_2 concentrations in terms of the global mean temperature rise. The notion of "acceptable risk" is directly linked to the "acceptable" global temperature change that would ensure the survival of humankind for the foreseeable future. Large uncertainties in climate sensitivity, i.e., amount of warming expected at different atmospheric CO_2 concentrations, can be attributed to the great number and complexity of the factors that shape climate. Currently, based on the overwhelming body of evidence including modeling studies and paleoclimate data, the majority of climate experts agree on the 2 °C change (above the preindustrial level) as an acceptable global mean temperature change target, which would require the stabilization of atmospheric GHG at about 450 ppm CO_2-equivalent level. Different CO_2 stabilization scenarios and roadmaps, as well as the implications of these scenarios for energy supply, GHG emissions, industry, transportation, and energy security, are discussed in this chapter.

4.1 Introduction

Society has to make important choices with regard to present and future GHG emissions. In order to make these choices, it is necessary to consider not only the current and projected GHG emissions but also the potential impacts of the stabilization of GHG (mainly, CO_2) in the atmosphere at a particular concentration level [1]. The information needed to evaluate such atmospheric CO_2 stabilization targets is multifaceted and would require answering a number of questions, e.g., to what extent would the amount of released anthropogenic CO_2 emissions affect the average atmospheric CO_2 concentration and, potentially, global climate, and what would be the major implications of the possible impact? This debate is not new: in 1896, based on hand-written calculations, Swedish scientist Svante Arrhenius stated that

N. Muradov, *Liberating Energy from Carbon: Introduction to Decarbonization*, Lecture Notes in Energy 22, DOI 10.1007/978-1-4939-0545-4_4,

reaching atmospheric CO_2 concentrations of roughly 560 ppm would likely increase the world average temperature by about 3 °C [2]. Since then, different estimates of the global temperature increase and its links to the rise in atmospheric CO_2 concentrations have appeared in the literature (amazingly, not far from the Arrhenius' estimate).

In the face of ever-increasing volumes of man-made carbon emissions released to the atmosphere, there have been attempts to estimate the tolerable or "safe" limits of atmospheric CO_2 concentrations in terms of the global mean temperature rise. Some climate experts (e.g., NASA's Hansen) believe that we have already passed the threshold of "safe" levels of CO_2 in the atmosphere (which is 350 ppm, according to Hansen [3]), while others argue that there is still some "reserve" left before we reach the dangerous point. In practical terms, the problem boils down to the following question: what is the threshold atmospheric CO_2 concentration at which it must be stabilized in order to avoid an excessive and uncontrollable increase in the Earth's surface temperature potentially leading to irreversible negative changes in its ecosystems and climate? The answer to this question is not easy because not only scientific but also some societal, judgmental, and, even, political factors are likely to be involved.

4.2 Link Between Atmospheric CO_2 Concentration and Global Mean Temperature

Since the beginning of the Industrial Revolution, atmospheric CO_2 concentration has risen by 35–45 % reaching 400 ppmv in 2013, the highest level in almost million years. The IPCC 2013 Assessment Report, the work of almost a thousand experts from around the globe, states with 95 % confidence that more than half of the observed increase in global mean surface temperature during the last six decades was caused by the human-induced increase in atmospheric GHG concentrations [4]. Another main conclusion of the report: "Cumulative total emissions of CO_2 and the global average surface temperature response are approximately linearly related. Any given level of warming is associated with a range of cumulative CO_2 emissions" [4].

A study reported by the US National Academy of Sciences also shows a link between the global mean temperature change and both atmospheric CO_2 concentrations and cumulative CO_2 emissions (Fig. 4.1a, b).

Figure 4.1a shows the relationship of the atmospheric CO_2 stabilization levels in the range of 320–1,490 ppm to equilibrium global average temperature change. According to the diagram, the increase in the equilibrium atmospheric CO_2 concentration from average 340 ppm to average 840 ppm would result in an increase in the probability of the temperature rise from about 1 °C to about 5 °C [1]. The diagram includes the "likely" range (i.e., 66 % chance) of atmospheric CO_2 concentrations associated with various global temperature changes based on climate models and paleoclimate data. Figure 4.1b depicting the relationship between the global mean

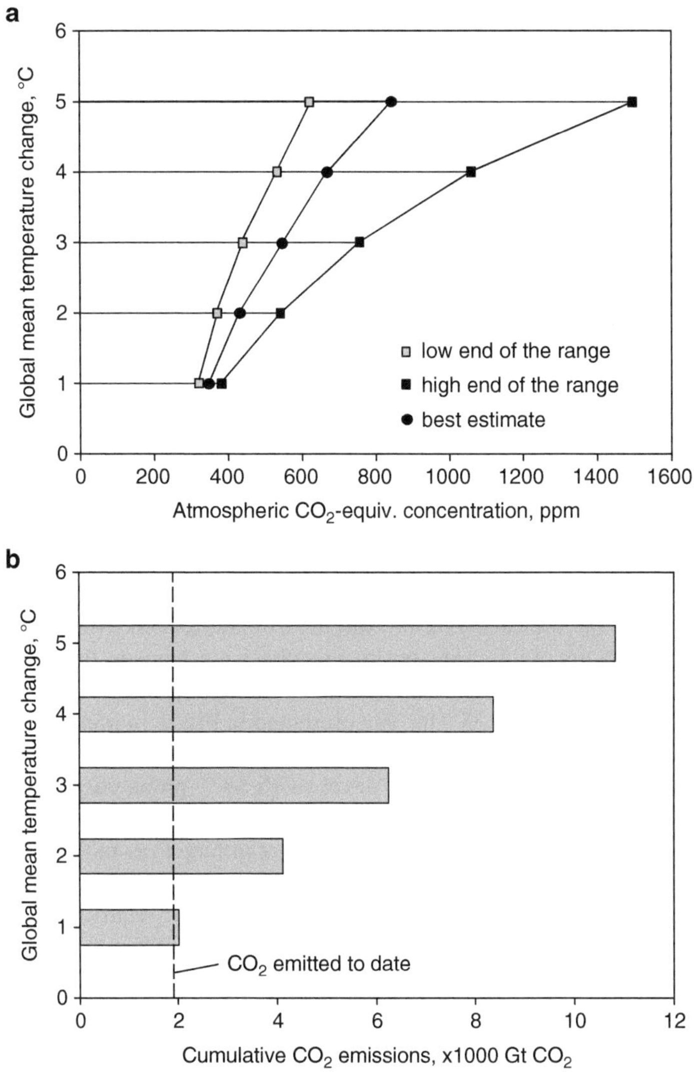

Fig. 4.1 Relationship between global mean temperature change and atmospheric CO_2 concentrations (**a**) and cumulative CO_2 emissions (**b**). *Source* [1]

temperature change and average cumulative carbon emissions shows a near-linear dependence between the temperature and the scope of emissions. The error bars (not shown on the diagram) in average amount to about 40 % in the high end of the range and about 30 % in the low end of the range (with regard to best estimates); they reflect uncertainties in carbon cycle and climate responses to CO_2 emissions due to observational constraints and the range of model results [1].

Large uncertainties in climate sensitivity, i.e., amount of warming expected at different atmospheric CO_2 concentrations, can be attributed to the fact that there are too many factors that shape climate. Because of these uncertainties, choices about atmospheric CO_2 stabilization targets depend upon value judgments with regard to the degree of acceptable risk. From this viewpoint, the cost of the judgment error could be enormous for our civilization, because of a fine balance between continuing moving along the path of economic growth and the devastating impact of the side effects of this "growth" on environment and climate. For all the practical purposes, the notion of "acceptable risk" is directly linked to the "acceptable" global temperature change that would ensure the survival of humankind for at least a millennium. Currently, based on the overwhelming body of evidence, including modeling studies and paleoclimate data, the majority of climate experts agree on the 2 °C as an acceptable global mean temperature change target (above the preindustrial level).

The 2 °C target was first adopted by the European Council in 1996 and later accepted by IPCC and The Group of Twenty (G20) [5]. In their documents, a temperature increase of 2 °C above the preindustrial levels was considered a critical point beyond which potentially catastrophic changes in our planet's ecosystems might occur [6]. The main concern is that beyond the 2 °C increase point there exists not only the elevated risk of extreme climate-related events but also the increased probability of the strong positive feedback mechanisms that would trigger even stronger climate impact with the potential of reaching a "tipping point" (see discussion in Sect. 2.5.3). The data presented in Fig. 4.1a imply that targeting the global temperature increase below 2 °C would require the stabilization of atmospheric GHG at 450 ppm CO_2-equiv.[1] level (with 54 % probability). Note that the CO_2-equiv. concentration of 450 ppm is not necessarily a universally accepted stabilization target: some countries support the 550 ppm target (as more realistic one), while other countries that are particularly vulnerable to an adverse climate impact are advocating for the much lower target of 350 ppm CO_2-equiv. [3]. In its 2007 assessment report, IPCC stipulates that in order to achieve 450–490 ppm CO_2-equiv. target, CO_2 emissions would need to globally drop to 50–85 % below 2000 levels by 2050 [7].

The link between the global temperature increase and cumulative carbon emissions implies that there is a quantifiable cumulative amount of CO_2 emissions that must not be exceeded in order to keep global mean temperature from rising more than 2 °C. Several recent analytical studies indicated that the total cumulative CO_2 emissions of about 3,700 Pg CO_2 would provide even odds of meeting the 2 °C target [1, 8] (Pg is petagram or 10^{15} g). To meet this target given already emitted carbon emissions would entail that the world has roughly half of the allowable carbon emission budget remaining. This would amount to about 50 years of carbon emissions at the current level, and it implies that the longer is the delay before starting to cut the emissions, the greater has to be the rate of these cuts in order to stay

[1] CO_2-equivalent (abbreviated CO_2-equiv.) concentration of a given mixture of greenhouse gases corresponds to the concentration of CO_2 that would have the same global warming potential.

within the allowable budget [8]. Oxford University's physicist Myles Allen estimated that the world could afford to put one trillion metric tons of carbon into the atmosphere by 2050 to have any chance of restraining global temperature increase below 2 °C [3]. Considering that nearly 570 billion tons of carbon has already been emitted to the atmosphere due to fossil fuel burning, deforestation, and other man-induced actions, Allen estimated that the trillionth ton of carbon will be released around the summer of 2041 at present rates of emissions.

Researchers at the Center for International Climate and Environmental Research-Oslo (CICERO) and the Norwegian Computing Center applied their model and statistics to analyze air and ocean temperature readings for the period starting from 1750 and ending in 2000 (their model included human-induced factors as well as fluctuations in climate caused by natural factors, e.g., volcanic eruptions, solar activity). The authors of the study determined that climate sensitivity to a doubling atmospheric CO_2 level would most likely be 3.7 °C, which more or less agrees with IPCC predictions [9]. But when researchers extended the time period to the year 2010 (i.e., adding data from 2000 to 2010), the modeling results indicated that with 90 % probability, the temperature rise due to a doubling of CO_2 concentration would vary in the range from 1.2 to 2.9 °C (an average of 1.9 °C). This is a substantial drop in the temperature rise estimates (almost by 2°) compared to previous reports. The explanation given by the authors of this study to this unexpected result boils down to the following. The Earth's mean surface temperature rose sharply during the 1990s, which may have caused the model to overestimate the climate sensitivity; since then, the temperature increase has leveled off at its 2000 level (likewise, ocean warming also appears to have stabilized), although CO_2 emissions and other man-made factors contributing to climate change are still on the rise. According to the researchers, natural variations that could occur over several decades may have caused the leveling off of temperature between 2000 and 2010 (which came on top of long-term warming) [9]. Despite the project's findings that somewhat allay the urgency of climate mitigation actions, the authors emphasize that it must not be construed as an excuse for the complacency in addressing human-induced global climate change. The results do provide some encouragement that achieving climate targets may be more within our reach than previously thought.

4.3 CO_2 Stabilization Scenarios: Paths to Different Energy Futures

Climate sensitivity is a measure of how much the global temperature is expected to rise if humans would continue releasing CO_2 and other GHG into the atmosphere; for the simplicity, the temperature rise is considered in reference to doubling CO_2 levels against preindustrial world (ca. the year 1750). According to many models, if GHG are being emitted at the current rate, there is a risk of doubling atmospheric CO_2 concentration by around 2050 [9].

Climate development is an extremely complex phenomenon for it is affected by a large number of different factors, among which various feedback mechanisms are the least understood factors. (Due to the feedback mechanisms some factors can greatly amplify or neutralize their impact on climate; more details on this phenomenon follow.) Due to uncertainties with regard to the impact of different factors and feedback mechanisms on climate, it is very difficult to prognosticate how high might the Earth's mean surface temperature rise due to anthropogenic GHG emissions. According to IPCC estimates, the climate sensitivity (in temperature rise units) to doubled atmospheric CO_2 concentrations varies between 2 and 4.5 °C warming (most probable being 3 °C) [9].

The IEA in its Energy Technology Perspectives 2012 report (ETP-2012) [10] unveiled three dramatically different energy future scenarios associated with the average global temperature increases of 6 °C, 4 °C, and 2 °C.

4.3.1 6 °C Scenario (6DS)

The 6DS pathway is essentially the extension of the current global carbon-intensive trends, and this scenario presupposes the absence of any tangible efforts to address climate change concerns and stabilize atmospheric GHG concentrations. In the 6DS, fossil fuel use and GHG emissions would almost double by 2050 (against 2009). In particular, coal use for electricity generation would increase more than twofold, and carbon capture and storage technology would not be deployed. The share of renewable energy sources would modestly increase from 19 to 24 %. Transport would almost entirely rely on petroleum-based fuels, with very slow improvement in fuel economy. Energy efficiency would slowly improve at about 1 % per year rate (comparable to that from 1971 to 2009). Oil prices would continue to rise and approach $150/barrel by 2050. To a large extent, the 6DS is consistent with Business-as-Usual Scenario (BAUS) or Reference Scenario (RS) described in relevant analytical studies. According to many experts, the 6DS path is clearly unsustainable in the long run and could have potentially devastating impacts on the Earth's ecosystems and its inhabitants [10].

4.3.2 4 °C Scenario (4DS)

The 4DS projects a long-term average global temperature rise of 4 °C and reflects concerted efforts to abandon current carbon-intensive trends and technologies. The 4DS takes into consideration recent pledges by major industrialized and developing countries to substantially reduce both energy demand and GHG emissions and improve energy efficiency. Policies and measures to achieve the 4DS include the boost in the use of renewable energy, shift away from fossil fuel dependency, and the increase in end-use energy efficiency. Under the 4DS, annual energy-related carbon emissions would rise by 27 % (against 2009) to 40 Gt by 2050, with fossil fuels still representing two-thirds of the total primary energy supply (TPES) [10]. Certain measures to decarbonize energy would start in the 4DS, but the transition will be rather slow, with only 2 % of total electricity capacity equipped with CCS and 35 % produced from renewable sources. In transport, in response to tighter fuel economy standards, fuel economy in passenger vehicles would improve in average by 30 % (over 2009), mostly due to an increased share of gasoline hybrid vehicles. Although the IEA analysis indicates that the 4DS is plausible, many experts already consider it an ambitious scenario, because it would require immediate and significant changes in energy policies and technologies and the associated cuts in carbon emissions.

4.3.3 2 °C Scenario (2DS)

The 2DS pathway presents a vision of a sustainable energy system; it is consistent with the widely agreed-upon target of limiting the mean global temperature increase to 2 °C. According to climate science research, the global temperature increase limit of 2 °C would stave off with high probability the worst consequences of the adverse impact of human activities on climate and ecosystems. Many experts believe that the 2DS would support a healthy economic growth and prosperity without catastrophic environmental consequences. The 2DS path is projected to deliver CO_2 emission savings of 24 Gt against 4DS scenario and 42 Gt compared to 6DS path. In both cases, about a quarter of CO_2 emission reductions would come from OECD countries and three quarters—from non-OECD countries. The 2DS path will be the focus of this chapter, where the energy and environmental implications of this energy future scenario are analyzed in detail.

4.4 Two-Degrees Scenario (2DS) and Its Implications

4.4.1 Implications of 2DS for Greenhouse Gas Emissions

According to the IEA's 2DS pathway, countries must take very strong coordinated actions in the energy, industry, and other sectors to drastically reduce the levels of GHG emissions into the atmosphere (i.e., cutting energy-related CO_2 emissions in half by 2050 compared to 2009). The IPCC's 2013 Report estimates that limiting the warming caused by anthropogenic CO_2 emitted since the 1850 s to 2 °C with probability of >33 %, >50 %, and >66 %, would require limiting cumulative CO_2 emissions from all anthropogenic sources to about 1,560 GtC, 1,210 GtC, and 1,000 GtC, respectively, over the same period [4]. Figure 4.2 depicts contributions of different technologies to carbon emission reductions in the 2DS pathway (relative to the 6DS pathway) according to ETP-2012 [10].

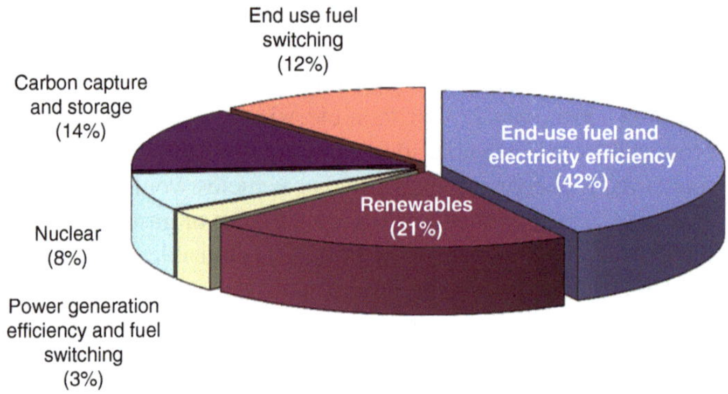

Fig. 4.2 Contributions of different technologies to carbon emission reductions in the 2DS pathway (relative to the 6DS) according to IEA's Energy Technology Perspectives 2012. *Source* [10]

Among carbon mitigation technologies, the end-use fuel/electricity efficiency is projected to be the main contributor (42 % of CO_2 savings). This can be attributed to the fact that energy efficiency improvement measures in the industrial and building sectors typically have short payback periods and very low net abatement costs [5]. An increased market share of renewables in power and heat generation would result in 21 % of total CO_2 savings in 2050 (relative to 6DS) [10]. The sizable share of CO_2 savings (14 % of total) could be achieved by a widespread introduction of carbon capture and sequestration technologies. The deployment of additional nuclear-based facilities for power generation and industrial applications would cut CO_2 emissions by about 8 % in 2050 [10]. According to the IEA's projections, the increased share of biofuels in the transportation sector would result in modest CO_2 savings of only 3 % [5]. Among different sectors, power generation and transportation hold the greatest potential to reducing carbon emissions (42 % and 21 % of total, respectively, relative to 4DS). Industry and residential/commercial buildings will also be making sizable contributions to the carbon emission reductions (18 % and 13 %, respectively, relative to 4DS). The emissions of CH_4, N_2O, halocarbons, and those from the land-use change and forestry (LUCF) will stay relatively flat until 2015–2020, and after that they will steadily decline and cumulatively amount to 5.1 $GtCO_2$-equiv. (or about 24 % of the total GHG emissions) in 2050 [5].

Figure 4.3 depicts IEA's long-term outlook (to 2200) for the atmospheric concentrations of CO_2 and all GHG according to the 450 ppm Scenario (450S) and reference scenario (RF) (Note that 450S and RF are the equivalents of the 2DS and 6DS in the ETP-2012.) [5].

It can be seen that the concentration of CO_2 and summarily GHG in the 450S pathway will be about half of that in the RS pathway, in which the atmospheric concentrations of CO_2 and GHG gases steadily increase during the current century and reach about 750 and 1,000 ppm CO_2-equiv., respectively, by 2150. In the 450S, the atmospheric GHG concentration peak at 510 ppm CO_2-equiv. in 2035 stay almost flat for about a decade and then slowly decline to the 450 ppm target level.

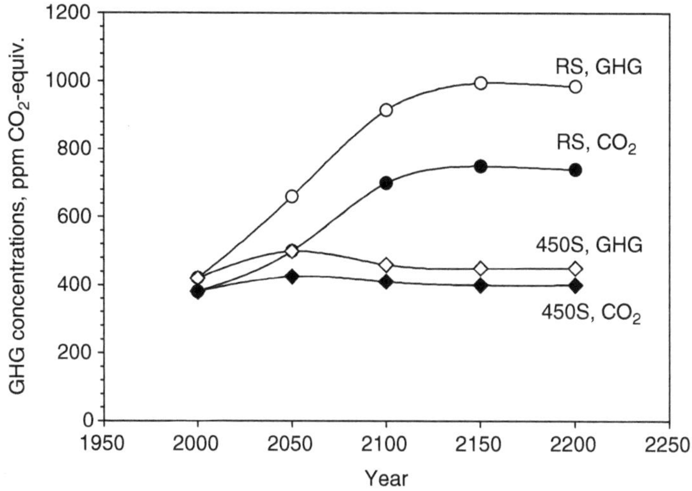

Fig. 4.3 IEA's long-term projections of the atmospheric concentrations of CO_2 and summarily all GHG according to 450S and RS (which are consistent with the 2DS and 6DS, respectively). *Source* [5]

4.4.2 Implications of 2DS for Total Energy Supply

It is realized that achieving targets would require the extensive global transformation of the existing energy system. To a large extent, success will depend on a substantial decoupling of the energy use from economic activity, which would require technological advancements, as well as the changes in economic structure and individual behavior [10]. The 2DS envisages the drastic measures and policies aiming at significantly diminishing the total energy supply growth through the mid-century. In particular, the total energy supply in OECD countries will slowly decrease over the entire period, while in non-OECD countries it will slow down until 2035 (compared to 2000–2010) after which it will start increasing again. The worldwide trend will almost repeat that of the non-OECD countries.

Although fossil fuels will continue being a major source of energy throughout the projection period, their share in the overall energy supply will significantly decline by the middle of the century. Decarbonization of electricity is the most important system-wide change in the 2DS. In 2009, fossil fuels (mostly, coal and NG) generated 67 % of global electricity [10]. According to IEA, policies stimulating the increased deployment of carbon-free energy sources, such as hydropower, onshore wind, and nuclear, as well as the significant expansion of emerging technologies (solar, offshore wind, geothermal) would bring the share of fossil fuels in power generation sector to less that 25 % by 2050 [10]. Along with the extensive deployment of CCS technology, these measures would result in an almost 80 % drop in the carbon intensity of electricity generation in the 2DS: from about 600 g CO_2/kWh in 2009 to 60 g CO_2/kWh [10].

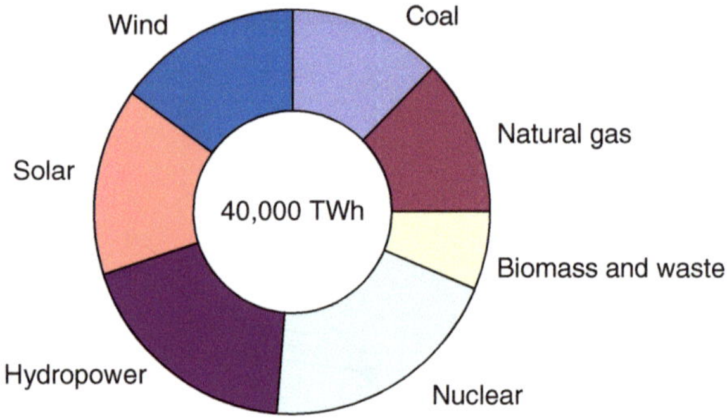

Fig. 4.4 Fuel mix in electricity generation in 2050, according to the IEA's 2DS pathway. *Source* [10]

In response to potential international carbon-restricting policies, energy security, and environmental concerns, the 2DS envisages a significant increase in the share of carbon-free fuels and energy sources in the overall energy demand. In the 2DS, global electricity demand will grow from about 20,000 TWh in 2009 to 40,000 TWh in 2050; however, compared with the 6DS/RS, this will constitute a drop of about 18 % (as a result of worldwide energy efficiency increase measures) [10]. Figure 4.4 depicts the composition of fuel mix in electricity generation in 2050, according to the 2DS.

IEA projects that carbon-free and low-carbon sources of electricity (e.g., hydro, nuclear, biomass, solar, and other renewables) will make up about three-quarters of the global total by 2050. In contrast, in the 6DS/RS, only about one-third of electricity will be produced from non- and low-carbon sources by 2050 [10]. The growth rate of electricity production from solar, geothermal, wind, and other nonhydro renewables is projected to outpace any other source of power generation. Wind (onshore and offshore) is projected to produce the lion's share of the renewable electricity by the mid-century. Due to the anticipated reduction in global primary energy demand and the introduction of CO_2 emission limits projected in the 2DS, fossil fuels will see a decline in production; their use will drop by about 20 % in 2050 compared to 2009 levels, but compared to the 6DS/RS path this represents 60 % reduction [10].

4.4.3 Implications of 2DS for Industry

To meet the 2DS goal in industry, it would be necessary to achieve significant technological advances in many sectors, optimize and increase energy efficiency of the majority technological processes. IEA projects that by using best available technologies (BAT) it would be possible to achieve reduction in industrial energy consumption by about 20 % in the mid-to-long term [11]. Between 2000 and 2010, the total industrial final energy use increased by 31 % from 103 to 135 EJ (associated with 8 $GtCO_2$ emissions) (EJ is exajoule or 10^{18} J). This growth was largely driven by China and India in response to an increased demand for industrial materials in these countries. In order to stay on the 2DS track, industry must rein in growth in its energy consumption to 162 EJ by 2020 with direct industrial CO_2 emissions peaking at around 9 $GtCO_2$ by that year.

Chemicals and petrochemicals is the largest energy-consuming industrial sector (37 EJ of energy used in 2010), accounting for 28 % of total industrial final energy consumption and 16 % of total direct industrial CO_2 emissions (1,292 $MtCO_2$) in 2010 [11]. The 2DS targets the reduction in the final energy consumption in this sector by an estimated 10 EJ using the BAT together with process integration, recycling, and energy recovery. This would allow reducing CO_2 emissions from chemical/petrochemical sector by 27 % in 2020 (compared to 2010 level), or by 350 $MtCO_2$ compared to 4DS levels.

The cement industry is the third largest consumer of energy in industry (13 EJ in 2010). To reach 2DS target by 2020, the cement industry will need to substantially raise thermal and electric efficiencies, use alternative fuels and clinker substitutes, and widely deploy CCS. The 2DS 2020 target is to reach thermal energy efficiency of 3.7 GJ per ton clinker by 2020 (in 2010, it was 3.9 GJ/t) and CO_2 emissions of 0.68 t CO_2 per ton cement (in 2010, it was 0.73 t CO_2/t cement). Iron and steel industry must limit growth in energy consumption to reach 32 EJ in 2020 and reduce CO_2 emissions by 247 $MtCO_2$ (relative to 4DS) through application of BAT.

4.4.4 *Implications of 2DS for Transport*

Transportation sector consumes a lion's share of oil (more than two-thirds of total oil consumption) and emits about a quarter of the total CO_2 emissions; therefore, it provides an excellent illustration of the potential impact of the 2DS path in terms of oil consumption savings and CO_2 emission reduction. The improvements in road transportation could be achieved through an improvement in fuel economy and the use of advanced types of vehicles. According to the 2DS, oil will be replaced in transport by a portfolio of three alternative fuels: electricity, biofuels, and hydrogen [10]. This will require significant technological advancements and price reduction in vehicle propulsion systems, in particular, with regard to electrification of light-duty vehicles.

4.4.4.1 Fuel Economy

In the 2DS, fuel economy accounts for 0.6 $GtCO_2$ (or about 60 % of total carbon emissions from road transportation) reduction in 2020 [11]. This represents a reduction in oil demand of approximately 2.4 million barrels per day in 2020 (excluding savings from penetration of hybrid-electric vehicles). In the near term (i.e., next decade), improving fuel economy of conventional ICE vehicles holds the greatest potential to reduce fuel consumption and CO_2 emissions in the road transport.

ICE vehicles are projected to represent more than 90 % of light-duty vehicles (LDV) sold worldwide between 2010 and 2020 (passenger LDV consume 60 % of total fuel used in road transport). 2DS targets to achieve global average fuel economy of new LDV equal to 5.6 l of gasoline equivalent (Lge) per 100 km in 2020 (compared to 7.6 and 8 Lge/100 km in 2008 and 2005, respectively) [11]. This represents an improvement in fuel economy of about one-third in 2020 against 2005 level (i.e., the annual improvement of 2.7 % during 2011–2020 decade). In the heavy-duty vehicles (HDV) category, 2DS targets an annual fuel economy improvement of 1.5 %. Although CO_2 emissions will drop by 25 % compared to 2009, still, overall CO_2 emissions in transport will amount to 5 $GtCO_2$ in 2050 [10], mainly due to the dramatic increase in the number of vehicles in developing countries.

4.4.4.2 Advanced Vehicles

This category of vehicles includes hybrid-electric (HEV), plug-in hybrid-electric (PHEV), full battery-electric (BEV), and fuel cell electric (FCEV) vehicles. Around 100,000 PHEV and BEV were sold globally in 2012 (twice the amount sold in 2011). This trend puts the sales growth rate on a track to meet 2DS targets [11]. In 2012, HEV broke the 1.2 million mark in annual sales. Recently, there have been dramatic improvements and cost reductions in battery and charging technologies. Cost of batteries dropped from US$800–1000/kWh in 2010 to around US$500–600/kWh by the end of 2012. In 2012, there were breakthrough developments in charging technology for BEV and PHEV, particularly a three-phase fast electric vehicle charger developed by Volvo Car Corp. The charger reduced a vehicle charging time by a factor of six (to about 1.5 h), and is scheduled to enter the market in 2013. These developments undoubtedly boost consumer confidence in the electric-driven vehicles by lessening anxiety over their driving range (which will approach the driving range of conventional vehicles).

The 2DS sees 20 million BEV and PHEV on the road by 2020, with yearly sales reaching 7 million vehicles [11]. The 2DS also projects the stronger displacement of ICE vehicles in the post-2025 period, with the share of BEV, PHEV, and FCEV increasing sharply to 50 % of new vehicle sales by 2050. HEV also plays an important role in the 2DS as a transitional technology aiming at a fuel economy improvement. According to 2DS, annual HEV sales will reach 10 million by 2020 (or 12 % of the global market share) and they will peak at 40 million (30 % of the global market), as the share of BEV, PHEV, and FCEV will steadily increase. It is projected that battery costs will be further reduced to an estimated US$300/kWh to reach cost parity with ICE. This would dramatically reduce the cost of ownership of electric vehicles to be attractive for consumer levels, and, hence, would boost their market penetration.

In March 2012, AutoNews Europe reported that by 2015, the auto industry must reduce CO_2 emissions from new cars sold in Europe to a fleet average of 130 g CO_2/km (for comparison, in 2011, 2009, and 2006, the averages were 141 g CO_2/km,

146 g CO_2/km, and 160 g CO_2/km) (Note that 130 g CO_2/km corresponds to fuel consumption of 5.6 l of gasoline or 4.9 l of diesel fuel per 100 km.) [12]. In 2011, Fiat and Toyota ranked first and second in Europe based on average CO_2 emissions of 125.9 g CO_2/km and 130.0 g CO_2/km, respectively. Toyota Motor Corp. only needs to cut its fleet CO_2 emissions by 4.2 % by 2015 to achieve their target of 20 % reductions by 2020. PSA/Peugeot-Citroen SA and BMW AG need to cut their overall fleet emissions by 7 % to comply with the emission regulations. The trend of shifting toward cheaper, smaller, and lighter vehicles would help automakers to meet the CO_2 reduction target. The next emission reduction target is 95 g CO_2/km by 2020, which, according to industry experts, would be difficult to achieve because this would require manufacturing of vehicles from exotic and expensive lightweight materials (Fiat reduced the average weight of its models from 1,337 kg in 2009 to 1,067 kg in 2011.) [12].

In the USA, transportation is the second largest source of CO_2 emissions, accounting for about 31 % of total CO_2 emissions and 26 % of total US GHG emissions in 2010 [12]. The US EPA proposed the target of achieving a fleet-wide level emissions of 155 g CO_2/km and 101 g CO_2/km for the model years 2016 and 2025, respectively (in the model year 2025, a passenger car target is 89 g CO_2/km and a light truck target is 126 g CO_2/km). The most extensive CO_2 emission control is mandated in the state of California (USA), where legislature passed a bill requiring 25 % reduction in state CO_2 emissions by 2020, with the first major control taking effect in 2012. The California mandate will result in cuts of about 174 million tons of CO_2 (which corresponds to cutting emissions to their 1990 level). Several northeastern states also signed regional agreement to reduce CO_2 emissions with the target reduction of 24 million tons CO_2. The US National Research Council released a report in 2013 that analyzed various alternative vehicles and fuels with two main objectives [13]: (1) reducing oil consumption by 50 % below 2005 levels by 2030 and (2) reducing both oil consumption and GHG emissions by 80 % below 2005 levels by 2050.

4.4.5 Economics of 2DS

The implementation of the 2DS on a global scale would require a substantial investment in carbon mitigation technologies including an across-the-board improvement in energy efficiency, energy conservation, and the widespread deployment of zero-, near-zero- and low-carbon technologies. The total investment needs in the 2DS between 2010 and 2050 are estimated at US$140 trillion or US$36 trillion more compared to the investments in the 6DS path (or about US$1 trillion additional investment each year until 2050) [10]. These additional investments would amount to approximately 1 % of cumulative GDP over this period and, according to the reported estimates, is unlikely to present a very large burden on the global economy.

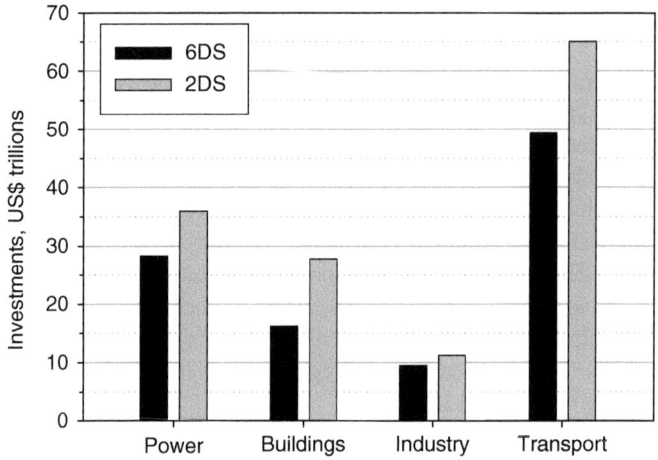

Fig. 4.5 Additional investment requirements by sector in the 2DS and 6DS pathways. *Source* [10]

The four key sectors where the most of the additional investments would be directed to are power generation, buildings (residential and commercial), industry, and transport. Figure 4.5 compares cumulative additional investments by sector in the 2DS and 6DS pathways.

Figure 4.5 shows that the largest investments in the 2DS and the incremental increase in the investments against 6DS are in the transportation sector (US$65 trillion and US$15.6 trillion, respectively), where money will be spent mainly on purchasing more efficient and less polluting vehicles, e.g., HEV, PHEV, and BEV. (Note that only the cost of powertrain is presented in the Fig. 4.5; for the full vehicle costs, the investments would have almost tripled.) Power generation sector is the second largest area of additional investments in the 2DS amounting to the total of US$35.9 trillion by 2050. These additional investments would help to expand the share of renewables, nuclear power, and to incorporate CCS technology, especially, in coal-fired power plants. In the building (residential, commercial, public) sector, the investments in the 2DS (totaling US$27.8 trillion) would mainly be directed toward the measures to improve energy efficiency, boost energy conservation, and the use of renewable resources. Industrial sector (iron and steel, chemicals, cement, pulp and paper, and aluminum) will need US$11.2 trillion of investments in the 2DS.

From 2010 to 2020, the investment requirements in the 2DS are expected to be modest (in average, US$2.4 trillion per year, or 25 % higher than in the 6DS), with money mostly directed toward improvements in energy efficiency and offsetting the high investment cost for low-carbon technologies [10]. The following years of 2020–2030 will see average annual investments rising to US$3 trillion (or 36 % increase over the 6DS), which will mostly go to renewable power and deployment of CCS in the power generation and industry sectors, and also improvements in energy efficiency of buildings. After 2030, investment needs in the 2DS will substantially increase with the annual investments reaching US$4.4 trillion (more than 50 % than that of 6DS), which could mainly be attributed to higher investment costs of decarbonizing transport and power generation sectors.

According to the IEA's estimates, the transition to a low-carbon energy sector through the 2DS path would require an additional investment of about US$130 per person per year (on average) from now until 2050 [10]. This value will widely vary depending on the country and region with higher investment costs required in the countries consuming more energy per capita (e.g., OECD countries), as opposed to countries with lower per capita energy consumption (e.g., developing and non-OECD countries). For example, additional investments in the 2DS in the USA are projected to amount to US$386 per capita per year, whereas this value will be only US$54 per person per year in developing countries (not including China and India).

One of the major findings of the IEA'a ETP-2012 study was that future savings from the 2DS would potentially outweigh its upfront investment costs, even without accounting for the value of preventing potential damages from climate change [10]. Investments in zero- or low-carbon technologies are projected to generate estimated US$103 trillion between 2010 and 2050, which represent undiscounted net savings of US$61 trillion (or in average US$1.5 trillion per year) (relative to 6DS or BAUS). Even with 10 % discount rate, it would still result in net savings of US$5 trillion, which points to the affordability of implementing low-carbon economy. According to IEA, most of the savings will come from lower fossil fuel usage. The projected increases in fossil fuels prices (especially, oil prices, in 6DS) would drive a demand for these fuels down. As a secondary effect, the lower demand may potentially dampen fossil fuel price increases in 2DS resulting in even more savings. If the impact of lower fuel prices is also taken into account, the total reduction in fuel purchases would amount to US$150 trillion [10].

The additional investments in low-carbon energy sector would result not only in environmental benefits but also in improved energy security, since the dependence on fossil fuels would be greatly reduced. These positive developments would be especially advantageous for the countries that import oil and gas, since this would free up foreign reserves for other uses. In addition, the investment in the 2DS would provide significant health benefits and additional employment opportunities [10].

4.4.6 Implications of 2DS for Energy Security

The energy security of a given country is determined by its ability to obtain the uninterrupted availability of its main energy sources at affordable prices and to react promptly to sudden (in many cases, unforeseen) changes in their supply. Historically, energy security was predominantly linked to oil (or petroleum products) supplies. While oil supply still remains a key issue for most countries, the increasing complexity of energy systems requires a systematic understanding of a wider range of vulnerabilities, because disruptions can affect not only oil but also other fuel sources and the infrastructure as a whole [10]. In response to this challenge, IEA has developed a comprehensive tool to assess energy security called MOSES (the MOdel of Short-term Energy Security) [10]. The MOSES examines both risks and resilience factors associated with short-term (days to weeks) disruptions of energy supply, and it extends beyond oil monitoring and analyzes several important energy sources and the entire infrastructure. The utility of MOSES is that it helps countries understand their energy security profiles in order to identify energy policy priorities. For example, analysis of vulnerability for oil and gas disruptions is based on the risk factors such as net-import dependence and the political stability of suppliers, and, sometimes, political climate in the region. The resilience factor is linked to the ability of a country to respond to disruptions by substituting with other suppliers and supply routes. For instance, the resilience factors may include the diversification of suppliers, the level of fuel stocks, and the number of entry points (ports, pipelines) [10].

The energy security portfolio of the 2DS promotes the diversification of energy sources (as a resilience factor) and lowering of a total energy demand. For example, as an important energy security measure, the use of fossil fuels is projected to decrease by about 50 % in both electricity generation and transport in the OECD countries [10]. Of particular importance is the security of electricity supply, because it will account for the larger share of a final energy demand in the 2DS. Electricity decarbonization policies must be coupled with efforts to ensure the reliability of electrical grids, making them more flexible in terms of transporting, storing, and trading electricity. The importance of reliability and flexibility factors will especially increase with the inclusion of large-scale solar and wind power plants into the interregional grids.

4.5 CO_2 Stabilization Roadmaps

4.5.1 Carbon Abatement Options

In the ETP-2012 report, the IEA outlined the vision of abatement options through mid-century [10]. Figure 4.6 depicts the global marginal abatement options and the associated abatement costs according to the report.

Marginal abatement costs represent the estimated cost for the last ton of CO_2 emissions eliminated via abatement measures [10]. The ETP-2012 analysis follows two guiding principles: (a) less costly technologies are applied before more expensive ones and (b) the margin (i.e., the most costly) abatement costs should be roughly equal across all sectors and regions that are considered. The values of the abatement costs presented in Fig. 4.6 represent the cost of the most expensive options applied to mitigate CO_2 emissions in the 2020–2050 time period. It is recognized that uncertainty surrounds each cost projection, increasing as the date of projection moves further into the future. Furthermore, the marginal abatement costs are dynamic by

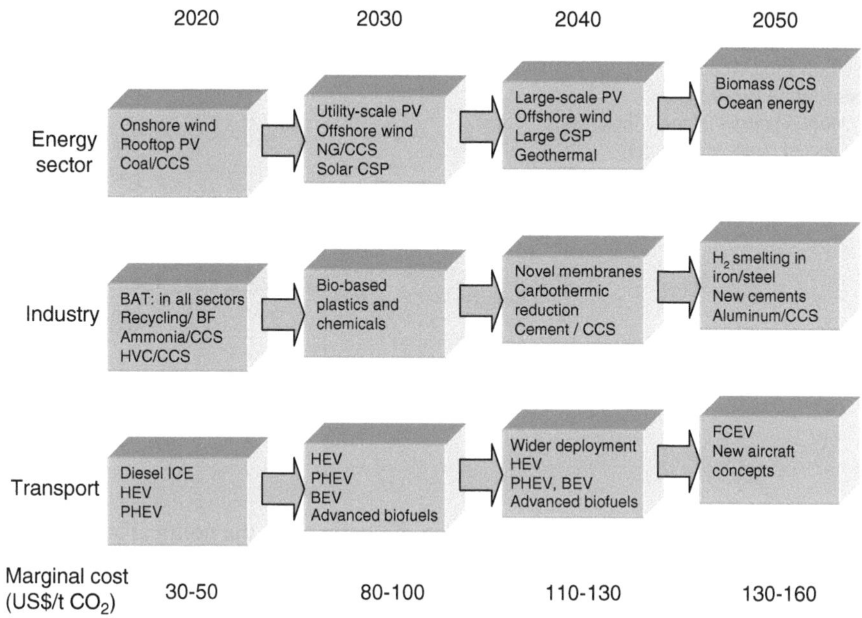

Fig. 4.6 The global marginal abatement costs and marginal abatement options according to the IEA's ETP-2012 report [10]. *BEV* battery electric vehicle, *BF* blast furnace, *CCS* carbon capture and storage, *CSP* concentrating solar power, *FCEV* fuel cell electric vehicle, *HEV* hybrid electric vehicles, *HVC* high-value chemicals, *ICE* internal combustion engine, *PHEV* plug-in hybrid electric vehicle, *PV* photovoltaic

nature: they evolve over time and affect each other. Two factors work in opposite directions: the abatement costs increase as emissions get deeper (assuming everything else being equal); however, as more clean energy technologies are deployed, their cost may decline as a result of learning [10]. The combined effect will depend on whether learning outpaces the move up along the cost curve.

The cost estimates of delaying carbon abatement measures could be found in the literature. For example, Mignone et al. examined the implications of the delay using a well-tested model of the ocean carbon cycle [14]. The authors of the study determined that when future CO_2 emissions are constrained to decline at the rate of 1 % per year, the peak atmospheric CO_2 concentration (the so-called stabilization frontier) increases at the rate of about 9 ppm per year (average). The simulation results indicated that the atmospheric CO_2 stabilization level below doubling a preindustrial CO_2 level (i.e., 550 ppm, which is cited as a target according to some climate policy assessments) would require dedicated carbon mitigation efforts to start within roughly the next decade. The study concludes that the delay of more than a decade would not guarantee the stabilization of atmospheric CO_2 concentration below 550 ppm. According to other reports, the minimum the humans can do to avoid serious consequences of climate change is to start from now cutting CO_2 emissions by roughly 2.5 % per year until the year 2050 [3].

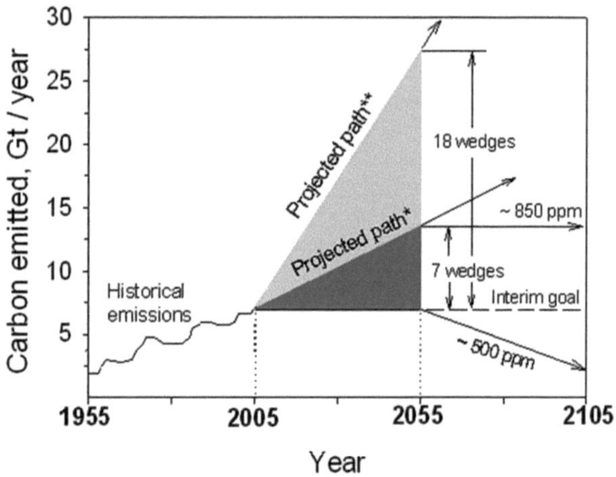

Fig. 4.7 Schematic representation of the Stabilization Wedges concept. (*Asterisk*) and (*Double Asterisk*) denote the projected paths according to refs. [17] and [18], respectively

Davis et al. emphasized the importance of overcoming the political and technological inertia in curbing man-made CO_2 emissions [15]. According to the authors, the existing century-old energy and transportation infrastructure is expected to contribute significant amounts of CO_2 emissions over the next half century due to an infrastructural inertia. Assuming no additional CO_2 point sources would be built, and all the existing CO_2 emitters would be allowed to live out their normal lifetimes, cumulatively, 282–701 Gt of CO_2 from combustion of fossil fuels would be emitted by the existing infrastructure between 2010 and 2060, resulting in CO_2 atmospheric level of about 430 ppm, and forcing mean temperature increase of 1.1–1.4 °C above the preindustrial level. Since nobody doubts that more CO_2-emitting devices will be built, this scenario cannot be considered realistic, but it offers a means of assessing the threat of climate change from existing sources relative to those that have to be built.

4.5.2 "Stabilization Wedges" Concept

In 2004, Pacala and Sokolow outlined the "Stabilization Wedges" (SW) concept as a useful tool for quantifying the actions that would be necessary to stabilize atmospheric GHG concentrations at acceptable levels within half a century [16]. Figure 4.7 provides a graphical representation of the SW concept.

In this diagram, the baseline (or a business-as-usual) case is represented by a linear growth trajectory leading to doubling of global CO_2-equiv. emissions by 2055: from about 7 GtC/year to about 14 Gt C/year by mid-century. The flat trajectory on the diagram represents "zero emission growth," with the target of achieving the removal of 7 GtC/year in 2055. The area between the flat and ramp trajectories form the so-called stabilization triangle, which corresponds to the policies required to stabilize the atmospheric CO_2 concentration below doubling the preindustrial levels (about 500 ppm) over the period of half a century. According to the authors, one "wedge" corresponds to about 1 GtC/year of the emission savings in 2055 due to the implementation of a single mitigation strategy; thus, to achieve the zero-carbon growth (or the flat trajectory), there will be a need to introduce roughly seven wedges. After achieving the interim goal of the zero emission growth by 2055, a more challenging carbon target could be mapped out that would further reduce the global GHG emissions. The authors suggested 15 potential wedges or carbon-mitigating technologies, including the implementation of different energy efficiency increase measures, decarbonization of electric power generation and fuel supplies (e.g., through shifting to low-carbon fuels), CCS, use of nuclear and renewable energy resources, and biological storage in forests and soils. The selection of a particular wedge will depend on a variety of factors and, in most cases, markets would help to determine the "winning" wedges. Although some technological wedges have already been deployed on a commercial scale and no fundamental breakthroughs would be required, many believe that these wedges could be difficult to implement due to a number of reasons of technical, social, environmental, and political nature.

Some experts do not agree with the quantitative aspects of the SW concept, suggesting that the original SW concept greatly underestimates needed CO_2 reductions to stabilize atmospheric CO_2 levels. Hoffert et al. [18] pointed to the fact that the original SW scenario was built on the business-as-usual emissions baseline based on an assumption that a shift in the fossil fuels mix would result in the reduction of a carbon-to-energy ratio (i.e., the amount of CO_2 released per unit of energy). Although the carbon-to-energy ratio indeed declined during prior shifts from coal to oil to NG, this trend may no longer be valid, because as oil and NG production peaks in the future, coal production will rise, effectively increasing the carbon-to-energy ratio (e.g., a large number of coal-fired power plants will be built in China, India, and other developing countries). Considering this trend, Hoffert estimated that not 7, but 18 new wedges would be required to curb carbon emissions by mid-century in order to keep future warming below 2 °C (see Fig. 4.7).

In general, despite some simplifications and scientific, economic, and political uncertainties shrouding certain aspects of the Stabilization Wedges concept, it could be a useful tool providing guidance on where to focus the limited resources in order to achieve the stabilization of atmospheric CO_2 concentrations at acceptable levels within the next 50 years, and how to prepare the grounds for further CO_2 emission rate reduction in the following half century.

4.6 Carbon Pricing: Status, Prospects, and Challenges

There is a strong consensus among many experts that carbon pricing should be at the core of an effective long-term climate policies, either through *a carbon tax* or an *emission trading scheme* [19]. The rationale for that is to make carbon emissions more expensive, while (judiciously) giving carbon-free alternatives (e.g., nonfossil energy and energy efficiency) a marginal advantage. Ideally, a carbon price should be equal to the net cost to the society caused by an additional ton of carbon emissions [10]. It is believed that carbon pricing is necessary to incentivize actions, and, if applied across the entire economy, it could deliver an efficient outcome, since all sectors would face equal marginal abatement costs.

Carbon tax has been implemented in a number of developed countries, and, currently, it is a subject of discussion in many (including developing) countries. Since Sweden introduced its carbon tax in 1991, another nine OECD countries have followed the suit [19]. In general, encouraging developments in the carbon market are now evident in many regions of the world. The Emission Trading System (ETS) has been in effect in New Zealand since 2009. Australia is in the process of implementing its carbon pricing law, with carbon tax evolving into a full-blown ETS by 2015 [10]. In the USA, the North-Eastern States' Regional Greenhouse Gas Initiative has been in operation since 2009 (albeit, at present, with low prices), and the state of California's system is to start in 2013. Canadian province of Alberta has a carbon price system in effect, with revenues directed to fund innovative GHG mitigation technologies. South Korea recently approved a law to implement ETS by 2015 [10].

In Europe, the carbon market presents a mixed picture. The European Union ETS (the largest of such systems in the world) is currently hampered by a large surplus of emission allowances, which stems from both the economic downturn and an overallocation to industrial sources early on [10]. The resulting carbon price of less than €10 per ton CO_2 is not enough to motivate using gas instead of coal in the European power generation sector, and it provides only limited incentives to renewables and nuclear options.

Among recent significant developments, China has launched six carbon market pilot programs covering four large cities, Beijing, Shanghai, Chongqing, and Tianjin, and two provinces [10]. If successful, these pilots would pave the way for a nationwide system by 2015. Other developing countries interested in implementing various types of carbon market mechanisms as part of the World Bank's Partnership for Market Readiness are Brazil, India, South Africa, Chile, Colombia, Costa Rica, Indonesia, Jordan, Mexico, Morocco, Thailand, Turkey, Ukraine, and Vietnam [10]. The Conference of the Parties to UNFCCC (COP-17, Durban, 2011) countries agreed to establish a new carbon market mechanism to support carbon mitigation actions in the developing countries [10]. Without underestimating the challenges of establishing and implementing carbon market mechanisms, the prospects for carbon pricing are positive.

Encouragingly, the available data show that energy and carbon taxes (in the countries where they have been implemented) clearly affected energy consumption behavior: countries with higher average effective tax rates on CO_2 tend to have lower carbon emissions per unit of GDP [19]. At the same time, in general, despite some regulatory actions with regard to carbon pricing around the world, the outcome is far from optimal, and the overall picture is still quite chaotic.

4.7 Is Reaching Zero-CO_2 Emissions a Realistic Goal?

According to the IEA's extended scenario described in the ETP-2012 report, net energy-related CO_2 emissions may need to reach zero by the year 2075 in order to stay on the 2DS trajectory (i.e., for the temperature rise to stay below 2 °C) [10]. Is this a realistic goal? Given the distant timeframe and inherent uncertainties, the simplest approach would be to extrapolate the trends in the 2040–2050 period into the future and adjust them using a number of realistic assumptions. But this approach may entail other questions:

- Would the continuation of the 2DS trends beyond the year 2050 be sufficient to reach zero-carbon emission target by 2075?
- What technologies would be the most crucial to achieve that target in the 2050–2075 timeframe?
- Would the improvement of existing technologies be sufficient or breakthroughs will be needed to achieve the zero-emission goal?

We stipulate here that the term "zero-CO_2 emissions" should not be taken literally in the sense that the last ton of CO_2 will be eliminated in the year 2075; many authors (e.g., [19]) use this term and lay down that a margin of a few percent would need to be preserved in order to be realistic.

Let's consider the range of critical technological developments that would be necessary to achieve the zero-carbon emission target by 2075. According to IEA, in the energy sector, by 2075, 99 % of electricity has to be produced from low- or zero-carbon technologies [10]. The share of renewables in the electricity generation mix has to increase from about 60 % in 2050 to more than 70 % in 2075. The remainder will be based on nuclear (19 %) and fossil-powered plants with CCS. The projected increases in the share of zero-carbon electricity sources (in % of total electricity generation) in order to stay on the 2DS trajectory during 2050–2075 transition timeframe are shown in Table 4.1.

Although the share of biomass energy post-2050 will remain mostly unchanged, it is projected to play very important role in achieving zero emissions from power

Table 4.1 Projected increases in the share of zero-carbon electricity sources (in % of total electricity generation) between the years 2050 and 2075, *Source* [10]

	Projected share of zero-carbon electricity sources (in % of total)	
Zero-carbon electricity sources	Year 2050	Year 2075
Concentrating solar power	8	13
Solar PV	7	9
Wind	15	20
Geothermal	3	5
Biomass[a]	10	10
Nuclear	18	19

[a]Biomass is a carbon-neutral (or net zero CO_2) energy source

generation and industrial sectors, because through coupling with CCS bioenergy produces negative carbon emissions (the subject of negative carbon emissions will be discussed in details in Chap. 10 of this book). In particular, 4.2 $GtCO_2$ will need to be captured through bioenergy-CCS technologies in biofuels and power production and industrial sectors [10]. To allow the continued use of coal-fired plants with CCS in the zero-emission trajectory, they will have to be combined with biomass co-firing; for example, through blending 10 % biomass with coal, the carbon intensity will drop from 120 g CO_2/kWh (without biomass) to 30 g CO_2/kWh [10]. It has been pointed out, however, that the usage of biomass in such enormous quantities might raise resource sustainability concerns. To stay on the required trajectory, 80 % of all gas-fired power plants will have to be equipped with CCS. Overall, the amount of CO_2 captured in the power generation sector will need to be increased from about 3.5 Gt/year in 2050 to about 5 Gt/year in 2075. This would result in overall emissions from the power generation of 0.2 Gt CO_2/year in 2075, which corresponds to the carbon intensity of electricity generation of less than 1 g CO_2/kWh [10] (for comparison, this value was 500 g CO_2/kWh in 2009).

Achieving zero emissions in industry could be challenging, due to steady growth in the production of energy-intensive materials; thus, additional breakthrough technologies (some of them are currently in R&D stage) may be required to achieve deep reductions in CO_2 emissions. In the transportation sector, the fairly even mix of electricity, hydrogen, and advanced biofuels will cover different niches and applications. By 2075, all new passenger cars will have to be either electric or fuel cell driven; trucks and heavy-duty transport will be dominated by biofuels and FC-powered systems; ships and aircrafts will be heavily dependent on biofuels (possibly, with the contribution of hydrogen fuel in later stages). Overall, to stay on the zero-emission trajectory, CO_2 emissions from transport will have to drop in half, from about 6 $GtCO_2$ in 2050 to about 3 $GtCO_2$ in 2075 [10]. Most of the carbon emissions will be produced by ships, heavy trucks, and aircrafts, which will use predominantly biofuels with appreciable share of petroleum fuels.

Economic and resource sustainability factors may introduce some uncertainties to the projected portfolio of zero-emission technologies. In some sectors, the cost of eliminating fossil fuels and, consequently, CO_2 emissions may be very high. For example, aircrafts can be powered, in principle, by three renewable carbon-neutral fuels: advanced biofuels (e.g., synthetic jet-fuel), noncarbon hydrogen (produced from water and renewables), and zero-carbon electricity. In order to reach zero-carbon emissions in transport, more advanced biofuels will be needed to power aircrafts and also ships and trucks, which may raise land constraints and other sustainability issues (especially, if there is large demand for biofuels in industrial and other sectors).

The use of hydrogen as aircraft fuel also faces some challenges and uncertainties. Although hydrogen-fueled aircrafts have been shown to be feasible (in 1988, the Soviet Union successfully tested Tu-155 commercial aircraft partially converted to liquid hydrogen fuel), the technology is still in its infancy [19]. Liquid hydrogen occupies four times the volume of petroleum jet-fuel, resulting in the much larger fuel tanks and, consequently, the greater air resistance and increased fuel consumption (by 9–14 %) [20]. There are also environmental concerns with regard to burning hydrogen at the altitudes of 10–12 km, where produced water vapor will create a significant greenhouse impact (about half of that of kerosene). The use of electricity to power aircrafts will require the development of breakthrough technologies. Thus, considerable trade-offs between economic, technological, environmental, and sustainability factors may become a serious issue on the pathway to zero-carbon emission target.

There is another important aspect to achieving zero-carbon emissions that should not be overlooked. Even if the zero-emission target is achieved by the 2050–2075 timeframe, there will be a likelihood of the accumulated impact of past GHG. (Note that 45 % of CO_2 emitted now will be in the atmosphere 100 years from now [21], see also Chap. 2.) Thus, the arguments could be heard that zero emissions may not be enough, and there may be a need to go beyond zero-carbon emissions and make the provisions to include carbon-negative technologies (that remove CO_2 from the atmosphere) in the portfolio of technological options. What will be these solutions, and will they be affordable by the time they are needed? It is difficult to foresee now.

What about the timeframe of transition to zero-emission economy: is it realistic? Would 4–5 decades be sufficient to make such drastic transformations in the energy systems considering a powerful inertia factor? Historical examples show that large transitions in the energy economy in a relatively short period of time are nothing new. For example, the US energy sector was thoroughly transformed during 4–5 decades at the turn of the twentieth century: from horses and coal-powered trains to petroleum-fueled cars and electricity [19]. France's electricity sector transition provides another example of nearly complete transformation of a large sector within only 2–3 decades. In 1973, 65 % of France's electricity was from fossil fuels and 8 % from nuclear [19]. As a result of the government's "energy independence" decision, within 23 years, the share of fossil fuels in the power generation dropped to about 9 % and that of nuclear increased to 77 % (with an associated drastic drop in CO_2 emissions). Although specific to each country, and not without certain challenges, these examples show that with the right policies and incentives in place, the transformations of carbon-intensive economy to low-to-near-zero-carbon economy, in principle, could be achieved in about five decades timeframe.

Summarizing, in general, achieving zero emissions by 2075 appears to be feasible, but will be extremely challenging even if the IEA's 2050 targets will be successfully met. Applying one of nature's fundamental laws, it gets more and more difficult the closer you approach perfection (i.e., zero emissions). The success will depend on a myriad of factors, most of which, given the timeframe of 50–60 years into the future, are highly uncertain. It raises concerns about whether simply extending the technology trends achieved by 2050 could lead to zero-carbon emission energy system by 2075. The use of zero-carbon fuels, such as noncarbon electricity, hydrogen, and biofuels, is expected to expand with the greater use of these fuels in increasingly broader range of applications. On the other hand, there are particular concerns around bioenergy and biofuels due to possible issues with the resource sustainability and their carbon-mitigation potential (more discussion on this topic is in Chap. 6). It is likely that most of the improvements and energy efficiency gains with regard to conventional technologies will already be achieved by 2050; so, in order to close the "carbon gap," additional technological innovations and breakthroughs will be needed in all sectors: power generation, industry, and transportation. Further research is needed to identify opportunities, challenges, and barriers on the pathway to the zero-emission target.

References

1. National Academy of Sciences (2011) Climate stabilization targets: Emissions, concentrations and impacts over decades to millennia. The National Academic Press, Washington, DC
2. Biello D (2013) A dirty business. ScI Am
3. Hansen J, Sato M, Kharecha P et al (2008) Target atmospheric CO_2: Where should humanity aim? Open Atmos Sci J 2:217–231
4. U.N. Intergovernmental Panel on Climate Change (2013) Working group I contribution to the IPCC fifth assessment report climate change 2013: The physical science basis. Summary for policymakers http://www.climatechange2013.org./images/uploads/WGIAR5-SPM_Approved27Sep2013.pdf. Accessed 27 Sep 2013
5. International Energy Agency (2009) World energy outlook 2009 edition. ISBN: 978-92-64-06130-9. Paris
6. Meinshausen M (2006) What does a 2 °C target mean for greenhouse gas concentrations? A brief analysis based on multi-gas emission pathways and several climate sensitivity uncertainty estimates. In: Schellnhuber H, Cramer W, Nakicenovic N et al (eds) Avoiding dangerous climate change. Cambridge University Press, Cambridge
7. Intergovernmental Panel on Climate Change (2007) Climate change 2007. Mitigation. Contribution of working group III to the fourth assessment report of the IPCC. Bosch P, Dave R, Davidson O et al. (eds) Cambridge University Press, Cambridge
8. Matthews D, Solomon S (2013) Irreversible does not mean unavoidable. Science 340:438–439
9. Amundsen B, Lie E (2013) The Research Council of Norway. Global warming less extreme than feared? http://www.forskningsradet.no/en/Newsarticle/Global_warming_less_extreme_than_feared/1253983445 35/p1177315753918?WT.ac=forside_nyhet. Accessed 20 Apr 2013
10. International Energy Agency (2012) Energy technology perspectives. Pathways to a clean energy system. IEA/OECD, Paris
11. International Energy Agency (2013) Tracking clean energy progress. IEA input to the Clean Energy Ministereal, IEA, Paris, www.iea.org/publications/TCEP_web.pdf. Accessed 10 Sep 2013
12. CO_2 capture in vehicles and home heating systems (2012) Carbon Capture J. http://www.carboncapturejournal.com/displaynews.php?NewsID=1043&PHPSESSID=18dkjsha7qaa6s815mlrslvtq4. Accessed 30 Nov 2012
13. National Research Council of USA (2013) The transitions to alternative vehicles and fuels. The National Academies Press, Washington, DC http://www.nap.edu/openbook.php?record_id=18264&page=R1. Accessed 12 Aug 2013
14. Mignone B, Socolow R, Sarmiento J et al (2008) Atmospheric stabilization and the timing of carbon mitigation. Climate Change 88:252–265
15. Davis S, Caldeira K, Matthews D (2010) Future CO_2 emissions and climate change from existing energy infrastructure. Science 329:1330–1333
16. Pacala S, Socolow R (2004) Stabilization wedges: solving the climate problem for the next 50 years with current technologies. Science 305:968–972
17. Socolow R (2006) Stabilization wedges: an elaboration of the concept. In: Schellnhuber H et al (eds) Avoiding dangerous climate change. Cambridge University Press, Cambridge, pp 347–353

18. Hoffert M (2010) Farewell to fossil fuels? Science 329:1292–1294
19. Makhijani A (2007) Carbon-free and nuclear-free. A roadmap for U.S. energy policy. IEER Press and DRD Books, Takoma Park, MD
20. Forsberg C (2007) Future hydrogen markets for large-scale hydrogen production systems. Int J Hydrogen Energ 32:431–439
21. Gurria A (2013) Carbon price vital to address climate change. Responding to climate change (RTCC). http://www.rtcc.org/2013/10/09/oecd-chief-carbon-price-vital-to-address-climate-change/. Accessed 14 Oct 2013

Chapter 5
Pathways to Decarbonization of Energy

Abstract History of industrial civilization is history of the progression of primary fuel substitution: wood → coal → oil → gas. This evolutionary trend of reducing carbon intensity of primary energy is referred to as *decarbonization*. During these historical transitions, human society moved to more convenient, efficient, and clean energy sources that enabled new technological advances in industry, transportation, and other areas. However, during the last couple of decades, this positive decarbonizing trend dramatically slowed down and practically ceased. In this chapter, the current trends in carbon intensity of global economy and prospective decarbonization options are analyzed using *Kaya Identity* (KI) modeling tool. The KI analysis indicates that the cessation of decarbonization of global economy can be largely attributed to a reversal of the evolutionary fuel substitution trend and "detour" to coal by populous rapidly developing countries. Dramatic reductions in both energy and carbon intensities of world economy would be necessary to stop and reverse this worrisome trend. Among proposed carbon mitigation policies, improvements in energy efficiency promise the largest near-term dividends and are central to achieving atmospheric CO_2 stabilization goals.

5.1 Decarbonization Concept: Historical Background

History of industrial civilization is history of the progression of primary fuel substitution: wood yielded its dominance to coal, and the latter successively to oil and natural gas. During this fuel progression process, which was mostly driven by considerations of convenience and versatility (see Chap. 1), the environmental merit of fuels was not among important considerations (if any); nevertheless, every posterior fuel was "cleaner" than its predecessor in terms of amount of CO_2 emissions per unit of energy produced. The "cleanness" of fossil fuel is determined by its H/C atomic ratio: the higher H/C ratio the cleaner fuel since it emits less CO_2 during combustion (per unit of energy). (Note: sulfurous and other impurities in fuels are

N. Muradov, *Liberating Energy from Carbon: Introduction to Decarbonization*, Lecture Notes in Energy 22, DOI 10.1007/978-1-4939-0545-4_5,

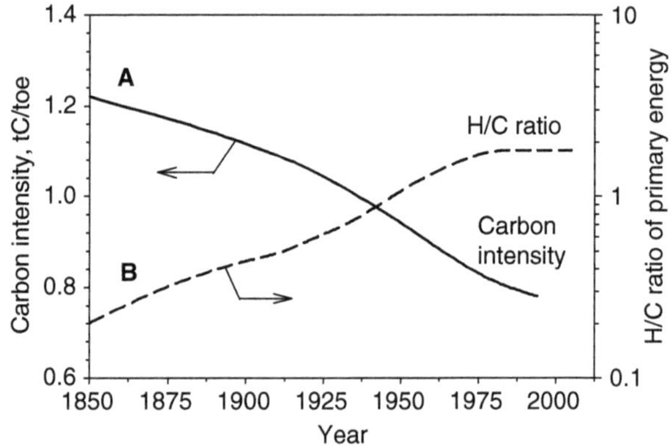

Fig. 5.1 Historical decarbonization trends. (*A*) The evolution of the carbon intensity of the world. (*B*) The evolution of H/C ratio of primary energy. *Source* [2, 3]

not considered in this analysis.) The historical trend of fossil fuel substitution is consistent with the increase in the H/C ratio of fuels as follows:

$$\text{wood}(\text{H/C} < 1) \rightarrow \text{coal}(\text{H/C} \approx 1) \rightarrow \text{oil}(\text{H/C} \approx 2) \rightarrow \text{natural gas}(\text{H/C} \approx 4).$$

This evolutionary trend of reducing carbon content of fuels or, in more accurate terms, carbon intensity of primary energy is referred to in the literature as *decarbonization*. Currently, the term "decarbonization" is often used in a broader context covering a wide range of policies aiming at removing carbon from the energy equation and diminishing the carbon footprint of economy.

In general, for any given fossil fuel (or primary energy source) to assume a dominant role in the global energy mix, it has to satisfy a number of the requisite conditions such as abundance, accessibility, the convenience of use, versatility, and cost-competitiveness. (Note that for any particular country, there could be local specifics intertwined with a variety of complex geopolitical considerations, but here only general trends are discussed.) During each of these historical transitions, human society moved to more efficient primary energy sources that enabled new technological advances in industry, transportation, and other areas (as an example, moving from coal to oil to gas allowed switching from steam engines to internal combustion engines to turbines, respectively). Advantageously, in most cases, more efficient also meant more clean and convenient.

In the 1980–1990s, Nakićenović, Grübler, Marchetti, and other researchers at the International Institute for Applied Systems Analysis (IIASA) in Austria analyzed the long-term decarbonization trends in the global energy system. In particular, they were tracking the evolution of the carbon intensity of primary energy since the onset of the Second Industrial Revolution marked by the emergence of steam-powered ships, engines, railways, etc [1, 2]. One of the main conclusions of their analyses was that since the1850s to early 1990s, decarbonization appeared to be a continuous and persistent global trend: carbon intensity of energy and GDP have been falling globally at the average rates of about 0.3 % per year and 1 % per year, respectively [2] (Fig. 5.1, curve A). The authors emphasized, however, that since the early 1970s there seemed to be a distinct deceleration trend or slowdown in the pace of decarbonization. This decelerating trend was underscored in more recent (2012) study conducted at the same Institute by Aguilera and Aguilera [3]. The authors of this analytical study concluded that after more than a century of a continuous rise, starting from the mid-1970s, H/C ratio of the global energy mix leveled up and stayed nearly constant at the H/C ratio of about 1.8 (Fig. 5.1, curve B), implying that, since that time, the positive historical decarbonization trend has practically ceased.

Many experts link the emergence of the slowing trends in decarbonization of global economy to accelerating tendencies in the growth of carbon emissions during last decades. For example, over the last two decades, global CO_2 emissions from fossil fuel combustion have been accelerating with the growth rate increasing from 1.2 % per year in 1990–2000 to 3.0 % per year in 2000–2010 to 3.2 % per year in 2010–2011, which is greater than the carbon emission growth rate projected by the IPCC in their most conservative scenarios [4–6].

These worrisome recarbonizing trends were also stressed by other energy analysts who warned of their potential consequences for climate policies [6–9]. It was emphasized that the IPCC assumptions for the near-term decarbonization are already inconsistent with the recent evolution of the carbon intensity of global economy. The current trends and ongoing developments on the energy market do not inspire optimism either: most of carbon emission scenarios (even "optimistic" ones) project that the world's economy will continue moving along the carbon-intensive path. What are the major global and regional drivers that pull the world economy along this path?

5.2 Kaya Identity and Decarbonization

Kaya Identity (*KI*), a simple yet powerful and versatile carbon quantification tool, was introduced in the early 1990s by a Japanese researcher Kaya for calculating CO_2 emissions by segregating different drivers responsible for the emissions [10]; since then, it has been often used for computing and forecasting CO_2 emissions on the regional and global scales [7, 8, 11]. At the basic level, the KI is composed of two primary factors that are linked to the changes in economy and technology. Each primary factor can be broken down into two subfactors: the economic factor driven by the changes in population and GDP, and the technology factor—by the changes in the energy intensity of GDP and the carbon intensity of energy. Figure 5.2 shows the schematic representation of the KI concept.

In a simple mathematical form, the KI expresses the global CO_2 emissions as a product of four driving forces:

$$F = P\left(\frac{G}{P}\right)\times\left(\frac{E}{G}\right)\times\left(\frac{F}{E}\right) = P\times g\times e\times f \tag{5.1}$$

where F is the global CO_2 emission flux from fossil fuel combustion and industrial processes, P is global population, E is global primary energy consumption, G is

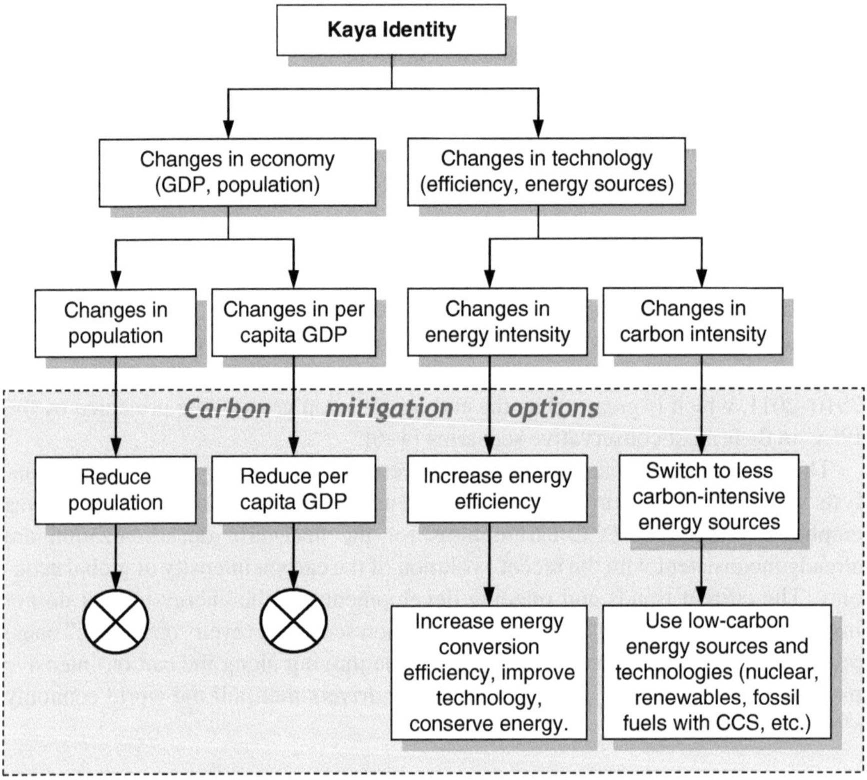

Fig. 5.2 Schematic representation of the Kaya Identity concept

world GDP, $g=(G/P)$ is global average per capita GDP, $e=(E/G)$ is the energy intensity of world GDP (i.e., energy consumed per unit GDP), and $f=(F/E)$ is the carbon intensity of energy (i.e., the amount of CO_2 emitted per unit of energy produced).

Equation (5.1) can be further simplified by combining two last factors $(E/G)\times(F/E)=(F/G)$ as follows:

$$F = P\left(\frac{G}{P}\right)\times\left(\frac{F}{G}\right) = P\times g\times h \tag{5.2}$$

where $h=e\times f=(F/G)$ is the carbon intensity of GDP.

Note that the upper- and lower-case symbols in (5.1) and (5.2) relate to extensive and intensive variables, respectively. Combined, the KI factors can describe the variety of economic and technological drivers contributing to the recent dramatic growth in global CO_2 emissions and the associated increase in atmospheric CO_2 concentrations.

Fig. 5.3 The evolution of Kaya Identity factors from 1971 to 2011. All data are normalized against the year 1971. (*A*) All KI factors are shown in one diagram, (*B*) factors *f*, *e*, and *h* are shown separately in the expanded *Y*-coordinate. The data on *g*, *f*, *e*, and *h* factors for the year 2011 are not available. *Source* [4, 5, 12]

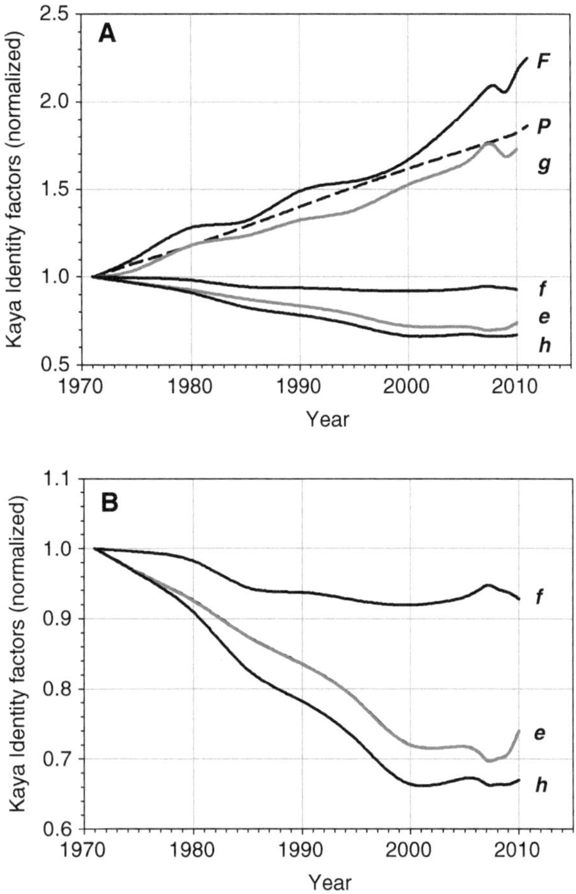

Figure 5.3A, B summarizes the evolution of global CO_2 emissions and all KI factors from 1971 to 2011 based on the IEA's 2012 statistical data [4, 5, 12] (some data for 2011 are not yet available).

Figure 5.3 indicates that from the beginning of the 1970s through the end of the 1990s, the KI factors *e*, *f*, and *h* were dropping at the average annual rates of about 0.8 %, 0.2 %, and 1.1 %, respectively, but at the end of the twentieth century they practically leveled off. The trends in the KI factors evolution show that the dramatic increase in anthropogenic CO_2 emissions, especially during the last decade, is driven not only by the steady increase in population (*P*) and per capita GDP (*g-factor*) but also by a reversal of positive century-long decarbonization trends, manifesting themselves in decreasing in the energy intensity of GDP (*e*-factor) and the carbon intensity of energy (*f*-factor).

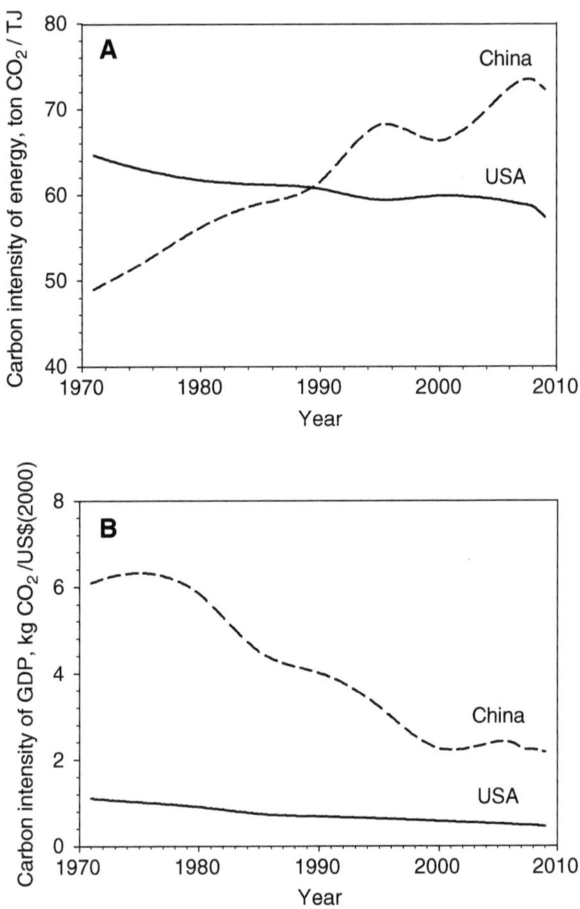

Fig. 5.4 The evolution of the carbon intensity of energy (*A*) and GDP (*B*) from 1971 to 2010 for the USA and China. TJ is terajoule or 10^{12} J. *Source* [12]

The data shown in Fig. 5.3 represent world-average values of the KI factors and, as such, do not reflect the specifics of each country. To make the above trends more elucidatory, Fig. 5.4 depicts the evolution of the carbon intensities of energy (*f*-factor) and GDP (*h*-factor) from 1971 to 2010 for the two most carbon-emitting countries in the world: the USA and China (note that in 2007, China overtook the USA as the world's greatest CO_2 producer).

Carbon intensity of energy in the USA (and in many EU countries) gradually decreased over last four decades. Carbon intensity of the US GDP decreased from 1.11 kg CO_2/US$ in 1971 to 0.46 kg CO_2/US$ in 2009, i.e., the reduction of 1.5 % per year (based on 2000 exchange rates). During the analyzed period, China's carbon intensity of energy grew by more than 40 %. The carbon intensity of China's GDP dropped over three decades from 6.09 to 2.25 kg CO_2/US$, but since 2000 this factor practically leveled off (based on 2000 exchange rates). In 2011, China was the world's largest emitter of carbon emissions, with the amount of emitted CO_2

rising by 9.3 %, mostly, due to coal consumption in the power sector [5]. Similar trends could be seen in India and other developing countries where economies are being heavily fueled by the abundant and cheap coal resources.

In the IEA's 2013 annual report to Clean Energy Ministerial (CEM), it was made clear that the drive to clean up the world's energy system had practically stalled (CEM brings together ministers from 22 countries representing more than 75 % of global GDP and energy consumption and 80 % of global CO_2 emissions) [13, 14]. The report underscored that despite much talk by world leaders and some progress in renewable energy, "*the average unit of energy produced today is basically as dirty as it was 20 years ago.*" To quantify this worrisome trend, the IEA introduced the Energy Sector Carbon Intensity Index (ESCII), which is defined as the average amount of CO_2 emitted per unit of energy produced (in toe). According to IEA's estimates, over a period of two decades, the ESCII had barely moved from 2.39 t CO_2/toe in 1990 to 2.37 t CO_2/toe in 2010. The IEA report emphasizes that while there is some progress in clean energy (e.g., in solar, wind, and advanced vehicle technologies), for the majority of technologies, the progress remains alarmingly slow. Although the revolution in shale gas technology has triggered significant switch from coal to much cleaner fuel—gas in the USA, this is still considered a regional phenomenon, because coal's use expanded elsewhere, especially, in China, India, and even in Europe, where the share of coal increased in the fuel mix at the expense of gas [14].

The results of the KI analysis and other developments in the energy sector indicate that the practical cessation of the decarbonization of global economy can be largely attributed to a reversal of the historical positive fuel substitution trend (wood → coal → oil → gas) and "detour" to coal by populous rapidly developing countries taking advantage of the most reliable and the lowest cost fuel. Unfortunately, this trend is projected to continue in the future if no policies are implemented to avert this negative trend.

Besides being an important tool for analyzing past, present, and future CO_2 emission scenarios and climate models, the Kaya Identity is very useful in identifying optimal decarbonization strategies and evaluating the carbon mitigation policies and technological options aiming at reducing CO_2 emissions to a specific CO_2 stabilization target. Equations (5.1) and (5.2) imply that the reduction in anthropogenic CO_2 emissions could be achieved, in principle, by reducing one or several Kaya Identity factors, namely, *P*-factor (population), *g*-factor (per capita GDP), *e*-factor (energy intensity of economy), and *f*-factor (carbon intensity of energy). Obviously, the first two options are non-starters because they would violate one of the undisputable ground rules of climate policies that limiting economic growth as a means of reducing CO_2 emissions is simply not an option [8]. Thus, realistically, there are only two decarbonization options left, and they are linked to the KI factors (*e*) and (*f*), i.e., reducing energy and carbon intensities of economy.

IEA in its ETP-2012 report identifies the portfolio of the least-cost low-carbon technologies that will have to be implemented to reduce in half CO_2 emissions by 2050 (compared to 2009) in order to comply with the projected atmospheric CO_2 stabilization target [15]. This portfolio includes a broad range of low-carbon technologies and solutions:

- Increase in end-use energy efficiency
- Increase in power generation efficiency
- Deployment of carbon capture and storage
- Nuclear
- Renewables
- Switching to zero-carbon and low-carbon fuels

From analysis of Fig. 5.2 it is evident that the first two technological options are linked to the *e*-factor, while the remaining options—to the *f*-factor in the Kaya Identity.

5.3 Technological Pathways to Reducing Energy Intensity

Reducing energy intensity of economy could be achieved through several technological pathways, such as:

- Decrease in the amount of primary energy consumed per unit of product
- Decrease in requirements for energy-intensive materials ("dematerialization")
- Increase in energy conversion efficiency
- Conservation of energy
- Energy demand management
- Recycling
- Integration (or decentralization) of energy systems

Increasing energy efficiency of an industrial process implies that producing a unit of a product (e.g., kWh of electricity, 1 t of cement, or plastic, or a fertilizer) would require less amount of energy. Energy that is not consumed does not have to be produced, transported, and converted, and, hence, it would not produce any GHG emissions. Thus, increasing energy efficiency by reducing energy consumption might ultimately reduce vulnerability to all the potential problems along the value chain, and doing this greatly contributes to achieving climate mitigation goals. Ultimately, the efficiency with which energy (both primary and secondary) is utilized does profoundly affect economics.

5.3.1 *Energy Efficiency: A Critical Target*

Dramatic improvements in energy efficiency is central to achieving the IEA's 2DS objectives, which call for a "transformative shift" in energy savings and CO_2 reductions [15]. According to the 2DS, the combined effect of the increase in fuel and electricity end-use efficiencies would contribute 45 % of the overall reduction target which would be necessary to reduce global CO_2 emissions in half by 2050 (compared to 2005 level) [15]. In particular, the IEA's analysis shows that it would be possible to achieve energy savings equivalent to nearly one-fifth of today's global energy use by 2035 through a combination of policies, regulations, technological developments, and energy investments [16]. The IEA stressed that recent trends toward US oil independence, in part profited from regulations requiring increased vehicle fuel efficiency; this tendency is likely to continue into the future, as increasingly tougher mileage standards go into the effect.

The IEA's WEO-2012 report puts a special emphasis on the importance of efficiency gains in the electricity generation sector [17]. Since nearly 1.3 billion people worldwide currently lack an access to electricity, the report projects the demand for electricity to grow by 70 % between 2010 and 2035, with more than 80 % of that growth coming from non-OECD countries, primarily from China (38 %) and India (13 %) [17]. One of the major energy-saving tools in the power generation sector is cogeneration, i.e., simultaneous production of electricity and high-grade heat or steam for industrial processes. Cogeneration has the potential to deliver generation efficiencies up to 90 % compared to only 45 % achieved by today's best coal-fired power plants [15]. Renewable electricity sources (solar, wind) would require additional efforts to improve their efficiency and deliver their energy to the consumer.

Building sector is another important target for increasing end-use energy efficiency to stabilize atmospheric CO_2 concentration. However, two diverse challenges have to be taken into consideration in the building sector. Many developing (non-OECD) countries are pursuing a rapid expansion of their building stock, and this process is expected to intensify through the mid-century. If these countries would take advantage of the advancements in building technologies and implement innovations, they could play a leading role in constructing highly energy efficient residential and commercial buildings. But if they miss this opportunity, they would risk "locking-in" the inefficient buildings that will stand for decades. Such a lock-in situation is evident in some OECD countries, where building stock is growing rather slowly, and, thus, the potential to improve overall efficiency by constructing more efficient buildings is rather limited [15]. In these countries, a major route to increasing efficiency in the building sector is through renovation of existing buildings. The greatest potential to decreasing carbon footprint of buildings is through the combination of measures aiming at lowering energy demand per square meter of floor space and using decarbonized electricity and very efficient heating/cooling systems based on heat pumps. In large cities and densely populated areas, district heat-and-power cogeneration systems could further reduce energy consumption in buildings.

ExxonMobil's energy analysts estimated that the improvements to energy efficiency would reduce a global energy demand growth by about 65 % through 2035 [18]. Without these improvements, the energy demand would almost double due to population growth and economic expansion, but because of the efficiency gains the global energy demand would increase by "only" 35 %. In terms of environmental benefits, improved efficiency would offset more than 75 % of carbon emissions that could have been produced without those efficiency gains. It is projected that in heavy industry and chemical sectors, the efficiency improvements will offset 60 % of the energy demand growth through 2030, whereas in the energy industry sector it will remain flat, even as the demand continues to grow.

In the transportation sector, the increase in energy efficiency is linked to the improvements in the fuel efficiency of vehicles, which includes the following tools:

- Improved vehicle design targeting reduction in an aerodynamic drag
- Advanced engines and transmissions
- Reduced vehicle weight through the use of composite light-weight materials
- Reduced rolling resistance through the use of improved tires
- Advanced powertrains including diesel or gasoline hybrids

Such advancements in vehicular technology could potentially save fuel and reduce carbon emissions without compromising safety and comfort [18]. Increase in energy efficiency is often referred to as a "win-win-win" solution because it reduces the energy demand growth, reduces CO_2 emissions, and extends the life of world's energy resources.

Lovins estimated that stabilizing CO_2 emissions would require the reduction of energy intensity of GDP at the rate of 2–3 % per year to meet the IPCC's climate stabilization target [19]. The USA has been steadily decreasing the energy intensity of its economy by an average of 1.5 % per year since the early 1970s [4], and during short periods (1981–1986, 2001, 2006) it reached and even exceeded 3 % decrease per year [19]. After oil crisis of 1973, for over a decade, the US economic growth occurred practically without an energy growth (in average). The UK, Germany, and other European countries have achieved comparable reductions in their energy intensity. The European Commission targets 20 % increase in energy efficiency by 2020 (about 2 % per year) [20].

China is striving to reduce the carbon-footprint of its coal-intensive economy by closing inefficient coal power plants and capping energy use [21]. In 2004, China set a new energy strategy (called "leap-frog technologies") aiming at rapid improvements in the energy efficiency of new power plants, buildings, and factories, and banning fuel-inefficient cars [22]. Although China overtook the USA as the major CO_2 emitter, its per capita emissions are much lower (in 2011, 6.8 tCO_2 per capita in China vs. 16.9 tCO_2 per capita in the USA). But, at the same time, China's per capita emissions tripled since 1990, and by this index the country has already overtook France, Spain, and other European countries, and it may reach the USA levels by 2017 [23]. Despite this track record, one Chinese government official in charge of climate policy pointed that China would not "*follow the path of the US*" and allow per capita emissions to rise high, which would be a "*disaster for the world*" [23]. As a basis for this statement, the officials indicated that they are making efforts to control GHG emissions and their carbon intensity is decreasing. China's current 5-year plan projects economic growth of about 40 % from 2010 to 2015, but also a 17 % drop in carbon intensity of GDP. A longer-term goal is to boost national energy efficiency by 40–45 % by 2020 from the 2005 levels.

The latest developments in the power generation and industrial sectors show that reducing globally energy intensity at the rate of 2–3 % per year is not an unrealistic goal, especially, considering that a significant part of power plants in developing countries are quite inefficient (30 % or lower), and very efficient combined heat and power (CHP) units started their conquest of the power generation market not long

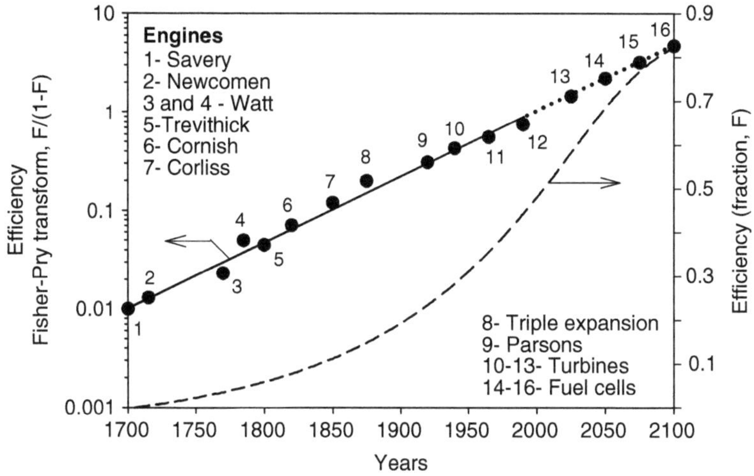

Fig. 5.5 Ausubel–Marchetti diagram for predicting energy conversion efficiencies and time of the deployment of prospective power generators. The data on the efficiency of existing power engines (points 1 through 12) forming a *solid line* are taken from the source [26]; the projected efficiencies (13 through 16) lie on a *dotted line* which is the extension of the *solid line*

ago (the overall efficiency of CHP reaches up to 90–91 % [24]). According to reported estimates, the increase in energy efficiency of coal-fired power plants from the current 30 % to the state-of-the-art 50 % efficiency can potentially reduce CO_2 emissions almost by third (32 %) per unit of energy produced [25].

5.3.2 *Trends in Fuel-to-Electricity Energy Conversion Efficiencies*

It is widely recognized that one of the key solutions to securing a stable power supply while reducing carbon emissions and conserving finite fuel resources is to boost the energy efficiency of power generation systems. Therefore, it is important to prognosticate the ensuing technological progress in this area in order to assess its technical potential to deliver further reductions in fuel consumption and CO_2 emissions per unit of electricity produced. In the mid-1990s, Ausubel and Marchetti developed a diagram that tracked the improvements in the efficiency of power engines over three centuries (from 1700 to 1997) [26]. They showed that the energy efficiencies of different power engines could be arranged in the form of a linear Fisher-Pry transform function (often used for technological forecasting). Presuming that the Ausubel–Marchetti's plot possesses a predictive power, Fig. 5.5 attempts to prognosticate the energy efficiency of prospect power generators and the proximate time of their market deployment by extrapolating the linear function into the future (until the end of this century).

It is evident from the diagram that the extrapolation of the Fisher-Pry transform line to the years 2025, 2050, 2075, and 2100 yields the energy conversion efficiencies of 60 %, 70 %, 76 %, and 82 %, respectively. These values (especially, last three) may seem quite high for conventional heat engines, but they are well within the range of electrochemical energy converters: fuel cells that are an object of intensive research, development, and commercialization efforts worldwide.

5.3.2.1 Fuel Cell-Based Power Generation

Unlike traditional power generation systems (e.g., turbines, diesel generators, ICE) that are controlled by the Carnot cycle's heat-to-work limitations, FC directly convert chemical energy of fuel into electricity via electrochemical reactions. Actual (practical) efficiency of FC can be defined as follows:

$$\eta = \frac{\Delta G}{\Delta H^{o}} \mu \frac{V}{V_{0}} = \frac{\Delta G}{\Delta G^{o} + T\Delta S^{o}} \mu \frac{V}{V_{0}} \tag{5.3}$$

where η is FC actual efficiency, ΔG, ΔH, and ΔS are free energy, enthalpy, and entropy of a fuel oxidation reaction (superscript "o" denotes standard conditions), μ is a fuel utilization coefficient (assumed 0.8–0.9), and V and V_0 are operating and open circuit voltages of FC, respectively.

Based on (5.3), practical efficiencies of 60 %, 65–70 %, and 80–85 % for hydrogen, direct methane, and direct carbon FC, respectively, could potentially be obtained [27], which are almost twice the efficiencies of the conventional power generators. An additional advantage of FC power generators is that fuel and air are not mixed (as in a turbine, or ICE), but are directed into separate compartments of FC: fuel is introduced to an anode and air to a cathode compartment. As a result, oxidation products are not diluted by nitrogen and anode outlet stream is rich with CO_2, which significantly simplifies its capture and storage.

The Ausubel–Marchetti's diagram predicts the widespread deployment of FC-based power generators with low carbon footprints in the marketplace in about two decades. The types of FC currently under development for portable, distributed, and centralized power generation are summarized in Table 5.1.

Two basic technological options for the FC-based power generation systems that utilize fossil fuels can be envisioned. In the first option, fuel is first reformed or gasified to reformate gas, which is introduced to the anode compartment of FC (in case of PEMFC, reformate has to be further processed to pure H_2 before introducing it to FC). In the second option, carbonaceous fuel (e.g., NG, methanol) is directly introduced to FC (without preprocessing) resulting in the H_2O–CO_2 stream, from which CO_2 is easily recovered (this approach is used in DMFC, SOFC, MCFC).

Although most of FC-related activities today are focused on hydrogen- and hydrocarbon-fueled FC, direct carbon FC (DCFC) is increasingly attracting attention of researchers due to its potential to achieve the highest practical fuel-to-electricity (FTE) efficiency of 80–85 % among all FC. It also features the highest

Table 5.1 Comparison of operational characteristics of different fuels cells and the status of their technological development

Type of FC	Temperature range (°C)	Efficiency (%)	Power range (MW)	Development status
Polymer electrolyte membrane (PEMFC)	60–100	35–50	0.01–1	Commercial
Alkaline (AFC)	70–200	60	0.01–0.1	Demonstration
Phosphoric acid (PAFC)	150–200	35–45	0.2–10	Commercial
Direct methanol (DMFC)	50–120	30	10^{-4} to 0.01	Demonstration
Molten carbonate (MCFC)	600–900	45–55	1–100	Demonstration
Solid oxide (SOFC)	650–1,000	50–60	1–100	Demonstration
Direct carbon (DCFC)	650–950	75–85		R&D

fuel utilization efficiency of 80–90 % (since fuel-carbon and the product-CO_2 exist in separate phases), allowing the full conversion of carbon in a single pass. Different types of DCFC utilizing a variety of electrolytes have been under development [28–30], and conversion efficiencies of up to 80 % have already been demonstrated on a laboratory scale DCFC using different types of carbon [31]. Despite potential advantages of DCFC over other types of FC, the practical implementation of DCFC is hindered by several factors mostly related to the system sustainability, and a need for the supply of clean (i.e., sulfur- and ash-free) carbon fuel for DCFC.

From the viewpoint of large-scale power generation systems, SOFC has received much attention lately due to a number of advantages, such as high efficiency, long-term stability, fuel flexibility, capacity to integrate with other power sources (e.g., turbines) and with CCS, thus, drastically reducing carbon emissions per unit of electricity produced. They have been used in a number of stationary applications, such as auxiliary power units in vehicles and stationary power generation units with the capacities ranging from 100 W to 2 MW [32]. Researchers at the US DOE Pacific Northwest National Laboratory (PNNL) reported that they have developed a 2 kW SOFC system capable of achieving efficiencies of 57 % [33]. This development could potentially lead to highly efficient, distributed power generation systems. Benefits, barriers, and perspectives of using FC for distributed power generation are reviewed and analyzed in [34].

5.3.2.2 Integrated FC-CCS Systems

Fuel cells hold great promise of drastically reducing CO_2 emissions through the integration with CCS systems. Although some types of FC have already been commercialized (or are close to commercialization) on niche markets, in general, the implementation of FC-based systems coupled with CCS is still in validation and demonstration stage. The largest existing FC-based demonstration units are at 1 MW_{el} scale, and it will take, probably, another 5–10 years before large-scale

(10–100 MW_{el}) units will be commercially deployed. Due to high CO_2 concentration in off-gas, integrating FC-based power generators with CCS systems could be relatively easy and, potentially, less costly compared to conventional fossil fuel-based power plants, and this may offer the advantages of reduced energy efficiency penalties and improved process economics. It has been reported that if CO_2 capture is incorporated to a high-temperature SOFC-based power generation system, the FTE efficiencies will drop to 60 % from 67 % without CO_2 capture [35] (i.e., penalty of only 7 %, compared to penalties of 30–50 % and higher for conventional power plants). Since the cost of CO_2 transport is a function of the amount of CO_2 transmitted, grouping of several small-scale FC units to achieve the total capacity of 100 MW_{el} and higher could substantially reduce the cost of CO_2 transport [36].

Among the most advanced integrated FC-CCS projects, three British companies, B9 Coal, PowerFuel Power Ltd (PPL), and AFC Energy, unveiled plans to build a fuel cell power station near Doncaster, UK [37]. PPL, which operates a colliery near Doncaster, is planning to build an 800 MW demonstration IGCC power station with carbon capture. Initially, an 800 MW combined cycle gas turbine, which can operate on coal-derived syngas will be built. In the next phase, this syngas, will be cleaned up of the impurities and introduced to alkaline fuel cell (AFC) to generate electricity with 60 % generation efficiency. If successful, the B9 Coal, PPL and AFC joint venture will develop high-efficient IGCC/CCS/AFC power stations in the UK and other countries.

5.3.2.3 Integrated Hybrid FC-GT Power Generator

Recently, R&D efforts in the area of advanced power generators focused on promising hybrid systems based on integrated FC and gas turbine (IFCGT) cycles. In a typical hybrid IFCGT system, thermal energy of FC exhaust gas is converted into additional electrical energy through a heat engine. A great variety of these systems powered by NG, coal, biomass, and other fuels have been proposed. Theoretical analysis and experimental results of testing of IFCGT cycles indicate that these integrated systems show synergies not present in conventional combined cycles; in particular, their FTE efficiencies are higher than those of either FC or a gas turbine (GT) alone, and the costs for a given efficiency and power output may become lower than for either one acting separately [38].

High-temperature FC (HTFC) such as MCFC and SOFC are especially well suited for the hybrid FC-GT systems because of high operational temperatures in a close range, which would make their integration relatively easy. Although, in principle, different heat engines can be coupled with HTFC, including reciprocal engines, steam turbines, and gas turbines, in practice, only microturbine generators (MTG) have been tested in IFCGT systems so far, because they are especially well matched to the requirements of HTFC. In particular, the MTG can operate at relatively low turbine inlet temperature and pressure ratios that are amenable to the direct combination with HTFC [39]. In this system, a HTFC (e.g., MCFC or SOFC) replaces the fuel combustor of a conventional Brayton cycle. As a result, the Carnot-limited chemical-to-thermal energy conversion in the combustor is substituted with the direct chemical-to-electrical energy conversion in the FC, with its waste heat being used in a GT to produce additional electricity. About 80 % of the total electricity output of the IFCGT system is produced by FC with the remaining 20 % generated in the GT. The integration of the two types of power generators significantly increases the overall energy conversion efficiency of the system and lowers carbon emissions per kWh energy produced.

IFCGT is not a new technology; the first patents describing hybrid FC-GT technologies were issued in the 1970s (e.g., [41]). In the late 1990s, the US DOE in collaboration with several companies (FuelCell Energy, Siemens-Westinghouse, M-C power, and McDermott) initiated several feasibility studies on the evaluation of various IFCGT schemes, including MCFC, SOFC, and state-of-the-art turbines, mostly, the 20 MW class power systems. The results of these studies demonstrated that nominal energy efficiencies in the range of 60–71 % (depending on the type of HTFC and particular configuration) could be delivered by the studied IFCGT systems.

5.3.2.4 Triple Combined Cycle Power Generators

Recent technological developments demonstrated that the addition of a steam turbine (ST) to the hybrid FC-GT system could further increase the overall efficiency of the power generator. Mitsubishi Heavy Industries, Ltd. (MHI) (Japan) is one of the leading companies involved in developing and commercializing a triple combined cycle (TCC) that integrates SOFC with GT and ST [40, 42]. In a project commissioned by the New Energy and Industrial Technology Development Organization (NEDO), MHI has initiated a joint research venture with Tohoku Electric Power Co., Inc. aiming at the full-scale development of TCC. Figure 5.6 depicts the schematic diagram of an integrated SOFC-GT-ST power generation system.

In the shown TCC diagram, SOFC is located upstream of a combined cycle gas turbine. Because the SOFC is placed in a high-pressure area upstream of the GT combustor, it must have a robust structure, and hence tubular ceramic SOFC are advantageous for this application. By generating power at three stages—FC, gas turbine, and steam turbine—the resulting fuel cell-combined cycle (FCCC) system could potentially achieve outstanding FTE efficiency. The FCCC system is expected to achieve the world's highest power generation efficiency exceeding 70 % for several hundred MW class power generation and over 60 % efficiency for several tens MW class power generation [42]. MHI considers FCCC-TCC combined cycle power generation a revolutionary technology that will result in 10–20 % improvements in power generation efficiency over existing gas-fired power generation plants.

Reported estimates indicate that upon successful implementation of the integrated technology, NG-to-electricity energy conversion efficiencies could increase from the current 57 % (for the base case NG combined cycle) to 75 % for SOFC-GT-ST system [42, 43]. Figure 5.7a, b provides the performance summary of existing and emerging advanced power generators, in terms of their energy efficiency, preferred power output range, and projected time of deployment.

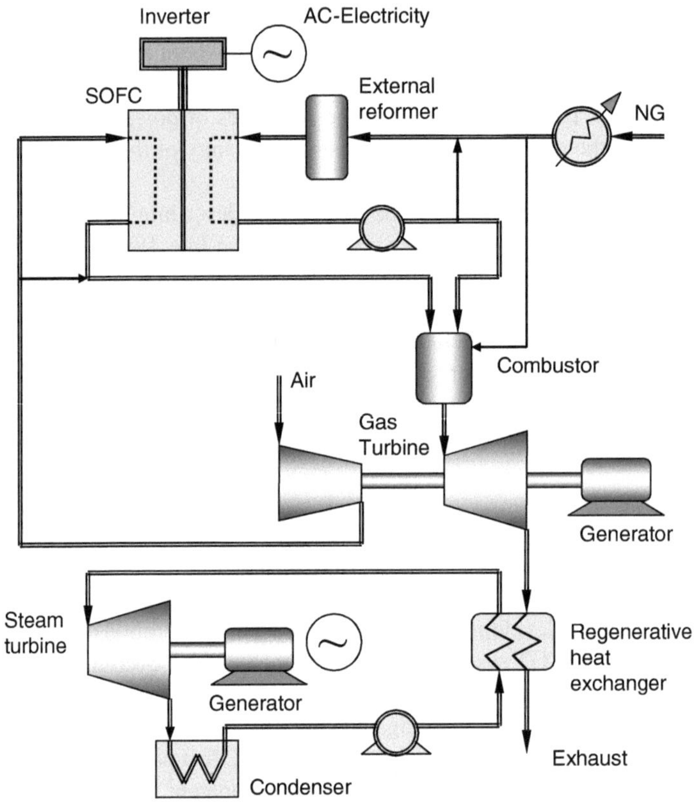

Fig. 5.6 Schematic diagram of a triple combined cycle including SOFC, gas, and steam turbines. *Source* [40]

Figure 5.7b shows the evolution of the efficiencies of power generating systems since 1900 in Fisher-Pry transform coordinates, including latest developments in conventional combined cycles and advanced integrated (hybrid) FC-based power generators, and projections to the end of the century. The current trends in the development of power generation systems seem to be in a good agreement with the projections of the Ausubel–Marchetti's diagram, especially for NG-powered systems.

5.3.3 *Energy Conservation: "A Low-hanging Fruit"*

Energy conservation, like energy efficiency, is one of the main energy reduction strategies and is an important part of current energy policy. The main difference between energy conservation and energy efficiency is that the former refers to reducing energy through using *less of an energy service*, whereas the latter refers to

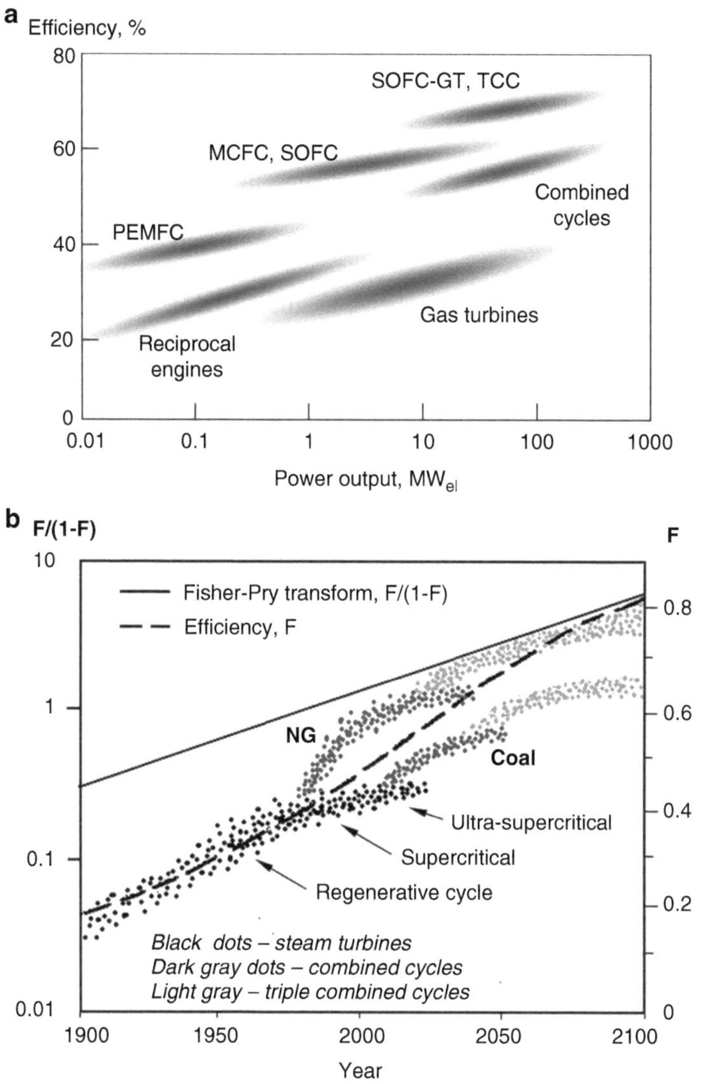

Fig. 5.7 The performance summary of existing and emerging advanced power generating systems. (**a**) Comparison of different types of conventional and FC-based integrated power generators. *ICE* internal combustion engine, *PEMFC* polymer electrolyte membrane fuel cell, *GT* gas turbine, *MCFC* molten carbonate fuel cell, *SOFC* solid oxide fuel cell, *TCC* triple combined cycle. (**b**) Evolution of the efficiencies of power generating systems starting from 1900 in Fisher-Pry transform coordinates and projections to 2100. *Source* [40]

using *less energy for a constant service*. Driving less and using incandescent bulbs for home lighting for fewer hours are examples of energy conservation. Driving the same distance using a vehicle with higher fuel efficiency or replacing incandescent bulbs with the same number of compact fluorescent bulbs are examples of the energy efficiency approach.

5.3.3.1 Conservation

Energy conservation is considered "a low-hanging fruit" in the arsenal of near-term options for drastically reducing energy intensity of the economy due to its potential to rapidly ameliorate adverse impacts of the energy sector on environment at relatively low cost. Available statistical data indicate that energy consumption per capita (E_{cap}) even in developed countries with similar Human Development Index (HDI) could differ by a factor three-four. (The HDI is a composite statistic combining three major indices: life expectancy, education, and income) For example, within the OECD-Europe countries with almost identical HDI, E_{cap} varies by as much as factor four; the USA's E_{cap} is twice as high as the average value for OECD-Europe countries [44]. It would be fair to assume that the USA's E_{cap} could be reduced to the level of OECD-Europe (i.e., by half) with little sacrifices via a combination of energy efficiency improvements and changes in the transportation infrastructure, and other policies [45]. This could result in enormous savings because the USA consumes almost a quarter of the total energy consumed by the world.

Currently, annual per capita energy/power consumption in the USA is equivalent to about 9 kW/person, whereas the world average is close to 2 kW/person [45]. Researchers at the Swiss Federal Institute of Technology (Lausanne, Switzerland) have developed a vision of "*a 2-kW-per-capita society*" to be established by the middle of the current century [44]. Many experts consider this vision to be technically feasible; however, its realization would require a combination of (1) increased energy efficiency in several sectors (power generation, industry, buildings), (2) policies that encourage energy conservation and the use of high-efficiency systems, and (3) structural changes in transportation systems [45].

According to a report by the United Nations Development Programme (UNDP), a reduction of 25–35 % in primary energy consumption in the industrialized countries without sacrificing the level of energy services could be achieved at acceptable cost in the next two decades [44]. Similar reductions of up to 40 % and 45 % are economically achievable in the transitional and developing economies, respectively.

5.3.3.2 Recycling

Recycling of energy-intensive materials is one of the key energy conservation measures for the realization of resource-efficient economy. Figure 5.8a, b shows the recycling rates and reuse percentages of several important categories of materials and industrially significant metals, respectively.

According to UN Environment Programme's International Resource Panel 2011 report, less than one-third of 60 industrially most important metals are recycled worldwide at the rates of 50 % and higher, and more than half of the surveyed metals are recycled at rates of less than 1 % [48]. The report underscored that global post-consumer recycling rates for many metals show room for improvement. These metals could be brought back into the economy by improving recycling rates.

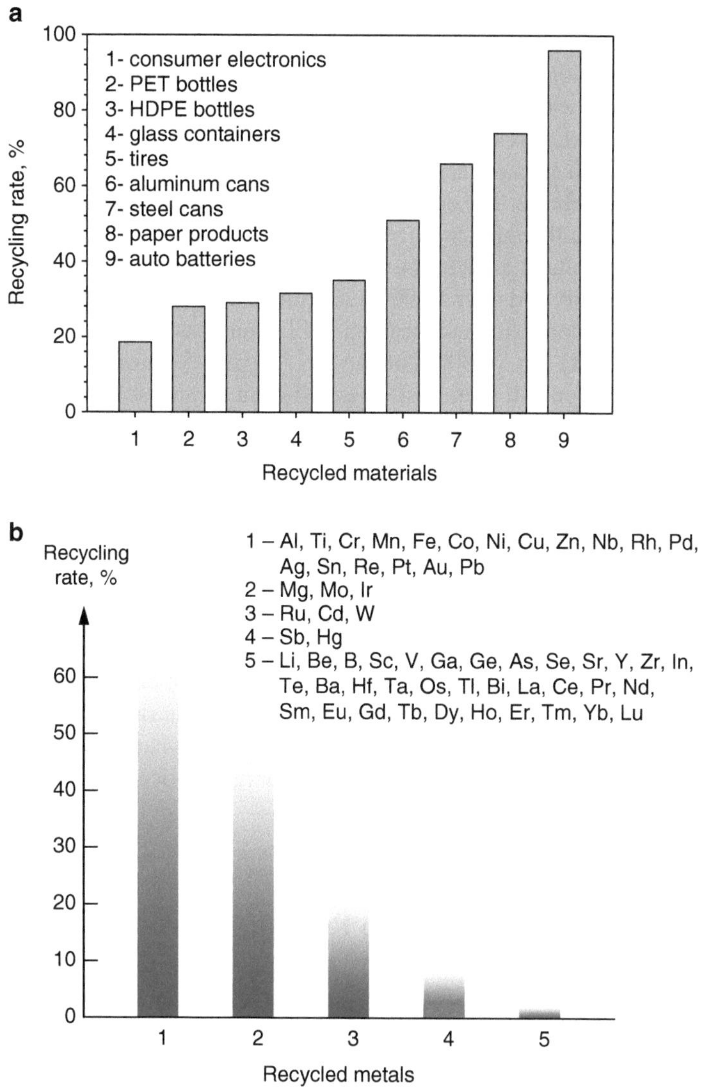

Fig. 5.8 Global post-consumer recycling rates of (**a**) selected materials and (**b**) industrially important metals. *HDPE* high-density polyethylene, *PET* polyethylene terephthalate. *Source* [46, 47]

The report indicates that whereas some metals enjoy high recycling rates, such as lead used in industrial and vehicle batteries (practically all lead is recycled) and iron, chromium, nickel, and manganese used in the manufacturing of steel (recycling rates higher than 50 %), other important metals are recycled at alarmingly low rates. For example, recycling rates of such strategically important metals as lithium (used in rechargeable batteries), cerium (used in electronic devices and catalysts), and indium (used in semiconductors and light-emitting diodes) fall below 1 % mark [46]. The low recycling rates of many metals can, at least partially, be explained by very low concentration of these metals in products and the immense complexity of their recovery from waste streams.

In discussing matters related to energy efficiency and conservation, it is important to touch upon a human factor that could markedly affect the efficacy of the carbon mitigation policies. The energy efficient technologies are pointless if nobody wants to use them, although very often, the problem simply is one of an inertia or perception. The results of a recent survey are a case in point. An average incandescent lamp costs $0.50, and over 10,000 h of using it will cost about $49 (in electric bill charges). A compact fluorescent lamp (CFL) may run at $3, but will result in electricity cost of $11 per 10,000 h of use (i.e., about 3.5 times less) [49]. But the majority of population still prefer using less efficient incandescent lamps, although they realize that not only they are more expensive in a long run, but are associated with production of more CO_2 emissions. However, when CFL are presented as a default option in a new home or a newly renovated home, 80 % of consumers accepted this option (as opposed to insisting to put incandescent lamps) [49]. Therefore, many analysts argue that understanding consumer behavior needs to be at the center of a meaningful energy policy.

According to UNDP, as a combined result of energy efficiency improvements, energy conservation measures, and structural changes in transportation and other areas, along with the increased recycling and substitution of energy-intensive materials, the energy intensity of economy could decline at the rate of 2.5 % per year over the next 20 years [44, 45]. More comprehensive information on the matters of energy efficiency, conservation, recycling and energy services can be found in an excellent book by Danny Harvey [50].

5.4 Technological Pathways to Reducing Carbon Intensity

The main approaches to reducing carbon intensity of economy most frequently discussed in the literature (e.g., [51, 52]) include:

- Expansion of nuclear energy
- Increase in the use of renewable energy resources
- Expansive deployment of carbon capture and storage technology
- Switching from high-carbon to low- and zero-carbon fuels and energy carriers

There are a number of daunting technical, economic, and political challenges and pitfalls associated with the introduction of these carbon remediation measures to the marketplace. For the majority of promising low-carbon technologies, competition with fossil incumbents is an uphill battle now. In the following chapters, the latest developments in low-to-zero-carbon technologies and the near-to-mid term outlook will be analyzed in the light of the current technological trends and advancements. The major existing and emerging innovative solutions in the pursuit of a low-carbon future that lie at the heart of a sustainable energy system will be discussed in detail.

References

1. Grübler A, Nakićenović N (1996) Decarbonizing the global energy system. Technol Forecast Soc Change 53:97–110
2. Nakićenović N (1996) Freeing energy from carbon. Daedalus 125:95–112
3. Aguilera R, Aguilera R (2009) Global possibilities of future methane and hydrogen economies. J Pet Technol June:34–39
4. International Energy Agency (2011) CO_2 emissions from fuel combustion. IEA report. 2011 ed. Paris, France
5. International Energy Agency (2012) Global carbon-dioxide emissions increase by 1.0Gt in 2011 to record high. http://iea.org/newsroomandevents/news/2012/may/name,27216,en.html. Accessed 15 Jun 2012
6. Olivier J, Janssens-Maenhaut G, Peters J, Wilson J (2011) Long-term trend in global CO_2 emissions. European Commission's Joint Research Center, PBL Netherlands Environmental Assessment Agency, The Hague, The Netherlands
7. Raupach M, Marland G, Ciais P et al (2007) Global and regional drivers of accelerating CO_2 emissions. Proc Natl Acad Sci U S A 104:10288–10293
8. Pielke R Jr (2010) The climate fix. Basic Books, New York
9. Pielke R, Wigley T, Green C (2008) Nature 452:531–532
10. Yamaji K, Matsuhashi R, Nagata Y, Kaya Y (1991) An integrated system for CO_2/Energy/GNP analysis: case studies on economic measures for CO_2 reduction in Japan. Workshop on CO_2 reduction and removal: measures for the next century. Intern. Inst. Applied Systems Analysis, Laxenburg, Austria, 19 Mar 1991
11. Hoffert M, Caldeira K, Jain A et al (1998) Energy implications of future stabilization of atmospheric CO_2 content. Nature 395:881–884
12. International Energy Agency (2012) CO_2 emissions from fuel combustion 2012. http://www.iea.org/publications/freepublications/publication/name,4010,en.html. Accessed 12 Mar 2013
13. International Energy Agency (2013) Tracking clean energy progress 2013. IEA input to the Clean Energy Ministireal, IEA, Paris. www.iea.org/publications/TCEP_web.pdf. Accessed 10 Jul 2013
14. Carbon Capture Journal (2013) IEA-progress towards clean energy stalled. 17 Apr 2013. http://www.carboncapturejournal.com/displaynews.php?newsid=1125&phpsessid=18dkjsha7qaa6s815mlrslvtq4. Accessed 15 Jun 2013
15. International Energy Agency (2012) Energy technology perspectives. Pathways to a clean energy system. IEA/OECD, Paris, France
16. Johnson J (2012) The new Saudi Arabia. Chem Eng News 90:30–31
17. International Energy Agency (2012) World Energy Outlook. OECD/IEA. Paris, France. http://www.iea.org/publications/freepublications/publication/English.pdf. Accessed 25 Jan 2013
18. Cohen K (2010) Our most powerful energy solution? Efficiency. ExxonMobil. http://www.exxonmobilperspectives.com/2010/11/29/our-most-powerful-energy-solution- efficiency/. Accessed 20 Oct 2011
19. Lovins A (2009) Profitable climate solutions: correcting the sign error. Energy Environ Sci 2:15–18
20. European Commission (2011) Strong impetus to energy savings and energy efficiency. http://europa.eu/rapid/pressReleasesAction.do?reference=IP/11/770. Accessed 12 Dec 2011
21. Jiao L, Stone R (2011) China looks to balance its carbon books. Science 334:886–887
22. Lovins A (2005) More profit with less carbon. Scient Am September:74–83
23. Black R (2011) China "won't follow US" on carbon emissions. BBC News. 25 Oct 2011. http://www.bbc.co.uk/news/science-environment-15444858. Accessed 8 Nov 2011
24. Mueller C (2013) Improving energy efficiency with CHP. Distributed Energy January/February:48–50
25. International Energy Agency (2011) World energy outlook 2011. France, Paris. ISBN 978-92-64-12413-4

26. Ausubel J, Marchetti C (1997) Elektron: electrical systems in retrospect and prospect. In: Ausubel J, Lanhford D (eds) Technological trajectories and the human environment. National Academy Press, Washington, DC, pp 110–134
27. Cao D, Sun Y, Wang G (2007) Direct carbon fuel cell: fundamentals and recent developments. J Power Sources 167:250–257
28. Saito H, Hasegawa S, Ihara M (2008) Effective anode thickness in rechargeable direct carbon fuel cells using fuel charged by methane. J Electrochem Soc 155:B443–B447
29. Cherepy N, Krueger R, Fiet K et al (2005) Direct conversion of carbon fuel in a molten carbonate fuel cell. J Electrochem Soc 152:A80–A87
30. Jain S, Lakerman J, Pointon K, Irvine J (2007) Carbon content in a direct carbon fuel cell. ECS Trans 7:829–836
31. Steinberg M, Cooper J, Cherepy N (2002) Proc. AIChE 2002 spring meeting, New Orleans. pp. 2112–2127
32. Patel S (2012) Major developments for solid oxide fuel cells. Power, 1 Aug 2012. http://www.powermag.com/renewables/wind/Major-Developments-for-Solid-Oxide-Fuel-Cells_4801_p2.html. Accessed 21 Jan 2013
33. Fuel Cell Today (2012) PNNL Develops new, high-efficiency solid oxide fuel cell system for community power. 1 Jun 2012. http://www.fuelcelltoday.com/news-events/news-archive/2012/june/pnnl-develops-new,-high-efficiency-solid-oxide-fuel-cell-system-for-community-power. Accessed 10 Oct 2012
34. Pehnt M, Ramesohl S (2003) Fuel cells for distributed power: benefits, barriers and perspectives. IFEU, Wuppertal Institut. WWF and FCEu report on fuel cells for stationary power-2003.pdf
35. Jansen D, Dijkstra J (2003) CO_2 capture in SOFC-GT systems. Second annual conf. carbon sequestration, Alexandria, VA, USA
36. International Energy Agency (2002) Transmission of CO_2 and energy, IEA greenhouse gas R&D programme, report PH4/6. IEA Greenhouse gas R&D Programme, Cheltenham, UK
37. The Hydrogen Journal (2010) Plan for UK generating plant using coal gasification, CCS, hydrogen and fuel cells production. 14 Oct 2010. http://www.h2journal.com/displaynews.php?newsid=525. Accessed 2 Feb 2012
38. Brouwer J (2006) Hybrid gas turbine fuel cell systems, chapter 4. In: Dennis RA (ed) The gas turbine handbook. U.S. Department of Energy, Morgantown, West Virginia, DOE/NETL-2006/1230
39. Brouwer J (2003) Hybrid gas turbine fuel cell systems, DOE, NETL, Coal Power Technologies. pp. 127–162. www.netl.doe.gov/technologies/coalpower/turbines/refshelf/handbook/1.4pdf.
40. Kobayashi Y, Ando Y, Kabata T et al (2011) Extremely high-efficiency thermal power system-solid oxide fuel cell (SOFC) Triple combined-cycle system, Mitsubishi Heavy Industries Technical Review 48:9–15. http://www.mhi.co.jp/technology/review/pdf/e483/e483009.pdf. Accessed 18 Jan 2012
41. Bloomfield D (1976) Pressurized fuel cell power plant with single reactant gas stream. US Patent 3,976,507
42. Net: Developers' Journal (2012) MHI to develop fuel cell triple combined cycle power generation system. http://dotnet.sys-con.com/node/2288456. Accessed 10 Dec 2012
43. Kvamsdal H, Jordal K, Bolland O (2007) A quantitative comparison of gas turbine cycles with CO_2 capture. Energy 32:10–24
44. United Nations Development Programme (2004) World energy assessment: energy and the challenge of sustainability. UNDP, New York, 2004
45. Goswami Y, Kreith F (2007) Global energy system. In: Kreith F, Goswami Y (eds) Handbook of energy efficiency and renewable energy, chapter 1. CRC, Boca Raton, FL, pp 1–24
46. Kemsley J (2011) Metals recycling falls short. Chem Eng News 89:9
47. Bomgardner M (2011) Taking it back. Chem Eng News 89:13–17

48. United Nations (2011) UN Environment program's international resource panel report "recycling rates of metals: a status report. unep.org/resourcepanel/publications/recyclingratesofmetals/tabid/56073/default.aspx. Accessed 20 Dec 2012
49. Powell E (2012) Earth, wind and water. Discover 11:50–53
50. Harvey D (2010) Energy and the new reality. Energy efficiency and the demand for energy services. Earthscan, London
51. Jacobson M (2009) Review of solutions to global warming, air pollution, and energy security. Energy Environ Sci 2:148–173
52. Moriarty P, Honnery D (2011) Rise and fall of the carbon civilisation. Springer, London, UK

Chapter 6
Carbon-Neutral Energy Sources

Abstract Among the main approaches to decarbonizing global economy, the switching to carbon-neutral energy sources such as nuclear and renewables (solar, wind, biomass, etc.) is mentioned most often. Nuclear energy is considered an important carbon mitigation option; despite the recent Fukushima accident, the majority of countries with nuclear power remain committed to its use. Renewables are no longer regarded immature technology; while the cost of some renewables has dropped significantly over the last decades (e.g., onshore wind, solar photovoltaic), the competition with fossil incumbents is still an uphill battle. There are a number of daunting technical and economic challenges and pitfalls associated with the expansion of the carbon-neutral energy sources in the energy market. This chapter analyzes the latest scientific, technological, and commercial developments in the area of carbon-neutral energy sources and fuels, as well as their carbon mitigation potential and outlook in the light of current technological trends.

6.1 Nuclear Energy as a Carbon Mitigation Option

Nuclear energy does not inherently involve any direct production of CO_2 or other GHG and, as such, is a major producer of carbon-free electricity [1, 2]. Although not without controversies, nuclear energy is considered an important carbon-free energy source, which, through displacing coal, oil, and NG, would substantially reduce overall CO_2 emissions and, thus, alleviate the potential power shortage problem without disturbing the Earth's fragile carbon balance.

Besides being practically carbon-free, another important advantage of the nuclear energy source is that it enjoys the highest power density among all electricity generating technologies, and, especially, in comparison with other non-carbon sources of electricity. The energy density of nuclear fuel varies from 3,456,000 MJ/kg (for 3.5 % enriched ^{235}U in a light water reactor) to 86,000,000 MJ/kg (uranium fuel in a "fast" breeder reactor), which is about million times greater than that of fossil

N. Muradov, *Liberating Energy from Carbon: Introduction to Decarbonization*, Lecture Notes in Energy 22, DOI 10.1007/978-1-4939-0545-4_6,

fuels (e.g., 46 MJ/kg for gasoline). Renewable energy sources, such as solar, wind, and biomass, are diffuse, and, in terms of energy density, they are by many orders of magnitude lower than the nuclear source. For example, an average nuclear plant of 1 GW capacity takes up about 6 km^2 of land space, whereas a wind farm of the comparable capacity would occupy an area of 609 km^2 [3].

There are many excellent reviews and books on the subject of nuclear energy and state-of-the-art nuclear reactors for production of power, heat, and fuels, and thus, a detailed discussion of this topic in this book would be superfluous; the interested readers could find an in-depth information on this topic in literature (e.g. 4). The objective of this chapter is to provide a brief introduction to the carbon mitigation aspects of the nuclear source and discuss the current challenges and post-Fukushima developments that may not have been reflected in other books and reviews.

6.1.1 Nuclear (Fission) Energy: Trends and Challenges

Nuclear (fission) power is a mature technology: it has been commercially practiced since the middle of the twentieth century, and, in many countries, it supplies a lion's share of the total electricity demand. Nuclear energy had ups and downs throughout the relatively short history of this technology. At the onset of this energy technology in the 1950s, it was advocated by its proponents as a miracle source of abundant "cheap to meter" power that would revolutionize the way people use energy (though the serious scientists such as Enrico Fermi, Robert Oppenheimer, and Glenn Seaborg warned of the "unwarranted optimism" [5]). Although things did not turn out to be exactly that way, nuclear energy enjoyed period of stable growth until the nuclear accidents at the Three Mile Island (TMI) plant (USA) and, especially, in Chernobyl (Soviet Ukraine, 1986). For almost three decades, no new nuclear plants have been ordered in the USA, and the share of nuclear electricity has been steadily declining as older plants were approaching retirement age.

6.1.1.1 Current Trends and Outlook

At the beginning of the twenty-first century, nuclear energy was increasingly favored as an important (practically) emission-free part of the energy mix (especially, in view of potential carbon pricing). The ground was broken for 16 new reactors (mainly, in non-OECD countries), and 67 reactors were under construction (26 in China alone) [2]. However, the so-called "nuclear renaissance" had been significantly hindered by the global financial crisis of 2008 and, especially, the Fukushima accident (2011), which cast uncertainty over the future of nuclear power.

According to the IEA's statistics, as of the end of 2012, there were 437 operational nuclear power plants worldwide, with the total capacity of about 392 GW (representing 23 GW increase from 2000 levels) [6]. Most currently operating nuclear reactors are known as Generation II reactors, which are based on the 1970s technology and the 1960s materials [7]. About 2,518 TWh of nuclear electricity was generated in 2011, accounting for over 12 % of the world's electricity mix (21 % of the US electricity and about 80 % of France's electricity). Since the middle of the last decade, an average of 2.4 GW in global capacity has been added each year, including 3 GW in 2012 [6]. Between 2000 and 2012, China, Japan, and South Korea completed the construction and grid connection of 22 Generation II and III reactors [6].

The cost and construction time for modern nuclear power plants vary significantly by the reactor type and country/region. The average costs of Generation II and III reactors vary in the range of US$1,560–3,000 per kW in Asia and US$3,900–5,900 per kW in Europe [2]. In terms of construction time, some plants were built in 4 years, and for some it has taken more than 20 years to complete construction (e.g., in Romania and Ukraine).

The analysis of current trends in the nuclear power sector shows that interest in small modular nuclear reactor with the net capacity of less that 300 MW and medium-size reactors (300–1,000 MW) is increasing. In the USA, nuclear energy R&D spending is projected to substantially increase and focus on the development and industrial deployment of small modular reactors with advanced designs and safety features (US$450 million was provided for the development of the small reactors targeting the construction of a first-of-a-kind reactor before 2022) [6]. The rational for the deployment of small reactors in the USA is that these physically small reactors would enjoy more flexible applications and incur lesser financial risk and construction cost than the nation's current fleet of 104 large and ageing nuclear power plants.

In 2012, the US DOE designated an engineering firm Babcock & Wilcox (B&W) to design, license, and commercialize a first small modular nuclear reactor for the US market [8]. The reactor will be compact enough to be manufactured in a factory, shipped to an end user, and installed about 50 m underground and covered by about 3 m thick concrete slab. The B&W is designing a 180 MW unit (which is a modification of a basic light water reactor) with the plans to install two units by 2022 (with expected life of at least 60 years). The potential markets for the small modular reactors in the USA include: (a) the replacement of coal-fired plants that are scheduled to close, (b) providing heat and power to remote communities, and (c) heat and power supply for industrial facilities, such as chemical plants, petrochemical and metallurgical complexes. The modular reactors could be grouped together for a large-scale utility application.

The advantage of the underground design is that it protects the reactors from weather-related threats (e.g., hurricanes, tornados) and terrorist attacks, but, at the same time, it may increase vulnerability to other hazards such as floods and earthquakes and makes it much harder for operators to gain access to equipment during emergency. A 180 MW module will cost around US$1 billion to build, and this size was selected because it is the biggest size the company can construct and still put on a railcar, and because it is roughly the size of a typical coal- or gas-fired power plant that the nuclear plant could replace. Opponents of this approach question whether the small reactors will be competitive with standard large units considering the apparent loss of the economy of scale. The B&W argues that there are several ways to reduce the cost per kilowatt to the level of large reactors, namely, by doing most of the manufacturing job in a factory environment (as opposed to doing the construction at the job site), and also by relying on passive safety systems and reducing back-up support and staffing for emergencies.

South Korea's SMART small modular reactor received a standard design approval in July 2012; a target application for the technology is a combined power and desalination field [6]. Two small modular reactors on floating barges (KLT-40S design) are being constructed in Russia. In 2012, China started construction of two 100 MW gas-cooled high temperature reactors (HTR-PM); this is a first step toward a Generation IV Very High Temperature Reactor (VHTR) [6].

Outlook

The IEA sees nuclear power playing a substantial role in the decarbonization of the electricity sector, contributing 8 % to overall CO_2 savings by 2050 [2] and reaching about 16 % of global generation by 2025 [6]. According to the IEA's 2012 scenario (see Chap. 4), to reach 2025 goal, nuclear capacity must increase by over 250 GW from 2012 levels [6]. In order to reach the targeted levels of deployment, in addition to construction of new reactors, the extended long-term operation of existing units will be required [9]. The extension of the operation of existing nuclear plants beyond their original design lifetime (which would require license extensions or renewals and substantial investments by utility companies) can help to prolong nuclear capacity until new reactors will replace the older ones. In its ETP-2012 projections, IEA assumes a 60-year lifetime for the US reactors, and 55-year lifetime elsewhere (at the end of 2012, over 70 reactors in the USA had received license extensions of up to 60 years) [6]. However, due to the changes in safety requirements and government policies and regulations after the Fukushima accident, getting the permissions for the extended operation is becoming more and more complex and difficult. Taking into account that the closure of existing nuclear capacity could be accelerated, the IEA projects the increase in the rate of introduction of new capacity by 25–50 % through 2030 in order to meet its CO_2 stabilization target by mid-century [9].

6.1.1.2 Controversies and Challenges

Nuclear energy is not without a controversy; but, recently, pros and cons voices have been getting increasingly louder with plenty of conflicting arguments from each side. Here we give a podium to both sides: the supporters and opponents of nuclear energy. The main argument of the supporters of nuclear energy is that it has to be an essential part of our energy supply if we want to prevent a major climate disruption due to continued massive carbon emissions, especially, considering that other alternatives are not yet ready. The opponents, however, have more cautious-to-negative assessments of the future role and scope of using nuclear energy source as a carbon mitigation option, putting forward the following arguments:

- There may not be adequate nuclear resources to support carbon mitigation policies in a first place. It was estimated that at the current rate of consumption, uranium reserves might last 50–100 years [10]; however, if nuclear plants were to provide the predominant share of electricity, uranium resources would last only 30 years (assuming the present-day types of the reactors and ultimately recoverable uranium resources) [11, 12] (there is also a reported estimate of 90 years for known reserves [13]). Taking into account the nuclear reactor lifetime of about 50 years, it would be imprudent to base energy policy on the nuclear source without knowing if there will be enough fuel to run the reactors.
- The growth in nuclear power at the required pace may not be feasible because of the severe constraints in the uranium fuel cost, power plant site availability, safety, public opposition, and waste disposal considerations [12]. (Note that the price of uranium fuel is about $4 per barrel of petroleum equivalent [14]).
- Nuclear reactors are very expensive: an average nuclear plant with a capacity of 2.2 GW would cost about $12 billion [15], and it would take almost a decade to construct (including all required permissions).
- Nuclear energy is not completely a "carbon-free" source since its life-cycle GHG emissions are on the order of 65 g CO_2-eqiv./kWh) [16].
- People fear radiation because they cannot see, control, or feel it: this natural fear cannot be dismissed and does not seem to be irrational.
- Three Mile Island, Chernobyl, Fukushima.
- Reasonably high degree of safety of nuclear reactors can never be achieved or could be achieved only at an exorbitant cost. And the insurance industry is keenly aware of this fact: for example, in the USA, they would never have given construction loans to build nuclear power plants without the Price-Anderson Act, which limits their liability to $375 million (a mere pittance compared to recent disastrous oil spill accidents) [17].
- Many existing nuclear reactors (in the USA, almost all of them) are approaching (or already have somewhat exceeded) their operational lifetime (in most cases, about 40 years). Longer these reactors run past that designated time, greater is the risk of structural and radiation-exposed material failure with extremely costly consequences to the society.
- The problem of safe long-term storage of nuclear waste has not been solved yet.
- Nuclear proliferation, possible terrorist attack issues, etc.

The proponents of nuclear energy are trying to allay the above and other concerns by putting on the table the following counter arguments:

- Every energy technology comes with associated risks: (1) NG is known to blow up houses and even whole city blocks; (2) coal mines are notorious for frequent and deadly accidents; (3) there were several battery-related explosions (including couple on Boeing's Dreamliner); and other examples are plenty.

- Mining fossil fuels is not less hazardous than nuclear fuel: a case in point is "Deepwater Horizon" oil spill, "mountaintop removal" method, radioactive gas radon from NG development, etc.
- Coal which provides almost half of world's energy is also a source of radioactivity. A series of studies on the environmental impact of coal produced a quite surprising conclusion: the solid waste produced by coal-fired plants is actually more radioactive than that generated by their nuclear counterparts. It turned out that the fly ash emitted by a coal power plant carries into the surrounding environment 100 times more radiation than a nuclear power plant per the same electrical output [18]. The sources of radioactivity are uranium and thorium; these elements occur in natural (or "whole") coal in trace amounts; however, when coal is burned into fly ash uranium and thorium are concentrated at up to ten times their original levels, which can no longer be ignored.
- Nuclear reactors are only as safe, reliable, and unfailing as man makes them—no different from space shuttles, deep-water oil rigs, or industrial robots. Current advances in nuclear technology and high-performance construction materials enable the development and deployment of much safer nuclear reactors with multiple levels of security (e.g., modular nuclear reactors with advanced design and safety features).
- The waste from burning fossil fuels is not less problematic than nuclear waste pointing to CO_x, SO_x, NO_x, mercury, and arsenic emissions from coal power plants.
- Nuclear waste problem, although technically and politically challenging, is not insurmountable and can be solved in the near future.
- Many worries about leaked radiation and its health hazards are exaggerated and there are far worse dangers out there. The death rate from the Three Mile Island accident is practically zero. A recent study conducted by University of Würzburg (Germany) found that quarter of a century after the Chernobyl, worst nuclear disaster in history, many children and teenagers in the affected areas of Ukraine and Byelorussia who developed thyroid cancer due to radiation are in a complete (64 % of patients) or near complete (30 % of patients) remission [19]. These findings suggest that the victims of the recent nuclear accident at the Fukushima plant might face a lower risk of developing advanced-stage thyroid cancer.
- Current death toll in Japan from earthquake/tsunami is about 50,000, but directly from radiation is zero.
- Humans are not going to give up the benefits of modern civilization, which is based on abundant energy supplies. For the next half century or so, renewables and/or other alternative clean energy sources will not be able to fully power civilization; so, the role of nuclear power will remain essential, and it is very unlikely to contribute to climate change [20].
- The natural fear of people toward radiation could be overcome, since low levels of radiation (e.g., from nuclear waste storage sites) can provide some health benefits.

The latter point may require some explanation. Many scientific reports indicate that small doses of ionizing radiation emanating from storage sites can provide some positive effect on human health through the so called, "radiation hormesis" phenomenon [21]. Radiation hormesis hypothesis, while accepting that high levels of radiation are hazardous, suggests that the radiation exposure comparable to and just above the natural is not harmful but, actually, beneficial. Supporters of the radiation hormesis claim that radiation-protective responses in a human's cells and the immune system not only counter the harmful effects of radiation but also act to inhibit spontaneous growth of cancer cells not related to radiation exposure [22]; they also point that the radiation levels were much higher in the past, and life on the Earth, including Homo Sapiens species, evolved coping with it. It has been reported that an optimum "healthy" level of ionizing radiation is about 50 times ambient levels [23]. The implications of the radiation hormesis notion is that an appropriate radiation supplementation (e.g., from a radioactive waste storage site) would provide abundant health benefits (e.g., the occurrence of cancer in this area would become rarer compared to radiation-free places) [24]. It should be noted that the radiation hormesis hypothesis has not been accepted by either the US National Research Council or the US National Council on Radiation Protection and Measurements.

6.1.2 Fukushima Accident's Implications and Lessons

Japan's 11 March 2011, powerful earthquake followed by a massive tsunami caused a severe damage to coolant systems and, subsequently, led to meltdowns in three operating reactors at the Fukushima-Daiichi nuclear power plant (440 km north of Tokyo). This was the worst nuclear accident since the Chernobyl disaster resulting in the release of large amount of radioactivity and forcing evacuation of hundreds of thousands of people within 40 km from the plant.

6.1.2.1 Fukushima's Global Implications

The Fukushima accident further exacerbated already heated debates about the future of nuclear power. In some countries, ambitions for nuclear power have been scaled back as they have reviewed their policies after the Fukushima. There was a sober realization that if such a technically advanced nation as Japan was vulnerable to such an accident, the nuclear technology might indeed carries too many inherent risks (the Fukushima plant was designed to withstand supposedly a worst-case scenario). Some countries announced that they would consider reducing their dependency on nuclear power, or even completely abandon it, while others stated that they would not make any changes to their nuclear energy policies and deployment targets. Here are some examples of the responses to the Fukushima accident from selected countries.

Germany—Europe's the largest economy—announced plans to abandon nuclear energy by 2022, effectively shutting down all 17 nuclear power plants and replacing them with renewable energy sources and high-efficiency gas- and coal-fired CHP plants. (As of 2011, Germany produced about a quarter of its electricity from nuclear power, about the same share as in the USA.). Chancellor Angela Merkel said "We believe that we can show those countries who decide to abandon nuclear power — or not to start using it — how it is possible to achieve growth, creating jobs and economic prosperity while shifting the energy supply toward renewable energies" [25].

Belgium also chose to phase out nuclear power (by 2025) by closing down or not extending the lifetime of existing plants. *Switzerland*, where nuclear power plants produce significant share (about 40 %) of electricity, also announced its plans to gradually shut down its reactors as they reach their average life span of about 50 years (this implies that the last nuclear power plant will be taken off the grid in 2034 or so [25]). *Italy* has already abandoned nuclear power (even before the Fukushima), which was voted down in a referendum after the 1986 Chernobyl nuclear disaster.

France questioned the Germany's move arguing that it would be impossible for the EU to meet CO_2 emission cutting targets without nuclear power. France remains in favor of nuclear energy source despite a recent explosion at Marcoule nuclear waste processing site in southeastern France in 2011, which, luckily, did not release any radioactive leak or waste [26]. The French government reviewed the country's policy after the nuclear disaster in Fukushima, and it is considering reducing the share of nuclear power from 79 % in 2011 to 50 % by 2025, and it has scheduled a closure of the country's oldest nuclear plant in 2016 (at the same time, the government supports the construction of the first European Pressurized Reactor at Flamanville) [6]. *Sweden* also did not support the Germany's decision to do without nuclear power and was particularly concerned that this could drive up electricity prices across Europe [25]. The *United Kingdom* has an active nuclear program; it intends to construct up to eight units by 2025, amounting to at least 10 GW, with plans to launch two projects in 2013 [6].

In the *USA*, the US Nuclear Regulatory Commission (US-NRC) task force commissioned after the Fukushima concluded that the risk of a similar accident in the USA is "very, very small," but any risk would be inherently unacceptable, and nuclear power regulators should learn lessons from the Japanese nuclear crisis [27]. In 2012, the US-NRC approved licenses to build two new reactors—the first authorized reactors after over 30 years, hiatus [28]. The reactors will be sited in Georgia at Vogtle nuclear power plant complex, about 270 km from Atlanta (the plant already houses two older reactors). The new nuclear reactors feature AP 1000 type reactor technology and are expected to have a price tag of $14 billion and provide 2.2 GW of power to one million homes. The AP 1000 is the newest modification of the Westinghouse-designed nuclear reactors approved by the US-NRC. The Vogtle reactor would be the first one built in the USA, although four of them are already under construction in China. The reactor features passive cooling design based on gravity and condensation, which makes it much safer than older electricity-driven designs (including one at Fukushima, which relied on electric power for cooling the fuel rods). In addition to the Vogtle nuclear plant, 16 other plants across the US have applications with the US-NRC to build 25 more nuclear reactors [28]. Most of the new reactors will be built at existing nuclear power plant sites, but two applications are submitted for brand new nuclear plants: one in Levy County, Florida, and another one near Gaffney, South Carolina.

Japan. In the wake of the Fukushima's accident, Japan in its "Innovative Strategy for Energy and Environment" announced plans to reduce the country's reliance on nuclear energy, which before the accident provided about quarter of all electricity [29]. In July 2011, while engineers were still struggling to restore normal cooling at the plant and find a solution to storing more than 100,000 t of radioactively contaminated water, Japan's Prime Minister Naoto Kan outlined a plan to gradually move away from nuclear power. "When we think of the great risk this nuclear energy imposes, our traditional position of ensuring safety will not be enough," Kan said. "We would like to seek to build a society which does not depend on nuclear energy" [30]. In 2011, nuclear power supplied about 18 % (102 TWh) of electricity, down from 26 % (or 288 TWh) in 2010, and only 2 reactors out of 50 operational reactors had been restarted by the end of 2012 [6]. But even if no new nuclear plants are built through 2035 (except for the two reactors at Shimane-3 and Ohma that are already at an advanced stage of construction) and existing plants will have shorter lifetimes than originally projected, nuclear generation in Japan could recover a 20 % share by 2020 and 15 % by 2035 (with the shortage picked up by renewables) [29].

China announced a new ambitious construction program, which will be based on Generation III reactor designs and installed on coastal sites only [2, 6]. *Russia* and *India* also have active nuclear programs; they confirmed plans to continue to build nuclear plants with the capacity additions of 15 GW and 20 GW, respectively, by 2025 [6]. Among countries that delay or make changes to first nuclear power plant introductions are *Thailand*, *Malaysia*, *Philippines*, and *Indonesia*. The remaining (about twenty) countries that have nuclear power plants have not changed their plans for nuclear energy as a result of the Fukushima accident.

6.1.2.2 Fukushima's Lessons

Different in many ways, all three major nuclear accidents, TMI, Chernobyl, and Fukushima, had one thing in common—the damage to the core of the reactors due to the loss of a coolant (water). Unlike TMI and Chernobyl accidents that were caused by human errors, the Fukushima accident was due to the "act of God;" nevertheless, it has brought into focus one of the most serious vulnerabilities of modern nuclear plants—the danger posed by spent fuel pools (that are also cooled by water). Although many details of the Fukushima accident are still investigated, there are several important lessons to be learned from the Fukushima-Daiichi nuclear accident that might help to prevent similar accidents in future in countries that use the same type of light water reactors (e.g., the USA and France).

Lesson 1

The interim storage of spent fuel at the nuclear plant site (before reprocessing or a permanent disposal) has to be completely rethought with regard to the location of the pools and their packing with spent fuel. In particular, the location of pools above ground (just below the roof) and outside the hardened containment should be now reexamined [31]. Another interim storage related vulnerability is concerned with densely packed spent fuel in pools, which poses an increased risk of radiation or radionuclide release (as a result of coolant loss due to, e.g., earthquake, tornado, and sabotage). Spent fuel at the Fukushima nuclear plant was stored in seven spent fuel pools (one at each reactor plus one large central pool) and, to a lesser extent, in dry cask storage [32]. All pools had high-density racks (though not fully loaded) and sustained some type of damage (the pool at the unit 4 was directly damaged by the earthquake). In contrast to water-cooled spent fuel pools, the dry casks suffered little damage.

According to an analysis of the US nuclear policy in the light of the Fukushima accident, in the USA, spent fuel pool racks have been redesigned to hold up to four times the originally intended amount of fuel [32]. After the pools are allowed to be filled to the maximum high-density capacity, older fuels are moved into on site dry storage. There were reports questioning the wisdom of this policy and recommending to revert to low-density open-cage racks in the pools by moving more spent fuel into dry storage casks, which are passively safe [33] (again, the US-NRC maintains that the present-day spent fuel safety systems are adequate). Sweden (which produces 35–40 % of its electricity from nuclear energy) has been practicing different (albeit admittedly more costly) approach to handling spent fuel: they move recently discharged fuel to a centralized underground spent fuel pool, thus, avoiding filling the reactor pools with hot spent fuel [32]. Regardless of which of these approaches is better or worse, the Fukushima accident clearly exposed the vulnerabilities of the current spent fuel interim storage policies and emphasized the importance of careful planning and efficient management.

Lesson 2

National nuclear energy programs need a robust well-thought long-term plan for the permanent disposal of the nuclear waste. In Japan, the long-term management strategy for spent fuel involves its reprocessing to extract plutonium (Pu) and uranium for reuse in new fuel with the subsequent disposal of highly radioactive waste products in a mined geologic repository [34]. But the nuclear waste problem was exacerbated by the fact that Japan's recently constructed waste reprocessing facility at Rokkasho (expected to start operation in 2007) is still inoperable due to unexpected technical difficulties, and, as a result, the spent fuel has been built up at nuclear reactor sites in Japan (note that before Fukushima, Japan did not have an interim spent fuel storage policy) [32].

Lesson 3

Currently, emergency plans for the majority of nuclear plants (e.g., for most of them in the USA) envision a single incident involving one reactor, but at the Fukushima plant, the enormous tsunami damaged all four operating reactors. Among the US-NRC recommendations were calls for ensuring a protection, enhancing accident mitigation, strengthening emergency preparedness, and, especially, rethinking each facility's disaster plans with regard to how strong an earthquake, tornado, hurricane, tsunami, or flood the plant is designed to withstand [30]. The US-NRC also recommended extending the battery back-up operation time to 8 h (twice the current standard) and enable the cooling system to operate 72 h during a blackout.

Lesson 4

Natural forces are unpredictable (this old lesson is being learnt over and over again), and there are limits to the information that seismologists can provide to the governments and nuclear engineers in terms of timing and magnitude of future catastrophic quakes. Although it is well known that Japan is located in a tectonically active region (part of the Ring of Fire), the science of evolution and movement of tectonic plates is relatively young, and seismologists are far from understanding all the lithospheric processes causing the occurrence of large earthquakes and associated processes such as liquefaction in the soil and tsunamis [35]. These uncertainties and the accounts of what is known and not known must be taken into serious consideration during a decision-making process with regard to the safety of nuclear plants (e.g., the height of the protection wall, and the location of emergency back-up power generators), or nuclear waste repository sites [32].

The Fukushima accident (despite its terrifying and appalling consequences) was an important test demonstrating the role of nuclear energy as an essential carbon mitigation policy. After the accident, Japan's government made a decision to take a significant number of nuclear reactors offline [6]. This measure caused a substantial energy shortage in the country, which was mostly compensated with the increased usage of imported fossil fuels in the power generation sector (along with conservation measures). As a result, in 2011, Japan's CO_2 emissions increased by 28 million ton, or 2.4 % of total [36].

6.1.3 Nuclear Waste: Problem Waiting for a Solution

Begin with the end in mind

(Stephen Covey: The 7 Habits of Highly Effective People) [37]

Nuclear waste storage problem is one of the major trump cards in the arsenal of the opponents of nuclear energy. They argue that it would be irresponsible to base a

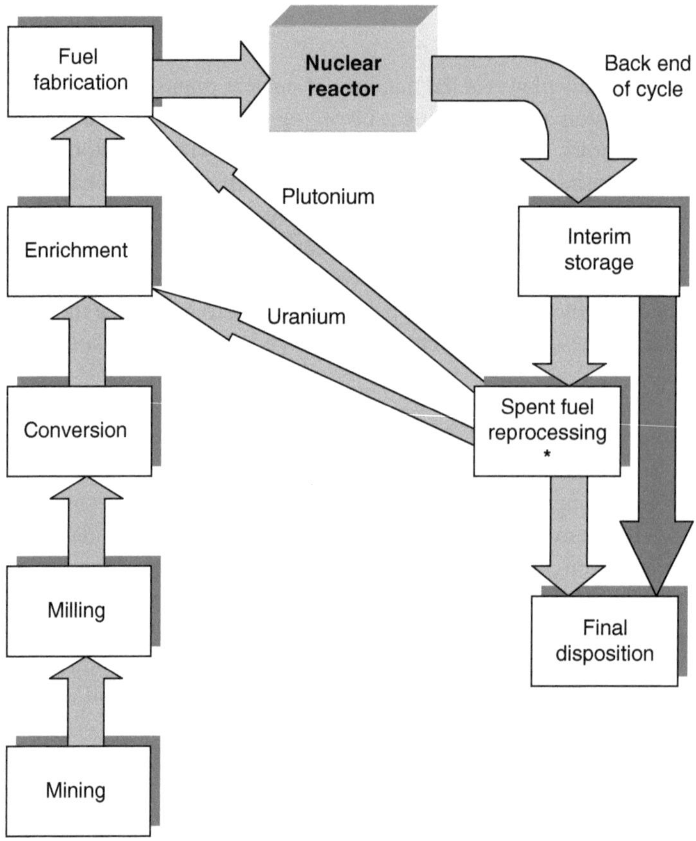

Fig. 6.1 Schematic diagram of a nuclear fuel cycle. (*) Currently, spent fuel reprocessing is omitted from the fuel cycle in most countries, including the USA *Source* [38]

global energy system on technology which does not have clear plans on what to do with the waste that will remain radioactive for thousands of years. Indeed, nuclear waste is an intractable problem for which no cost-effective and a reasonably safe long-term solution has been developed so far. Currently, the spent nuclear fuel is being stored in temporary repositories such as pools and dry casks (in most cases, within the perimeter of nuclear power plants) under continuous control and monitoring. Figure 6.1 shows the diagram of a typical nuclear fuel cycle with the back end of the cycle including interim and final disposal of waste products.

The back end of the nuclear fuel cycle, mostly spent fuel rods, contains a variety of fission products, such as ^{234}U, ^{237}Ne, ^{238}Pu, and ^{241}Am that emit alpha, beta, and gamma radiation. Certain radioactive elements, such as ^{239}Pu, ^{99}Tc, and ^{129}I, in the spent fuel will remain hazardous to humans for thousands to millions of years, which will be of a particular concern in the nuclear waste management.

Several methods of the permanent radioactive waste disposal have been or are currently under consideration by nuclear nations [39]:

- Outer space disposal (not yet realized).
- Deep borehole disposal (not yet realized).
- Rock-melting (not yet realized).
- Ocean disposal (or ocean dumping) (implemented in 1954–1993 by USSR, UK, Switzerland, USA, Belgium, France, The Netherlands, Japan, Sweden, Russia, Germany, Italy and South Korea; this option was banned in 1993 and is no longer permitted by international agreements).
- Sub-seabed disposal (currently, not permitted by international agreements).
- Disposal in ice sheets (banned by an Antarctic Treaty).

There is a general international agreement that high-level nuclear waste (e.g., spent fuel or the remains of reprocessing) would require a geologic repository for a final disposal [38]. To launch such a repository, an institution (a government agency, a private, or an industry-backed entity) will need to be established, which will determine the location of a repository site and repository operations. A successful siting strategy is a staged process, in which most management decisions are not made at the outset but along the way [40]. Currently, this process is being practiced in Canada that began with a survey of the public's attitudes to repositories, in general [32].

The geological disposal of a high-level radioactive waste and spent fuel is now being actively developed in several countries. The basic principle is to locate a sufficiently large and very stable geologic formation and using state-of-the-art mining technology to drill a shaft 500–1,000 m below the surface where vaults can be excavated for the disposal of nuclear waste. The objective is to permanently isolate the high-level radioactive waste from human environment. To date, only very few countries have established programs for the long-term geological disposal of nuclear waste. In January 2013, Finland's POSIVA company (a subsidiary of the two nuclear utility companies) submitted an application to the government to build a geological repository and waste encapsulation plant [6]. In Sweden, a similar application has been submitted by the Swedish nuclear fuel and waste management company in 2011. In France, its nuclear waste management agency is planning to submit an application to build a geological repository in 2015, to start operations in 2025 [6].

In the USA, the nuclear waste disposal problem has been a controversial and highly sensitive issue having as many political as technical and economical facets, which delays its practical implementation. Until recently, the official US nuclear waste disposal policy has been to forgo reprocessing (for economic and nuclear non-proliferation reasons) and send spent fuel directly to the specially designated geologic repository at Yucca Mountain, Nevada [32]. However, in 2010, the Obama Administration reconsidered the original plans for the disposal at this site and decided to pull the license application and stop further construction at the $15 billion radioactive waste repository citing a lack of public support and the political stalemate surrounding the issue [41, 42]. The Administration established a Blue

Ribbon Commission for America's Nuclear Future with the objective of rethinking policy for the country's high-level nuclear waste disposal [38]. Following this decision, there were proposals to create two temporary storage sites by 2025 that would centralize the storage of the spent fuel rods now scattered at over hundred sites across the country and create a permanent facility by 2048 [43].

Sweden went through at least three iterations of its nuclear waste policy before finding a technically and politically acceptable site for its high-level nuclear waste repository (which happened to be a site with suitable geology and already existing nearby nuclear facilities) [44]. The repository in Sweden is planned to open in 2023, provided no further political impediments will arise [32]. Canada, Finland, France, Switzerland, and the UK have also gone through several iterations of their high-level nuclear waste siting and disposal policy. It is realized that during decision-making process, purely technical or economical considerations concerning reasonable nuclear repository sites would not be enough, and they should be balanced with public acceptance.

6.1.3.1 Recycling of Nuclear Fuel

Recycling of nuclear fuel potentially provides the following advantages [45]:

- It reclaims valuable fuel for future power production (about 96 % of the nuclear energy remains in the spent fuel).
- It is a well-developed technology utilized by most of the nations using nuclear power (the UK, France, Germany, The Netherlands, Switzerland, Japan, Russia, India).
- It reduces the volume of waste needing long-term storage (fivefold compared to original waste volume).
- It reduces the half-life of some radioactive elements in the exhausted fuel by about 20 % (i.e., from thousands to hundreds of years).
- Its cost is gradually decreasing (whereas the long-term storage cost is increasing).

The recycling was halted in the USA in the 1970s, but could be restarted because it currently appears to be economically competitive with the long-term storage.

Another often-overlooked benefit of recycling is concerned with the recovery of some high-value fission products from a nuclear waste. For example, one of the fission products found in spent fuel is rhodium (Rh) (its price fluctuates between $65 and $390 per gram). One ton of spent fuel contains 400 g of Rh, which makes the "waste" worth $25,000 per ton just for its Rh content [46]. Advantageously, the Rh radioactive isotopes are relatively short-lived (about 30 years), which is comparable with the age of spent fuel in most nuclear power plants. The process, however, would unlikely be economically justified because the quantity of the "ore" is small (the total production of spent fuel in the USA is about 42 t per week).

Yet another angle to the nuclear waste utilization that is rarely discussed concerns with its potential to be a source of energy in the future. Uranium ore not yet mined contains decay products that have accumulated over the past four or so billion years, and in several hundred years the discharged nuclear fuel will have decayed to the point that it will no longer be significantly more dangerous than pure uranium in a transient equilibrium with four billion years of decay products [47]. While current fuel processing plants are expensive and tend to (sometimes) pollute the environment with leaked fission products, it is a fact that only about 3 % of potentially available fission energy in that fuel can be extracted by modern (conventional) technology. Attempting to dispose of that fuel such that it will be locked from humanity for thousands of years might not make sense, because in a few hundred years our descendants will be seeing it as an immense source of energy—and, of course, they will have much greater technological capabilities for processing the nuclear fuel than we do. Thus, it might be the best solution to put the nuclear waste in steel or concrete canisters, similar to those now used at nuclear power plants to store aged discharged fuel, and safely store them in suitable repositories until technology will become available to efficiently recover remaining energy from the waste.

6.1.4 Advanced Nuclear Cycles

It is generally understood that in order for nuclear reactors to continue to be a producer of carbon-neutral energy in the foreseeable future, they have to move beyond today's conventional a once-through nuclear fuel cycle. Recently, Grimes and Nuttall [48] proposed a "Two-stage Nuclear Renaissance" concept, where they suggested six potential complementary routes to adopting a long-term sustainable nuclear energy strategy:

- Unconventional uranium resources.
- Reprocessing spent fuel for multiple mixed U-Pu oxide fuel recycle.
- Breeder or "fast" reactors.
- Thorium fuel cycle.
- Accelerator-driven subcritical reactors.
- Nuclear fusion energy.

According to the authors, the first stage of this process would include replacing or extending the life of existing nuclear power plants, with continued incremental improvements in efficiency, safety, and reliability. In the second stage, which will start after 2030, new fuel cycles including fuel reprocessing will be introduced, significantly contributing to decarbonization of economy and sustaining carbon-neutral power production for more than thousand years.

In recent decades, new technological advancements in nuclear reactor technology led to the development of Generation IV reactors (GEN-IV) with the objectives of improving economics, safety, reliability, and security (including proliferation resistance) of the reactors and the fuel cycle. Due to high operational temperature, GEN-IV reactors will significantly expand the range of technological options by producing not only electricity and heat (as conventional reactors) but also carbon-neutral fuels such as hydrogen and synthetic fuels; this will improve the sustainability of the nuclear source to meet the needs of present and future generations.

The US DOE has led an international consortium of ten countries in the Generation IV International Forum, which has identified six advanced reactor designs to be given a priority for further development, as shown below (the operational temperature range of the reactors is shown in brackets) [49–52]:

Very high temperature reactor	(900–1,000 °C)
Supercritical water reactor	(400–600 °C)
Fast gas-cooled reactor	(850 °C)
Heavy metal (Pb) cooled reactor	(540–650 °C)
Fast sodium-cooled reactor	(550 °C)
Molten salt-cooled reactor	(700–850 °C)

The US DOE projects that GEN-IV nuclear reactors will be deployed beyond the year 2025 timeframe [51].

6.1.4.1 Fast-Neutron or Breeder Reactor

Fast-neutron or breeder reactor technology can potentially alleviate the uranium resource constraint problem and significantly extend the uranium reserves (by a factor of about 30). "Fast" reactors offer the significant margin of safety and would allow meeting most of electricity needs with virtually no CO_2 emissions [15]. A number of experimental breeder reactors have been built and tested, e.g., Superphenix in France and Enrico Fermi plant in the USA (both are not currently operational). There are estimates in the literature that it might take two to three decades before large-scale commercial breeder reactors would contribute to the world's energy needs [12].

6.1.4.2 Thorium (Th)

Thorium was recently "reintroduced" as fuel having a tremendous potential to provide energy without causing environmental and proliferation-related problems [53, 54]. The proponents of the thorium-based nuclear energy point to the following advantages of the technology: (a) Th is more abundant than U (the estimated resources of Th—4.5 million ton), (b) Th is potentially less expensive to process than uranium, (c) Th-fueled reactors are not conducive to making and collecting materials that can be used to make nuclear bombs, and (d) radiation toxicity of waste products from the Th usage persists for just tens of years compared with thousands of years for the uranium waste. However, some experts are taking more cautious approach to thorium as a "miracle" nuclear fuel; among major concerns are that an enormous investment of time and resources would be required before any new type of Th-based nuclear reactor could be licensed for a commercial operation [55]. There are also reports that some claims of relatively "benign" waste products from the thorium cycle seem to be somewhat exaggerated. ^{233}U as well as ^{231}Pa (the fission products contaminating the fuel) have half-lives of 162,000 and

32,500 years, respectively [56], and, besides, ^{233}U fission products' distribution resembles those of ^{235}U and ^{239}Pu, namely it includes ^{90}Sr, ^{131}I, and ^{137}Cs. Meanwhile, India is developing its own Th-fueled nuclear industry to take an advantage of the country's large reserves of Th minerals [53]. Some energy analysts believe that the twenty-second century is much more likely to see thorium cycle power plants than fusion plants [47].

6.1.5 Nuclear Fusion Energy

Nuclear fusion energy that powers the Sun and other stars can potentially provide clean and virtually limitless energy. In many respects, it is a perfect energy source: seawater can provide millions of years of fuel, and fusion reactions are safe because they emit neither radioactive waste nor GHG. But its realization is extremely technically challenging: two hydrogen isotopes (deuterium and tritium) must be held at 200 million degrees Celsius until they react to produce helium. But this does not mean it is impossible; the European experimental facility JET (located in the UK) claims that it has accomplished it (for just a couple of seconds), and the objective now is to extend that duration and improve the technology [57].

In general, there are currently two main approaches to the realization of nuclear fusion: through the magnetic containment of plasma and a laser pulse ignition. A multinational nuclear fusion consortium is building the $12 billion International Thermonuclear Experimental Reactor (ITER) in Cadarache (France) with the objective of demonstrating controlled nuclear fusion by 2030 (the seven countries involved in the project are the USA, EU, Japan, China, Russia, India, and South Korea). The project targets harvesting energy from magnetically contained plasma heated to extremely high temperatures. A supporting project dubbed "The Broader Approach" funded by Japan and six European nations is located in Rokkasho, Japan [58].

In the USA, $3.5 billion National Ignition Facility (NIF) aims at igniting fusion fuel (hydrogen) by explosively compressing it with powerful NIF laser that produces the world's highest energy pulses [59]. Ignition is considered being achieved when the fusion reaction is not only self-sustained but produces more energy than the laser pulse that sparked it. Achieving the ignition at NIF facility, however, proved to be more challenging than expected. When NIF went into full operation in 2009, the project managers confidently predicted achieving the ignition before the end of 2012 [60]. This, unfortunately, did not happen, despite the fact that models predicted that ignition should have been materialized at the facility. The disagreement between NIF experimental data and models apparently reflects an inadequate understanding of the key physics issues. At the end of 2012, NIF asked for at least 3 more years to investigate those key scientific issues and identify what has prevented the giant laser fusion laboratory from achieving ignition (annual operating budget of NIF is $450 million). The work will possibly involve exploring alternatives to the indirect drive approach now used at NIF (in which the laser beams target a gold cylinder containing a fuel capsule), and focus on the direct drive approach (the beams directly hit the capsule), which promises delivering a jolt of much higher energy density. Another possible alternative will deal with using immense electrical pulses to crush fusion fuel magnetically.

Researchers at Osaka University (Japan) also utilize powerful lasers, but attempt a different approach known as a fast ignition fusion [59]. In contrast to the NIF system that uses the same laser to accomplish two necessary functions the compression and ignition of the fusion fuel, the Japanese researchers use two types of lasers each optimized for separate tasks, which would result in significant economic benefits. In the experiments, a hydrogen-filled capsule was compressed to the density 600 times greater than that of a liquid material, and with the second laser the researchers increased temperature to about ten million degrees Kelvin. Although the researchers were able to detect fusion reactions, the nuclear ignition itself has not been achieved. The team now is planning to use more powerful heating laser capable of achieving 60 million degrees Kelvin as part of the project called the Fast Ignition Realization Experiment (FIREX) [59].

But achieving the nuclear ignition is only one part of the story. There will be a need to advance in parallel an engineering agenda into key reactor technologies that will enable commercial fusion power plants to reliably deliver electricity in a highly competitive market. This means technological advancements will be needed in areas such as structural and functional materials, power conversion, and reliability [57]. Due to the host of still persisting technical challenges, nuclear fusion is considered by many as a long-term option: most experts agree that, in all likelihood, it will contribute to overall energy supply by the end of this century.

6.1.5.1 Concluding Remarks

It would be interesting to see the real-world competition of nuclear power with other carbon-neutral technologies in the near future. Unlike the 1970s, which saw a real boost in nuclear power in response to energy crisis, the current global energy field has new dynamic players, e.g., solar and wind industries and CCS technologies. These competitors offer the advantage of relatively quick installation, compared with about 10 years of construction typical of nuclear power plant installation [61]. The outcome of this competition over the next decade will, most likely, define future global energy interplay.

6.2 Renewable Energy Sources

6.2.1 Renewables: No Longer Immature Technology

According to a widely accepted definition, *renewable energy sources* (or for brevity, "renewables") are primary energy resources that are replenished by nature at a sufficiently high rate to enable humans to use them indefinitely. Currently, the term "renewables" covers a wide variety of energy sources, such as solar, wind, geothermal, hydropower, ocean thermal, tidal, wave energy, and various forms of biomass energy, including liquid biofuels, biogas, landfill gas, sludge, municipal solid waste, energy crops, and agricultural waste. The major advantage of renewable-based energy systems is that they can provide secure energy with predictable future costs, largely unaffected by geopolitics and global energy markets, because sunshine, or wind, or biomass are widely available to virtually any country in the world and they are "free" for everyone to use it.

Due to recent major technological advancements, renewables are no longer considered a very expensive, immature technological option for curbing carbon emissions; they are becoming increasingly cost-competitive with traditional energy sources in many countries around the globe. In its recent Tracking Clean Energy Progress 2013 report, IEA pointed to a boom in renewable energy over the last decade as one of the few bright spots in an assessment of global progress toward low-carbon energy [6]. Despite ongoing economic and policy turbulence in the sector, renewable energy technologies have achieved impressive growth levels, especially, in emerging economies that are stepping up efforts in clean energy and enhancing policy support for the renewable electricity sector.

Globally, renewables continued to grow strongly in the last several years in both OECD and non-OECD countries. From 2000 to 2011, global renewable generation grew by 1,620 TWh (4.1 % annually), of which non-hydropower amounted to 680 TWh (13.6 % growth annually) [6]. In 2011, renewable non-hydropower capacity increased by 77 GW (+19 %). Global investments in new renewable power plants (excluding large hydropower) reached US$240 billion in 2012 [6]. While several governments (e.g., Germany, Italy, Spain) reduced economic incentives for renewable energy technologies as their competitiveness improved, others upgraded economic incentives and policies to boost the deployment of renewables. For example, Japan has introduced a feed-in tariff scheme for a wide range of renewables. China introduced measures to facilitate the grid connection of distributed PV solar plants with the target of 10 GW deployment in 2013, and Korea has introduced renewable energy certificates and tax incentives [6].

Some renewables are becoming increasingly cost-competitive not only with more traditional sources like hydropower, but also with nuclear and fossil-based energy sources in a broad set of markets. Figure 6.2 summarizes available data on the cost of renewable electricity (for comparison, the costs of fossil-based and nuclear electricity are also included in the diagram).

Wind already competes successfully with new fossil fuel power plants in several countries: Brazil, Turkey, and New Zealand [6]. Solar power is competitive in the markets with high peak prices for electricity (e.g., from oil-fired generation). Decentralized solar photovoltaic generation costs tend to be lower than grid-electricity prices in some countries. In 2011, the number of countries with installed renewable energy (including onshore and offshore wind, bioenergy, and solar photovoltaic) capacity above 100 MW grew significantly compared to 2005 [6].

The notion that countries and even entire regions could be completely powered by renewable energy seems like a pipe dream to some people; but to others not only

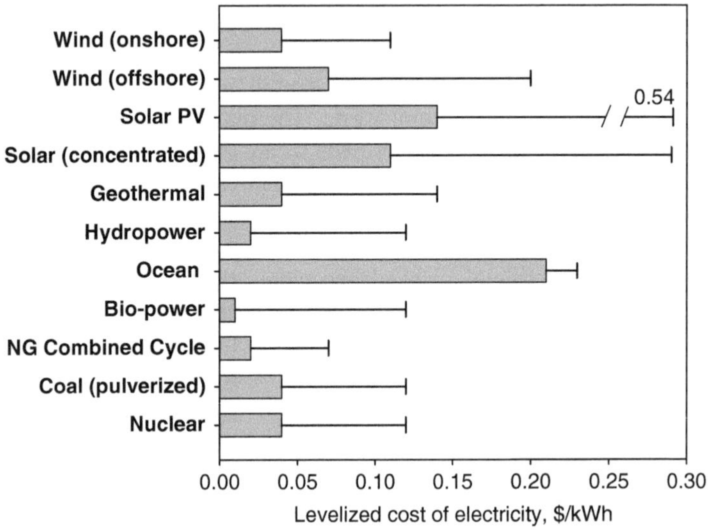

Fig. 6.2 Levelized cost of renewable electricity. *Source* [62]

it is a fully achievable goal but it also makes an economic sense, and they point to several places where 100 % renewable energy threshold has already been reached. One of the most inspiring examples is Rhein-Hunsruck in southern Germany, where it is projected that 100 % of the energy needs of 100,000 inhabitants will be covered by solar, wind, and biomass energy by the end of 2013; by 2014, this rural community will be providing 236 % of its own energy needs from renewable sources and will generate substantial revenue by selling excess carbon-free electricity on the open market [63]. Besides shifting to renewables, another main component of the Rhein-Hunsruck project success includes aggressive energy efficiency programs, which allowed reducing overall electricity consumption by a quarter. In May 2012, Germany's solar industry set a world record producing 22 GW of electricity, which met a third of electricity needs on a work day and almost half of the demand on a weekend (it is important to note that this electric power output is equivalent of about 20 nuclear power plants) [64]. Currently, Germany gets about 20 % of its overall annual electricity from renewables: solar, wind, and biomass. Due to switching to renewables combined with efficiency increase measures, Germany aims to cut its GHG emissions by the year 2020 by 40 % from the 1990 level.

In the USA, a small town of Greensburg (state of Kansas) is 100 % powered by wind. The town also has the highest per capita LEED platinum-certified green buildings in country (LEED stands for Leadership in Energy and Environmental Design) [63]. The examples of Greensburg and Rhein-Hunsruck emphasize the possibility of synergy between high energy efficiency and renewable energy that in combination can deliver the 100 % carbon-free energy to people. Denmark is striving to reach the 100 % renewable energy target for electricity, heat, and transportation. Development of modular microgrids would boost the penetration of renewables and meeting carbon reduction goals [63]. The specific portfolios of different combinations of wind, water, and sunlight resources that could power entire states such as California and New York have been mapped by Jacobson and Delucci [65].

Even though the role of renewables is projected to increase across all sectors, the development and progress in the renewables area are becoming more complex and face many challenges—especially in the policy arena. Many renewables no longer need high economic incentives, but they still need long-term policies that provide a predictable and reliable market and a regulatory framework. Policy uncertainties could be a strong deterring factor for investors. Worldwide subsidies for fossil fuels remain six times higher than the economic incentives for the renewables. There are also experts who argue that although in some areas solar and wind power is cheaper than fossil fuels, the cost of shifting to 100 % renewable energy is too high and the shift itself may not be necessary [66]. The competition from NG—abundant and relatively clean fossil fuel—could be another potential challenge that may negatively impact the march of renewables by putting pressure on some governments to alter their renewable energy policies in favor of gas [67].

Because of a large diversity of the renewable resources, they are typically classified according to either intermittent nature, or carbon content, e.g.,

- Non-intermittent sources: biomass, hydro, geothermal, and ocean thermal energy.
- Intermittent sources: solar, wind, tide, wave energy.
- Non-carbogenic sources: solar, wind, geothermal, wave, tide, ocean thermal.
- Carbogenic sources: biomass and biomass-derived fuels and energy sources (biofuels, biogas, landfill gas, bio-methane, MSW).

Since the main tenor of this book is concerned with the role of carbon in energy systems, the following discussion of the current status of renewable energy technologies is arranged according to the carbon content of renewables, i.e., non-carbogenic vs. carbogenic sources.

6.2.2 Non-carbogenic Renewable Sources

6.2.2.1 Solar

According to many reports (e.g., [13]), the global available and technical potential of solar energy greatly exceeds that of other renewable sources. Two major energy technologies taking advantage of the solar resource are solar photovoltaic (PV) and concentrating solar power (CSP). The former utilizes the photonic (or quantum) component of solar spectrum (e.g., ultraviolet and visible light), whereas the latter uses thermal component (or infrared radiation) of solar light.

Solar Photovoltaic

Solar PV technology is expanding very fast in many regions of the world from marginal levels in 2000 to an estimated 32 TWh in 2010, to 65 TWh in 2011 (the global average annual growth of 47 %) [6]. In 2012, the growth in the solar PV capacity remained strong, despite increased turbulence in the upstream manufacturing industry and incentive cuts in some countries (e.g., Germany and Italy). In sunny locations with moderate or high electricity costs, PV units now provide electricity at prices at or below parity with grid electricity. This "grid parity" is becoming more geographically widespread, accelerating the deployment of PV throughout the world. Until recently, most solar PV growth was concentrated in countries with strong policy support (e.g., USA, Germany, Italy); however, improving competitiveness of the PV technology facilitates its spreading into new markets in south Asia, Latin America, Middle East, and Africa.

Electricity generation by solar PV has grown into $100 billion per year global industry [68]. This growth is fueled by the availability of increasingly efficient and durable PV modules at rapidly falling prices. Due to economic advantages, currently, most PV systems are installed directly in buildings rather than at centralized PV power plants, since in the latter case PV competes against grid-electricity prices; this trend is likely to continue in future. Commercially available amorphous and polycrystalline silicon PV devices (the predominant type of PV systems on the market) nowadays have efficiencies in the range of 10–18 %. Some state-of-the-art PV cells including single-crystal silicon, GaAs, and $CuIn_{1-x}Ga_xSe_2$ (CIGS) enjoy solar-to-electric power conversion efficiencies up to 25 % under full sunlight, which approaches the theoretical energy conversion limit of 32 % for single band gap devices (the theoretical limit for multi-gap PV cells under full sunlight is about 65 %) [69, 70].

Currently, solar PV generation contributes about 0.1 % of the US electricity, and it is projected to reach 3 % penetration in California by 2015 [71]. The US DOE's SunShot Program supports research and development efforts to reduce cost and improve sunlight-to-electricity conversion efficiencies of the PV modules. However, despite the impressive technological advances made by solar PV, the high cost of electricity storage will likely limit PV penetration to less than about 5 % of the US primary energy, unless breakthrough technologies would enable to cost-effectively store many hours of electricity generated by PV units [68]. The models of solar resource-rich California's electricity grid show that PV curtailments will begin at penetrations as low as 12 %, while about one-third of PV electricity would be curtailed at penetrations of about 28 % [72]. The marginal economic value of installing PV is projected to fall by half when PV meets more than 20 % of the California utility load [72].

In Germany, in 2012, about 5 % of annual electrical energy generation comes from solar resource, and in a summer day, solar energy can contribute more than 20 % of the required grid power [73]. This level of solar penetration was made possible mainly through a combination of a demand-side management, trading of surplus power at low cost to neighboring jurisdictions, and the dispatch of expensive NG peaking or load-following plants. The higher levels of penetration of solar energy would require either reducing generation from base-load plants (a function that they are not designed to perform efficiently) or dumping of surplus solar electricity. A recent analytical study found that Germany's new program offering up to 660€/kW subsidy for energy storage tied to PV will not lower battery payback periods enough to induce new investments [74].

China is currently the world's largest exporter of solar panels, and it recently quadrupled the goal for solar installations to 21 GW by 2015 [75]. A global glut of cheap Chinese-made mono- and polycrystalline silicon PV cells during 2010–2013 caused prices for traditional crystalline solar modules to crash, and it was especially devastating for the manufacturers of thin-film cells. Due to shrinking profit margins, many companies in thin-film PV business (e.g., Solyndra, Abound Solar, MiaSole, Nanosolar, Solibro, Uni-solar) went bankrupt. (In general, thin-film cells such as CIGS and CdTe are more efficient than crystalline silicone cells, but currently are significantly more expensive than latter per Watt of output.) In 2012, thin film technology accounted for only 11 % of the PV product sales, down from 21 % in 2009 [76]. But some companies, e.g., First Solar (USA), Solar Frontier (Japan) are still in the business and optimistic about the future. First Solar makes CdTe-based thin film modules and as of mid-2013 still remains profitable, although its quarter-to-quarter results hinge on uneven revenues from its solar project.

Along with explosive growth of massive solar plants in some of the world's sun-drenched desert areas, there are new technological developments involving floating offshore solar panel arrays. Researchers at the Norwegian foundation Det Norske Veritas (DNV) are developing the concept of a dynamic floating offshore solar field especially well suited for powering congested urban regions such as coastal megacities. The so-called SUNdy project involves a floating hexagonal array of thin-film flexible 560 W PV panels that are grouped together to generate 2 MW of power, and if multiple islands are further combined together, up to 50 MW could be generated by the solar field [77]. Due to its unique dynamic compliant design, the structure is capable of withstanding considerable external mechanical loads. 30 kV electrical transmission lines connect separate islands of the solar farm to form a closed loop and continue to the electrical substation onshore for the grid connection. The floating solar fields are being developed by other companies, such as Solaris Synergy (Israel) and Sky Earth (France). The drawbacks of the floating solar plants include cumbersome maintenance and repair, high cost, and some ecological concerns.

In the long-term future, space-based solar power systems may play a significant role in the overall energy supply; the solar flux in space is eight times greater than that on the Earth [11]. It is proposed that solar electricity will be transmitted to the Earth by microwave energy with the efficiency of 50–60 %. A PV array the size of Manhattan on a geostationary orbit (800 km) would transmit power to a surface rectenna (area of 10×13 km^2) with 5 GW power output [11].

Concentrating Solar Power

CSP systems take advantage of sunlight beams concentrated by different types of mirrors on an absorber of solar radiation which is heated to high temperatures (depending on the concentration ratio, from 200 to 2,000–2,500 °C) to power an electricity-generating turbine. CSP is a proven commercial technology for electricity and heat generation. Available and technical potential energy for CSP systems are estimated at 9,250–11,800 PWh/year and 1.05–7.8 PWh/year, respectively [13] (PWh is peta-Watt hour or 10^{15} Wh) . These values are less than the estimates for solar-PV systems, which can be attributed to the fact that the land area required to produce unit electricity is about one-third greater for CSP compared to PV.

CSP systems did not recently experience the same explosive growth as solar PV. During 2000–2011 period, the total growth in CSP was just over 3 TWh (about 20 % annually) reaching an estimated 2 TWh in 2010 and 4 TWh in 2011 [6]. The main challenge to the widespread CSP technology deployment comes from the competition with lower-cost solar PV (because of the economic factor, some projects in the USA have been converted from CSP into solar PV). However, to its advantage, CSP is well suited for the integration with gas-fired power plants (see Chap. 8) and other high-temperature industrial processes, which can potentially boost its market penetration.

Until recently, commercial CSP units have been concentrated in a few sunny areas where they can better compete with conventional technologies, mostly, in Spain and the USA (California, Arizona). Currently, numerous CSP projects are being developed in the Middle East, North Africa, Australia, India, China, and South Africa. For example, in Morocco, the first phase of the Ouarzazate project (160 MW of the 500 MW) secured financing in 2012, and it is expected to start operation by 2015 [6]. In the USA, CSP systems including solar heating and solar hot water applications together contribute less than 0.1 % of the US primary energy, and their deployment is growing slowly [68].

Today, CSP technology is more expensive than solar PV: the levelized cost of electricity (LCOE) from CSP is (in average) roughly twice that of PV electricity. However, advantageously, CSP systems have a low incremental cost of heat storage that enables the dispatch of electricity when it is most needed [68]. The solar heat collected is typically stored as sensible heat in molten salts, from where the stored thermal energy could be extracted when needed with almost no loss. Today's cost of the thermal energy storage is approximately \$30/kWh$_{th}$ [78], which is equivalent to about \$75/kWh$_{el}$ (assuming an average CSP steam Rankine turbine efficiency of 40 %), and it is already lower than the aggressive \$100/kWh$_{el}$ research goals for short-term electrical storage [79] (kWh$_{th}$ and kWh$_{el}$ correspond to kWh-thermal and kWh-electric, respectively).

6.2.2.2 Wind

About 2 % of incoming solar energy is converted into wind energy through atmospheric circulations. The globally available and technically achievable potential of wind energy are estimated at 630–700 PWh/year and 410 PWh/year, respectively (the theoretically available wind power is about 72 TW) [13]. Currently, wind is the most commercially significant source among non-hydro renewable energy sources. Top five commercial wind turbine manufacturers are: Vestas (Denmark), GE (the USA), Sinovel (China), Enercon (Germany), and Goldwind (China).

Onshore Wind

Recently, onshore wind power has been experiencing very high growth rates: from 2000 to 2011, onshore wind generation increased by 400 TWh (27 % annually), reaching estimated 335 TWh in 2010 and 435 TWh in 2011 [6]. In 2010, China overtook the USA as the country with the largest installed wind power capacity (42.3 GW vs. 40.2 GW, respectively) [80]. The cost of kilowatt-hour of wind-generated energy has dropped from $0.40/kWh in the early 1980s to less than $0.05/kWh, and, in some locations, to $0.03 per kWh [81]. In Denmark and some of the northern regions of Germany, wind is now providing 14–19 % of the total electricity output [13, 81]. Spain has installed 15,000 MW of wind power which provides about 28 % of its electricity needs [82]. In Brazil and Turkey, wind projects are successfully competing against fossil fuels in wholesale electricity markets without economic incentives [6]. In the USA, wind farms annually produce 100 TWh of electricity, or about 2.5 % of total demand [83].

The ongoing trends show, however, that, globally, wind power growth rates started slowing down, mainly due to grid integration challenges (e.g., in China), and uncertainties over key policy incentives in some countries (e.g., USA and India) [6]. Besides, wind energy faces several scientific and technological challenges to overcome. One of the major issues is that the wind patterns are not well understood, and the reason for that is that wind behaves differently at the height of 60–120 m (where most turbines operate) compared to near ground (about 10 m, where it is typically measured). Because of the lack of data, energy projections at wind farm sites, especially those built over an uneven terrain, can be highly inaccurate (errors up to 20 %). Furthermore, complex airflows over hills and mountains could create wind shear and turbulence that produce a significant stress on the turbine's blades, gearboxes, and bearings. Since wind turbines are designed and manufactured without a complete understanding of such complex wind airflow patterns, their failure rate is higher than expected [83]. Of particular concern are sudden wind gusts, because they can damage the turbines and could cause sudden transmission line overloads. More research work needs to be done to be able to predict average wind speed with greater accuracy and better understand wind flow patterns in order to reap the full benefits of wind power.

Offshore Wind

Compared to onshore wind power, the contribution of offshore wind generation is now relatively small. Although its growth has accelerated in the past few years, offshore wind has reached the capacity of only12 TWh in 2011 [6]. The technology is still emerging and requires further development to reduce the cost. The world's largest offshore wind farm has opened off the British coast with 100 wind turbines capable of supplying enough electricity (300 MW_{el}) for 200,000 homes a year [84]. Denmark also has substantial offshore wind capacities.

Despite continued progress in reducing the cost of wind power, the intermittent nature of the wind resource (both onshore and offshore) remains a serious drawback. There is a need for an expanded transmission capacity that could deliver wind power from remote rural and offshore sites to major load centers and for the deployment of highly efficient technologies to store wind energy in large quantities. The explosive growth of wind power may also face some material shortage challenges, in particular, rare-earth metals—neodymium (Nd) and dysprosium (Dy) used in permanent magnets (e.g., as $Nd_2Fe_{14}B$ alloy).

6.2.2.3 Geothermal Energy

The available resources of geothermal energy are immense (1,390 PWh/year): they are second only to solar energy [85]. However, most of this energy lies deep under the Earth's crust, making it very difficult and costly to extract; because of this, the estimated technical potential of the geothermal source is only 0.57–1.21 PWh/year [13]. Based on the capacity estimates of currently known geothermal basins, the resources of geothermal energy are equal to roughly 80 times the world's oil resources. The geothermal resources in the USA alone are estimated to exceed 70 million quads (quad is a unit of energy equal to 1.055×10^{18} J; for reference, one quad is an equivalent to the amount of power annually produced by 34 nuclear plants) [81].

From 2000 to 2011, geothermal generation grew by over 19 TWh (2.9 % annually) reaching 70 TWh [6]. Currently, the geothermal resource represents the significant portion of electricity production in several countries: Iceland—27 %, El Salvador—26 %, Kenya—19 %, and Philippines—15 % (Japanese government has approved development of 12 GW of geothermal power) [6]. In the USA, approximately 2,300 MW of geothermal capacity is installed, most of which is in California. The world's largest geothermal plant located in The Geysers (California) has a total installed capacity of 1,224 MW_{el}, and it generates 2.2 % of the state's electricity needs [86].

The geothermal systems showed a significant cost reduction during last two decades approaching that of the wind resource (see Fig. 6.2). The cost of geothermal energy has come down from US$0.15–0.16 per kWh in 1985 to US$0.04–0.06 per kWh now thanks to a gained experience, improved drilling technology, and the economy of scale; it is projected to further drop to less than US$0.04 per kWh by 2020 [65].

Hot Dry Rock Technology

"Hot dry rock" technology makes use of another form of the geothermal energy resource: it utilizes the increase in the rock temperature with the increase in depth (The Earth's continental crust temperature increases 30–35 °C per kilometer depth on average, and significantly more in geothermally active regions.) [87]. At the depth of about 8 km, the rock temperature exceeds that of the boiling point of water. Due to this feature, the hot rock technology has a capacity to produce base-load electricity at any flat location on the Earth. For example, the hot dry rock resource could be tapped to provide as much as 200,000 EJ of electrical energy in the USA—about 2,000 times the total US energy consumption [88, 89]. However, no commercial system is in practical use anywhere in the world, and the estimated cost of electricity produced by this method is several times higher than that of conventional geothermal systems [12].

6.2.2.4 Hydroelectric Power

Although both available and technical potential energies for hydroelectric source are rather low (16.5 PWh/year) compared to solar, wind, and geothermal sources, today, it is the greatest source of renewable electricity in many countries. It grew from 2,700 TWh in 2000 to an estimated 3,640 TWh in 2011 (3 % annually). Hydroelectric power plants today produce a significant fraction (about 10 %) of the electrical power generated in the USA (most of it is concentrated in the Pacific Northwest). China, Canada, Brazil, Russia, Norway, Venezuela, and Egypt also have very large hydropower capacities. Despite its present-day importance as a source of carbon-free electricity, hydropower will unlikely see a significant growth in its future share of the renewable energy market.

6.2.2.5 Ocean Thermal Energy

Ocean thermal energy conversion (OTEC) systems take advantage of the difference between the water temperatures at the surface and the depth of warm tropical regions of the ocean (for example, on average, there is about 20 °C temperature difference between surface and 1,000 m depth). Although the overall heat-to-work energy efficiency of the OTEC system is relatively low (due to the small temperature gradient, it is only 2–3 %), the immense mass of warm water would potentially allow the OTEC plant to economically produce electricity at suitable locations [90]. At a commercial level, OTEC power generation remains small, at less than 1 TWh in 2011 [6]. The largest commercial ocean power project started operation in South Korea in 2011. There are large ocean power generation plants in Canada and France.

6.2.2.6 Wave and Tidal Energy

Wave and tidal energy sources are examples of underdeveloped renewable resources. Their available potentials reach about 23.6 and 7 PWh/year, and the technical potentials amount to 4.4 and 0.18 PWh/year, respectively [13]. For the tidal source to be economical, it has to have the tide height of at least 3 m at a suitable collection area, where turbines can produce electricity from the streams flowing in both directions. Wave energy potential is estimated based on the consideration that 2 % of our planet's coastline (800,000 km) has or exceeds wave power density of 30 kW/m, which translates to about 480 GW (or 4.2 PWh/year) of technically feasible power output [12]. It is projected that in 2020, the cost of wave energy would drop to US$0.04 per kWh [65].

6.2.3 Carbogenic Renewable Energy Sources

6.2.3.1 Biomass and Bioenergy

Biomass is a product of the photosynthesis reaction where solar energy drives the reaction of CO_2 with water to produce starch and other high-energy content compounds (see reaction 1.1). Due to the fundamental role and abundance of the photosynthesis reaction on our planet, biomass is available on a renewable (or recurring) basis. As a source of energy, biomass can be produced in natural (e.g., trees, grasses, algae, crops, plants, aquatic plant, agricultural products, forestry wastes) or industrial settings (e.g., microalgae production from power plant off-gases, biogas production from animal wastes and agricultural residues, municipal solid waste). Biomass is considered *carbon-neutral* or *zero net* CO_2 energy source: although CO_2 is released during the energy use of biomass, the equivalent amount of CO_2 is captured from the atmosphere during its growth, in a so-called "closed carbon loop."

As carbon mitigation technology, biomass growth is limited by the relatively low solar-to-chemical energy conversion efficiency,[1] typically 0.4–1 % for agricultural biomass, which is an equivalent of about 1 W/m^2. Some types of aquatic biomass (e.g., specially designed microalgae and aquatic plants) can reach higher energy conversion efficiencies of 2 % (the theoretical maximum efficiency is about 11 %) [10]. Due to the low solar-to-biomass energy conversion efficiency, the area covered by biomass for CO_2 utilization applications (based on existing technology) would be immense (e.g., the estimated area covered by algae absorbing CO_2 from 100 MW coal power plant would be about 50 km^2 [91]).

Biomass can be converted into different forms of energy: heat, electricity, and fuels. Production of energy from biomass (*bioenergy*) has gained a significant interest worldwide recently, which can be attributed to three factors, namely, bioenergy is:

- A domestic resource (which would potentially alleviate the dependence on imported energy resources, particularly oil).
- A carbon-neutral resource (it does not add CO_2 to the atmosphere).
- Compatible with the fossil fuel infrastructure (e.g., biomass can be co-processed with coal, and biofuels can be delivered, distributed and used in existing engines with minimal changes).

From 2000 to 2011, electricity generation from solid biomass (plants, grasses, etc.), biogas, MSW, and liquid biofuels grew by over 170 TWh (in average, 8 % annually) reaching 280 TWh in 2010 and 310 TWh in 2011 [6].

[1] Solar-to-chemical energy conversion efficiency is defined as a ratio of chemical energy of a product (biomass) to energy of incident solar irradiation.

As an energy source, biomass is unevenly distributed across the globe: some countries have abundant resources (e.g., Russia, Canada, USA, North Europe), and some not (e.g., in North Africa, Middle East). On the other hand, some bioenergy feedstocks, e.g., wood pellets, are internationally traded, and, besides, MSW can contribute to renewable energy production anywhere in the world [6]. The major players in bioenergy utilization are: USA (the largest current capacity), Nordic countries (e.g., Norway, Denmark, Sweden produce both electricity and heat in cogeneration plants for district heating systems), and UK, which is widely practicing co-firing biomass with coal. In Sweden, for example, seven combined heat and power plants using pulp mills with the average output of 130 MW_{el} (equiv.) are under operation [91]. In the USA, the sustainable annual biomass potential for bioenergy production is estimated at about 1.366 billion of dry tons, of which 0.998 billion tons comes from agriculture (crop residues, perennial crops, grains-to-biofuels, process residues) and 0.368 billion tons from forest residues (manufacturing residue, logging debris, fuel-wood, urban wood waste) (Note that except for grains-to-biofuels, almost all of the biomass relates to cellulosic biomass) [92]. Large bioenergy developments are also underway in China, Brazil, and Japan.

Utilization of biomass for heat and power production has been practiced in many countries for decades. These bioenergy plants are relatively small (the typical capacity of about 30 MW), and they generate about 0.2–0.3 Mt CO_2 per year [91]. There are more than 200 bioenergy plants operating in North America and Brazil emitting about 73 Mt CO_2 per year (CO_2 concentration in the off-gases of bioenergy plants varies in the range of 3–8 vol.%) [93]. In general, the local availability of biomass feed (mainly, crop and forestry residues) determines the size of the plants. Future projections point to the increasing role of large bioenergy plants utilizing dedicated (possibly, genetically engineered) energy crops. Reported analytical studies indicate that the plant capacities of several hundred megawatts are feasible for bioenergy plants using dedicated energy crops, and, in many cases, the economy of scale would outweigh the additional cost of biomass transportation [94].

There is, however, some skepticism of the future role of biomass as a major energy source, mostly, due to the fact that photosynthesis (fundamentally) has a very low power density (less than 1 W/m^2) for biomass to significantly contribute to the world energy market. According to a recent study, production of 10 TW power (the amount of power that would be needed to cover the most essential energy needs of humankind) from biomass would require more than 10 % of the Earth's land surface, which is comparable to overall agricultural area [11]. Besides, biomass production requires significant amount of water (about 1,000–3,000 t of water per ton of biomass) and nutrients [10]. These factors may potentially result in resource constraints and a significant environmental impact that need to be carefully considered.

6.2.3.2 Liquid Biofuels

Broadly, the term "biofuel" covers a wide range of gaseous (e.g., biogas, landfill gas, sludge gas, biohydrogen, biomethane) and liquid (bioethanol, biodiesel, "green" gasoline, biomass-to-liquids, etc.) fuels produced from biomass via a

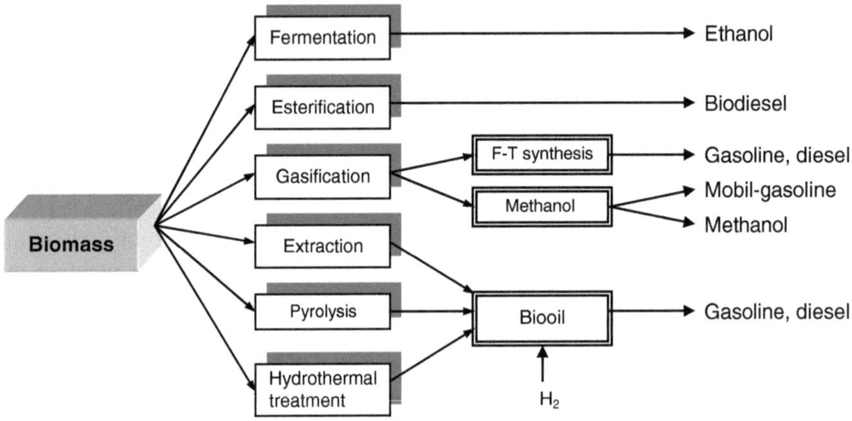

Fig. 6.3 Major technological pathways to production of liquid biofuels for transportation applications

variety of fermentative, photobiological, and thermochemical processes. Liquid biofuels, in general, are compatible with the existing fuel infrastructure and vehicle technology and can be conveniently blended with conventional transportation fuels, thus, making a considerable contribution to reducing the dependence on imported fuels and curbing GHG emissions.

The biofuel sector grew the fastest in the last decade. Driven by government policy support and subsidies in more than 50 countries, global production of biofuels grew from 16 billion Lge in 2000 to more than 100 billion Lge in 2011 and is projected to reach almost 250 billion Lge in 2020 (Lge is liter of gasoline equivalent) [2, 6]. Worldwide, biofuels accounted for about 3 % of road transport fuels with significant variations in different countries, e.g., as high as 21 % in Brazil, and an increasing share in the USA (4 %) and the EU (about 3 %) [2].

Main technological (biological and thermochemical) routes to producing liquid biofuels are summarized in Fig. 6.3.

Conventional Biofuels

Conventional biofuels, also known as *First-Generation biofuels*, are produced from food crops, e.g., cereal crops (wheat, maize), sugar crops (sugar beet, cane), and oil crops (vegetable oils). Crops such as wheat and sugar are the most widely used as a feedstock for bioethanol production, whereas oil seed rape has been found to be very effective in the production of biodiesel fuel. Other types of conventional biofuels include biobutanol, biomethanol, "green diesel" (or hydrotreated vegetable oil), ethyl tertiary butyl ether, and straight vegetable oils. Conventional biofuels are produced via well-established biological (e.g., fermentation) and chemical (trans-esterification, hydrolysis, oil hydrocracking) processes. Currently, bioethanol and biodiesel are two main commercially produced biofuels.

Industrial bioethanol production process involves fermentation of sugar/starch-based feedstocks (e.g., sugar cane, beet, corn) into ethanol and CO_2, whereby two-thirds of carbon from the feed ends up in the biofuel and the remainder in the near-pure CO_2 by-product. The concentrated CO_2 stream is easily separated via gas–liquid separation, whereas ethanol is separated from water via distillation. Bioethanol, as transportation fuel, is typically used in a mixture with gasoline in a wide range of proportions: from 5 to 10 % in the USA and Europe to up to 85 % in Brazil. Recently, the US EPA suggested that it was considering increasing the allowable content of ethanol in gasoline in the US market to 15 % (called E15), which would add about 7 billion additional gallons of ethanol to the domestic ethanol market [95].

Currently, relatively high cost of biofuels is a major barrier to their broader introduction to the market. Without a government support and subsidies, only sugarcane-derived ethanol produced in Brazil is competitive with petroleum-based hydrocarbon fuels [96]. Several countries have either mandated or promoted biofuel blending standards in order to diversify fuel supplies for transportation. For example, in Brazil, gasoline contains 20–25 % of ethanol, and cars purchased after 2008 can run either on 100 % ethanol or on ethanol–gasoline blends [96]. The USA, the EU, Canada, and Australia are also mandating the use of biofuels.

Fermentation plants for bioethanol production are the sources of appreciable amounts of CO_2 emissions: an average plant produces about 0.2 Mt CO_2 per year [93]. Advantageously, bioethanol plants produce off-gas in the form of almost pure CO_2 which makes it easy to capture and store. The scale of future global production of bioethanol and related CO_2 emissions will depend on many factors such as improvements in biomass conversion technologies, the land use factor, water availability, and the competition with other alternative fuels [91]. There are concerns, however, about the potential impact of the high bioethanol production levels to food and livestock prices, as well as water quality. Recently, the US National Research Council reported that health and non-climate-related damages from corn/grain-based ethanol are similar or slightly worth than those from gasoline because of energy required to produce and convert corn into fuel [97].

Advanced Biofuels

Advanced biofuels (also called *Second-Generation biofuels*) have been developed to overcome the limitations and deficiencies of the conventional biofuels in that they are produced from sustainable biomass feedstocks, have higher "net energy gains," and can (in many cases) utilize existing fossil fuel infrastructure, thus, facilitating their penetration to the market place. The sustainability of a biomass resource is defined by many factors, such as availability of the (nonfood) biomass feedstock and water, impact on GHG emissions, land use changes, and impact on biodiversity.

The advanced biofuels are typically produced from nonfood crops, wood, plant residues, organic wastes, food-crop wastes; most of the feedstock sources are lignocelluloses-rich materials. The examples of advanced biofuels include: cellulosic

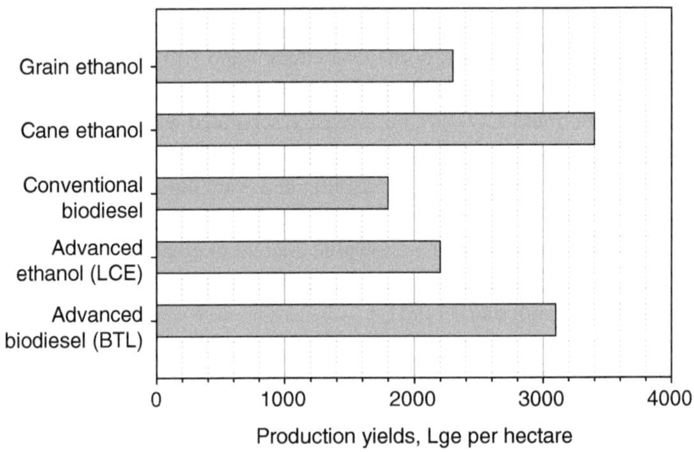

Fig. 6.4 Comparison of production yields of conventional and advanced biofuels in liters of gasoline equivalent (Lge) per hectare. *Source* [2]

ethanol (CE), biomethanol, biodimethyl ether, biocrude, biomass-to-liquid (BTL), and Fischer–Tropsch diesel. (Note that such gaseous fuels as biohydrogen, biosynthetic NG also belong to advanced biofuels category.) The advanced biofuels are produced via biological (fermentation, with and without pretreatment) and chemical (hydrolysis, gasification, aqueous reforming, catalytic synthesis, hydrothermal treatment, etc.) routes. Figure 6.4 compares advanced and conventional biofuels in terms of their production yield in Lge per hectare.

Several types of the advanced biofuels (e.g., CE, BTL) have been commercialized or are close to the commercialization stage. Currently, installed capacity of advanced biofuels is less than 200 million Lge, and another 1.9 billion Lge per year production capacity is under construction [2].

Cellulosic Ethanol. The production of lignocellulosic ethanol requires a pretreatment step where cellulose is separated from lignin. This is followed by chemical or enzymatic hydrolysis step, where complex cellulose chains are converted into simple sugars that can be further fermented to ethanol via conventional fermentation process. Recently, there has been a surge in a number of companies worldwide involved in the commercialization of CE technology. Dutch chemical firm DSM and Poet (a major US-based corn ethanol producer) formed joint venture and launched Project Liberty, a 25 million-gallon-per-year facility in Emmetsburg, Iowa [98]. The plant is projected to start operation in the second half of 2013. DuPont has recently broken ground on its first large-scale facility for CE production from the leaves and stalks of corn. When fully operational, the plant will produce close to 30 million gallons (or about 120 million liters) of ethanol per year [99]. The facility will cost more than $200 million to build, and it is expected to be completed by 2014. Other projects owned by American Process, Ineos Bio, Abengoa, Masco, Fiberight, and ZeaChem are scheduled to start commercial production of CE in 2013 [98]. The US Department of Agriculture

(USDA) will guarantee a $233 million loan to ZeaChem to build one million liters per year CE facility in Boardman, Oregon [100]. The 25 million-gallon-per-year facility will produce both CE and biochemicals from farmed trees and other biomass sources.

Raizen, the world's largest producer of sugarcane ethanol, plans to build CE plant in São Paulo, Brazil, that will use technology developed by Iogen Energy (Canada) [101]. Novozymes (the world's largest enzyme-manufacturing company) and Italian cellulosic biofuels company Beta Renewables are combining their technologies for manufacturing CE [102]. The plants are targeting the production of about 60 million to 160 million liters of ethanol per year, depending on the type of biomass selected. As of the end of 2012, Beta Renewables was commissioning the world's first commercial-scale CE plant in Crescentino, Italy [102].

Biomass-to-Liquid Fuels. Currently, the interest in BTL technology is rapidly increasing due to recent improvements in the technology and the realization that it can provide economical means of converting biomass into liquid fuels with desirable characteristics. The technology is similar to coal-to-liquid process in that it consists of two stages: gasification (where biomass is gasified to syngas) and Fischer–Tropsch (FT) stage, where syngas is catalytically converted into a wide range of liquid hydrocarbons (typically, C5 to C30 and higher) (sometimes, BTL fuels are also called *FT liquids* or *FT hydrocarbons*).

BTL fuels offer important advantages over bioethanol, biodiesel, and even petroleum-based fuels. While during manufacturing of bioethanol and biodiesel only part of the biomass feedstock is utilized (e.g., starch, sugar, oil, cellulose), in the case of BTL fuels the whole plant is utilized (since it is completely gasified to syngas). Furthermore, in contrast to alcohols that contain significant amount of oxygen (which dramatically reduces energy density of the fuel) BTL fuels contain only carbon and hydrogen (e.g., gravimetric energy densities of ethanol and BTL gasoline are 30 MJ/kg and about 45 MJ/kg, respectively). BTL fuels are most similar to conventional petroleum-based fuels by physical and chemical characteristics; thus, no changes in the fuel infrastructure or car engines will be necessary. Furthermore, compared to petroleum-based fuels, BTL are much cleaner since they contain practically no sulfurous or other harmful impurities, and, barrel for barrel, they emit less CO_2 and lower levels of NO_x emissions and particulate matter than petroleum fuels [103]. More detailed information on the technological and economic aspects of BTL fuel production from lingocellulosic biomass can be found in an excellent review by Schaub and Pabst [104].

There are alternative thermochemical routes to the production of liquid hydrocarbon-based biofuels from biomass. Universal Oil Products (UOP, USA) has invested in the development of "green" gasoline, diesel, and jet-range hydrocarbon fuels by integrated thermochemical processing of biomass [105]. The technology is based on the integration of fast pyrolysis of biomass feedstock to bio-oil followed by its catalytic hydrodeoxygenation to liquid fuels. UOP is planning to build a demonstration unit with the capacity of 1 t per day in Hawaii (USA) to convert cellulosic biomass and algae residues into hydrocarbon fuels. If the project proves to be successful, UOP will expand the operation to a commercial unit with the capacity of 190 million liters of liquid fuels per year.

Algal Biofuels. Although algal biofuels also belong to advanced biofuels category, sometimes they are differentiated from the main group and called *Third Generation biofuels*. The fast growth rate of algal feedstock, high oil yields, efficiency in utilizing CO_2, widespread availability, and, most importantly, the use of land and water unsuitable for food production and other purposes make algae and aquatic biomass a promising bioenergy resource. Many believe that algae farms might provide a salvation from the energy starvation [106]. In principle, the algal biofuel production can utilize the biotechnological pathways similar to biodiesel production (extraction, transesterification). Thermochemical routes can also be used for the conversion of algal feedstock into biofuels, e.g., gasification followed by FT synthesis could produce a wide range of fuels such as diesel, gasoline, and jet fuel.

Recently, there has been a significant boost in algae-related R&D and technology validation and demonstration projects. Oil and petrochemical companies after years of sitting on the sidelines, entered the biofuel game. ExxonMobil have announced that it would invest US$600 million to develop algae-derived biofuels with California-based Synthetic Genomics [107]. Another major oil company, Total (France), invested in biofuels start-up company Gevo, which opened a pilot facility in St. Josef, Mo, USA [108]. Besides biofuels, Gevo is also developing chemical coproducts such as isobutanol, polyethylene terephthalate and polymethyl methacrylate. More information on algae-based fuels can be found in Chap. 9.

Carbon-negative Biofuels. Carbon-negative biofuels (also referred to as *Fourth-Generation biofuels*) have all the characteristics of sustainable advanced biofuels (i.e., second- and third-generation biofuels) with an additional feature of capturing CO_2 at all stages of their production, followed by its transport and sequestration using conventional technologies such as Post-CCS and geosequestration. This approach not only captures CO_2 from atmosphere and locks it away, but also reduces CO_2 emissions by replacing fossil fuels that would otherwise produce CO_2 emissions. Carbon-negative systems are discussed in Chap. 10.

6.2.3.3 Concluding Remarks

In the times of increasing oil prices and concerns about environmental damage done by fossil fuels, bioenergy and biofuels are gaining worldwide attention as a possible remedy and, at least, a partial solution to these problems. There are some features that distinguish bioenergy from other carbon-free and carbon-neutral sources that have to be taken into consideration. As an energy resource, biomass growth is limited by a relatively low solar-to-chemical energy conversion efficiency (even the fastest growing biomass species do not exceed energy conversion efficiencies of 2 %), which will translate into very large areas to be covered by biomass. It is recognized that the widespread use of conventional biofuels (bioethanol, biodiesel) is associated with "fuel vs food" dilemma, which has already been blamed for the global increase in food prices over the last decade (3–30 % of the increase in food prices in 2008). Therefore, most of R&D, demonstration and commercialization efforts worldwide are currently focused on advanced (i.e., second- and third-generation) biofuels. There are also some questions raised recently on the role of biomass and biofuels as an effective carbon mitigation option. A recent US National Research Council report on national biofuels policy concluded that the USA will unlikely meet its 2022 target on the amount of cellulosic ethanol and biodiesel that had been mandated by the US Congress in the Renewable Fuel Standard (RFS) [109]. The Council pointed out that the RFS may be an ineffective policy for reducing global GHG emissions because of uncertainties in the environmental impact of biofuel production and the associated land use. These developments add more heat to conflicting assessments of biofuels as genuine carbon-neutral fuels.

6.2.4 Storage of Renewable Energy

A fundamental limitation associated with the efficient utilization of solar and wind energy sources relates to their intermittent nature (i.e., daily and seasonal variations) and a nonuniform distribution over the land. Thus, the practical realization of reliable and affordable energy storage systems is a prerequisite for the efficient use of these intermittent renewable energy sources and their integration into a nationwide energy system. Therefore, the energy storage systems are projected to play an extremely important role and become an integral part of the future renewable-based infrastructure. The existing energy storage options have different performance characteristics with regard to their response time, maximum storage capacity, lifetime, etc. In general, intermittent renewable energy can be stored in three forms: electrical, thermal, and mechanical (or kinetic) energy.

6.2.4.1 Intermittent Electricity Storage

Figure 6.5 summarizes available energy storage technologies by their rated capacity and discharge time.

Attempts to store large amounts of renewable energy and, thus, smoothen their supply have had a limited success so far [110]. While most conventional batteries have rather limited capacity and are applicable to powering vehicles and small-scale end users, some advanced batteries, such as sodium-sulfur battery, can store up to 100–200 MWh electricity, but even that would be sufficient to store less than one hour of the electric output from a mid-size wind farm [111]. As evident from Fig. 6.5, pumped hydro technology has the highest capacity and discharge time. A 130 GW of pumped hydro capacity has been installed worldwide, accounting for over 99 % of the global storage capacity [2]. Most of the pumped hydro units were installed from the 1970s to mid-1990s; however, since the late 1990s with the advent of gas turbines for covering peak-loads the incentives to build more pump hydro have greatly diminished. It is widely believed that pumped hydro will unlikely be the answer to intermitted renewable energy storage for a number of technical, environmental, and economical reasons.

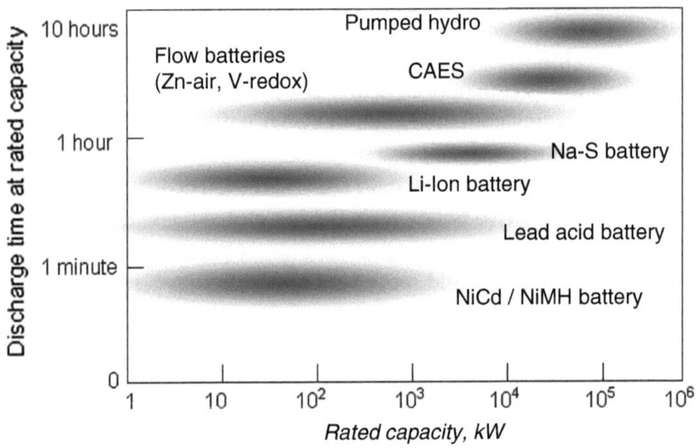

Fig. 6.5 Comparison of existing energy storage technologies by their rated capacity and discharge time. NiCd: nickel–cadmium, NiMH: nickel–metal hydride, Na–S: sodium sulfur, Li-Ion: lithium ion, V- redox: vanadium-redox, Zn-air: zink–air, CAES: compressed air energy storage. *Source* [2]

Compressed air energy storage (CAES) is the second largest storage capacity connected to the electricity distribution systems (400 MW installed capacity worldwide) [2]. The technology has successfully been operated in Huntorf (Germany) and McIntosh (Alabama, USA) for about couple of decades (Note that rather than expanding the compressed air, these systems supplement the stored energy with NG combustion.) For example, CAES facility in McIntosh, Alabama, can produce 110 MW of power for up to 24 h [111]. Both the US and Germany's plants have demonstrated admirable rates for reliability and availability, 90–99 %. A modification of the traditional compressed air storage system: underwater-CAES (UW-CAES) has been recently developed by the University of Windsor researchers in Canada [111]. In contrast to the CAES, UW-CAES utilizes distensible underwater air storage reservoirs. Anchored to the bottom of a seafloor or lakebed, these balloon-like reservoirs would expand or contract in response to the amount of stored compressed air. Since the hydrostatic pressure is constant, UW-CAES releases energy at a constant rate. But the downside is that the reservoirs have to be at suitable depths (at least 80 m deep). After successfully completing the pilot project, the researchers in collaboration with Hydrostor company have begun the development of a 4 MWh demonstration facility near Toronto on Lake Ontario.

Siemens engineers consider hydrogen to be the only viable option to store energy in quantities larger than 10 GWh [112]. In this energy storage option, water is electrolyzed to hydrogen and oxygen at times of low energy demand using electricity from an intermittent renewable energy source (e.g., solar, wind). Hydrogen is stored in a suitable storage reservoir, and at times of a high demand, hydrogen is used to produce electricity. By means of electrolyzers, hydrogen can link a variety of power generation systems, both intermittent and steady (renewables, nuclear, fossil-based) to energy storage via re-electrification, i.e., through combustion in gas turbines. Siemens is addressing the large-scale electrical energy storage problem by developing high-capacity polymer electrolyte membrane (PEM) electrolyzers for splitting water to hydrogen and oxygen. The advantages of PEM electrolyzers are that they are very robust, flexible, and best suited for operating under fluctuating conditions of intermittent power sources (e.g., wind turbines or PV panels) compared to traditional alkaline electrolyzers, which are typically used for the continuous steady-state industrial production of hydrogen. PEM electrolyzers are capable of jumping from a standby to a full load mode in less than 10 s, the feature that is extremely important from the power management viewpoint. Siemens plans to deliver Generation I electrolyzers with the capacity of 1–10 MW around 2015, which will be followed by 100 MW Generation II electrolyzers.

The future market for the hydrogen-based storage technology is immense: converting only 10 % of globally generated wind energy into hydrogen via electrolysis would store immense quantities of electricity. In this scenario, large-scale electrolyzers (100 MW capacity) will be located close to commercial wind farms, with the excess wind electricity to be converted to hydrogen. According to a market research company Frost & Sullivan, only in Europe, the market for large-scale energy storage projects will reach US$2 billion by 2017 [112]. The company emphasized that implementing smart grids and achieving complete automation would be possible only if large-scale energy storage is introduced to the marketplace.

6.2.4.2 Thermal Energy Storage

Thermal energy storage systems are designed for storing high-temperature heat from solar concentrators or other sources. Conventional thermal storage systems involve several types of storage media, predominantly thermal oil and molten salts. The use of thermal oils is limited to 200–250 °C (due to relatively low decomposition temperature of oil at about 300 °C and its inflammability). Molten salts (e.g., $NaNO_2$–$NaNO_3$–KNO_3 mix) have been used as an efficient high-temperature (250–1,000 °C) heat transfer medium since the 1930s (e.g., at refineries). The main problem with the molten salt storage medium relates to its high melting point (about 220 °C), which requires to always keep the salt medium preheated. The first commercial solar thermal plant (50 MW) incorporating molten salt storage operates in Spain.

The role of large-scale energy storage systems is projected to significantly increase mostly due to smart grid applications and may reach storage capacities in the magnitude of hundreds to thousands of MWh. Thermal energy stored at very high temperatures will help improve the thermodynamics of Rankine and Brayton cycle applications. More detailed information of state-of-the-art energy storage options can be found in [113].

6.2.5 *Outlook and Challenges for Renewables*

In its 2DS pathway, IEA sees renewables dominating power generation by the mid-century; the scenario assumes an increase in the renewable energy share of global electricity generation from 20 % in 2010 to 28 % by 2020 to 57 % by 2050 [6]. IEA projects that in 2020, 7,500 TWh of renewable electricity will be generated (of total generation of 27,165 TWh) with hydropower being the largest contributor (17 % of total electricity generation), followed by wind (6 %), biomass and waste (3 %), and solar (2 %). The largest proportion of global renewable electricity in 2020 will come from China (24 %), followed by OECD-Europe (19 %), the USA (11 %), Brazil (7 %), and India (5 %) [6].

The positive outlook for renewable power generation can be attributed to two main factors [6, 114]:

- Investments and widespread deployment of renewables are accelerating in emerging economies, where they help address fast-rising electricity demand, provide energy diversification, and alleviate local air pollution concerns, at the same time, contributing to climate change mitigation. Non-OECD countries are projected to account for two-thirds of the global renewable power generation increase between now and 2018. Such rapid progress is expected to compensate for slower growth and smooth out volatility in OECD countries, notably the USA and Europe.
- Renewables are becoming increasingly cost-competitive not only in such traditional areas as hydropower but also in new markets (e.g., wind, solar) where they already competes with new fossil fuel power plants in several countries with favorable conditions.

In its *Medium-Term Renewable Energy Market Report* (*MTRMR*) 2013 report, IEA sees renewable power increasingly cost-competitive with fossil fuel generation and, especially, nuclear power in the near future [114]. The rapid expansion of renewables, however, will most likely heavily rely on subsidies. In 2011, these subsidies globally amounted to $88 billion per year. However, to meet the IEA's target, worldwide support for renewables over the projected period needs to increase to $200 billion annually reaching $4.8 trillion by 2035 [115].

According to the IEA's projections, the gains for biofuels in transportation and for renewable sources for heat supply will be at somewhat slower growth rates compared to renewable electricity. For example, biofuels output, adjusted for energy content, is expected to account for about 4 % of global oil demand for road transport in 2018, only one-percentage growth from 3 % in 2012. The share of renewable sources in final energy consumption for heat (excluding traditional biomass) will rise from about 8 % in 2011 to 10 % in 2018. Thus, the potential of renewable heat remains largely unexploited [6, 114].

In 2012 report "Renewable Electricity Futures Study," a panel of the US experts analyzed the implications and challenges of the generation of 80 % of all the US electricity from renewable technologies by 2050 [116]. The study concluded that within the limits of the scenarios assessed, the estimated US electricity demand in 2050 could be met with 80 % of generation from renewable electricity together with a mix of flexible conventional generation and grid storage, the additions of transmission lines, more responsive loads, and changes in power system operations. Although many studies emphasize that the widespread penetration of the intermittent renewable sources such as solar and wind will unlikely occur without fossil fuel backup (e.g., gas turbines), some authors point to the possibility of providing nearly all (99.8 %) of electricity through a combination and efficient management of renewable sources in places like California [117].

The reductions in CO_2 emissions due to the expansion in renewable energy sources are expected at 17 % of total (or by 7.3 Gt CO_2) by 2050 [2]. To achieve this target, the renewables should provide 40 % of the primary energy supply and 48 % of power generation by 2050, which may require large investments in power transmission networks and technological advances, especially in large-scale storage of intermittent wind/solar electricity, further cost reductions in PV electricity, development of intercontinental super-grids, etc. There are estimates in the literature that the cost of replacing 70 % of fossil fuels is about $170–200 billion per year over the next 30 years; however, a carbon tax of $45–50 per ton of CO_2 would pay for the investment and provide incentives for implementing renewable technologies [118].

Despite the impressive growth in the recent years and "sunny" long-term outlook, renewables may face some challenges in the years ahead. The competition from NG, as a relatively clean and cheap fossil fuel, is a serious potential challenge, and it could potentially negatively impact the pace of penetration of renewables to the global marketplace. Fatih Birol, chief economist of the IEA, underscored this problem as follows: "If gas prices come down, that would put a lot of pressure on governments to review their existing renewable energy support policies ... We may see many renewable energy projects put on the shelf." [67]. Some advanced renewable technologies, such as onshore wind, would continue to prosper but offshore wind and solar energy could be the worst affected technologies. Birol warned that if the world fails to invest in renewables, a new generation of gas-fired power stations would have a lifetime of at least 25 years, effectively "locking in" billion tons of carbon emissions a year.

Although public opinion, in general, is in favor of renewables, it would be a mistake to discount the voices of concerned citizens. There are many examples of the Not-In-My-Back-Yard (NIMBY) factor intervening and sometimes killing many important initiatives with significant potential benefits to the society based on the sentiments like: "we strongly support clean alternative energy sources, but, please, put those wind turbines somewhere else." For example, citizens in Germany and other countries initially enthusiastic about wind energy, are now complaining that their energy is more expensive, their homes are less tranquil (due to proximity of wind-turbine motors), and the environment is suffering due to deforestation and wildlife deaths [119]. More public outreach activities emphasizing the benefits of renewable energy sources and pointing to what could be possible alternatives to them might help in these situations.

References

1. Schultz K, Gorensek M, Herring J et al (2012) Nuclear power for production of carbon-free energy and fuels. In: Muradov N, Veziroglu V (eds) Carbon-neutral fuels and energy carriers. CRC, Boca Raton, pp 203–289
2. International Energy Agency (2012) Energy technology perspectives. Pathways to a clean energy system. IEA/OECD, Paris, France
3. Percopo B (2008) US nuclear power's time has come-again. Power 1 Jan 2008. http://www.powermag.com/u-s-nuclear-powers-time-has-comeagain/. Accessed 8 Mar 2009
4. Murray R (2009) Nuclear energy: an introduction to the concepts, systems and applications of nuclear processes. Butterworth-Heinemann, Oxford, ISBN: 978-0-12-370547-1, p 513
5. Makhijani A (2007) Carbon-free and nuclear-free. A Roadmap for US energy policy. IEER, MD, USA
6. International Energy Agency (2013) Tracking clean energy progress. IEA input to the Clean Energy Ministerial, IEA, Paris. www.iea.org/publications/TCEP_web.pdf. Accessed 10 Sept 2013
7. Jacoby M (2009) Coming back to nuclear energy. Chem Eng News 87:14–18
8. Johnson J (2012) Snap-together nuclear power. Chem Eng News 90:42
9. Nuclear Energy Agency (2012) The role of nuclear energy in a low-carbon energy future. OECD/NEA, Paris. www.oecd-nea.org/nsd/reports/2012/nea6887-role-nuclear-low-carbon.pdf. Accessed 5 Dec 2013
10. National Research Council and National Academy of Engineering (2004) The hydrogen economy: opportunities, costs, barriers and R&D needs. National Academies Press, Washington, DC
11. Hoffert M, Caldeira K, Benford G et al (2002) Advanced technology paths to global climate stability: energy for the greenhouse planet. Science 298:981–987
12. Moriarty P, Honnery D (2007) Intermittent renewable energy: the only future source of hydrogen? Int J Hydrogen Energ 32:1616–1624
13. Jacobson M (2009) Review of solutions to global warming, air pollution and energy security. Energ Environ Sci 2:148–173
14. Cowan G (2010) Letter to editor. Chem Eng News 88:7
15. Sobolev I (2011) Pursuing nuclear power. Chem Eng News 89:2
16. Lenzen M (2008) Life cycle energy and greenhouse gas emissions of nuclear energy. Energ Convers Manag 49:2178–2199
17. Lawton E (2011) Saying no to nuclear. Chem Eng News 89:4
18. Hvistendahl M (2007) Coal ash is more radioactive than nuclear waste. Sci Am. http://www.scientificamerican.com/article.cfm?id=coal-ash-is-more-radioactive-than-nuclear-waste. Accessed 12 Jul 2012
19. Science Daily (2013) Chernobyl follow-up study finds high survival rate among young thyroid cancer patients. http://www.sciencedaily.com/releases/2013/04/130424132630.htm. Accessed 12 Jul 2013
20. Baum R (2011) Nuclear power. Chem Eng News 89:5
21. Feinendegen L (2005) Evidence for beneficial low-level radiation effects and radiation hormesis. Br J Radiol 78(925):3–7. doi:10.1259/bjr/63353075
22. Allison W (2009) Radiation and reason: the impact of science on a culture of fear. York Publishing Services, York. ISBN 0-9562756-1-3
23. Luckey T (2011) Much more on nuclear power. Chem Eng News 89:6
24. Chen W, Luan Y, Shieh M (2004) Is chronic radiation an effective prophylaxis against cancer? J Am Phys Surg 9(1):6–10
25. Baetz J (2011) Germany decides to abandon nuclear power by 2022, Associated Press. http://news.yahoo.com/s/ap/20110530/ap_on_bi_ge/eu_germany_nuclear_power. Accessed 31 May 2011

26. Namasivayam D, Vandoorne S, Bittermann J et al. (2011) Explosion at French nuclear site kills one, CNN online, http://www.cnn.com/2011/world/europe/09/12/france.nuclear/index.html?hpt=hp_t2. Accessed 12 Sept 2011
27. Glenn H (2011) Regulators deem US reactors safe. Chem Eng News 27:14
28. Hargreaves S (2012) First nuclear reactors OK'd in over 30 years, CNN News. http://money.cnn.com/2012/02/09/news/economy/nuclear_reactors/index.htm?hpt=hp_t3. Accessed 9 Feb 2012
29. Patel S (2013) World energy outlook foresees distinct generation shift. Power 8–12. 1 Jan 2013. http://www.powermag.com/world-energy-outlook-foresees-distinct-generation-shift/. Accessed 5 Feb 2013
30. Smith M, Ahlers M (2011) US calls for stepped-up safety at nuclear plants. CNN. http://www.cnn.com/2011/US/07/13/nuclear.safety/index.html?hpt=hp_t2. Accessed 13 Jul 2011
31. US Nuclear Regulatory Commission (2010) Spent fuel storage in pools and dry casks. Key points and questions and answers. www.nrc.gov/waste/spent-fuel-storage/faqs.html. Accessed 15 Jul 2011
32. Macfarlane A (2011) The overlooked back end of the nuclear fuel cycle. Science 333: 1225–1226
33. Alvarez R, Beyea J, Janberg K et al (2003) Reducing the hazards from stored spent power-reactor fuel in the United States. Sci Glob Secur 11:1–51
34. Japan Atomic Energy Commission (2005) Framework for nuclear energy policy. AEC, Ibaraki, Japan
35. Goldfinger C (2011) Disaster preparedness and the ring of fire. www.pdxcityclub.org/content/risks-and-potential-nuclear-energy. Accessed 10 Oct 2011
36. International Energy Agency (2012) Global carbon-dioxide emissions increase by 1.0Gt in 2011 to record high. http://iea.org/newsroomandevents/news/2012/may/name,27216,en.html. Accessed 20 June 2012
37. Covey S (1989) The 7 habits of highly effective people. Free Press, New York
38. US Department of Energy (2011) Blue Ribbon commission on America's nuclear future. Disposal subcommittee report to the full commission. 1 June 2011. Washington, DC http://brc.gov/sites/default/files/documents/draft_disposal_report_06-01-11.pdf. Accessed 5 Dec 2011
39. World Nuclear association (2010) Storage and disposal options. www.world-nuclear.org/info/inf04ap2.html. Accessed 10 Feb 2012
40. US National Research Council (2003) One step at a time: the staged development of geologic repositories for high level radioactive waste. National Academic Press, Washington, DC
41. US Department of Energy (2011) Letter from S. Chu, Secretary of Energy, to co chairs the Honorable B. Scowcroft and the Honorable L. Hamilton of the Blue Ribbon Commission. Washington, DC. 11 Feb 2011
42. Hess G (2011) Future unclear for Yucca mountain. Chem Eng News 89:24
43. Cho A, Malakoff D (2013) Chu's legacy at DOE: some fields gain, others falter. Science 339:635–636
44. Macfarlane A, Ewing R (2006) Uncertainty underground: Yucca mountain and the nation's high-level nuclear waste. MIT Press, Cambridge, MA
45. Moore E (2011) Much more on nuclear power. Chem Eng News 89:4
46. Norbeck E (2011) Much more on nuclear power. Chem Eng News 89:6
47. Read J (2011) (Much) more on nuclear power. Chem Eng News 89:6–7
48. Grimes R, Nuttall W (2010) Generating the option of a two-stage nuclear renaissance. Science 329:799–803
49. National Research Council and National Academy of Engineering (2004) The hydrogen economy. Opportunities, costs, barriers and R&D needs. National Academies Press, Washington, DC
50. Lee S, Speight J, Loyalka S (2007) Handbook of alternative fuel technologies. CRC, Boca Raton, USA
51. U.S. Department of Energy (2002) A technology road map for Generation IV nuclear reactor systems. U.S. Department of Energy, Washington, DC
52. Evans R (2007) Fueling our future. Cambridge University Press, Cambridge, 180

53. Jacoby M (2009) Reintroducing thorium. Chem Eng News 87:44–46
54. Wetzel J (2010) Letter to editor. Chem Eng News 88:6
55. Swanson J (2010) Letter to editor. Chem Eng News 88:6
56. Cooper E (2011) Thorium cycle hype. Chem Eng News 89:4–5
57. Cowley S (2013) Nuclear fusion is the "perfect energy source". CNN. http://edition.cnn.com/2013/03/12/opinion/fusion-nuclear-energy-future/index.html?hpt=hp_c4. Accessed 10 Apr 2013
58. Normile D (2010) Fusion "Consolation Prize" gears up for show time. Science 328:1464–1465
59. Clery D (2010) Laser fusion energy poised to ignite. Science 328:808–809
60. Clery D (2012) NIF report asks for more time to achieve ignition. Science 338:1519
61. Johnson J (2010) New jolt for nuclear power. Chem Eng News 88:31–33
62. US Department of Energy (2012) The transparent cost database. National Renewable Energy Laboratory. http://en.openei.org/apps/TCDB/. Accessed 8 Aug 2012
63. Asmus P (2013) All renewables, all the time. Forbes. http://www.forbes.com/sites/pikeresearch/2013/05/16/all-renewables-all-the-time/?goback=.gde_77156_member_242495376. Accessed 1 Jul 2013
64. Kirschenbaum E (2012) Germany sets new solar power record, institute says. Reuters. http://www.reuters.com/article/2012/05/26/us-climate-germany-solar-idUSBRE84P0FI20120526. Accessed 12 June 2012
65. Jacobson M, Delucchi M (2009) A path to sustainable energy by 2030. Sci Am. p 58–65. Nov 2009 Issue. http://www.scientificamerican.com/magazine/sa/2009/11-01/. Accessed 10 Jan 2011
66. Moriarty P, Honnery D (2011) Rise and fall of the carbon civilisation. Springer, London, UK
67. Harvey F (2011) Natural gas is no climate change 'panacea', warns IEA. The Guardian. http://www.guardian.co.uk/environment/2011/jun/06/natural-gas-climate-change-no-panacea. Accessed 20 June 2012
68. US Department of Energy (2013) ARPA-E. DE-FOA-0000949: Full spectrum optimized conversion and utilization of sunlight (Focus). https://arpa-e-foa.energy.gov/#Foaldfca200cc-bf46–4789-a8c1-e2be7e41509e. Accessed 20 Sept 2013
69. US Department of Energy (2005) Basic research needs for solar energy utilization, Report of the U.S. DOE basic energy sciences workshop on solar energy utilization, Argonne National Laboratory, Argonne, IL
70. Licht S (2001) Multiple bandgap semiconductor/electrolyte solar energy conversion. J Phys Chem B 105:6281–6294
71. U.S. Energy Information Administration (2013) Annual energy outlook. U.S. Energy Information Administration, Washington, DC
72. Mills A, Wiser R (2012) Changes in the economic value of variable generation at high penetration levels: a pilot case study of California. US Department of Energy, Technical report: LBNL-5445E
73. Burger B (2013) Electricity production from solar and wind in Germany in 2012. Fraunhofer Institute for Solar Energy Systems ISE, Freiburg
74. Bloomberg New Energy Finance (2013) Will Germany's energy storage subsidy spur investment? Bloomberg New Energy Finance, London, p 1
75. Bomgardner M (2012) US solar firms try China. Chem Eng News 90:22
76. Bomgardner M (2013) Odds narrow for thin-film solar. Chem Eng News 91:20–21
77. Patel S (2013) Floating solar on water. Power 8–12. http://www.powermag.com/floating-solaron-water/. Accessed 5 Mar 2013
78. Kolb G, Ho C, Mancini T et al. (2011) Power tower technology roadmap and cost reduction plan. US Department of Energy. Technical report SAND2011–2419
79. US Department of Energy (2013) ARPA-E GRIDS Funding opportunity announcement. https://arpa-e-foa.energy.gov/FileContent.aspx?FileID=98c6222e-471c-4216-a377-024dbdb45549

80. Olivier J, Janssens-Maenhaut G, Peters J et al (2011) Long-term trend in global CO_2 emissions. European Commission's Joint Research Center, PBL Netherlands Environmental Assessment Agency, The Hague
81. Rifkin J (2002) The hydrogen economy. J Ptarcher/Putnam, NY
82. Evans-Pritchard A (2008) Spain's gain from wind power is plain to see. www.telegraph.co.uk/money/main.jhtml?xml=/money/2008/04/07/ccwind107.xml. Accessed 12 Nov 2008
83. Powell E (2012) Earth, wind and water. Discover 11:50–53
84. Cable News Network (2010) Wire staff. World's largest offshore wind farm opens in UK. http://www.cnn.com/2010/world/europe/09/23/uk.largest.wind.farm/index.html?hpt=c2. Accessed 23 Sept 2010
85. Tester J, Drake E, Driscoll M et al (2005) Sustainable energy. MIT Press, Cambridge, MA, 846pp
86. National Commission on Energy Policy (2004) Ending the energy stalemate. Report of the National commission on energy policy, Washington, DC
87. Pollack H, Hurter S, Johnson J (1993) Heat-flow from the Earth's interior: analysis of the global data set. Rev Geophys 31(3):267–280
88. Tester J, Murphy H, Grigsby C et al. (2006) The future of geothermal energy: impact of enhanced geothermal systems (EGS) on the United States in the 21st century, Mass. Inst. of Technology. Report INL/EXT-06–11746, MIT, Cambridge. p 372
89. CO_2 storage with geothermal energy production (2012) Carbon Capture J. http://www.carboncapturejournal.com/displaynews.php?NewsID=973&PHPSESSID=up5ge1lr524iq0ita9tgcok406. Accessed 1 Aug 2012
90. O'Bockris J, Veziroglu N (2007) Estimates of the price of hydrogen as a medium for wind and solar resources. Int J Hydrogen Energ 32:1605–1610
91. Intergovernmental Panel on Climate Change (2005) IPCC special report on carbon dioxide capture and storage. Prepared by Working Group III of the Intergovernmental panel on climate change. In: Metz B, Davidson O, de Coninck H et al. (eds) Cambridge University Press, Cambridge
92. Forsberg C (2009) Meeting US liquid transport fuel needs with a nuclear hydrogen biomass system. Int J Hydrogen Energ 34:4227–4236
93. International Energy Agency (2002) IEA GHG. Building the cost curves for CO_2 storage, part 1: sources of CO_2, PH4/9, p 48
94. Dornburg V, Faaij A (2001) Efficiency and economy of wood-fired biomass energy systems in relation to scale regarding heat and power generation using combustion and gasification technologies. Biomass and Biomass Energy 21:91–108
95. C&EN Concentrates (2009) Chem Eng News 87: 32. 10.1021/cen-v087n049.p032
96. IEA Statistics (2009) International Energy Agency Statistics. CO_2 emissions from fossil fuel combustion, 2009th edn. IEA, Paris, France
97. Johnson J (2009) Fossil-fuel costs. Chem Eng News 87:6
98. Bomgardner M (2012) DSM joints Poet in biofuels push. Chem Eng News 90:11
99. Bomgardner M (2012) Advancing ethanol. Chem Eng News 90:10
100. McCoy M (2012) US guarantees loans for cellulosic fuels. Chem Eng News 90:28
101. Raizen M (2012) Iogen target cellulosic ethanol. Chem Eng News 90:22
102. Scott A (2012) Cellulosic collaboration. Chem Eng News 90:10
103. Kintisch E (2008) The greening of synfuels. Science 320:306–308
104. Schaub G, Pabst K (2012) Synthetic hydrocarbon fuels from Lignocellulosic biomass. In: Muradov N, Veziroglu TN (eds) Carbon-neutral fuels and energy carriers. CRC, Boca Raton, pp 465–490
105. US Department of Energy (2010) EERE. Biomass. www1.eere.energy.gov/biomass/pdfs/ibr_arra_uop.pdf. Accessed 2 Jul 2010
106. Manzanera M (2012) Biofuels from Oily Biomass. In: Muradov N, Veziroglu TN (eds) Carbon-neutral fuels and energy carriers. CRC, Boca Raton, pp 637–666
107. Alternative energy flourishes (2009) Chem Eng News 87: 28
108. Reisch M (2009) Chemicals from biorefineries. Chem Eng News 87:28–29

109. U.S. National Research Council (2011) Renewable fuel standard: Potential economic and environmental effects of US biofuel policy, National Academies Press, ISBN 978-0-309-18751-0. http://www.nap.edu/catalog.php?record_id=13105. Accessed 5 Dec 2011
110. Ng Y (2012) Natural gas key to a renewable energy future. Power Eng 116:84–86
111. Cheung B (2012) Storing energy. Mech Eng 38–41
112. Hoffmann P (2011) Electric energy giant Siemens launches big electrolyzers for utility scale H2 storage. Hydrogen Fuel Cell Lett 26:4
113. Huggins R (2010) Energy storage. Springer, New York, p 406. ISBN 978-1-4419-1023-3
114. International Energy Agency (2013) Renewables to surpass gas by 2016 in the global power mix. Medium-term renewable energy market report 2013 – market trends and projections to 2018, p 217, ISBN 978-92-64-19118-1
115. Johnson J (2012) The new Saudi Arabia. Chem Eng News 90:30–31
116. National Renewable Energy Laboratory (2012) Renewable electricity futures study. In: Hand M et al. (eds) NREL/TP-6A20-52409. Golden, CO. http://www.nrel.gov/analysis/re_futures/. Accessed 5 June 2012
117. Hart E, Jacobson M (2011) A Monte Carlo approach to generator portfolio planning and carbon emissions assessments of systems with large penetrations of variable renewables. Renew Energy 36:2278–2286
118. Shinnar R, Citro F (2006) A roadmap to US decarbonization. Science 313:1243–1244
119. Johnes O (2013) Germans revolt against once-promising wind energy. Big Think. http://bigthink.com/ideafeed/Germans-revolt-against-once-promising-wind-energy. Accessed 25 Jul 2013

Chapter 7
Carbon Capture and Storage: In the Quest for Clean Fossil Energy

Abstract The main objective of carbon capture and storage (CCS) is to prevent CO_2 from entering the atmosphere by capturing CO_2 from large industrial sources and securely storing it in various carbon sinks. CCS is considered a critical component of the portfolio of carbon mitigation solutions, because global economy heavily relies and will continue to rely on fossil fuels in the foreseeable future. Currently, there are close to 300 active and planned CCS-related projects around the world—an indication of a growing commitment to this technological option. However, despite significant progress in CCS technology, the pace of CCS commercial deployment is rather slow. The major challenges facing the large-scale CCS deployment worldwide relate to a very high financial barrier and limited economic stimuli or regulatory drivers to encourage investments in the technology. This chapter highlights scientific and engineering progress in all three major stages of the CCS chain, CO_2 capture, transport, and storage, and the current status of existing and planned commercial CCS projects. Technological, economic, environmental, and societal aspects of the large-scale CCS deployment and its prospects as a major carbon abatement policy are analyzed in this chapter.

7.1 Introduction to Carbon Capture and Storage (CCS)

7.1.1 An Overview of CCS

The objective of CCS is to prevent CO_2 from entering the atmosphere by capturing it from large industrial sources and (practically) permanently storing it in various carbon sinks. CCS is considered and will remain a critical component of the portfolio of carbon mitigation options as long as fossil fuels and carbon-intensive industries continue dominating the energy sector and the global economy, as a whole. CCS is particularly popular among the proponents of "clean" coal technology who believe that fitting new (or retrofitting existing) coal-fired power plants with carbon

N. Muradov, *Liberating Energy from Carbon: Introduction to Decarbonization*, Lecture Notes in Energy 22, DOI 10.1007/978-1-4939-0545-4_7,

capture units would allow to continue the usage of coal in ever-increasing amounts without being accused of impacting our planet's environment and climate.

The importance of CCS as a carbon mitigation strategy was reinforced by a number of intergovernmental bodies and international committees. The Group of Eight (G8) at its 33rd convention in Hokkaido, Japan (July 2008), expressed its support for initiating 20 new industrial-scale (larger than 1 Mt/year CO_2) CCS demonstration projects with the view of a widespread commercial deployment of the CCS technology by 2020 [1]. The 2012 UNFCCC conference held in Doha (Qatar) established and refined the arrangements for CCS in the Clean Development Mechanism (CDM) that would facilitate the creation of the institutional arrangements necessary for CCS introduction as a mitigation option and enhance industry and public confidence in CCS [2]. In particular, two issues were discussed: (a) further exploring the eligibility of the CCS projects involving geological storage sites located in more than one country, and CO_2 transport from one to another country, and (b) the establishment of a global reserve of certified carbon emission reduction units for CCS projects. Other issues related to transboundary CCS projects were also considered at this convention.

CCS technology has been commercially proven since 1996 (the Sleipner project, Norway) and is considered a mature technology. In the beginning of 2013, there were eight large-scale CCS projects in operation around the world and nine projects under construction with the total CO_2 capture and storage capacity of 37 million tons per year [3]. Worldwide, there are close to 300 active or planned CCS-related activities—an indication of a growing commitment to this technological option. Most of CCS projects are in North America (the USA, Canada), Europe (Norway, the UK, Netherlands, and others), Australia, and China. Despite the progress in the development and industrial deployment of CCS technology, it is not currently used in the power generation market, mostly, because of considerable energy penalties and associated additional large capital and operational costs for capturing CO_2 at fossil fuel-based power plants. Today, all the existing large-scale CCS projects include CO_2 capture as part of an already established industrial process, e.g., in gas-processing facilities or enhanced oil recovery. In the IEA's ETP-2012 scenario, the CCS share of overall CO_2 reductions will amount to 14 % by 2050 [4].

The CCS technology, however, is not without a controversy; it faces a resistance from many members of scientific and environmental communities, who put forward the following points:

- CCS would provide only a temporary relief and make humankind even more dependent on fossil fuels, thus, making the necessary changes later even more difficult
- There are concerns about its efficacy, permanence, potential CO_2 leakage problems, and long-term ecological uncertainties
- Due to the relatively high lifecycle emission rate of the coal-CCS technology (255–440 g CO_2-equiv./kWh [5]), it does not have any advantages over other carbon mitigation options
- There is a widespread public opposition toward CCS in Europe and other places [6]

- Due to abundant and possibly long-lasting supply of cheap low-carbon fuel, NG, the need for the deployment of rather expensive CCS systems may not be that necessary or, at least, is not as urgent as previously thought

Regarding the last point, the supporters of CCS have a counterargument that the overreliance on NG as a carbon mitigation option may serve to delay the commercialization of CCS technology and, in fact, hold back rather than speed the reductions in GHG emission from power generation sector in the long run [7]. According to IEA, coal and gas will remain dominant fuels in the power sector for at least half a century; therefore, to meet atmospheric CO_2 stabilization targets, fossil power will need major reductions in CO_2 emissions that NG alone cannot provide. Because of that, the technology proponents argue that CCS needs to be widely implemented in the power and industrial sectors as soon as realistically possible to make a genuine progress in reducing GHG emissions. The cost gap could be reduced if stricter CO_2 emission limits on fossil power plants are imposed, incentives on beneficial CO_2 reuse (e.g., enhanced oil recovery) are expanded, and if government support for CCS demonstration or commercial projects is increased, rather than diminished [7].

The supporters of CCS are concerned that if the above controversies further persist, this may hinder the dynamics of the market penetration of the CCS technology. For example, IEA warns that if CCS is not widely deployed in the 2020s, this will put an extraordinary burden on other low-carbon technologies with potentially negative implications for carbon reduction policies [4]. The IEA's 2013 report "Tracking Clean Energy Progress" pointed to relatively slow progress of the worldwide deployment of CCS systems and underscored that while the CCS technologies are mature in many applications, they are unlikely to be deployed commercially (especially, in power generation sector) until governments make strong commitments and enforce appropriate policies [8]. The Agency pointed out that while 38 projects that apply CCS to power generation are required by 2020 to reach climate stabilization goals, none are operating now [9].

Technologically, CCS is a complex set of industrial processes and operations encompassing three major steps: CO_2 capture, transport, and storage, and a manifold of technological options and variations associated with each of these steps. Figure 7.1 provides a general outline of the complete CCS system including existing and emerging CO_2 capture, transport, and storage technologies.

7.1.2 Carbon Capture Strategies

CO_2 capture is a first step in the multistage CCS process. In general, CO_2 capture is an energy-intensive and rather costly process; capturing CO_2 from diluted streams, purifying and compressing (or liquefying) it to a state that makes its transport economically feasible, in most cases, represents the largest component of the overall cost of the CCS chain. Therefore, CO_2 capture technology is a focus of the intensive worldwide efforts aiming at improving the process efficiency and significantly

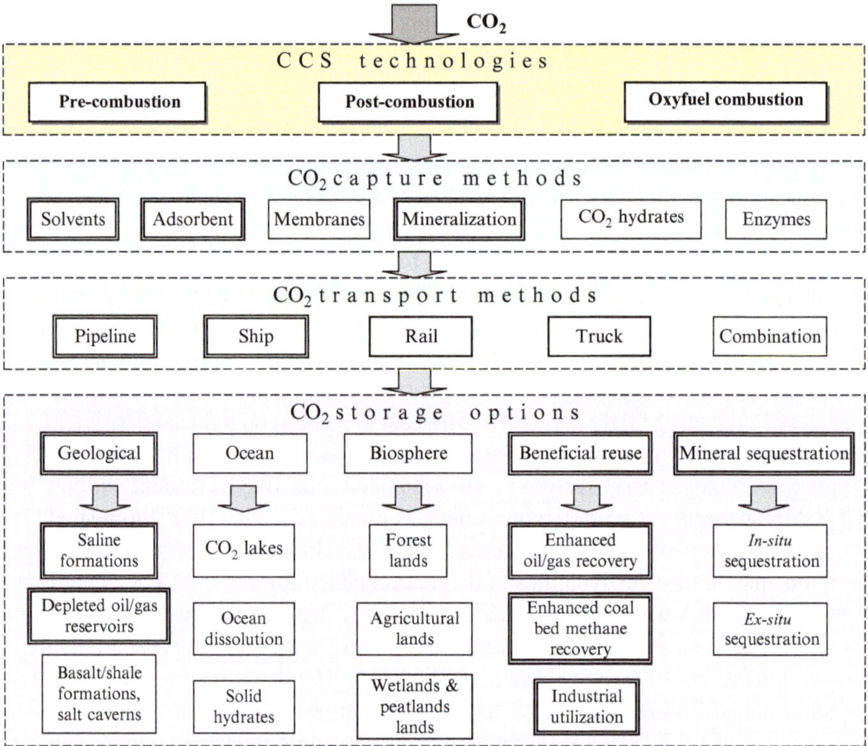

Fig. 7.1 General outline of complete CCS system including CO_2 capture, transport, and storage. (existing commercial technologies are *circled* by a *double-line*)

reducing its cost. The overwhelming majority of these efforts are targeting CO_2 capture from coal-fired power plants because they are the largest stationary sources of CO_2 emissions and, in all likelihood, will remain as such in the foreseeable future. The existing CO_2 capture technologies are also applicable to NG-fired power plants and the majority of large industrial sources of CO_2 emissions, such as synthetic fuel plants, cement manufacturing and metallurgical plants, hydrogen production plants, and refineries.

Currently, the main technological approaches to CO_2 capture from fossil fuel usage are:

- Pre-combustion carbon capture
- Post-combustion capture
- Oxyfuel combustion

There are multiple technological routes within each of these major categories that best suit specific fuel types, geographical locations, climate conditions, and economic development level. In order to better characterize the level of the technical development and maturity of CCS technologies, a widely accepted

Technology Readiness Level (TRL) format [10] is used in this book. The nine technology readiness levels (as applied to CCS technologies) are as follows (the TRL values are shown in *italics*) [11]:

Basic principles observed	*1*
Application formulated	*2*
Proof of concept	*3*
Laboratory component testing	*4*
Prototype development	*5*
Prototype demonstration	*6*
Pilot plant	*7*
Demonstration plant	*8*
Full-scale commercial deployment	*9*

The first five TRL levels deal with the technology advancement from basic principles to component validation in relevant environment. TRL 6, 7, and 8 relate to process development units (PDU) with the capacity of 0.1–5 %, greater than 5 %, and greater than 25 % of the full commercial scale, respectively [12]. TRL 9 indicates a full-scale commercial deployment (Note that with regard to the assessment of advanced coal-fired power plants, TRL 9 would be achieved by a plant with the capacity range of 400–800 MW (net)) [12]. Currently, pre-combustion capture is the only CO_2 capture technology that has reached the TRL of 9, while post-combustion and oxyfuel combustion options are less developed and have reached the TRL level of 7.

7.2 Pre-combustion Carbon Capture

Pre-combustion carbon capture (Pre-CCC) is the most mature CO_2 capture technology; the main elements of this technology have been practiced at the commercial scale for decades for separation of CO_2 from naturally occurring and industrial CO_2-containing gaseous streams. Many NG fields around the world produce gas with a high content of CO_2, which necessitates its removal to meet NG transport specifications or the purity requirements specified by customers. A number of industrial processes (e.g., coal gasification, steam methane reforming, and synthetic fuels production) also involve separation of CO_2 from gaseous streams using Pre-CCC technology. Currently, the industrial applications of Pre-CCC technology are being expanded to power generation sector. A simplified block diagram of the Pre-CCC technology for power generation application is shown in Fig. 7.2a.

In the gasification (or reforming) reactor, fuel (e.g., coal or NG) is gasified (or reformed) in the presence of oxidants (steam, oxygen, or air) into syngas (comprising mainly H_2 and CO, and small amounts of CO_2). The syngas is directed to a water gas shift (WGS) reactor where CO is converted (or shifted) into CO_2 and additional hydrogen in the presence of steam:

$$CO + H_2O \rightarrow H_2 + CO_2 \quad \Delta H^{\circ} = -41 \text{ kJ/mol} \tag{7.1}$$

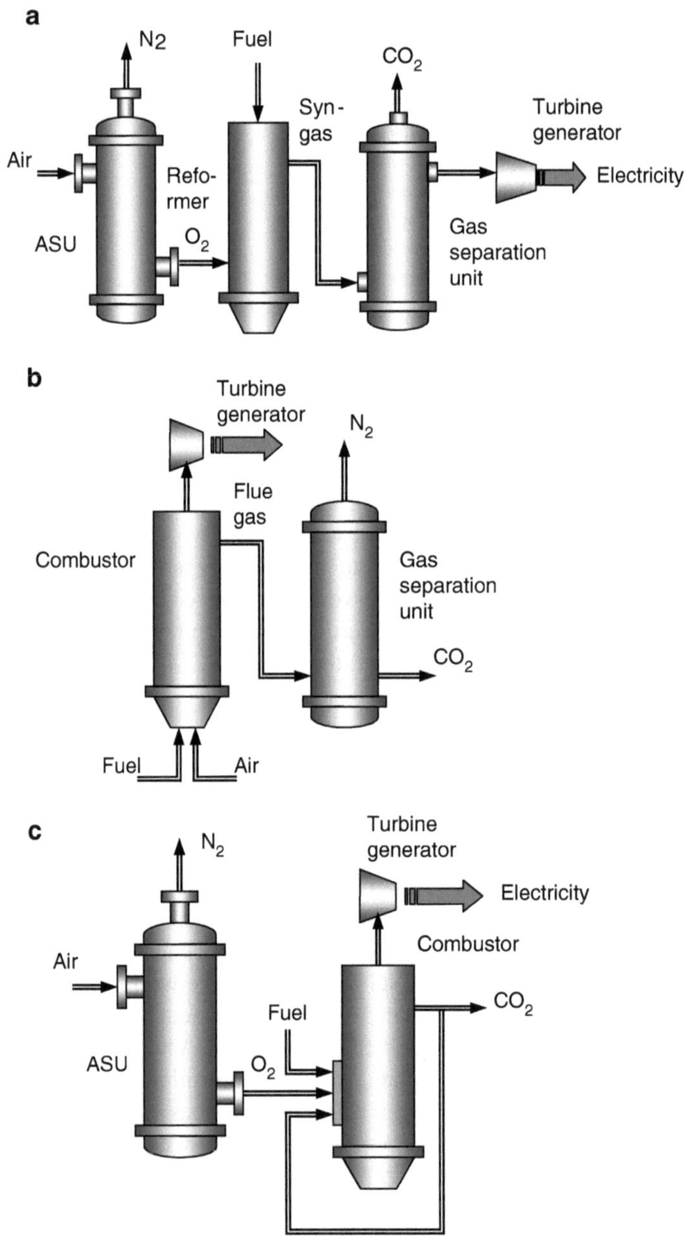

Fig. 7.2 Summary of CO_2 capture strategies. (**a**) Pre-combustion capture, (**b**) post-combustion capture, and (**c**) oxyfuel combustion

The CO_2 concentration in the resulting gaseous mixture varies in the range of 20–60 vol.% (on dry basis), with balance being mostly H_2 and minor quantities of CH_4, CO, and N_2, depending on the feedstock composition and operational parameters. Separated CO_2 is then dried, compressed, and readied for transportation and storage. Practically carbon-free hydrogen fuel can be fired in a gas turbine generating electricity; additional electricity can be produced by a steam turbine utilizing hot flue gas (alternatively, heat could be used in a variety of thermal applications, e.g., heaters and boilers). In most cases, the fuel processing (gasification, reforming) section is thermally integrated with the combined cycle section to increase overall fuel-to-electricity energy conversion efficiency. More detailed description of the fuel processing technologies is presented in Sect. 7.2.2.

Although many components of the Pre-CCC (as applied to power generation) are mature technologies, most of the power plants equipped with Pre-CCC are still in the planning stages of the development; only one of them (a power plant with planned output of 582 MW in Kemper County, Mississippi, USA) is in the advanced stage and is projected to reach TRL 9 in 2014 [12]. The major challenge associated with the addition of Pre-CCC to a power plant is to substantially reduce inevitable energy penalties due to the incorporation of additional processing steps (e.g., air separation, gasification, water gas shift, and gas cleanup) that could result in the loss of net power output (or thermal efficiency) up to 20 %; along with the added system complexity and capital cost, these factors are considered a major detriment of the CCS technology deployment [12]. For this reason, Pre-CCC is more likely to be applied to newly built power plants rather than retrofits to avoid substantial additional cost and performance penalties. Due to relatively high concentration of CO_2 in the gaseous stream and high operational pressure, a Pre-CCC process incurs less energy penalty (about 20 %) than post-combustion capture technology (about 30 %) at the 90 % CO_2 capture level [12]. Pre-CCC systems are flexible and strategically important due to their capacity to deliver a suitable mix of electricity, hydrogen, and low-carbon containing fuels or chemical feedstocks with a relatively high efficiency.

7.2.1 CO_2 Capture Technologies: Status and Prospects

Pre-CCC technology involves CO_2 removal from gaseous streams with relatively high CO_2 concentration (up to 40–50 vol.%). The search for efficient CO_2 capture materials has been of great practical and theoretical interest for many decades, which has recently intensified due to the development of commercial CCS systems. Various materials have been used for CO_2 capture: zeolites, amines, activated carbons, alkali and alkaline earth metal oxides, ionic liquids, polymeric membranes, microporous polymers, amine-modified mesoporous silica, metal-organic frameworks (MOF), etc. [13]. The most important requisites with regard to the selection or design of CO_2 capture materials are:

- CO_2 sorption capacity
- Sorption/desorption kinetics

Table 7.1 CO_2 capture technologies for Pre-CCC applications

Processes, methods	Compounds, materials, processes
Chemical solvents	Diethanolamine (DEA)
	Methyldiethanolamine (MDEA)
	Potassium carbonate (Benfield process)
Physical solvents	Glycol: Selexol
	Methanol: Rectisol
	Propylene carbonates
	N-Methyl-2-pyrolidone (Purisol)
Hybrid physical/chemical solvent absorption	Sulfinol (mixture of diisopropanolamine and tetrahydrothiophene oxide)
	Flexsorb® PS
	Ucarsol® LE: Amisol
Physical sorbents	Zeolites
	Activated carbon
	Metal-organic frameworks
Membranes	Polymer membranes
	Ceramic membranes
	Hollow fiber membrane supports
Cryogenic distillation	
Hydrates	

Source [1, 14]

- Sorption/desorption temperature and pressure
- Interference with common flue gas components or contaminants (e.g., SO_x, NO_x, H_2S)
- Stability and regenerability
- Cost and economic feasibility

The main existing technological options for CO_2 capture from Pre-CCC gaseous streams are summarized in Table 7.1.

7.2.1.1 Absorption

Current commercial Pre-CCC technologies are mostly based on the absorption processes, i.e., the use of selective solvents that can accomplish greater than 90 % CO_2 separation and removal. In general, absorption is an uptake of gas (CO_2) into the bulk phase of another material (e.g., an aqueous solution, an organic solvent). Depending on the nature of the interaction between CO_2 and the molecules of the bulk material, absorption could be purely a physical process (e.g., the dissolution of CO_2 in cold methanol) or a chemical process (e.g., the reaction of CO_2 with amine compounds). Both chemical and physical absorption methods are widely applied to Pre-CCC.

Chemical Absorption

Since CO_2 is an acidic gas, most of the absorbing media for its capture are basic solvents, and the efficiency of CO_2 removal is controlled by acid–base neutralization reactions. Commercially significant chemical absorption systems include aqueous amines, chilled ammonia (NH_4OH), and hot potassium carbonate (K_2CO_3). The most commonly used amine-based solvents for CO_2 capture in commercial systems are alkanolamines such as monoethanolamine (MEA), diethanolamine (DEA), and methyldiethanolamine (MDEA). Although the absorption process capacity is equilibrium limited, alkanolamines are capable of achieving the CO_2 recovery levels of 90 % and higher from the flue gases due to fast kinetics and high chemical reactivity.

The use of off-the-shelf amine-based solvents, however, incurs high energy penalties due to their regeneration via steam stripping (for the MEA solvent, the energy penalty is 1.9 MJ/kg CO_2 captured). The minimum work required to separate CO_2 from coal-fired flue gas and compress it to 150 atm is 0.11 MWh/t CO_2 [15]. The presence in flue gases of such contaminants as SO_x, NO_x, hydrocarbons, and a particulate matter could be another challenge, since these impurities may eventually reduce the absorption capacity of amine-based solvents and cause equipment corrosion problems. To avoid these problems, some commercial units practice various pretreatment options, which may increase the cost of CO_2 recovery. There are other operating problems encountered in the amine solvent systems, such as foaming, vapor entrainment of the solvent, and replenishment of the solvent, but these factors have a relatively small effect on the overall cost of the process. Other R&D efforts are focused on potassium carbonate promoted by piperazine, integrated vacuum carbonate absorption process, and novel oligomeric solvents [16].

Physical Absorption

Physical absorption methods are based on the preferential absorption of CO_2 from gaseous mixtures by inorganic or organic solvents. Since physical absorption systems are governed by the Henry's law (i.e., low temperature and high pressure favor CO_2 capture), this method is preferred for CO_2 capture from the mixtures where CO_2 partial pressure is relatively high (greater than 500 kPa). Advantageously, the regeneration of physical solvents is less energy intensive than that of chemical solvents. Due to a high-pressure requirement, this technology is considered particularly preferred for CO_2 capture from coal gasification gases in Pre-CCC. Commercial processes for physical absorption of CO_2 include glycol-based compounds (e.g., dimethyl ether of polyethylene glycol), cold methanol, propylene carbonate, and others [1]. The state-of-the-art process *Selexol* for CO_2 capture from shifted syngas is based on glycol-based solvent.

Advantageously, in the glycol-based systems, CO_2 recovery requires very small heat input for regeneration, and CO_2 and H_2S capture could be combined. However, the glycols have a relatively low carrying capacity, which requires circulating substantial volume of the solvent (e.g., more than 20 kg of the glycol solution per kg of CO_2 captured).

Other drawbacks of the glycol solvent use are that CO_2 pressure is lost during flash recovery, leading to the loss of hydrogen with the CO_2 stream. In the methanol-based absorption systems (commercial process: *Rectisol*, by Lurgi), a CO_2-rich stream is cooled and contacted with liquid methanol, which readily dissolves CO_2. The process is capable of capturing in excess of 90 % of CO_2 from gaseous streams; however, the high cost of refrigeration hurts the process economics (e.g., compared to glycol-based and other solvents). A techno-economic analysis conducted at NETL indicates that Selexol-based CO_2 capture raises the cost of electricity from a newly built coal-fired power plant by 30 % [17].

7.2.1.2 Adsorption

CO_2 physical adsorption and separation processes are based on the selective adsorption of CO_2 on high-surface area solids such as zeolites and activated carbons. The adsorption kinetics and capacity of adsorbents are controlled by a number of factors including their surface area, pore size, and volume, and the affinity of the adsorbed gas for the adsorbent. A pressure swing adsorption (PSA) process is the most widely used commercial technology in this category. PSA is the method of choice for the separation of H_2–CO_2 mixtures where high purity of hydrogen is required (e.g., 99.999 % and higher). In the PSA cyclic operation, a plurality of adsorbent beds (loaded with zeolites or activated carbons) adsorb gases at high pressure and then desorb them at lower pressure. The major technical challenges facing the adsorption capture systems involve the development of advanced adsorbents with (a) tolerance to higher temperatures in the presence of steam, (b) an increased adsorptive capacity and selectivity for CO_2 capture, and (c) improved kinetics and stability over thousands of cycles.

Emerging physical adsorption processes involve metal organic frameworks (MOF) and nanostructured carbon-based sorbents. The main advantage of MOF is that they have very high porosity and adjustable chemical functionality that can be tailored to increase the CO_2 adsorption capacity. A wide variety of MOF have been synthesized and tested at a laboratory scale. For example, MOF $Zn_3O_9(BTB)_2$ exhibited CO_2 sorption capacity of 1.4 g CO_2 per gram of a sorbent material, which is an improvement over conventional zeolite-based sorbents [18]. Some modifications of MOF have exceptional capacity to capture and store CO_2 and release it when MOF is exposed to sunlight. This development could help to overcome the problems of expensive energy-intensive recovery of CO_2 [19]. The MOF technology, however, is still expensive and at an early stage of development.

Although most of the CO_2 capture systems are envisioned for centralized applications (e.g., large CO_2 capture units attached to centralized CO_2 point sources such as power plants, cement manufacturing plants, and ethanol plants), recently, there have been reports on the small CO_2 capture systems that could be applied to moving vehicles (cars and trucks) and home heating systems. For example, the developers of *Strataclear*® technology claim that their CO_2 capture technology could remove CO_2 from exhaust gases: up to 25 % in automobiles, 40 % in trucks, and 50 % in home heating furnaces [20]. The CO_2 capture system includes removable sorbent cartridges that replace the resonator and muffler in a vehicle's exhaust system. The exhaust gases pass the catalytic converter then enter the Strataclear® exhaust treatment system where CO_2 is captured. The cartridges are stored along the perimeter of the car trunk until they are easily replaced at a gas station or another exchange facility. The technology produces a solid residual material that can be sold for use in many industrial applications.

7.2.1.3 Gas Separation Membranes

The membrane separation of CO_2 utilizes the permeable or semipermeable materials that selectively transport CO_2 across a membrane in response to a partial pressure gradient. Membranes have a promise of simplicity (no moving parts, a passive operation), compared to capital- and maintenance-intensive solvent-based separation systems; therefore, they are expected to eventually become more reliable and more cost-effective. A key challenge to developing the membranes for CO_2 capture is to design a material that has high selectivity for CO_2 combined with a high permeability. Most existing commercial membranes are held back by a trade-off between these two properties.

In general, gas separation could be accomplished via either physical or chemical interaction between a membrane and CO_2-containing gas. For example, polymer-based membranes transport gases by a solution-diffusion mechanism, which explains the relatively low gas transport flux. Polymeric membranes are quite effective (due to a large surface to volume ratio) and inexpensive; however, they are less selective and suffer from a gradual degradation. (In most cases, the increase in membrane permeance decreases the separation factor and, conversely, the increase in the separation factor reduces the membrane permeance.)

Considerable R&D efforts are required to implement the large-scale membrane separation of CO_2 from industrial gaseous streams [16, 17, 21]. Recently an international team of researchers discovered a class of novel membrane materials made of a porous organic polymer featuring a nitrogen heterocycle, tetrazole, exhibiting an exceptionally high gas permeability and CO_2 separation selectivity [22]. The novel membrane material has CO_2/N_2 selectivity and CO_2 permeability that far exceeds those of the existing top performing membranes. Hydrogen bonding coupled with Lewis acid–base interactions between CO_2 and tetrazole groups imparts this porous polymer a prominent CO_2-capturing capacity. As a result, CO_2 molecules strongly adsorb to the polymer's nanopores, thus, blocking adsorption and transport of other gases (e.g., nitrogen), which can be vented through other channels.

7.2.1.4 Cryogenic Separation

In cryogenic gas separation methods, a low-boiling temperature liquid is separated from a high-boiling temperature liquid via evaporation and condensation.

This technology is most efficient and cost effective when components in the gaseous mixture have significantly different boiling points. The cryogenic process is widely used commercially in the processes involving liquefaction and purification of CO_2 from streams with relatively high CO_2 content. The advantages of the technology are the good economy of scale and the possibility of direct production of liquid CO_2 that can be transported via liquid pumping or stored at high pressure. The main disadvantages of the cryogenic process are that it is energy intensive, and it requires the removal of the components having relatively high freezing point (e.g., water, NO_2) prior to cooling (which adds significantly to the cost of separation).

7.2.1.5 Separation Through CO_2 Hydrates

The method is based on the formation of solid CO_2 hydrates when liquid water is exposed to CO_2 at relatively low temperature (0–4 °C) and high pressure (10–70 atm) depending on CO_2 partial pressure in a gaseous mixture. The solid hydrates are easily separated from the liquid stream and then heated to release CO_2. The separation technique is particularly advantageous for CO_2 separation from high-pressure gaseous streams (e.g., pre-combustion streams) with minimal energy losses. The process' shortcomings include high refrigeration energy requirements to counteract the heat of hydrate formation (up to 3.3 MJ per kg CO_2 captured) [23] and the possibility of ice formation (in the reactor's cold spots) that may cause some operational problems. Currently, research efforts are underway to develop special additives to speed up hydrate formation while enabling 90 % CO_2 capture and hydrate-forming reactors with the improved heat exchanger (to avoid cold spots formation in the reactor).

Among recent developments, DOE–NETL in a partnership with Nexant and Los Alamos National Laboratory (LANL) has been working on a low temperature and high pressure *SIMTECHE* process for removing CO_2 from shifted syngas containing H_2 (60 %) and CO_2 (40 %) via formation of CO_2 hydrates [17]. The developers of the technology have demonstrated the technical feasibility of the continuous production of CO_2 hydrates from shifted syngas.

7.2.2 *Fuel Processing Technologies*

The first step in the Pre-CCC technological approach is to convert fuels into a gaseous mixture from which CO_2 could be extracted. The nature of fuels dictates the type of the conversion technology applied: typically, light hydrocarbons (e.g., NG, LPG, naphtha) are processed through a *reforming* process, whereas coal, petroleum coke, and residual oil through a *gasification* process (each of these technologies has the great variety of modifications depending on the process design and type of oxidant used: steam, oxygen, air, or their combination).

7.2.2.1 Reforming

Steam methane reforming (SMR) is by far the most important and widely used fuel reforming technology; it is used for industrial manufacture of hydrogen, amounting to about 80 % of hydrogen produced in the USA and 40 % worldwide [24]. SMR is a multistep process that can be represented by the following chemical equation (for the overall process):

$$CH_4 + 2H_2O_{liq.} \rightarrow 4H_2 + CO_2 \quad \Delta H^\circ = 253.1\,kJ/mol \tag{7.2}$$

SMR technology has been commercially practiced for many decades, and, currently, it is available in a wide range of hydrogen production capacities, from less than 1 t/h H_2 (small decentralized units) to 100 t/h H_2 and higher (large centralized H_2 production plants for ammonia manufacturing). The strongly endothermic reaction of methane with steam is carried out over Ni-based catalyst at high temperatures of 800–900 °C with heat provided by combusting of part (about a third) of the NG feed. The produced syngas undergoes water gas shift (reaction 7.1) followed by gas separation stages. Present-day SMR plants use physical adsorption technology (in particular, PSA) for the gas separation stage. It should be noted that, in most cases, the PSA unit does not selectively separate CO_2 from other waste gases (CH_4, CO); thus, the off-gas from the PSA unit contains CO_2 (about 40–50 % by volume), CH_4, CO, and small amounts of H_2, and it is used as fuel in the reforming reactor with CO_2 typically being vented to the atmosphere. Since the resulting off-gas is heavily diluted with N_2, the capture of CO_2 from SMR plant emissions would require one of the post-combustion CO_2 capture processes described in the following section.

In an alternative approach, which is more suited for the Pre-CCC option, the PSA process could be modified to recover both H_2 and CO_2 in pure form, e.g., by including an additional PSA section to remove CO_2 prior to the H_2 separation step (e.g., *Gemini* process developed by Air Products) [25]. However, the modification of the SMR plant to produce pure CO_2 stream comes with an energy penalty. The overall efficiency of H_2 production (at the pressure of 6 MPa) at an SMR plant with the capacity of 720 t/day H_2 without CO_2 capture is estimated at 76 % (on a lower heating value, LHV, basis) with overall CO_2 emissions of 9.1 kg CO_2 per kg H_2 [26]. If the process is modified to produce nearly pure CO_2 (e.g., via combination of amine solvent scrubbing with PSA), the efficiency is reduced to 73 %, with reduction in CO_2 removal rate down to 8.0 kg CO_2 per kg H_2.

Besides SMR, there are a number of other technological routes to conversion of light hydrocarbons to syngas and hydrogen, e.g., via partial oxidation, autothermal reforming, dry reforming, and combined reforming. A comprehensive overview of the state-of-the-art hydrocarbon-to-hydrogen technologies can be found in [27].

7.2.2.2 Gasification

Among the existing fuel gasification technologies, coal gasification is the most commercially important technology. The overall process can be expressed by the following generic equation (For the simplification, coal is presented as elemental carbon, and a minor CO_2 by-product is not shown.):

$$3C + H_2O + O_2 \rightarrow H_2 + 3CO \tag{7.3}$$

The amount of O_2 has to be carefully controlled such that only portion of fuel is burnt providing enough heat to run steam gasification of coal to syngas. The composition of the syngas produced heavily depends on operational conditions: the nature of coal (e.g., anthracite or lignite), the type of a gasifier (e.g., entrained flow, fluidized bed, or moving bed), process parameters (pressure, temperature, H_2O/C, and O_2/C ratios in the feed), syngas cooling method (water quench vs heat exchangers), a gas cleanup system, etc. There are three main types of commercial gasifiers differing primarily by coal/gas interface and coal particles movement patterns [28]:

- Moving bed (e.g., Lurgi process)
- Fluidized bed (e.g., KRW process)
- Entrained flow bed (e.g., ChevronTexaco, Shell processes)

Depending on the gasifier type, operational temperature and pressure could reach up to 1,600 °C and 85 atm, respectively, with typical thermal process efficiencies varying in the range of 51–63 % [29]. Shell and Texaco (GE)-type gasifiers produce syngas with the H_2–CO composition of 26.7–63.3 vol.% and 34–48 vol.%, respectively [30]. The gasification stage is followed by water gas shift (reaction 7.1) and gas separation stages. Besides main components (H_2, CO_2, CO), the gasifier output contains appreciable amounts of impurities (depending on the type and composition of coal: COS, H_2S, NH_3, HCN, N_2, Hg, volatile minerals); these harmful impurities have to be captured and dealt with.

In 2010, there were 128 gasification plants operating worldwide with 366 gasifiers producing 42,700 MW_{th} of syngas, and about 24,500 MW_{th} of syngas capacity were under construction (with 4,000–5,000 MW_{th} of syngas capacity added annually) [31]. Present-day commercial coal gasification systems are mainly focused on the following five application areas:

- Hydrogen and ammonia production (many plants are being operated in China)
- Integrated gasification combined cycle power plants
- Substitute natural gas production (e.g., North Dakota plant in the USA)
- Fischer–Tropsch liquids production (e.g., Sasol technology in South Africa)
- Polygeneration (production of electricity, steam, and chemicals)

7.2.2.3 Integrated Gasification Combined Cycle (IGCC)

IGCC is an advanced coal gasification technology for the production of syngas and hydrogen at power plants. The technology was initially demonstrated in the 1980s, and since then several IGCC power plants fueled by coal and petroleum coke have been constructed. For example, the US DOE's *Clean Coal Power Initiative* envisions

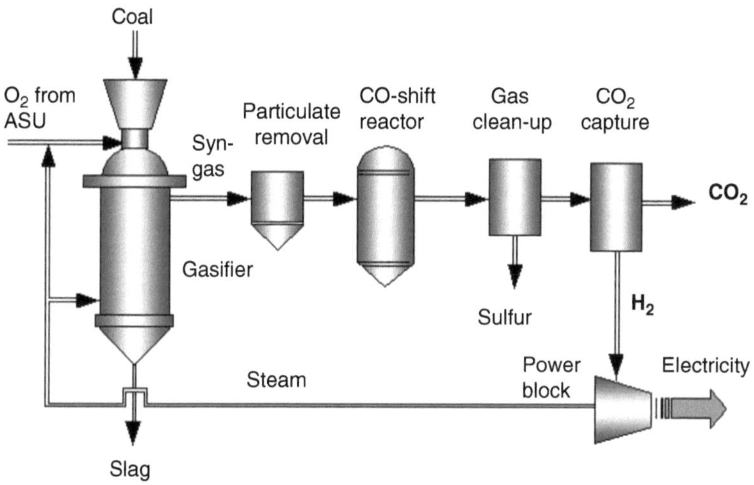

Fig. 7.3 Simplified block diagram of a modern IGCC plant. ASU—air separation unit

the development of a "clean" coal processing technology with substantially reduced CO_2 emissions through IGCC with CO_2 capture and sequestration. Recently, IGCC has evolved as an ultra-low emission power generation technology integrating advanced coal gasification processes with highly efficient combined power generation cycles. Under the US DOE *FutureGen* program, several IGCC plants are planned or under construction aiming at a full-scale demonstration of this technology. Figure 7.3 provides a simplified block diagram of an IGCC plant.

In this process, coal feedstock is gasified with an oxygen-steam mixture to syngas, which is cleaned of particulate matter and further subjected to CO shift in sulfur-tolerant shift reactors to produce H_2–CO_2 mixture. Typically, a double-stage Selexol unit is utilized for the removal of H_2S and CO_2 from the syngas (H_2S is preferentially scrubbed in the first absorber using a physical solvent) [28]. The resulting gas is treated in a Claus-process unit, where H_2S is oxidized to sulfur. The H_2S-free ("sweet") syngas enters the second absorber, where remaining CO_2 is removed from the syngas. Fuel gas after the Selexol unit consisting mainly of H_2 is sent to a gas turbine, and the CO_2 by-product is released at a relatively high pressure of 3.5 atm (it may require further pressurization before pipeline transport).

The advantages of the IGCC technology are:

- Syngas and hydrogen are produced and converted into electricity at the same site to avoid the high cost of pipeline transport
- Additional electrical power can be produced onsite from steam generated during syngas cooling

- Relatively low energy penalties due to high pressure and relatively high CO_2 concentration in the process gas
- Significant reduction in the emission of criteria air pollutants such as SO_x and NO_x
- Production of value-added by-product—elemental sulfur

The shortcomings of the IGCC system include: the complexity of the system compared to conventional power plants and high capital and investment costs. Besides, there is a limited operational data on the effect of coal quality on the process efficiency. (So far, most of the process development has been conducted using high-rank coals; the potential impact of widely available low-rank coals is not well known.)

The widespread deployment of coal-fueled IGCC technology is currently hindered due to a strong competition from NG combined cycle (NGCC) plants and pulverized coal-fired steam-electric plants. However, because of its environmental advantages over conventional coal-fired power plants, IGCC technology will be economically more attractive if carbon-mitigation policies will be reinforced in the power generation sector.

7.2.3 Enabling Technologies: Hydrogen-Fired Turbines

Gas turbines with their long history of efficient and reliable performance in the industrial and utility sectors are prime candidates for hydrogen-fueled power generation in the Pre-CCC systems. Modern power generators utilize a combined cycle (CC) configuration, which integrates gas and steam turbines based on Brayton and Rankine cycles, respectively. In the CC, high-temperature exhaust heat from a gas turbine enters a heat recovery system and a steam generator, which powers a steam turbine, thus, significantly enhancing the overall fuel-to-electricity energy conversion efficiency. The current trends in the CC design are to use a single-shaft configuration, whereby both the gas and steam turbines drive a common generator to reduce the capital cost, operating complexity, and the system footprint [32].

The development and practical implementation of hydrogen-fired turbines having comparable performance to NG-fueled turbines faces significant challenges in combustion technology. Hydrogen has lower volumetric energy density compared to NG (10.8 kJ/L H_2 vs 36.4 kJ/L for NG, in a gaseous form); thus, its use would require a higher volumetric flow rate through the advanced gas turbine to achieve the same power output as NG, which may result in a larger gasifier, air separation unit (ASU), and a plant, in general. Furthermore, stable, efficient, and low-NO_x combustion requires the rapid and homogeneous mixing of hydrogen fuel with air, which is very challenging [32].

Gas turbine efficiency is the key parameter in minimizing the cost of electricity, since higher efficiency translates into lower fuel consumption (for given power output) and lower capital cost of the plant. Thus, the major thrust in the development of advanced turbines (including, hydrogen turbines) is to increase combined cycle efficiency (and, hence, lower plant's cost) and reduce NO_x emissions (which indirectly reduces the capital cost by minimizing gas cleanup expenses). However, increasing the efficiency of gas turbines and reducing NO_x emissions are conflicting targets (To achieve higher efficiency, the temperature of turbine's working fluid has to be increased, but that would increase the rate of nitrogen reaction with oxygen in the combustion zone generating more NO_x.) [32, 33]. Besides, the increased firing temperature, pressure ratio, and mass flow rate due to the use of hydrogen (or syngas) as fuel would introduce significant challenges to the turbine aerodynamic, thermal, and mechanical designs [34].

The US DOE's two key initiatives, FutureGen and the Clean Coal Power Initiative, pursue the development of F-frame and G-frame gas turbines that can operate on syngas and hydrogen at an efficiency equivalent to that of NG with near-zero emissions. In particular, a major thrust of the US DOE's Turbine Program is to demonstrate 50–60 % (on a higher heating value, HHV, basis) efficient coal-based power generation at a capital cost of $800–900/kW with near-zero emissions (i.e., 2 ppm NO_x, with 99 % SO_2 removal and 95 % mercury removal) and 100 % CO_2 management by 2020 [32]. The US DOE's FutureGen 275 MW near-zero emission coal-based power and hydrogen production project is designed to demonstrate advanced systems and components, including the efficient gas turbines operating on hydrogen fuel with carbon storage [32].

Future progress in the high-efficiency turbines is linked to the development of ultra-high temperature turbines. The DOE's Turbine Program is supporting development of an alternative near-zero coal-based *Oxyfuel System* (OFS) (not to confuse it with the Oxyfuel Combustion system). In the OFS, water is injected into a reactor where clean gaseous fuel (syngas or hydrogen) is combusted in oxygen producing supercritical steam (pressure higher than 200 atm, temperature of 1,650–1,760 °C) and relatively small amount of CO_2 (or no CO_2 if H_2 was used as fuel) [32]. Due to the absence of nitrogen in the feed, no NO_x is formed in the combustor. Sulfurous and other potential pollutants can also be completely eliminated by cleaning gaseous fuel before combustion. The high-energy steam produced in the combustor drives a steam turbine resulting in a highly efficient power generation system with an exhaust that could be easily separated into water and sequestration-ready CO_2 (or it would produce only water if hydrogen was used as fuel). The US DOE's Turbine Program targets the development of 300–600 MW OFS with an efficiency of 50–60 % (HHV) operating on coal-derived syngas by 2015 [32]. In principle, it would be possible to further climb on the temperature ladder and utilize maximum possible temperatures that can be produced by fuel combustion in the order of 2,000–2,500 °C, which could potentially result in the efficiencies as high as 65–70 %, LHV [33]. However, this goal would involve a host of technical challenges and require technological breakthroughs in many areas such as advanced materials, cooling technologies, aerodynamic improvements, efficient catalysts, and NO_x mitigating techniques.

7.2.3.1 Summary of Pre-CCC

The TRL status of the Pre-CCC technologies is as follows [12] (the TRL values are shown in *italics*):

Chemical absorption (solvents)	*9*
Physical absorption (solvents)	*9*
Cryogenic distillation	*9*
Physical sorbents	*9*
Chemical sorbents	*6*
Gas separation membranes	*6*
Gasification	*9*
Reforming	*9*
Gas cleanup	*9*
Air separation unit	*9*
Hydrogen-fired turbines	*8*

Most of the Pre-CCC technologies are at the advanced stages of development (TRL 9) and have been commercially practiced for years. Among the least developed areas are membrane separation and chemical sorbents for CO_2 separation.

7.3 Post-combustion Carbon Capture

7.3.1 Current Status of Post-combustion Carbon Capture

In the post-combustion carbon capture (Post-CCC) option, fuel is combusted in air resulting in the flue gas containing N_2 and CO_2 as main components mixed with small amounts of NO_x and SO_x (Fig. 7.2b). The flue gas is first treated by existing pollution control technologies to remove SO_x, NO_x, and ash impurities, followed by CO_2 capture by a variety of gas separation techniques. The process is applicable not only to coal- and gas-fired power stations, but also to large industrial emitters of CO_2 (e.g., cement manufacturing, chemical, and metallurgical plants). The key advantage of the Post-CCC approach is that it can be retrofitted to existing large stationary point sources of CO_2 (e.g., coal-fired power and cement plants), where its end-of-pipe nature provides the potential flexibility from the design and operation perspectives, e.g., a flexibility to operate with or without carbon capture depending on the market conditions and/or existing regulations. Because the CO_2 concentration in the post-combustion flue gas is much lower than that in shifted syngas, in general, the post-combustion CO_2 capture is more expensive compared to pre-combustion technology. The main technical challenges facing Post-CCC technology are as follows:

- CO_2 concentration in flue gases is relatively low: 13–15 vol.% and 3–4 vol.% for coal- and gas-fired power plants, respectively [17], which makes its separation from the flue gases capital and energy intensive

- Due to reliance on air to burn fuel and the low pressure of the exit gas (close to atmospheric), the process requires a large-capacity equipment to accommodate very high gas volumetric throughputs, which would translate into significant energy penalties for CO_2 capture, high capital expenditure, and land availability problem, especially, when existing plants are retrofitted with CCS systems
- CO_2 needs to be compressed from about atmospheric pressure to the pipeline pressure of about 138 ata, which represents a significant auxiliary power load
- Gas separation units typically operate under oxidizing atmosphere, which might shorten the life of chemical solvents used in the CO_2 capture process
- The presence of impurities in the flue gas (e.g., SO_x, NO_x, and particulate matter) could potentially degrade some physical and chemical sorbents and adversely affect the performance of CO_2 capture systems.

Currently, Post-CCC technology has not reached full commercial status: there are currently no Post-CCC based large-scale integrated plants operational in the power generation sector; in most cases, its use in power plants has been restricted to slipstream applications. It was reported that some Post-CCC projects will soon reach TRL 8 (e.g., Boundary Dam 110 MW coal-fired plant in Saskatchewan, Canada) and start the operation in 2014 (there are also 16 Post-CCC projects applied to power plants in the planning stages) [12]. A preliminary analysis conducted at the US DOE National Energy Technology Laboratory (NETL) indicates that Post-CCC and compression of CO_2 to 152 ata could raise the cost of electricity by about 65 % [17]. Despite currently lagging in a technical readiness behind the pre-combustion capture, Post-CCC technology is projected to eventually dominate the field, especially coal-fired power plants [1, 25].

The developers of the technology will face several operational challenges to the integration of Post-CCC to existing power plants. Recent studies indicated that the retrofitting existing power plants with Post-CCC would allow effectively utilizing waste heat. An opportunity exists even for the low-efficiency power plants to significantly reduce their parasitic energy use, for the capture process provides a sink for low-temperature waste heat which otherwise is uneconomic to recover (e.g., in the power plants without carbon capture) [12]. Advantageously, such modifications can utilize existing heat exchange technology. Channeling certain amount of steam for solvent regeneration would reduce its flow to the low-pressure turbine and, thus, would have an operational impact affecting the plant's power output. Furthermore, the addition of Post-CCC will cause a significant increase in water usage, particularly, for water-cooled plants. The impact of these factors will be abated with the improvement in the efficiency of CO_2 capture processes.

7.3.1.1 Industrial (Non-power) Applications of Post-CCC

Although most of Post-CCC R&D efforts and demonstration projects are concerned with power generation, there are a few non-power related Post-CCC projects. The majority of the Post-CCC industrial applications deal with CO_2 capture from process heaters, boilers, and other fossil fuel-combusting equipment. There has been limited development with regard to Post-CCC application in oil refineries. One example involves Grangenmouth refinery project in Scotland with the capacity of 196,000 barrel/day, where feasibility studies were conducted on capturing CO_2 from fuel-fired heaters and boilers [12].

Another important and promising application area of the Post-CCC technology relates to cement manufacturing industry. Cement industry is one of the major CO_2 producers, emitting annually over two billion tons of CO_2 (from both the calcination of limestone and fuels used) (about 1 t CO_2 per 1 t of cement) [35]. The advantage of Post-CCC integration with the cement manufacturing plants is that it would be an "end-of-pipe" option that would not require fundamental changes in the clinker-burning technology; therefore, it could fit both new kilns and be retrofitted to existing plants. CO_2 capture technologies that are applicable to cement plants include absorbents (solvents), membranes and, particularly, CaO-based chemical sorbents (to be discussed in the next section). The latter technology is currently being assessed by the cement industry as a potential retrofit option for the existing cement manufacturing kilns [12]. Despite a great interest in capturing CO_2 from cement plants, CCS-related projects in this industrial sector are still at an early stage of development.

7.3.2 *CO_2 Capture from Diluted Streams*

Although many CO_2 capture methods discussed in the Pre-CCC section, in general, could be applicable to the Post-CCC, there are some limitations to the economic feasibility of their use due to significantly lower CO_2 concentrations in the flue gases of Post-CCC units. Main technological options for CO_2 capture from diluted streams (3–15 vol.%) are presented in Table 7.2 and discussed further in the text.

7.3.2.1 Chemical and Physical Solvents

Similar to the Pre-CCC systems, amine-based solvents are quite popular among the Post-CCC technological options. In view of the possible considerable increase in the capacity of CCS systems in the near future, the amine-scrubbing technology developers such as Fluor Corp., Mitsubishi Heavy Industries, and NETL are optimizing the chemical scrubbing technology. In particular, the focus is on the improvement of solvent formulations, lowering stripping steam requirements, the thermal integration of CO_2 capture system with a power plant, etc. A series of advanced amines with improved properties and stability have been developed, such as sterically hindered amines KS-1, KS-2, and KS-3, 2-amino-2-methyl-1-propanol (AMP), Cansolv, HTC Purenergy.

The main challenges to the amine-based CO_2 capture systems as applied to Post-CCC are associated with the relatively large parasitic loads originating mainly

Table 7.2 CO_2 capture technologies for Post-CCC applications

Processes, methods	Compounds, materials, processes
Chemical solvents	Monoethanolamine (MEA)
	Diglycolamine (DGA®)
	Sterically hindered amines KS-1, KS-2, and KS-3
	Cansolv
	HTC Purenergy
	2-Amino-2-methyl-1-propanol (AMP)
	Aqueous ammonia
	Chilled ammonia
Chemical sorbents	CaO
	Amine-enriched sorbents
Membranes	N_2/CO_2 polymer membranes
	Membrane-amine hybrids
	Gas absorption membranes
Enzymatic CO_2 capture process	Carbozyme
Physical solvents	Ionic liquids
Physical sorbents	Metal-organic frameworks (MOF)

Source [1, 14]

from three sources: (1) heating required to drive CO_2 off of an absorbing solution (needed for the absorbent regeneration), (2) pumping solutions, and (3) compressing the purified CO_2 to pipeline pressure. It was estimated that the parasitic loads due to equipping coal-fired power plants with CCS would reduce thermal efficiency of the plant from 38–39 % to about 27 % [12]. Even using advanced amine-based solvents would still keep the efficiency below 33 %. To reduce these parasitic losses, a team of researchers from Massachusetts Institute of Technology (MIT) have developed novel electrochemically mediated amine regeneration process for the post-combustion applications [36]. The technology shows the potential to exploit the excellent CO_2 removal efficiencies of thermal amine scrubbers while reducing parasitic energy losses and capital costs. The additional advantages of the developed process include higher CO_2 desorption pressures, smaller absorbers, and lower energy demands. The technology is still in early R&D stage of the development.

The US DOE/NETL is investigating advanced solvents that are more resistant to flue gas impurities (e.g., NO_x, SO_x) and could potentially incur lower energy penalties for the regeneration step compared to MEA. The aqueous ammonia-based CO_2 capture system (which converts ammonia into ammonium carbonate) showed some promise for capturing CO_2 in laboratory-scale studies. This technology exhibits the reduced heat requirement compared to amines, and it can potentially produce a fertilizer by-product by co-capturing SO_x and NO_x impurities in the flue gas. On the negative side, the system requires relatively low temperatures (26.8 °C) for ammonia carbonate to remain stable, and, besides, the reaction cycles involving ammonia reacting with CO_2 do not offer energy savings compared to amines.

CO_2 Technology Centre Mongstad (CO_2-TCM) in Norway has been developing Post-CCC technology to capture 85 % of the CO_2 contained in the flue gas from a nearby combined heat and power plant and a refinery cracker [37]. After a comprehensive evaluation, the CO_2-TCM developers selected two CO_2 capture processes: a chilled ammonia process from Alstom and an amine process from Aker Clean Carbon (ACC). Designed to capture about 100,000 t of CO_2 per year, the project will be the largest demonstration of CO_2 capture to date. The objective of the project is to increase knowledge on the chosen capture technology in order to reduce the technical and financial risk uncertainty and facilitate the wide-scale deployment of the technology.

New solvent chemistries and new process designs aiming at reducing the parasitic loads are currently underway. Of particular interest is a new class of solvents based on reversible ionic liquids (IL) that are capable of capturing CO_2 from low-pressure flue gases. Typically, IL contain an organic cation and either an inorganic or organic anion. Advantageously, the IL have very low vapor pressure and are thermally stable at temperatures up to several hundred degrees Centigrade (which would help to minimize the solvent loss during a high temperature operation). Also, as a physical solvent, IL require a relatively low heat input for CO_2 recovery. It has been reported that such IL as 1-hexyl-3-methylpyridinium bis(trifluoromethylsulfonyl) imide [hmim][Tf2N], trihexyltetradecyl phosphonium prolinate [P66614][Pro], trihexyltetradecyl phosphonium methionate [P66614][Met], especially those functionalized with amine groups, are promising solvents for the post-combustion CO_2 capture [38]. By tethering the amine group to the IL anion, the stoichiometry of the CO_2 capture reaction can be significantly increased (e.g., from one CO_2 for every two amines, as is the case with aqueous MEA, to one CO_2 for each amine). On the negative side, however, most IL are very viscous liquids, which may make pumping of these solvents in a power plant application very difficult and energy intensive. Furthermore, since the IL are not manufactured commercially, they are still very expensive ($350–2,000/kg), and their toxicity is unknown. The large-scale production of IL would make the process more economically attractive.

7.3.2.2 Chemical Sorbents

CO_2 removal from post-combustion flue gases by means of regenerable chemical sorbents offers the advantages of reducing the efficiency penalties compared to wet-absorption systems. In this method, flue gas is put in contact with a solid sorbent at high temperatures to allow the conversion of CO_2, typically, to carbonates. The solid sorbent then can be easily separated from the gas stream and sent to a regenerator-reactor (alternatively, the gas streams can be switched between the reactor and regenerator apparatuses). The key factors in the development of these systems is the adsorption capacity of sorbents and their cost.

CaO, as a typical representative of the family of regenerable CO_2 sorbents, has been used for many decades in a variety of industrial applications. The carbonation step involving CaO sorbent (reaction 7.4) is a very fast reaction occurring at temperatures above 600 °C, and its regeneration by calcining $CaCO_3$ to CaO and CO_2

(reaction 7.5) is thermodynamically favorable at temperatures above 900 °C and partial pressure of CO_2 of 0.1 MPa:

$$CaO + CO_2 \rightarrow CaCO_3 \quad (7.4)$$

$$CaCO_3 \rightarrow CaO + CO_2 \quad (7.5)$$

The main shortcoming of this method is that naturally occurring carbonate sorbents, such as limestone and dolomite, rapidly deactivate, which requires a large make-up flow of the sorbent to maintain the performance of the CO_2 capture-regeneration loop. On the positive side is that the CaO sorbent is not expensive, and the spent sorbent could find applications (e.g., in cement industry). The use of the CaO-sorbent method was successfully tested at a pilot plant with the capacity of 40 t/day utilizing two interconnected fluidized bed reactors (e.g., in *Acceptor Coal Gasification* process) [25].

Recent technological developments in this area include the chemical enhancement of physical sorption capacity of the sorbents. For example, researchers at the US DOE/NETL have developed amine-enriched sorbents that are produced from high-surface area sorbents (e.g., zeolites) by treating them with various amine compounds, which increases the surface contact area of the system and facilitates CO_2 capture [16]. The elimination of a water carrier in these systems offers another advantage since it improves the energy efficiency and economics of the process relative to the MEA scrubbing technology. The amine-enriched sorbents have demonstrated at least 8 wt% CO_2 uptake and stood up to more than 250 operating cycles [39]. The system's drawbacks include the difficulty of lowering and raising temperature of the solid material (compared to liquid solvents), which may cause the reduction in desorption rates.

In the *Dry Carbonate* process developed by the Research Triangle Institute (USA), supported sodium carbonate (Na_2CO_3) is used for scrubbing CO_2 from post-combustion flue gases. In this process, sodium carbonate reacts with CO_2 and water forming sodium bicarbonate ($NaHCO_3$) via a reversible reaction that requires temperature swing from 60 to 120 °C for the sorbent to be regenerated. The economic advantages of this process over conventional amine scrubbing are the reduced capital costs, lower auxiliary power load, and lower material costs. However, the process faces the challenges related to the continuous circulation of large quantities of solids and sensitivity to contaminants.

7.3.2.3 Membranes

Considerable R&D efforts are focused on the development of highly selective and permeable membranes for separating CO_2 from low partial pressure (i.e., diluted CO_2) flue gas streams. The latest developments in this area include the novel type of hybrid gas absorption membranes, where CO_2 separation is achieved by a hybrid system combining a CO_2-permeable membrane with an absorption solvent (e.g., MEA) to selectively remove CO_2 from flue gases [16]. In this membrane–liquid sorbent hybrid system, flue gas is contacted with a membrane, and a sorbent solution on the permeate side absorbs CO_2 and creates a partial pressure differential to draw CO_2 across the membrane. The advantages of the system include reduced attrition and shielding of the amine compound from contaminants in flue gas by the membrane. The shortcomings of the hybrid system relate to high additional costs associated with the membrane and the inability of the membranes to keep out all unwanted contaminants. The hybrid membrane/liquid sorbent CO_2 capture system is still at laboratory-scale stage of the development and would require pilot-scale testing to prove the commercial potential of the technology.

MTR company (USA) is developing thin-film composite polymer membranes to increase the CO_2 flux across the membrane by using a novel countercurrent flow design, where a portion of incoming combustion air is utilized as a sweep gas to maximize the driving force for membrane permeation. The company's Polaris™ membrane system uses a CO_2-selective polymeric membrane (microporous films, which act as semipermanent barriers to separate two different mediums) designed to capture CO_2 from flue gases of power plants. According to preliminary estimates, 90 % CO_2 capture at a 600 MW coal-fired power plant would require about 700,000 m^2 of membrane surface with a total footprint of about 2,024 m^2 [21].

Los Alamos National Laboratory (USA) in a partnership with Idaho National Energy and Engineering Laboratory, Pall Corp. and Shell Oil Co. are developing a new approach to CO_2 separation using thermally optimized membranes. In this concept, a desirable combination of high selectivity, high permeability, and mechanical stability is achieved at temperatures significantly higher than that of conventional polymeric membranes [17]. A polymeric-metallic membrane that is selective toward CO_2 at temperatures as high as 350 °C is under performance evaluation. The developers of the high temperature polymeric-metallic composite membranes target separating CO_2 at temperatures up to 450 °C and pressures of 10–150 atm to improve process economics.

7.3.2.4 Enzymatic CO_2 Capture Systems

CO_2 capture and release by enzyme-based CO_2 capture systems mimic a mammalian respiratory mechanism. The enzymatic sorbents feature fast CO_2 capture kinetics at the lower system size, and they can produce CO_2 at above atmospheric pressure. The system's shortcomings include low temperature resistance (temperature limited to below 38 °C to avoid the degradation of the enzymes) and the requirement to cool the flue gas before sorption (since the CO_2 sorption process is exothermic). The enzymes could be deactivated by SO_x, NO_x, and other acid gases; therefore, a preliminary gas cleanup would be required. In the enzymatic system developed by *Carbozyme* company, carbonic anhydrase enzyme catalyzes the conversion of CO_2 to bicarbonate at the flue gas interface and reverses the process at the CO_2 product side. The Carbozyme membrane system consists of two hollow-fiber microporous membranes separated by a thin liquid membrane. The laboratory-scale enzyme-facilitated membrane was validated recently on a 0.5 m^3 permeator demonstrating 85 % removal of CO_2 from a feed stream containing 15.4 vol.% CO_2 [40].

7.3.2.5 Cryogenic Methods

Until recently, cryogenic CO_2 capture was not considered a realistic option for the post-combustion CO_2 capture, mostly, due to expected high cooling costs. Indeed, in order to remove CO_2 from flue gases through the cryogenic gas–liquid separation, it would be necessary to compress the gas to pressures above the triple point of CO_2 (pressure of 5.2 bar and temperature of −56.6 °C for pure CO_2), which is too energy intensive and expensive [41]. Alternatively, flue gas could be cooled to below CO_2 sublimation point at atmospheric pressure, but this would result in CO_2 solid phase formation, which cannot be easily handled in a standard process equipment. The major advantage of the cryogenic concept is that deep CO_2 removal can be achieved: at minus150°C the vapor pressure of CO_2 is as low as 8 Pa, resulting in more than 99.9 % CO_2 removal (most of other existing separation technologies can only achieve up to 90 % of removal). This feature would generate both a very "clean" flue gas and high-purity CO_2 product (the process will simultaneously remove sulfurous and other impurities).

A novel cryogenic CO_2 capture technology based on the dynamic operation of packed beds (DOPB) that enables capturing CO_2 at atmospheric pressure has been recently developed in the Netherlands [41]. The process consists of three consecutive steps, cooling, capture, and recovery, and requires cooling down to temperatures below −120 °C. The techno-economic analysis of the process indicate that the economic feasibility of cryogenic CO_2 separation using DOPB method strongly depends on the availability of a low-cost (or free) cold source (e.g., the evaporation of LNG at a regasification terminal) and CO_2 concentration in the feed gas. In the case of using LNG evaporation as a source of cooling, 1 kg of CO_2 avoided corresponds to the evaporation of 2.7 kg of LNG [41]. If no LNG option is available and the entire required cooling capacity is to be provided by cryogenic refrigerators, the electricity consumption will be comparable to that of the power plant output, making the whole concept unrealistic.

Another recent advancement in the cryogenic post-combustion CO_2 capture technologies includes *Cryogenic Carbon Capture* (CrCC) process developed in the USA [42]. In this process, flue gas is cooled to desublimation temperatures (−100 to −135 °C) to form solid CO_2 that is separated from light gases; the cold products cool the incoming gases in a recuperative heat exchanger, and the solid/liquid CO_2 is compressed to final pressure of 100–200 atm, resulting in the separation of the compressed CO_2 stream from the light gas stream at atmospheric pressure. The developers of the technology claim that the overall energy penalties and economic costs would be at least 30 % lower than that of the most competing processes that involve ASU, amine solvents, and membranes. Additionally, the CrCC process could provide the highly efficient removal of most impurities (Hg, SO_x, NO_x, HCl, etc.) and has potential water usage savings. The CrCC process is supported by the US DOE and it is still in an early stage of technological development.

7.3.2.6 Metal-Organic Frameworks

Due to their exceptional capacity to capture and store CO_2, the unique porous materials, MOF have been extensively studied for CO_2 capture from Post-CCC flue gases. A team of researchers from Nottingham and Newcastle universities has designed MOF named NOTT-202a to adsorb and release CO_2 at lower temperatures than existing capture methods [43]. The material adsorbs CO_2 under pressure and releases it as the pressure is decreased, while allowing other components of the gaseous mixture such as hydrogen, nitrogen, and methane to pass through. In contrast to conventional amine solutions, NOTT-202a does not require heating to release the CO_2 and is not toxic. The material consists of two interpenetrating networks attached to a central indium metal atom, but with holes large enough to hold CO_2.

Summary of Post-CCC

The TRL ranking of main CO_2 capture technologies applicable to Post-CCC is shown below [12] (TRL values are shown in *italics*):

Absorption	*7*
Adsorption	*6*
Membrane separation	*6*
Enzymatic capture	*5*
Other methods	*4*

The post-combustion CO_2 capture option is less developed compared to the pre-combustion capture: most of the technologies have reached TRL of 5–7. The majority of the Post-CCC projects utilize amine-based absorption processes, albeit, at a relatively small scale (i.e., slipstreams from coal-fired power plants in 5–25 MW range). The adsorbent and membrane technologies promise lower energy consumption against absorption, but they are at the early phases of development (kW-range pilot units). The pilot test results of CO_2 capture from coal-fired flue gas by physical adsorption processes show that the energy consumption for carbon capture has significantly improved from original 708 kWh per ton CO_2 to 560 kWh per ton CO_2 [25]. Other methods are still in R&D and pilot stages.

7.4 Oxyfuel Combustion Capture

7.4.1 Current Status of Technology

Oxyfuel combustion (OFC) is a relatively new concept in the family of CCS technologies. OFC uses nearly pure oxygen instead of air for fossil fuel combustion resulting in flue gas consisting mainly of water vapor and CO_2 (see the process scheme in Fig. 7.2c). A portion of cooled flue gas is recycled back to the combustor to moderate very high temperature in the fuel combustion zone resulting from the use of pure oxygen. This process practically eliminates nitrogen from the technological chain, thus, significantly simplifying the downstream gas separation, purification, and compression stages. After condensation of water (through the use of cooling and desiccant systems and compression to a dew point of minus 40 °C), approximately 90 % pure CO_2 can be transported and stored directly without further purification (if regulations and geochemical conditions of the storage site would permit that) [12]. If specifications for CO_2 transportation and storage would require more pure grade CO_2, the flue gas impurities (mainly, O_2, N_2, and Ar) have to be further removed.

The OFC technology offers several important advantages as follows:

- Fuel flexibility: OFC can be applied to solid (coal, petroleum coke, biomass) as well as to liquid (residual oil) and gaseous (NG, refinery gas) fuels
- Applicability to both existing and new coal-fired power plants using conventional steam cycle technology
- Significant (60–70 %) reduction in NO_x emissions compared to air-fired combustion [17]
- The ultra-low emissions of conventional pollutants (SO_x, NO_x) can be achieved largely as a concomitant result of the selected CO_2 purification processes at little or no additional cost [12]
- The possibility of the co-capture of SO_2 (and its storage if the co-disposal becomes technically and economically feasible)
- Enhanced capacity to remove mercury (due to oxidation of mercury followed by its capture in an electrostatic precipitator)
- The added process equipment is largely familiar to power plant operators; no chemical operations or significant onsite chemical inventory is required
- Potential cost savings (due to more compact units and the elimination of certain gas cleanup devices)

The technology, however, is facing several technical challenges, in particular:

- The significant amount of pure oxygen is required, which increases CO_2 capture cost (oxygen is typically produced by energy-intensive cryogenic air separation or by using adsorption techniques)
- Operation of O_2-fired boilers requires high temperature resistant materials; there is also a possibility of air leakage
- The recycle of flue gas (about 70–80 %) is required to approximate the combustion and heat transfer characteristics of air in order to use a currently available combustion equipment (through the so-called synthetic air approach); this would result in an increased auxiliary power usage at OFC plants and, consequently, in the reduction in the net power production (by about 23 %) and decrease in net efficiency compared to air-fired plant with comparable output [12]
- If raw CO_2 is targeted for the storage, potentially, there could be a CO_2–NO_x–SO_x co-disposal problem (compressibility, corrosion issues, etc.)
- Environmental issues related to the emissions of carbon monoxide (CO) and unburned carbon, high concentration of acidic gases in the condensate, injection of toxic substances to sequestration sites, etc.

As of 2012, no full-scale commercial OFC-based power generation plants were being operated anywhere in the world, although there are currently several integrated pilot plants and demonstration projects utilizing OFC technology in Europe, the USA and Australia. Since 2009, Vattenfall company has operated a lignite-fueled 30 MW_{th} pilot plant at their Schwarze Pumpe power plant in Germany [12]. Total company (France) has been operating Lacq project since 2010 that involves an integrated NG-fueled 30MW_{th} boiler. CIUDEN's coal-fired OFC test facility in Spain includes a 20 MW_{th} pulverized coal boiler and a 30 MW_{th} circulating fluidized bed boiler. Babcock & Wilcox Co. (Ohio, USA) has successfully completed pulverized coal oxy-combustion testing at a 1.5 MW_{th} scale unit, and the technology is currently being evaluated at a 30 MW_{th} pilot scale unit. The project demonstrated 80 % reduction in the flue gas volume while achieving CO_2 concentration of 80 vol.% [16]. The pilot scale testing has also demonstrated the possibility of a smooth transition between air- and oxygen-firing modes.

Alstom Power (USA) has conducted the pilot-scale (3 MW_{th}) testing of the oxy-combustion process using a circulating bed fluidized bed combustor with coal and petroleum coke as fuels. The oxyfuel combustion technology is also being actively pursued in Australia, where several existing coal power plants are planned to be retrofitted with the OFC and amine-based CO_2 capture systems. For example, the CS Energy company is planning to convert a 30 MW_{el} pulverized coal power plant into OFC technology in Queensland, Australia. Besides the above-mentioned pilot plants, five larger plants (at TRL 8 stage) are in the planning and engineering stages of development worldwide [12]. Among planned full-scale commercial oxyfuel combustion units is a 200 MW coal-fired power plant in Meredosia (Illinois, USA) [44].

It is recognized that the technology development path for the OFC option may be more costly compared to Pre-CCC or Post-CCC because the former requires the commitment of the whole plant to the technology, whereas two other carbon capture options can be developed on the slip streams of existing plants [12]. While different retrofitting or repowering schemes have been proposed and are under consideration, it is not yet clear whether they will be economically justified. More studies are to be conducted to prove that the cost of oxyfired retrofit plant is lower than that of an optimized newly built plant.

7.4.2 Improvements to Technology and Future Directions

The increased auxiliary power use for air separation and recycling of flue gas results in a significant reduction in the net power output of OCF plants. A number of technological improvements to increase the overall efficiency of OFC plants are under consideration, including:

- An advanced ultra-supercritical (USC) steam turbine cycle with temperature of 680–700 °C and pressure of 352 bar
- Gas pressurized OFC with reduced recycle fan auxiliary power use and improved boiler efficiency

- Chemical looping combustion option to dramatically reduce the auxiliary power usage due to air separation
- Ion transport membranes

Employing an advanced USC turbine cycle would improve the plant's efficiency by 3.5 %, while using gas pressurized oxyfuel combustors would add another 1.4 % [12]. Employing chemical looping combustion technology promises the greatest (~5 %) improvement in thermal efficiency.

7.4.2.1 Chemical Looping Combustion

Chemical looping combustion (CLC) is a potentially attractive alternative to conventional OFC since it avoids direct contact between fuel and air. The main principle of the chemical looping combustion (CLC) process is to oxidize carbonaceous fuels (coal or gas) by an oxygen carrier, which circulates between the two reactors designed to combust fuel and regenerate the oxygen carrier. The advantage of CLC over conventional combustion approaches is that it obviates the need for an air separation plant and produces CO_2 in a highly concentrated sequestration-ready form (since it is not diluted with nitrogen), and it has a relatively simple gas purification stage (due to the absence of NO_x formation). Although the integration of CLC with gas pressurized oxyfuel combustion will be challenging, the combination of CLC with an advanced USC steam turbine cycle is technically feasible, and it can deliver an improvement that will make up and even exceed the losses due to added auxiliary power in CO_2 purification and recycling stages. This combination could result in OFC plant with 98 % CO_2 capture and near zero emissions of NO_x and SO_x pollutants with the efficiency comparable to state-of-the-art power plants currently being built [12].

The oxides of iron (Fe_2O_3), copper (CuO), nickel (NiO), and manganese (Mn_2O_3) are among suitable oxygen carriers for the CLC process [44]. Figure 7.4 depicts a simplified schematic diagram of a CLC system coupled with turbines.

In the metal oxidation reactor (MOR), the reduced form of the oxygen carrier, e.g., metallic Ni, exothermically reacts with air yielding NiO:

$$2\,Ni + O_2\,(+N_2) \rightarrow 2\,NiO\,(+N_2) + heat \tag{7.6}$$

The exothermic reaction increases the temperature of air, which enters a downstream expansion turbine producing electricity. The oxidized form of the oxygen carrier (NiO) is transported to the fuel oxidation reactor (FOR), where it is reacted with carbonaceous fuel (e.g., NG) resulting in the reduction of metal oxide to its original (metallic) form with the release of heat:

$$4\,NiO + CH_4 \rightarrow 4\,Ni + CO_2 + 2H_2O + heat \tag{7.7}$$

This strongly exothermic reaction produces a high-temperature gaseous stream that is expanded in a turbine for power generation. In a typical CLC process, the

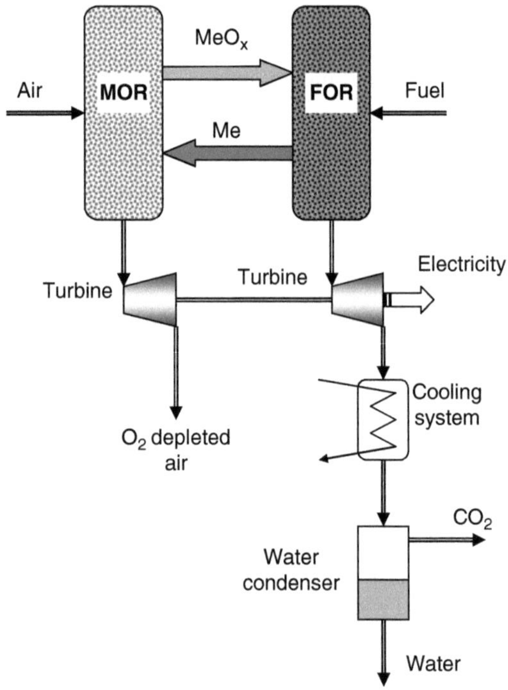

Fig. 7.4 Simplified schematic diagram of chemical looping combustion system. MOR—metal oxidation reactor, FOR—fuel oxidation reactor, Me—metal, MeO_x—metal oxide

oxygen carrier particles (approximately, 100–500 μm in size) circulate between the two reactors in a fluidized state, which is conducive to an efficient heat and mass transfer. Operational temperatures in the CLC process vary in the range of 800–1,200 °C. After water condensation, high purity CO_2 can be compressed to necessary pressure for the pipeline transmission. The thermodynamic analysis of the CLC system indicates that due to the lack of energy-intensive stages (e.g., air separation), relatively high overall efficiencies could be achieved by the CLC-based power stations [45].

A number of studies report the results of CLC technology testing. The fuel-to-electricity energy conversion efficiency of the CLC system running on NG fuel was estimated at the range of 45–50 % [46]. The application of the CLC technology to the NG-fired power plant utilizing combined cycle with CO_2 capture was reported by Naqvi et al. [47]. The system included a single CLC reactor coupled with the air and CO_2 turbines and steam cycle that was shown to achieve a net plant efficiency of about 52 % at the oxidation reactor temperature of 1,200 °C. The authors of the above studies concluded that the net plant efficiency of the CLC-based systems with close to 100 % carbon capture is superior to that of Post-CCC using amine solvents.

Currently, the CLC process is still in the pilot/demonstration stages of the development. There are indications that some companies (e.g., ALSTOM and CES) are currently building CLC-based power plants [48, 49]. Major challenges facing the CLC technology relate to the development of oxygen carriers that possess adequate long-term mechanical and chemical stabilities.

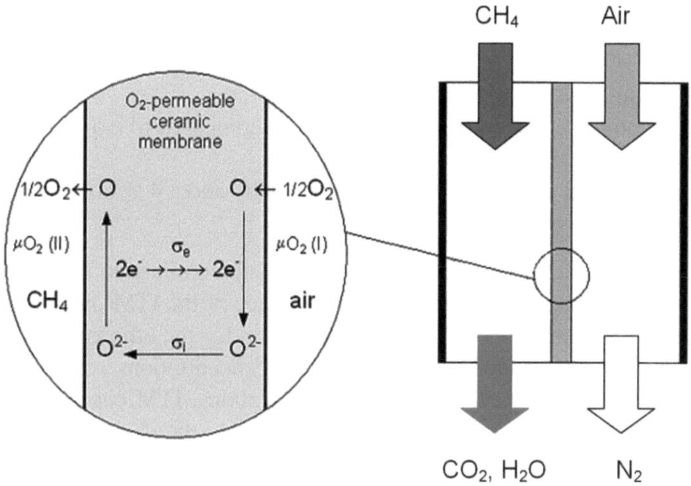

Fig. 7.5 Schematic diagram explaining the function of ceramic ion transport membranes. μ is chemical potential, σ_i and σ_e are ion and electron conductivities, respectively

7.4.2.2 Ion Transport Membranes

The use of O_2-separating membranes is another promising approach to avoiding energy penalties and high cost associated with the use of expensive air separation plants for OFC applications. Although the different types of membranes can be used for air separation, ceramic membranes are of particular interest due to their high selectivity, thermal stability, and, most importantly, their capacity to be integrated into an high-temperature oxy-combustion process. The O_2-separating dense ceramic membranes are referred to as *ion transport membranes* (*ITM*) or *mixed ionic-electronic conducting membranes (MIECM)* because two types of conducting species, ions, and electrons, participate in transporting oxygen through the membranes.

Typically, the ceramic membranes operate at the temperature range of 700–1,000 °C. In the ITM reactors, a ceramic membrane (i.e., MIECM) separates air and fuel streams, as shown in Fig. 7.5 for the case of methane fuel.

Oxygen molecules in contact with the ceramic surface are transformed into O^{2-} ions that are driven from the higher O_2 pressure (air side) to the lower O_2 pressure (fuel side) of the membrane through oxygen vacancies. In order to maintain charge balance within the ceramic membrane, electrons (supplied by fuel) are driven in the opposite direction (see the inset in Fig. 7.5). At the fuel side of the MIECM, oxygen ions are transformed to O_2 molecules that react with hydrocarbon molecules producing oxidation products (CO_2 and H_2O), thus, decreasing oxygen pressure and facilitating the O_2 flux across the membrane. The chemical composition and thickness of the membrane and interfacial processes are key parameters affecting O_2 permeation flux.

The advantages of the ITM technology over conventional O_2-fired fuel combustion approaches include [48]:

- The potential to achieve nearly 100 % CO_2 capture
- Reducing power consumption by about 70 % compared to existing methods of air separation
- Increasing the net power generation efficiency by about 4 % compared to a conventional OFC process

In some cases, the ITM is coupled with catalytic combustion of gaseous fuels, where a layer of catalyst is deposited on the surface of the ITM on the fuel side of the membrane. The presence of the catalysts allows for the efficient combustion of fuels at reduced temperatures and at lower fuel concentrations compared to those used in the regular flame combustion of hydrocarbons. ITM can be fabricated in tubular or planar configurations, which may facilitate the design of compact and efficient ITM reactors.

Key requirements to ITM materials are [48]: (a) fast oxygen-ion diffusion rate in the ceramic lattice, (b) high electronic conductivity, (c) thermodynamic stability under reducing conditions, and (d) stability under conditions favorable for carbon deposition. Among different classes of ITM, perovskite-based MIECM exhibit relatively high O_2 permeation fluxes. Currently, the acceptor-doped perovskites with the general chemical formula of $La_{1-x}A_xCo_{1-y}B_yO_{3-\delta}$ (where A = Sr, Ba, Ca and B = Fe, Cu, Ni) are ranked among the most promising materials for O_2 separation membranes. In particular, efficient O_2 transport has been reported for the perovskites with formula $La_{1-x}Sr_xCo_{1-y}Fe_yO_{3-\delta}$; more specifically, $La_{0.6}Sr_{0.4}Co_{0.2}Fe_{0.8}O_{3-\delta}$ was often used as an ITM material in experimental studies [50–52]. Significantly improved O_2 permeation flux has been demonstrated using another perovskite formulation: $Ba_{0.5}Sr_{0.5}Co_{0.8}Fe_{0.2}O_{3-\delta}$ [53].

In general, the oxygen permeation rate through membranes with mixed ionic and electron conduction is controlled by two factors: the oxygen ions bulk diffusion and the interfacial oxygen exchange on both sides of the membrane [54]. If the O_2 permeation rate is controlled by bulk diffusion, the O_2 permeation flux can be expressed by the Wagner's equation:

$$J_{O2} = -\frac{RT}{16F^2 L} \int_{\ln p'_{O2}}^{\ln p''_{O2}} \frac{\sigma_e \sigma_i}{\sigma_e + \sigma_i} d \ln p_{O2} \tag{7.8}$$

where J_{O2} is oxygen permeation flux; σ_e and σ_i are electronic and ionic conductivities, respectively; p_{O2}' and p_{O2}'' are high and low oxygen partial pressures on each side of the membrane; L is the thickness of the membrane; R and F are gas constant and Faraday constant, respectively; and T is absolute temperature.

The main limiting factor in the commercialization of ITM is a relatively low O_2 flux through the membranes and their high cost. For perovskite-type membranes, the O_2 flux density typically varies in the range of 1–8 μmol/cm^2 s, which may translate into very large surface areas required for practical systems. Among other ITM

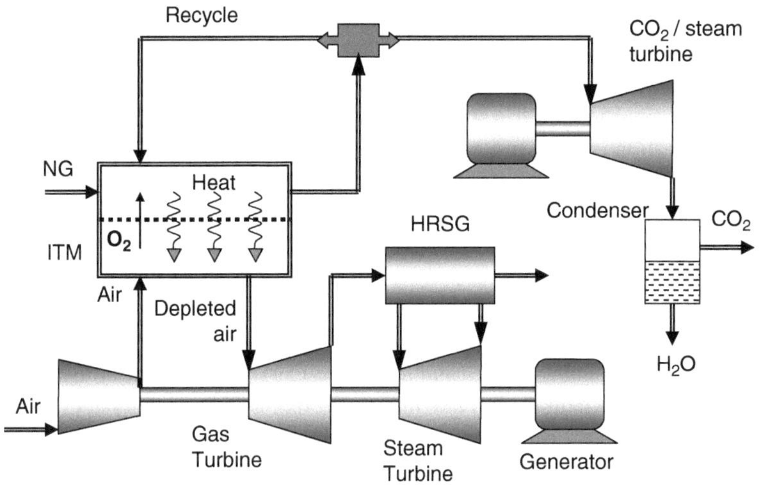

Fig. 7.6 Simplified diagram of the Advanced Zero-Emission Power plant (AZEP) concept. HRSG is heat recovery steam generator. *Source* [59]

challenges are demanding requirements for the high chemical and mechanical stability of the membranes under elevated temperatures (up to 1,000 °C) and pressure and oxidizing or reducing environment that are typical of actual operating conditions of the OFC systems. There have been reports that some membrane materials were subjected to a degradation under exposure to CH_4, CO_2, and CO, which adversely affected O_2 permeability through the membrane [55, 56]. $La_{0.6}Sr_{0.4}Co_{0.2}Fe_{0.8}O_{3-\delta}$ and La_2NiO_4 membranes, on the other hand, demonstrated a negligible degradation in CO_2 atmosphere [57, 58].

The application of ITM to OFC systems has been reported by a number of researchers [48, 59, 60]. The Advanced Zero-emissions Power Plant (AZEP) concept has been proposed as an advanced OFC process where a combustor was replaced with an ITM reactor. Figure 7.6 depicts the simplified schematic diagram of the AZEP concept as applied to NG fuel.

In the AZEP concept, the ITM reactor has three main functions: (a) the separation of O_2 from air, (b) the near-stoichiometric combustion of fuel, and (c) an efficient heat exchange (the transfer of combustion heat to O_2-depleted air). The AZEP system consists of a Brayton cycle combined with a bottoming steam cycle, CO_2-steam turbine, and a heat recovery steam generator. Pressurized NG and air enter two separate sections of the ITM reactor; in addition, a fraction of the exhaust gas from the fuel side of the ITM is recycled back to moderate the reactor temperature. Oxygen stream permeates through the ITM into the fuel section, where it oxidizes fuel producing heat and CO_2-steam exhaust gas, part of which is recycled to the ITM reactor and part is expanded into a specially designed CO_2-steam turbine. The substantial amount of heat generated in the fuel combustion section is transferred to O_2-depleted air, which is expanded in the main gas turbine. CO_2 is easily separated from water by cooling and condensing the latter, and is ready for compression and transport to a storage site. It was estimated that the deployment of AZEP would reduce the CO_2 capture costs by about 50 % compared to conventional NGCC with Post-CCC, at significantly lower investment cost [61].

The application of ITM to coal-fired power plants is a very active area of development. Praxair Inc. is conducting R&D work on the integration of an efficient ITM with the combustion process to enhance coal-fired boiler efficiency. According to the reported estimates, the deployment of ITM could reduce a parasitic power consumption required for O_2 production by 70–80 % (compared to a cryogenic ASU) [62]. Praxair in collaboration with the University of Utah (USA) researchers are designing and constructing a bench-scale ITM to evaluate its performance in a coal-based power system using single- and multitube ITM reactors. The construction and testing of a development-scale ITM-integrated oxy-combustion system is targeted for 2014–2016.

TRL rankings of the main OFC components are shown below [12]:

Air separation unit	9
Oxyfired boiler	7
CO_2 purification	8
CO_2 compression and drying	9
Ion transport membrane	5

ASU and CO_2 compression/drying are established commercial processes, whereas some components of the OFC system are similar to those of air-fired systems (e.g., heat transfer, gas quality control, thermal power utilization, and material handling) and would not require significant efforts in bringing them to the TRL 9 stage. The greatest remaining technical challenge is to integrate these subsystems and components into a complete power generation facility with carbon capture.

7.4.3 Carbon Capture Technologies: Challenges and Outlook

Although all the carbon capture technologies have the same objective of preventing fossil fuel-generated CO_2 from reaching the atmosphere, each of them faces a specific set of problems, challenges, and barriers to overcome.

- Pre-CCC is integrated (or "prewired") by the nature of this technological approach; thus, the operational problems in CO_2 capture could potentially impact the plant's performance through a lower reliability and availability. There is also a need to significantly improve the water-gas shift (or CO-shift) stage with new more active and durable catalysts, as well as the CO_2 capture stage with new adsorbents and absorbents having better capacity and durability. An optimization and integration on component and system levels should also be pursued.
- Post-CCC, due to its "end-of-the-pipe" nature, can be retrofitted to power plants, thus, providing greater flexibility and significantly reducing a capital investment risk.

Technological improvements are needed in the CO_2 absorption area, in particular, increasing the loading capacity of existing (first-generation) solvents and reducing their loss via chemical modifications or use of catalysts. There is also a need for a new (second) generation of efficient solvents capable of removing CO_2 and SO_2 from flue gases. The longer term Post-CCC technological developments would involve ionic liquids, phase change solvents, and high capacity adsorbents that are now at R&D stages. Reducing the cost of the Post-CCC technology and its impact on the plant's performance and environment are essential for its commercial success. According to many analyses, the Post-CCC will unlikely be applied to older plants due to the high energy penalties increasing dispatch costs, thus, impacting their capacity factor and, consequently, reducing revenue [12]. On the other hand, the ability of the Post-CCC to retrofit to newly installed high efficiency plants will be a critical aspect in ensuring that the technology will be able to operate and be highly competitive in the increasingly carbon-constrained world.

- OFC (similar to Pre-CCC) also operates in an integrated mode; thus, it may potentially have the same issues as pre-combustion systems (see above). For OFC, there is a persistent need for lowering the cost of oxygen production and associated energy penalties, and improving a boiler design and performance. Additionally, the development and demonstration of more efficient cycles (e.g., chemical looping combustion) and materials (e.g., ion transport membranes) will further improve the economics of OFC and its competitiveness. ITM is a promising technology that could potentially mitigate the penalties associated with O_2 separation; however, it is still mostly in R&D stage of the development. The significant enhancement in O_2 permeation flux and the reduction in its installation costs would be necessary for practical implementation of this technology.

7.5 Transport of CO_2

Transport of CO_2 is an intermediate operation (between CO_2 capture and its storage) in the overall CCS technological chain. Although in some projects, CO_2 capture and storage sites may be in a close proximity to each other; in most cases, CO_2 has to be transported from point-to-point for tens to hundreds of kilometers. Naturally, the longer distances translate into the higher cost and, in some cases, the additional challenges of technological and nontechnical nature, e.g., need for recompression and monitoring, unfavorable terrain, and public acceptance.

In principle, CO_2 can be transported in three physical states: gaseous, liquid, and solid. In order to transport CO_2 economically its volume should be substantially reduced: this can be done either by pressurization (compression), or liquefaction, or solidification, or hydration (to crystalline hydrates). CO_2 solidification is a more energy-intensive and costly option compared to CO_2 liquefaction, and, currently, it is not practiced for large-scale CO_2 transport (in a smaller scale, it is used in food and other industries). Transport of CO_2 in the form of hydrates is still at an early R&D stage.

7.5.1 CO_2 Compression and Dehydration

Captured CO_2 has to be compressed to the required pressure and dehydrated in order to meet pipeline transport specifications. CO_2 is converted into a dense-phase supercritical fluid by increasing the pressure above 7.4 MPa (the critical point pressure). CO_2 pipelines typically operate at the range of pressures 13.8–20.7 MPa, which allows for CO_2 to be pumped through the pipeline without further compression with the associated energy savings [1]. Before entering a pipeline, CO_2 has to be dried and cleaned of H_2S impurities in order to avoid a possible pipeline corrosion problem. During the staged compression of CO_2, its moisture content is reduced by cooling below its dew point and knocking out water. Finally, CO_2 stream is dehydrated by a treatment with solvents (typically, triethylene glycol, TEG) or solid adsorbents (e.g., molecular sieves). It has been reported that the CO_2 moisture content could be reduced to 20 ppm via TEG-based dehydration process [1]. Molecular sieves can achieve even higher levels of CO_2 dehydration.

To raise the pressure of 1 t of CO_2 from atmospheric to 10.34 MPa about 82 kWh of compression/pumping energy would be required. The selection of the type of a compressor depends on the volumetric flow rates, starting and final pressures, and gas composition (e.g., for the amine absorption process, starting pressure is 0.18 MPa). Currently, three compressor types are considered for CO_2 compression and pumping: (1) a reciprocating compressor, (2) a multistage, integrally geared centrifugal compressor, and (3) a single-shaft, multistage centrifugal compressor. MAN Turbo AG is one of the leading suppliers of CO_2 compressors.

Substantial R&D efforts are needed to improve the CO_2 compression technology for CCS applications, including (1) the development of large-scale semi-isothermal and high-pressure-ratio adiabatic CO_2 compressors, (2) the design of advanced, axial-flow CO_2 compressors that would allow the recovery of high temperature heat in the compressor after-coolers and, thus, improve the overall efficiency of power plants, (3) the optimization of the integration of CO_2 capture/compression systems together with plants, and (4) the establishment of the allowable levels of contaminants in CO_2 pipeline and/or compressors [1, 63].

7.5.2 Pipeline Transport of CO_2

Currently, the use of pipelines is the most economical method of CO_2 transportation in large quantities over long distances. It is a well-developed technology: building and operating long-distance CO_2 pipelines have been safely practiced for many decades by the oil and gas companies in the USA, Europe, and elsewhere. Currently, the CO_2 pipeline network extends over 6,000 km carrying about 50 millions of tons of CO_2 per year predominantly from industrial CO_2 sources to enhanced oil recovery sites.

Typically, pipelines are made of carbon-manganese steel, and CO_2 does not corrode them as long as relative humidity is below 60 %. It has been reported that

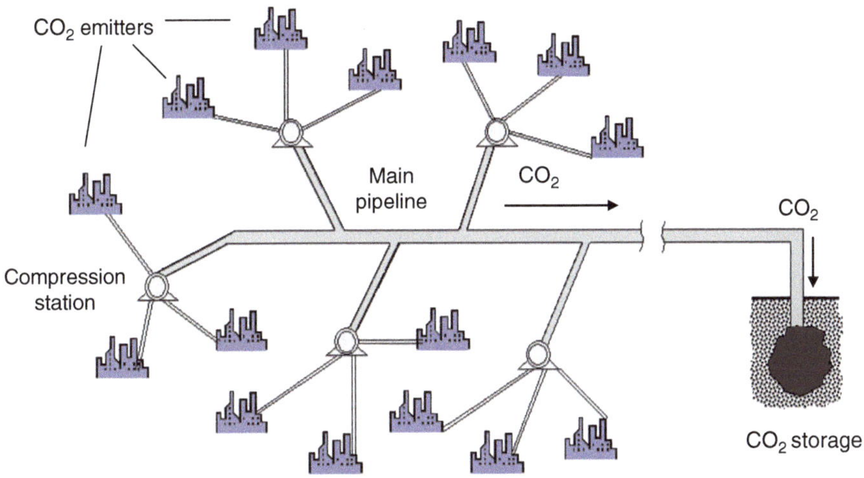

Fig. 7.7 Schematic diagram of an example of CO_2 transport network. *Source* [66]

at temperature of 3–22 °C, CO_2 pressure of 140 atm, and H_2S concentration of 800–1,000 ppm, the corrosion rate for X-60 carbon steel was less than 0.5 μm per year [64]. These and other factors put restrictions on the permissible levels of impurities in CO_2 to be transported through commercial pipelines. If corrosion-resistant pipelines from advanced inexpensive materials will be developed in the future, the cost of CO_2 transportation may be reduced as CO_2 streams with water vapor, H_2S, and other impurities could be safely transported to a storage site.

In many respects, the operation and maintenance of CO_2 pipelines is similar to those transporting NG, and the prior field experience indicates very few problems with the transport of high-pressure dry CO_2 in the carbon steel pipelines. The long-term operation data indicate that the incidence of CO_2 pipeline failure is relatively small: 0.0002–0.001 per km per year [65]. Despite its low probability, CO_2 leaking from pipelines may cause a potential physiological hazard, especially if CO_2 pipelines run in the vicinity of densely populated areas. This may add to the NIMBY sentiments among general public and result in a significant barrier to the implementation of the CCS systems. Despite some similarities with NG transport, one should take into account that CO_2 pipelines operate at much higher pressure than NG pipelines, and the CO_2 pipeline technology has not been developed to the same extent as gas pipelines.

The future large-scale deployment of CCS will necessitate the development of CO_2 transport clusters, hubs and networks that will be based on a comprehensive analysis and matching of CO_2 emitters and sinks (i.e., storage sites) [66]. In such systems, clusters of proximate CO_2 emitters will be linked through a hub to clusters of sinks by trunk pipelines. A representative network will resemble a "tree," where branches will represent feeder pipelines from CO_2 sources (e.g., power plants and cement manufacturing facilities), the trunk of the tree will be the main pipeline, and the roots will be the pipelines linking to storage sites. Figure 7.7 illustrates one

Table 7.3 Pipeline length required to meet the plans for future CCS deployment

		Year		
Region	Distance	2020	2030	2050
USA	km	8,000–21,000		35,000–58,000
Europe	km	2,300	15,000	22,000

Source [12]

example of the CO_2 transportation network (only one carbon sink is shown). Note that, in most cases, the existing system of oil/gas pipelines (e.g., the extensive network of oil and gas pipelines under the North Sea), in future, could be used for transporting CO_2.

The development of CO_2 transport networks has already been undertaken in many regions of the world (the USA, Europe, Australia). In Europe, the most advanced efforts in developing the CCS transport network relate to *Rotterdam Climate Initiative* (RCI) project. The project, which started in 2006 and now includes 18 participating companies, is aiming at handling as much as 20 Mt/year CO_2 by 2025 [12]. In North America, several regional partnerships are working on the development of CO_2 transport networks, e.g., the Plains CO_2 Reduction Partnership (PCOR) in the USA, the Integrated CO_2 Network (ICO2N) in Canada. In Australia, the Collie Hub and the CarbonNet CCS network are planning to integrate multiple CCS projects and industrial partners across the entire CCS chain within next decade to support transporting of close to 30 Mt/year CO_2 [12].

The projected lengths of the pipeline infrastructure at the scale needed to support full-scale commercial CCS deployment in the USA and Europe over the period of 2020–2050 are shown in Table 7.3.

To comply with the scale of the development reflected in Table 7.3, the rate of construction would need to be in the range of 1,200-1,500 km/year, which seems achievable based on history of pipeline construction by NG industry in the USA and Europe (e.g., 33,521 km of pipelines was built in the USA during decade of 1998–2007) [12]. On the other hand, due to such a large scale of pipeline construction the CCS will compete for resources (e.g., steel) with other pipeline construction needs, which may affect the rate of the construction.

7.5.3 *Transport of CO_2 by Shipping*

CO_2 is currently routinely transported by marine tankers (e.g., transporting CO_2 for enhanced oil recovery in the North Sea), but on a relatively small scale compared to pipeline transport. This option is particularly preferred when an emission source is within a reasonable distance to suitable seaport facilities that can be equipped to load CO_2 to a ship for injection in offshore storage sites.

The physical properties of L-CO_2 are not drastically different from those of liquefied light hydrocarbons (methane, ethane, and propane); so, the technology could be adapted and scaled up to extremely large L-CO_2 carriers serving large-scale CCS projects. L-CO_2 can only exist at a condition of low temperature and elevated pressure. A semi-refrigerated CO_2 tank (temperature from −50 to −54 °C, pressure 6–7 atm) is preferred for CO_2 transport by ships. Since these conditions are close to that of LPG carriers, in principle, CO_2 tankers could be constructed using the same technology as existing LPG carriers. Large L-NG carriers reach the capacity of 200,000 m^3, which could potentially transport 230,000 t of L-CO_2 [25].

Currently, worldwide, there are only a few relatively small ships specifically designed to transport L-CO_2, e.g., Coral Carbonic tanker with the capacity of 1,250 m^3 built in 1999 and operated by Anthony Veder [1]. These vessels are designed to transport food-grade L-CO_2 from ammonia plants in northern Europe to coastal distribution terminals, from where CO_2 is delivered to customers by tanker trucks or in pressurized cylinders [25].

Although most of marine CO_2 transport is currently serving enhanced oil recovery and other industrial users, there is a growing interest in using the ship tankers to transport CO_2 specifically for CCS applications. An intensive research and design work is ongoing in Norway and Japan to adapt the knowledge gained during decades-long operation of L-NG transport by ships to L-CO_2 transport [63]. In particular, the companies in Norway and Japan are working on the design of large L-CO2 carrier-ships and associated infrastructure, i.e., CO_2 liquefaction plants and intermediate onshore and offshore storage facilities. Statoil (Norway) is planning to transport CO_2 extracted from flue gases in pressurized tanks at temperature of −50 °C to offshore oil and gas fields (techno-economic analysis indicated that tanker-based CO_2 transport will be more cost effective than pipeline one) [1]. In one planned project, a 22,000 m^3 tanker will carry LPG from an oil field to a shore terminal, where it will discharge the cargo and replace it with CO_2 for the return journey to an offshore field. Major shipping companies, e.g., Maersk Tankers (Denmark), expressed an interest in shipping captured CO_2 [1]. The Maersk company estimates that ships with a capacity of about 25,000 t of CO_2 would be best suited for the job.

It is noteworthy, however, that transport of CO_2 by shipping may result in more CO_2 emissions than pipeline transport due to an additional energy-intensive step of CO_2 liquefaction and fuel usage in ships. According to IEA estimates, about 2.5 % and 18 % extra CO_2 emissions are produced during the marine transport of CO_2 over the distance of 200 km and 12,000 km, respectively (i.e., 1–2 % extra CO_2 emissions are produced for each 1,000 km of pipeline transport) [67]. CO_2 could potentially leak during shipping (about 3–4 % per 1,000 km), which could be reduced to 1–2 % per 1,000 km by the carrier tank design optimization.

7.5.4 *Land Transport of CO_2*

The land transport of CO_2 by means of railroad, truck tankers, and trailers can be justified in certain circumstances, for example, when the expected volumes of captured CO_2 do not warrant the construction of a new pipeline, or the point of CO_2

capture does not have an access to a pipeline facility. This choice would also be particularly attractive when (a) relatively small quantities of CO_2 are transported over short distances; so, building a pipeline infrastructure would not be cost effective and (b) an existing railway system is in close proximity to a CO_2 point source. Although CO_2 transport by either rail or truck tankers has been utilized for many years by industry, it is recognized that land transport of CO_2 is unlikely to contribute to large-scale CCS projects [1].

7.6 CO_2 Storage Technology

CO_2 storage is a final step in the multistep CCS process. Currently, the following technical options for CO_2 storage are under consideration:

- Geological storage
- Ocean storage
- Mineral sequestration
- Biological storage
- Industrial use of CO_2

7.6.1 Geological Storage

Geological carbon storage (sometimes called *geosequestration*) as a means of locking away man-made CO_2 and, thus, mitigating climate change was proposed in the 1970s in the works of Marchetti [68] and other authors; later, this concept gained credibility through the works of Kaarstad [69], Koide et al. [70], and others. In 1996, Statoil company of Norway initiated the world's first commercial-scale geological CO_2 storage project at the *Sleipner Gas Field* in the North Sea. This was followed by a number of research programs in the USA, Europe, Canada, Japan, and Australia (more details on carbon storage projects can be found in Sect. 7.8.2).

The first geological CO_2 storage projects were carried out by gas companies and dealt with natural gas fields with high CO_2 content (e.g., Statoil, *In Salah* in Algeria); as the level of confidence in the technology increased, electric utility companies and other industries started showing an interest in geological storage as a carbon mitigation option.

The geological storage of CO_2 is carried out by the injection of CO_2 in the deep geological formations (e.g., porous rocks, basalt, and saline formations) that are isolated from the atmosphere by the thick layers of an impermeable rock (a caprock). The density of injected CO_2 increases with depth, and it becomes practically constant at depths below 1.5 km. At a depth of about 800 m, CO_2 turns to a liquid-like "supercritical" or "dense" phase that is much denser than gaseous CO_2 (for this reason, most reservoirs that are considered for geological CO_2 storage have a depth of at least 800 m) [1]. Advantageously, the dense-phase CO_2 displays low viscosity close to that of gases, which allows for the efficient utilization of an underground storage space (e.g., the porous structure of sedimentary rocks). Similar to oil, the dense-phase CO_2 is of hydrophobic nature and the bulk of it not mixed with water forms a separate layer. (Note that small amount of CO_2 is dissolved in water producing carbonic acid that may react with other elements within the formation and form rock minerals such as limestone.) The injection of CO_2 into the reservoirs causes the displacement of naturally occurring fluids such as water, crude oil, and natural gas.

In order for geological CO_2 storage to be a viable option for meeting the objectives of climate change policies, it is expected that once CO_2 is injected into the selected geological formations, no less than 99 % of the injected CO_2 would be retained for at least 1,000 years [25]. The important criteria to be considered with regard to the suitability of a particular geological CO_2 storage formation are:

- Sufficient storage capacity
- Reliable confining unit (e.g., a satisfactory sealing caprock)
- Stable geological environment and acceptable tectonic activity (to ensure the long-term integrity of the storage site)
- Suitable geothermal and hydrodynamic characteristics of the formation
- Other technical, economical, environmental, and societal factors (industry maturity and infrastructure, the level of development, local economy, public acceptance, etc.) [71]

Several types of geological formations are suitable to safely store CO_2:

- Deep saline formations (DSF)
- Depleted oil and gas reservoirs (DOGR)
- Unminable coal beds
- Salt caverns
- Abandoned mines
- Basalts
- Organic-rich shale and other geological media

Once CO_2 is injected in a geological formation, it remains confined underground as a result of one or combination of several trapping mechanisms, such as (1) trapping below an impermeable layer (a caprock), (2) trapping in the pores of the storage formation as an immobile phase, (3) dissolution in naturally occurring fluids, (4) adsorption onto the surface of solid matter (e.g., minerals), and (5) chemical reactions with minerals to form carbonates.

The storage sites such as DSF and DOGR are relatively well understood, and, most likely, will be the main carbon storage options to be considered within the timeframe and scale required by the objectives of carbon abatement policies. The selection and characterization of appropriate carbon storage sites is one of the key limiting factors for the introduction of commercial CCS projects to the market, because, regardless of the CO_2 capture or transport methods, the capacity to safely store the extremely large quantities of CO_2 is the most critical issue underpinning the entire CCS value chain [63].

Analytical studies indicate that each of the storage sites will require data, time, and resource analysis to understand its capacity, containment, injectivity, cost, ease of monitoring, and potential conflicts with other land users [72]. A significant fraction of these sites (which is not yet known) are likely to prove unviable upon close analysis. Among European countries, Norway has the most extensive and long-standing "hands-on" experience with CCS technology. The recent estimates by the Norwegian Petroleum Directorate indicate that the country's geological formations could store 6.06 billion tons of CO_2 beneath the Norwegian Sea, compared to the country's current annual emissions of about 50 million metric tons [73].

7.6.1.1 Deep Saline Formations (DSF)

Saline formations represent the layers of a porous rock saturated with highly mineralized brines. Fluids in the deep saline formations flow at an extremely slow rate (centimeters to meters per year), and the formations can cover areas extending hundreds to thousands of kilometers [1]. Thus, when CO_2 is stored in a DSF, it is expected to be isolated from the near-surface layers for thousands of years. Due to the high concentration of minerals in the brine, CO_2 is likely to react with them (e.g., forming solid carbonates), thus, dissolving and trapping CO_2 within the formation making it especially well suited to serve as long-term storage sites.

DSF are the most abundant and geographically diverse potential sinks for CO_2 storage widely occurring throughout the world. The geographical distribution of CO_2 storage capacity in the DSF is as follows (in Gt): Canada—4,000, the USA—160–800, Europe—30–577, Australia—740, Japan—1.5–80, and China—2,300 [25, 74]. The examples of geological CO_2 storage in DSF include the Statoil Sleipner Project (Norway), In Salah Gas Project (Algeria), Statoil Snøhvit Project in the Barents Sea (Norway), and several pilot-scale or demonstration-scale projects such as the Ketzin Project (Germany), and the Lacq CCS project (France). Besides CO_2 sequestration, the saline formations, as a storage medium, have found very limited applications: just a few cases of chemical waste storage.

7.6.1.2 Depleted Oil and Gas Reservoirs (DOGR)

Depleted (or disused) oil and gas reservoirs are considered excellent candidates for CO_2 storage. Generally, they represent the most well-understood CO_2 storage option because they have proven history of gas and naturally occurring CO_2 containment for millions of years. The advantages of using DOGR as a CO_2 storage medium are: (1) the infrastructure (e.g., wells and pipelines) is already in place and can be easily readjusted for CO_2 storage, (2) a proven and reliable natural trapping mechanism (i.e., high confidence that these formations will be able to contain CO_2 ever extended period of time), (3) typically, large storage volumes due to the previous commercial-scale extraction of oil or gas from the reservoirs. DOGR represent attractive development opportunities as they have already undergone extensive site analysis during oil and gas exploration, which minimizes the upfront development costs for modeling and characterization of the formation's suitability for long-term CO_2 storage [63]. The storage potential of economical DOGR in Europe is estimated at 10–15 billion tons of abated CO_2, which is adequate for the lifetime of about 50–60 large CCS projects [66]. Most of these fields, however, are located in offshore northern Europe, which makes them twice as costly to access and operate as onshore fields.

7.6.1.3 Unmineable Coal Seams

The significant advantage of CO_2 sequestration in coal beds over conventional gas reservoir storage is that the coal seams can store 6–7 times more CO_2 than a reservoir of an equivalent volume because solid coal contains natural fractures or "cleats," pores, and micropores where CO_2 can diffuse and get tightly adsorbed on the surface of pores. A coal seam becomes a suitable site for CO_2 storage if it is no longer economical to be mined for coal (which is determined by its geological conditions and world energy prices, among many other factors). Typically, unmineable coal seams are likely to be several hundreds of meters or more in depth [1].

7.6.1.4 Shale and Basalt Formations

Shale and basalt have recently gained interest as potentially the most robust and stable CO_2 storage options in several regions around the world. Shale comprises thin layers of rock that in many cases contain about 1–2 % of organic matter capable of absorbing CO_2. Basalt formations represent the ancient volcanic rocks (lava) that have porosity and permeability in the fractures or cavities between blocks of a solid rock [1]. Depending on the chemical composition of basalt it can react with CO_2 producing solid stable products—carbonates. There are, however, many technical challenges facing storage in the basalt formations, including (a) difficulty of CO_2 injection into these heterogeneous formations, (b) the significant degree of porosity and permeability of the formations that will make the sealing unreliable, and (c) the lack of information about CO_2 storage properties of basalt formations. Although basalts are widespread around the world, to date, there are no large-scale projects on CO_2 storage in the basalt formations, and most of the activities are limited to laboratory-scale experimentation.

7.6.1.5 Storage in Salt Caverns

Naturally occurring underground salt caverns are also considered for storing large quantities of CO_2 provided they have an adequate geological sealing layer and they can support high pressures required for storage of dense-phase CO_2. Salt caverns differ from other geological formations in that they have an ability to deform and change volume until CO_2 pressure equalizes with surrounding pressure.

The volume of salt caverns has to be adequate in order to accommodate large industrial CO_2 sources, e.g., a 500 MW coal-fired plant equipped with CO_2 capture and producing about 3 Mt/year CO_2 would require a cavern with the equivalent volume of a spherical cavern 150 m in diameter to store CO_2 generated by a power plant just in 1 year [1]. Although the salt caverns have been used in the past for the temporary storage of natural gas and proved to be quite effective, currently, there are no CO_2 storage projects that make use of the salt caverns.

7.6.1.6 Geological CO_2 Storage Security

CO_2 storage permanence and security are among the major issues determining whether a geological CO_2 storage site is suitable for the storage of large amounts of man-made CO_2 emissions. The security and effectiveness of geological CO_2 storage is determined by a combination of a variety of physical and geochemical trapping or confinement mechanisms and processes. The most reliable storage mechanisms would involve CO_2 conversion to solid stable minerals or represent a thick, impermeable (or low permeable) seal (a caprock) under which the immobile CO_2 phase is permanently trapped.

Depending on the nature of a geological storage site, the physical trapping of CO_2 can be accomplished via stratigraphic, structural, and hydrodynamic mechanisms. The first two mechanisms occur when CO_2 is trapped below low-permeability seals, whereas hydrodynamic trapping mostly takes place in the deep saline formations where fluids migrate very slowly over long distances. After CO_2 is injected in a saline formation, it dissolves in saline formation water and migrates with groundwater. Since in most cases, the distance from a CO_2 injection site to the edge of the impermeable formation is in hundreds of kilometers, it would take millions of years for the CO_2 phase to reach surface [75]. A weak acid formed during CO_2 dissolution reacts with silicates of K, Na, Mg, Ca, Fe present in the rock matrix forming stable carbonate minerals (the process is called *mineral trapping*). Although the mineral trapping is the slowest process (thousands of years) among all geological storage options, it is considered the most preferred form of CO_2 storage due to the exceptional permanence it can provide coupled with a large CO_2 storage capacity.

According to existing models, physical trapping mechanisms are effective in the relatively short timeframe of tens to hundreds of years, whereas the geochemical trapping mechanisms dominate from thousands to millions of years. For example, modeling studies applied to CO_2 storage at the Weyburn Oil Field site (USA) indicated that over 5,000 years, all injected CO_2 would be dissolved and converted into carbonate minerals [76]. In another study, authors estimated the probability of a CO_2 release from the Weyburn Oil Field at 1 % in 5,000 years [77]. It is estimated that well-selected, designed and managed geological storage sites are likely to release only 1 % of injected CO_2 over 1,000 years period [25]. Experience with NG fields and trapped CO_2 accumulations can provide an indirect evidence of the potential storage permanence. For example, about 200 Mt CO_2 is trapped in the Pisgah Anticline (Mississippi, USA) for more than 65 million years.

7.6.2 *Beneficial CO_2 Reuse Applications*

The term "beneficial CO_2 reuse" covers enhanced oil, gas, and coal bed methane recovery applications, where captured CO_2 is used for generating a revenue from the sale of crude oil or natural gas that is obtained as a result of the CO_2 storage. Although, in general, a main motivation for the enhanced oil recovery and other beneficial CO_2 reuse applications is not concerned with climate change mitigation objectives; in practice, it results in the geological storage of CO_2.

7.6.2.1 Enhanced Oil Recovery (EOR)

EOR is the most widely used beneficial CO_2 reuse application; it has been practiced on a commercial scale for almost four decades by oil industry, where CO_2 injection technology and associated operations (e.g., monitoring of its subsurface behavior) have a proven track record. EOR is a generic term for the techniques aiming at increasing the amount of extracted crude oil that would otherwise remain stranded. It should be noted that there are many different types of EOR, e.g., chemical, microbial, and thermal EOR; however, in this book, the EOR term will be used exclusively for CO_2-induced EOR. Figure 7.8 depicts a simplified sketch of the CO_2-EOR technology.

Captured CO_2 is compressed to dense-phase CO_2, and injected into an oil reservoir through an injection well (CO_2 and water are pumped into the unit in alternating cycles). The injected CO_2 is mixed with crude oil in the reservoir making it less viscous and more mobile and forcing it to flow to the series of production wells (typically, there are several production wells per each injection well). Typically, the production stream consists of a wide range of petroleum components, water, CO_2, and other gases (methane, ethane, propane). At the surface, this stream is separated into various components to recover oil as well as water and CO_2. After the treatment, water and CO_2 are reinjected into the formation. The formation gas is separated into methane and LPG that are treated and sold in the market.

The EOR offers a potential to substantially increase the oil production through CO_2 flooding of an oil well. It was estimated that a conventional primary production recovers only 5–40 % of original oil, and additional 10–20 % of oil can be produced by a secondary recovery operation that utilizes water flooding [79]. EOR represents a tertiary oil recovery operation. Several estimates of the efficiency of the oil recovery by CO_2 flooding have been reported: they vary in the range of 6.7–23 % of original oil in place (the average for the USA is 14.6 %) [80]. For the USA, this translates into an additional 87.1 billion barrels of oil (based on the estimated 595.7 billion barrels of oil in place) [1].

Due to the technological imperfections, about half of CO_2 used in the EOR operation is retained in the reservoir after oil production is ceased [81]. Upon further improvements in the EOR technology, essentially all the injected CO_2 would remain in the reservoir other than minor losses from the operations, or intentional

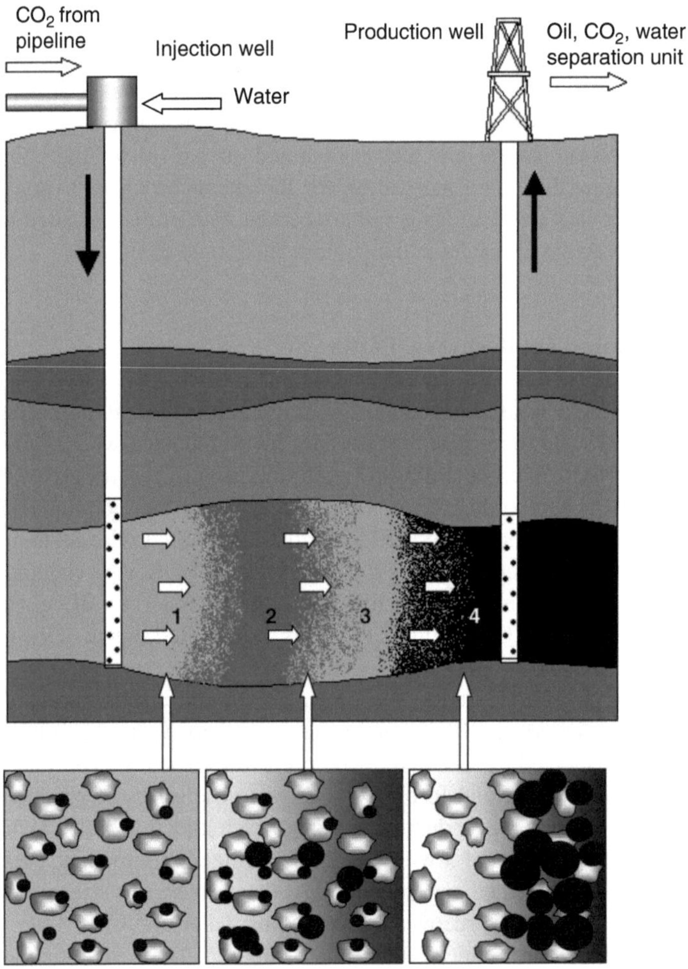

Fig. 7.8 Simplified scheme of CO_2-EOR. *1*—drive water, *2*—CO_2 and water zone, *3*—oil bank/miscible front, *4*—residual oil zone. *Source* [78]

flaring (if required). Typical EOR projects are planned to last decades, and when a field is no longer producing oil to justify continuation of the EOR operation, it can be converted into a dedicated carbon storage site.

At present, more anthropogenic CO_2 is geologically stored globally through EOR projects than through any other methods. There is an increasing interest in both developed and developing countries in the EOR for enhancing domestic oil production. Although the vast majority of the EOR projects have been practiced in North America, many other countries including China, Brazil, Hungary, Trinidad, and Turkey have a history of EOR operations [12]. The examples of the large EOR projects are the Rangely Project in Colorado (USA) and the Weyburn-Midale Project in Saskatchewan (Canada) (see discussion in Sect. 7.8).

It is important to emphasize that, today, more than half of the CO_2 used in EOR operations is coming from naturally occurring CO_2 sources produced from the subsurface (in the USA, approximately 80 %). Although the commercial EOR projects utilizing natural CO_2 sources allow gaining very valuable information on their potential and capacity for long-term CO_2 storage, they do not contribute in any meaningful way to the CO_2 abatement objectives, because they merely transport CO_2 from one underground reservoir to another. For the EOR to be considered a carbon abatement measure, the CO_2 should originate from human activities that would otherwise end up in the atmosphere.

Summarizing, the EOR provides opportunities for the easiest and most cost-effective way to initiate CCS projects, especially, for an onshore option, where there are existing regulations and infrastructure in place. The experience and knowledge gained from four decades of commercial EOR practice provide the understanding of the subsurface response to CO_2 injection; of particular value are the lessons learned from the long-distance CO_2 pipeline transport and from monitoring, measuring, and verification methodologies. Economic estimates indicate that EOR is justified in a large commercial-scale CCS applications, where CO_2 has a value. However, today, that value is perhaps only 25–50 % of the CO_2 capture costs for a coal power plant equipped with CCS [63]. Enhanced gas recovery (EGR) projects are similar to the EOR method; currently, there are no large-scale operations utilizing this approach.

7.6.2.2 Enhanced Coal Bed Methane Recovery

Enhanced coal bed methane recovery is another commercially important beneficial CO_2 reuse application. Most coal seams contain naturally occurring methane, and its content typically increases with the coal bed depth, coal rank, and pressure in the coal bed. When CO_2 is injected into coal seams, it readily displaces methane from coal surface, which is the basis of the operation, which can potentially increase the amount of recovered methane to about 90 % of the total gas, compared to the conventional recovery of only 50 % (by a pressure depletion method) [82]. The enhanced coal bed methane recovery is discussed in more detail in Sect. 9.2.1.

7.6.3 Ocean Storage of CO_2

7.6.3.1 Technical Background

The oceans cover about three quarters of the Earth's surface and they represent a natural sink for CO_2. As part of the global carbon cycle, they absorb significant amounts of anthropogenic and naturally occurring CO_2 (in fact, the oceans contains about 50 times more CO_2 than the atmosphere). Since the beginning of the Industrial Revolution, the oceans have absorbed about 500 Gt of the anthropogenic CO_2 emissions of the total of 1,300 $GtCO_2$ [25]. Dissolved CO_2 is mainly concentrated in the upper ocean layers resulting in a slight pH drop of about 0.1 pH units, compared to the preindustrial level (no changes in the pH of deep ocean layers have been detected so far) [83].

The ocean CO_2 storage approach is intended to inject CO_2 deep into the ocean (at least 1 km deep) where it is supposed to be retained for centuries isolated from the atmosphere. The concept was first proposed by Marchetti in the mid-1970s, who reasoned that if liquefied CO_2 is injected into the waters flowing from Mediterranean Sea to the mid-depth Atlantic Ocean, it would remain isolated from the atmosphere for centuries [67]. Due to the enormous volume of the Earth's oceans, it is widely assumed that the practically unlimited amount of anthropogenic CO_2 can be stored in the ocean for at least a millennium. The majority of modeling studies indicate that CO_2 injected into the ocean will eventually become part of the global carbon cycle, with deeper injections resulting in longer retention times. For example, it was estimated that 30–85 % of injected CO_2 will be retained after 500 years if stored at depths of 1,000–3,000 m [25].

Atmospheric CO_2 is in chemical equilibrium with carbonate ions in seawater according to the following equation:

$$CO_{2(gas)} + H_2O \leftrightarrow H_2CO_{3(aq)} \overset{pK_a=6.4}{\leftrightarrow} HCO^-_{3(aq)} + H^+ \overset{pK_a=10.3}{\leftrightarrow} CO^{2-}_{3(aq)} + 2H^+ \quad (7.9)$$

The equilibrium of this reaction is governed by several factors: CO_2 concentration in the atmosphere, seawater temperature, the rate of air/ocean exchange (mixing), the presence of other ionic species, and chemistry of seawater. Due to the dissolution of minerals (e.g., $CaCO_3$) in seawater, the ocean pH is slightly above 7 (i.e., it is slightly alkaline), which favors the dissolution of CO_2 in seawater. The rise in CO_2 concentration in the atmosphere will cause the equilibrium to shift favoring the dissolution of additional CO_2 in the ocean and forming more bicarbonate ions (as a result, the ocean pH and carbonate ion concentration in seawater will drop):

$$CO_{2(gas)} + H_2O + CO^{2-}_{3(aq)} \leftrightarrow 2HCO^-_{3(aq)} \quad (7.10)$$

The physical and thermodynamic properties of pure CO_2 and the CO_2–H_2O system determine the fate of CO_2 upon its release into the sea environment, in particular, the conditions at which CO_2 would exist as gas, liquid, or solid (hydrate) and their relative density against seawater. Figure 7.9 depicts different physical states of CO_2 in seawater as a function of depth.

At a typical temperature range in the ocean, CO_2 exists as gas at the depth from 0 to 500 m; below about 500 m, pressure is high enough to convert gaseous CO_2 into a liquid form. In the liquid phase, CO_2 is more compressible than seawater, the property that greatly affects the relative density of liquid CO_2 against seawater at different depths. At the depth shallower than about 2,400–2,500 m, L-CO_2 is lighter than seawater and would float upward, whereas in the ocean layers deeper than 3,000 m L-CO_2 is denser than seawater and it will sink. At temperatures below 8–9 °C and depth of about 400 m and deeper, CO_2 reacts with water forming crystalline CO_2

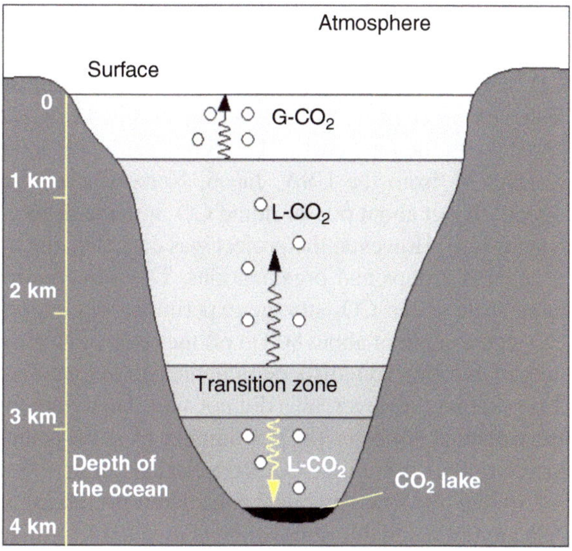

Fig. 7.9 Physical states of CO_2 in seawater as a function of depth. G-CO_2 and L-CO_2 are gaseous and liquid states of CO_2, respectively

hydrate ($CO_2 \cdot nH_2O$, where $6<n<8$), which is denser than seawater, and it sinks (CO_2 hydrates are not shown in Fig. 7.9). The formation of CO_2 hydrate can potentially slow down the dissolution of CO_2 and, from this viewpoint, it is beneficial for the ocean CO_2 storage. On the other hand, the formation of solid CO_2 hydrate may also create some technical difficulties during the CO_2 release into the ocean, in particular, by clogging pipelines and diffusers and impeding the CO_2 flow.

Based on the above theoretical considerations, currently, there are five different approaches to ocean CO_2 storage, in particular, through:

- Dispersing dense-phase CO_2 into the ocean at relatively shallow depths (about 1 km) and dissolving CO_2 in ocean layers (the so called *water column* option)
- Injecting liquid CO_2 into deep (about 3 km) ocean water, where it could form a plume of liquid CO_2 sinking to the bottom of the ocean
- Depositing the dense-phase CO_2 onto the sea floor at depths below 3 km where it would form a separate liquid phase ("CO_2 lake")
- Injecting dense-phase CO_2 at the depth where it would form CO_2 hydrate crystals sinking to the ocean floor
- Enhancing the ocean CO_2 storage capacity via increasing seawater alkalinity by dissolving carbonate minerals in seawater (via "carbonate mineral neutralization").

7.6.3.2 Current Status of Development

The ocean CO_2 storage technology is much less developed compared to the geological CO_2 storage option. Most activities are in the R&D phase involving modeling studies, laboratory, and small-scale field experiments. There are at least known two attempts to conduct large-scale in situ experiments involving the injection of liquid CO_2 into the ocean, but both of them did not even start due to a fierce opposition from concerned groups. The first project called *CO_2 Sequestration Field Experiment* (2001) with the total cost of US$5 million involved an international consortium of researchers from the USA, Japan, Norway, and Canada who were intending to directly inject about 60 t of liquid CO_2 into the deep ocean from a ship near Hawaii islands [25]. However, the project was canceled due to a stiff opposition from environmental groups and organizations. The second attempt to carry out a relatively large-scale ocean CO_2 storage experiment was supposed to release 5.4 t of liquid CO_2 at the depth of about 800 m off the coast of Norway. The researchers were planning monitoring CO_2 dispersion in the Norwegian Sea over the long period of time. This attempt, however, also did not start due to the opposition from the Norwegian Environment Ministry [84]. A number of small-scale experiments (at the scale of hundred liters of liquid CO_2) were carried out in different marine sanctuaries [85].

Among important practical considerations related to the ocean CO_2 storage is the selection of a suitable site, which is dictated by a number of factors, such as environmental and economical considerations, safety issues, and laws and international regulations. The oceans with the depth of more than 1 and 3 km amount to about 88 and 75 % of the total ocean surface area, respectively. Large CO_2 point sources located near these areas would be the most cost effective settings for the direct CO_2 injection into the ocean. The experience gained by oil and gas industries in the offshore operations (e.g., drilling platforms, deep-sea pipelines) can be applied for the large-scale ocean CO_2 storage projects.

The United Nations Convention on the Law of the Sea (UNCLOS) restricts the projects dealing with ocean CO_2 storage. Because of the restrictions and also opposition from concerned organizations worldwide, there is a very limited experience and practical knowledge with regard to the ocean CO_2 storage. Among the majority of experts, ocean CO_2 storage, as a carbon mitigation strategy, is not as favored as geological storage. For more detailed information on different technical aspects of ocean CO_2 storage, we would recommend IEA-released sourcebook [86].

7.6.4 Mineral Sequestration as CO_2 Storage Option

Mineral sequestration (also known as "mineral carbonation," or "enhanced weathering") is a CO_2 storage option dealing with the fixation of CO_2 in the form of insoluble stable carbonates. According to this relatively new approach, CO_2 reacts with complex metal oxides (preferably, silicates of Ca and Mg) producing corresponding carbonates and silica. The mineral sequestration approach is considered by its supporters as a promising *environmentally "benign" sequestration* (EBS) technology which could potentially provide an ecologically safe and geologically stable CO_2 disposal in the form of mineral carbonates.

Several industrially important mineral sequestration reactions involve the naturally occurring silicates of Ca and Mg such as olivine (Mg_2SiO_4), serpentine ($Mg_3Si_2O_5(OH)_4$), wollastonite ($CaSiO_3$) as follows:

$$Mg_2SiO_4 + 2CO_2 \rightarrow 2MgCO_3 + SiO_2 \quad \Delta H^o = -89\,kJ/mol\,CO_2 \tag{7.11}$$

$$\frac{1}{3}Mg_3Si_2O_5(OH)_4 + CO_2 \rightarrow MgCO_3 + \frac{2}{3}SiO_2 + \frac{2}{3}H_2O \quad \Delta H^o = -64\,kJ/mol\,CO_2 \tag{7.12}$$

$$CaSiO_3 + CO_2 \rightarrow CaCO_3 + SiO_2 \quad \Delta H^o = -90\,kJ/mol\,CO_2 \tag{7.13}$$

These reactions are spontaneously occurring in nature on a timescale of thousands of years (the so-called silicate weathering). Thermodynamically, the mineral carbonates are the most stable forms of carbon (more stable than CO_2 itself), which ensures the permanent fixation of carbon.

The material basis for the practical realization of the mineral sequestration process comprises both natural silicate rocks and industrial wastes, e.g., a fly ash (with CaO content of up to 35 %), a slag from steel production (CaO and MgO content of about 65 %), and waste cement. [25]. The experimental studies of the mineral sequestration technology were conducted by NETL researchers who found that the finely ground serpentine or olivine reacted with supercritical CO_2 to form magnesium carbonate ($MgCO_3$) [17]. At the temperature range of 150–250 °C and pressure of 85–100 atm, 84 % conversion of olivine was achieved in 6 h.

Depending on the nature of minerals and specifics of the process, mineral sequestration can be carried out in in situ and ex situ modes. The in situ mineral sequestration approach is very similar to the geological (mineral trapping) CO_2 storage option, and it involves the injection of a CO_2 stream into a silicate-rich formation, where the carbonation reaction will take place resulting in permanent storage of CO_2 in the form of carbonates. In the ex situ approach, mineral sequestration is carried out in a specialized processing facility, where the reaction between CO_2 and silicate minerals (or suitable industrial wastes) takes place producing carbonates, silica and other solid products that have to be disposed.

Main selling points for the mineral sequestration as a CO_2 storage option are twofold: the abundance and low cost of natural silicates suitable for this application and the exceptional CO_2 storage permanence, since CO_2 is permanently "locked" in the form of very stable insoluble carbonates. However, mineral sequestration faces many technical challenges as an energy-intensive process requiring the number of steps, such as the preparation of reactants (mining, grinding, and transport), the chemical processing, and the disposal of carbonates and other solid products with the associated energy penalties. There have also been significant difficulties with the engineering arrangement of the process, stemming from very slow reaction kinetics and thermodynamic limitations that tend to lower the product yield. In many cases, the mass of mineral matter required for CO_2 mineralization is many times larger than the mass of CO_2 to be stored adding to the process energy penalties (e.g., the fixation of 1 t of CO_2 in the form of carbonates would require processing of about 1.6–3.7 t of silicates and disposal of 2.6–4.7 t of solid products [25]).

The US DOE provides some preliminary data on the energy consumption and economics of a wet mineral sequestration process using several types of minerals (olivine, lizardite, antigorite, wollastonite with ore costs in the range of US$15–48 per ton) [87]. The reported estimates show that the energy input requirements for conducting the mineral sequestration process varied in the range of 180–2,300 kWh/ton CO_2 stored, depending on the mineral used and pretreatment procedure. As a result, 30 to 50 % of the total energy output would be used to carry out mineral sequestration reducing power plant efficiency from the original 35 to 25 % and 18 %. The authors of the study estimated the cost of CO_2 storage via mineral sequestration at US$55–250 per ton CO_2 stored.

The perspectives of mineral sequestration as a CO_2 mitigation strategy will be dictated by a number of trade-offs between the costs of the mineral mining, processing, carbonate disposal, and the benefits of permanent CO_2 storage. Considering the significant material intensity of the mineral sequestration process, the large-scale implementation of the technology might require the expanding of a mining industry to the scale comparable with coal industry [25]. The possibility of getting credits for by-products could improve the economics of the mineral sequestration processes. For example, during the pretreatment stage, certain high-value minerals can be extracted from the mineral ores, e.g., magnetite from olivine, minerals containing Cr, Ni, and Mn, and platinum group metals from periodite rocks. Due to the technical challenges, mineral sequestration is still in a relatively early stage of development and more studies are necessary to thoroughly evaluate its commercial potential and its environmental impact.

7.6.5 *CO_2 Sequestration in Biosphere*

The objective of CO_2 sequestration in the Earth's biosphere as a carbon mitigation option is to dramatically increase the rate of CO_2 intake while taking into consideration all the possible ecological, economic, and social implications of this action. The enhanced CO_2 sequestration in the biosphere can achieve the withdrawal of CO_2 from the atmosphere over the next several decades to gain some time for the technological development and practical implementation of other CO_2 mitigation options. Unlike the geological and ocean sequestration options that are mainly limited to fossil-based large industrial point sources of CO_2, the biospheric sequestration of CO_2 applies to all CO_2 sources including natural sources and atmospheric CO_2. CO_2 can be sequestered in the terrestrial, aquatic, and ocean ecosystems as parts of the Earth's biosphere.

7.6.5.1 CO_2 Sequestration in Terrestrial Biosphere

Terrestrial ecosystems, which include vegetation and soil, are major natural sinks for removing CO_2 from the atmosphere annually sequestering about 2 Gt of carbon [88]. The Earth's terrestrial biosphere consists of a broad diversity of ecosystems (peatland, forest, grassland, wetland, savanna, tundra, etc) that cumulatively store about 2,542 Gt of carbon. Peatland, tropical forest, and tropical savanna are the major carbon-storing terrestrial ecosystems of the world. The net primary productivity (NPP) of all ecosystems is about 59 GtC per year with variations from 11 $gC/m^2 \cdot year$ for extreme deserts to 1,180 $gC/m^2 \cdot year$ for wetlands (NPP is the rate at which all plants in the ecosystem produce net useful chemical energy, i.e., biomass) [88]. Carbon stock in soil exceeds that of a plant matter by a factor of about 4. Thus, enhancing carbon storage in soil may potentially have a more profound impact on carbon sequestration in biosphere compared to plant matter.

Carbon sequestration in terrestrial biosphere can be achieved via either enhancing net removal of CO_2 from the atmosphere, or reducing the net CO_2 emissions from the terrestrial ecosystems into the atmosphere, or by the combination of both these options. The list of possible approaches to CO_2 storage in terrestrial biosphere include:

- Increase in photosynthetic CO_2 fixation
- Reduction in decomposition of organic matter and other soil-derived CO_2 emissions
- Protection of ecosystems that store carbon
- Reverse land-use change trends that contribute to CO_2 emissions
- Create energy offsets by using biomass for production of fuels and other products
- Manipulate ecosystems to increase carbon sequestration beyond current conditions
- Apply new approaches based on biotechnology, molecular genetics, and species selection

Carbon sequestration in the terrestrial ecosystems can potentially provide the significant near- and mid-term (over the next 25–50 years) benefits as a carbon mitigation policy. The estimated potential for carbon sequestration in the terrestrial biosphere is in the order of 5–10 GtC per year. However, this value is rather optimistic since it does not adequately reflect many economic, energy, and environmental implications of biological carbon storage at such high rates. The uncertainties and complexities in estimating the potential for increasing biospheric carbon sequestration stem from many factors that are difficult to predict, e.g., cost of petroleum fuels and fertilizers, availability of water, various socioeconomic issues, and competition with other carbon management strategies.

7.6.5.2 Biological CO_2 Sequestration by Aquatic and Ocean Systems

Biospheric CO_2 sequestration in aquatic systems mostly relate to micro- and macroalgal systems, with the former being much more technologically advanced compared to the latter. The capacity of photoautotrophic organisms (algae) to efficiently capture CO_2 from both the atmosphere and CO_2-containing industrial streams (e.g., from power plants, cement manufacturing plants) and produce value-added products is in the focus of a research and commercial attention worldwide. Numerous types of microalgae strains have biomass productivity superior to that of terrestrial plants (trees, grasses, etc.) by at least one order of magnitude. To achieve high CO_2 uptake rates and, consequently, biomass production rates, microalgae has to be grown under the optimal conditions of light, temperature, pH, nutrient, and CO_2 concentrations [89, 90]. Compared to microalgae, macroalgal systems including some types of aquatic plants such as duckweed and hyacinth are much less developed as applied to CO_2 sequestration objectives. More detailed information on the use of microalgae for the industrial utilization of CO_2 is presented in Chap. 9.

The ocean CO_2 sequestration approach is based on the enhancement of the natural processes of CO_2 fixation by the ocean ecosystems. Although the ocean's biomass amounts to only about 0.05 % of the terrestrial biomass, it transforms about as much CO_2 to organic matter (about 50 GtC/year) as the land ecosystems [88]. A primary mechanism by which biological carbon sequestration occurs in the ocean is through the so-called biological pump, which represents a complex multistage process involving decomposition, gravitational settling, and the burial of biogenic debris formed in the upper layers of the ocean. CO_2 dissolved in the ocean is initially fixed by phytoplankton via photosynthetic mechanism. In the surface layers of the ocean, phytoplankton species are rapidly consumed by zooplankton, which, in turn, are grazed by higher trophic organisms, e.g., fish. Organic carbon in the form of decaying organisms and other organic matter sinks to seafloor. About 70–80 % of the fixed carbon is recycled back to CO_2 in the surface layers of the ocean, and the remaining part settles as particulate organic carbon in the ocean depths, where it is slowly mineralized by bacteria [88].

7.6.5.3 Advanced Biological Carbon Sequestration Systems

The development of advanced biological processes aims at enhancing natural biological processes for CO_2 uptake from the atmosphere in terrestrial and marine ecosystems via the use of novel organisms, designed biological systems, and genetic improvements in microbial, plant, and animal species. Among the currently explored approaches are [88]:

- Development of faster-growing and more stress-resistant crops and plants
- Design of biological methods to enhance geological carbon sequestration via the use of microorganisms
- New processes to enhance carbon sequestration in the ocean ecosystem through transgenic and genetic manipulation of members of the food chain
- Alternative microbial polymers or genetically improved plants as durable materials
- New microorganisms that do not rely on photosynthesis or carbon-based sources of energy

Through the genetic engineering, the metabolic networks of terrestrial and aquatic plants and algae could be designed to increase the share of products with desired characteristics. Specifically, new plant species could be designed to have the higher percentage of biomass below ground (making them more resistant to decay), and, at the same time, the structure of above ground plant biomass (e.g., its cell walls) could be engineered to facilitate the plant bioconversion processes and to make nonharvested biomass less degradable in the environment. The genetic modification of plants could significantly enhance their photosynthetic activity and ability to fix CO_2. In general, plants operate at rather low light-conversion efficiencies; For example, the photosynthetic efficiency of forests is in 0.05–0.1 % range, marsh grasses in 2–4 % range, and for corn and sugar cane it could be as high as 3.5–4 % [88]. The advanced bioengineering approaches could significantly improve the photosynthetic CO_2 fixation efficiency by enhancing the efficiency of two processes: conversion of captured light to chemical energy and primary carbon fixation catalyzed by the enzyme *Rubisco*.

Due to the high cost of CO_2 capture and separation from point sources emitting diluted CO_2 streams, the concept of integrated energy production and carbon capture by means of biological "energyplexes" has been proposed. These energyplexes have the potential to produce energy, treat waste, sequester CO_2, and produce value-added products with a minimal environmental impact [91]. For example, an energyplex will be able to capture CO_2 from flue gases of power plants by means of photosynthesis and store the reduced carbon in the form of algal biomass. Other approaches involve the integration of power production with sewage and other waste treatment processes with the benefit of using nutrients and carbon in biological processes at the site.

Carbon fixation in biomass with simultaneous production of durable and stable materials is another active area of development. Novel materials (e.g., biopolymers and bioplastics) with highly specialized properties such as biodegradability, biocompatibility, and chemical functionality have been produced using enzymes. In some areas, these biomass-based materials are replacing petroleum-based polymers (e.g., polyethylene, polystyrene). For example, polylactic acid-based resins were produced by conversion of starch to sugar followed by its fermentation to lactic acid (the process has been commercialized by Dow Chemical). The production of microbial cellulose is another example of the successful implementation of the biochemical approach. Since carbon originates from atmospheric CO_2 (via biomass) the net result of using these products is carbon sequestration.

7.7 Economics of CCS Systems

7.7.1 Economics of CO_2 Capture

7.7.1.1 Efficiency Penalties Due to CO_2 Capture

CO_2 capture operation is the greatest contributor to the overall cost of the entire CCS chain. Practically all CO_2 capture technologies are energy-intensive processes resulting in considerable energy efficiency and fuel usage penalties and, consequently, having an impact on the economics of power generation coupled with CCS.

For example, for the chemical absorption systems, the energy efficiency penalties stem from the following requirements: (1) heat input to regenerate solvents, (2) steam input for stripping operations, and (3) electricity input for fluid pumping and CO_2 compression. The reduction in the energy and fuel penalties is a focus of several R&D and demonstration projects worldwide [92, 93]. Davison [93] estimated additional fuel usage due to CO_2 capture per kWh electricity produced by coal- and gas-fired power plants and compared it to the same plants without capture. The results of analyses indicated that the increase in the fuel usage for CO_2 separation is greater for the post-combustion capture option compared to the pre-combustion capture technology because in the latter case CO_2 is removed from more concentrated, higher pressure stream; so, a physical absorption rather than a chemical absorption method could be utilized.

Technology Center Mongstad, the world's largest CO_2 capture research facility based in Norway, announced that they are able to run gas electric power plant equipped with carbon capture at 51 % efficiency [94]. Considering that the record efficiency of a state-of-the-art gas-fired power plant without CCS is 59 %, one arrives at 16 % energy penalty for running the carbon capture plant: $(59\text{–}51)/51 \approx 0.16$. Although this constitutes the substantial share of the overall electricity output (which otherwise would be available to consumers), it is still lower compared to about 30 % power requirement estimates of early CCS-equipped power plants. The Technology Center Mongstad engineers project that the plant efficiencies (including CCS) as high as 52–54 % are potentially achievable (in this case, the carbon capture unit will only draw about 9 % of electrical power). If this target is realized, it could lead to massive reduction in the overall cost of CCS deployment and improvements to its viability and consumer confidence in the technology. It has been estimated that if the coal price would drop 9 % below the gas price, it could be equally viable to construct a coal-fired power plant with CCS, as a gas-powered plant without CCS [94].

7.7.1.2 Cost of CO_2 Capture

The economics of carbon capture is in the focus of intensive research efforts (e.g., [95]). It has been shown that the cost of CO_2 capture is dependent on a variety of factors including the capture method, the process application, land availability, environmental factors, and prevailing regulations and existing policies, particularly those which impose an economic value on CO_2 emissions. Currently, the high cost of carbon capture combined with its impact on the performance of a power plant is among key challenges facing CCS technology; therefore, both capital and operating cost reductions for all types of carbon capture remain a main focus of technology development efforts worldwide.

The cost of CO_2 capture depends not only on the CO_2-generating process and capture technology but also on underlying assumptions, which explains why the reported cost estimates vary in such a wide range. Four common measures of the CO_2 capture cost are [25]:

- Capital cost
- Incremental product cost (e.g., cost of electricity, hydrogen, fuels)
- Cost of CO_2 avoided
- Cost of CO_2 captured

The entirety of these four measures reflects the added cost of capturing CO_2 in a particular application (industry or power generation).

Capital cost is a major contributor to the overall cost of CO_2 capture. It has been reported that the addition of carbon capture system to pulverized coal (PC) power plant would increase the capital cost by almost 70 %, whereas in case of IGCC and NGCC, the increases in capital costs are about 30 % and 100 %, respectively [96].

Incremental product cost reflects the effect of CO_2 capture on the product cost (e.g., cost of electricity, hydrogen, cement, steel). Product cost in power generation systems is often expressed as *levelized cost of electricity (LCOE)*. LCOE is the price at which electricity must be generated from a specific source to break even over the lifetime of the project (It includes all the costs over its lifetime: initial investment, operations and maintenance, and cost of fuel). LCOE can be defined as follows [25]:

$$\mathrm{LCOE} = \frac{(\mathrm{TCR} \times \mathrm{FCF} + \mathrm{FOM})}{(8{,}760 \times \mathrm{CF} \times \mathrm{kW})} + \mathrm{VOM} + \mathrm{HR} \times \mathrm{FC} \tag{7.14}$$

where LCOE is the levelized cost of electricity (e.g., US\$/kWh), TCR—total capital requirement (US\$), FCF—fixed charge factor (fraction per year), FOM—fixed operating costs (US\$/year), VOM—variable operating costs (US\$/kWh), HR—net plant heat rate (kJ/kWh), FC—unit fuel cost (US\$/kJ), CF—capacity factor (fraction), 8,760—total hours in a year), and kW—net plant power output.

Note that LCOE defined by (7.14) only relates to the power plant and CO_2 capture technologies and does not account for additional costs of CO_2 transport and storage (which are discussed in the next sections). Reported data indicate that the deployment of current CO_2 capture systems would increase the cost of electricity by 61 %, 26 %, and 30 % for PC, IGCC, and NGCC plants, respectively (not accounting for CO_2 transportation and storage) [96].

Cost of CO_2 avoided (CCA) is one of the most widely used measures in the economic evaluations of CCS systems. The CCA indicates the average cost of reducing CO_2 emissions by one unit while providing the same amount of useful product as a reference plant without CO_2 capture [25]. For the case of a power plant, the CCA can be defined as follows:

$$\mathrm{CCA}\left(\$\mathrm{t}^{-1}\,\mathrm{CO_2}\right) = \frac{\left[(\mathrm{LCOE})_{capture} - (\mathrm{LCOE})_{reference}\right]}{\left[(\mathrm{CO_2 kWh^{-1}})_{reference} - (\mathrm{CO_2 kWh^{-1}})_{capture}\right]} \tag{7.15}$$

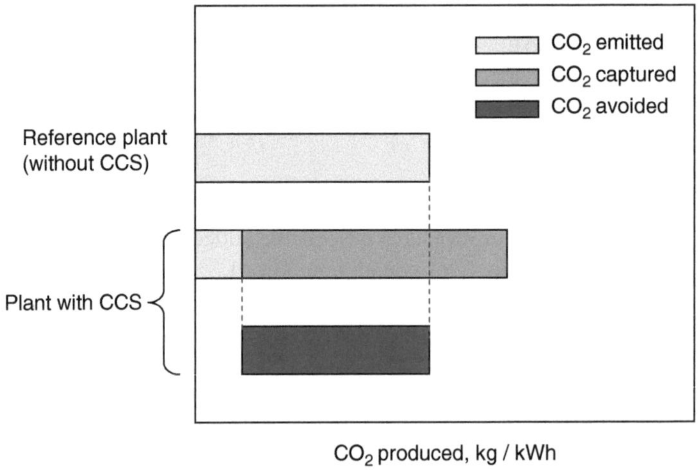

Fig. 7.10 Comparison of CO_2 emitted, captured, and avoided with and without CCS for a reference power plant. *Source* [25, 66]

where LCOE is the levelized cost of electricity ($/kWh) (as defined by 7.14), CO_2 kWh^{-1} is CO_2 emission rate (in tons per kWh generated). The subscripts "reference" and "capture" relate to the plant without and with CO_2 capture, respectively (the reference plant is assumed to be of the same type as the plant with CO_2 capture).

Cost of CO_2 captured (CCC) is defined as the mass of CO_2 captured (or removed) from an emission source (rather than CO_2 avoided). Again, for an electric power plant the CCC measure can be defined as follows [25]:

$$\mathrm{CCC}\left(\$\mathrm{t}^{-1}\,\mathrm{CO_2}\right)=\frac{\left[(\mathrm{LCOE})_{capture}-(\mathrm{LCOE})_{reference}\right]}{(\mathrm{CO_2})_{captured}\ \mathrm{kWh}^{-1}} \tag{7.16}$$

where $(CO_2)_{captured}$ kWh^{-1} is total amount (in tons) of CO_2 captured per net kWh for a plant with CO_2 capture.

CCC reflects the economic viability of the CO_2 capture system at a given market price for CO_2. For example, if CO_2 captured at a power plant can be sold to an end user (e.g., EOR or food industry), the LCOE for the plant with CCS would approach that of the reference plant (although the latter will have much higher emissions).

Figure 7.10 compares avoided and captured emissions for a reference power plant (assuming, a state-of-the-art plant) without CCS and a plant of the same type as the reference plant but equipped with CCS system. Unless the energy requirements for CO_2 capture are close to zero (which is practically impossible), the amount of CO_2 avoided would always be less than that of CO_2 captured. Thus, the CCC would be lower than the CCA, since the energy consumed during operation of CO_2 capture system would increase the amount of CO_2 emitted per kWh produced [25].

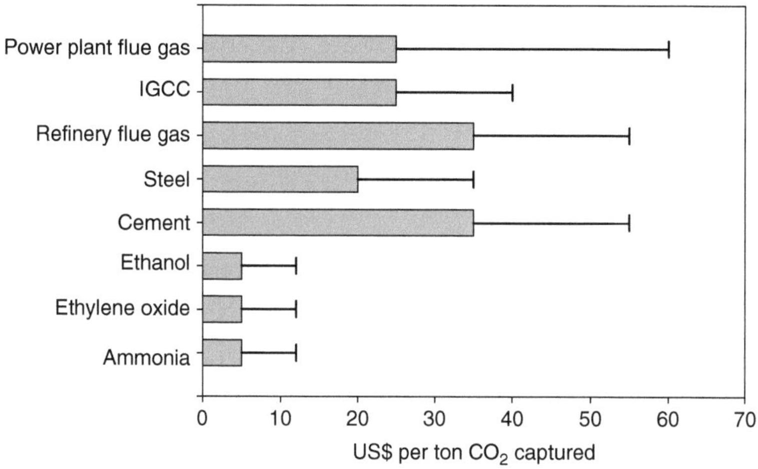

Fig. 7.11 Cost estimates for CO_2 capture and compression from different industrial sources. *Source* [1]

According to reported analytical studies, capturing and compressing CO_2 accounts for about two-thirds of the cost of the entire CCS chain [97]. The CCA estimates for early commercial-scale coal-fired power plants expected in the year 2020 are projected at \$50–70/ton CO_2 abated. A power plant has to generate in average about 30 % more power to cover the energy drain due to capturing 90 % of CO_2 emissions. The analyses of historical trends for the related energy and environmental technologies suggest that improvements to the current carbon capture technologies can potentially reduce the cost of CO_2 capture by at least 20–30 % over the next decade [25]. Despite numerous studies examining different technological options for carbon capture, there still exists a considerable uncertainty with regard to the magnitude of future cost reductions.

The costs of capturing and compressing (but not storing) CO_2 from a variety of power generation and industrial sources are summarized in Fig. 7.11.

The results of the analysis of carbon capture costs from different sources demonstrated that CO_2 could be captured from some industrial (non-power) facilities at costs less than that from coal-fired power plants (In the USA, those sources would amount to less than 25 % of the total emissions, where CCS could potentially be applied.) [1].

Tuinier et al. conducted comparative techno-economic assessment of different CO_2 capture technologies: amine absorption, membrane separation, and cryogenic capture [41]. The analysis showed that in each case, the preferred technology depended heavily on the availability of utilities. The amine scrubbing method requires low-pressure steam (to strip the solvent), while cryogenic method requires a source of cooling (e.g., L-NG evaporation at a regasification terminal). If steam and L-NG are not available at low costs, membrane technology could be the technology of choice. When low-cost steam is available, amine (especially, advanced amine) scrubbing shows advantage. The cryogenic capture could be attractive, if L-NG (or other cold source) is available at low costs.

7.7.2 Cost of CO_2 Transport

As CCS technology will be getting more matured and becoming an important industrial sector, the transport of ever-increasing volumes of CO_2 will require significant planning and investment over a relatively short period of time. It has been estimated that over 200,000 km of pipelines at a cost of US$2.5–3 trillion would need to be constructed to support CO_2 transport from the projected 3,400 industrial scale CCS projects according to the IEA's Blue Map scenario [12].

CO_2 transport via pipelines in a dense phase (pressure of 7.58 MPa at ambient temperature) is by far the most cost effective means of moving the large quantities of CO_2 over long distances [1]. The cost of pipeline transport depends on several factors: (1) construction, operation and maintenance costs, (2) the pipeline length and diameter, (3) the quantity and quality of transported CO_2, and (4) a route and terrain (onshore pipelines running through heavily populated areas could be 50–100 % more expensive compared to less urbanized areas) [25]. Offshore pipelines, in general, are 40–70 % more expensive than onshore ones. McCoy and Rubin reported the cost of transport of five million tons CO_2 per year (the approximate output of an 800 MW coal-fired power plant equipped with CCS) through 100 km pipeline as follows (in US$ per ton CO_2): 1.16, 0.30, and 0.20 for the pipelines constructed in the Midwest, Central, and Northeast USA, respectively [98].

Many analysts reported that there are significant economies of scale resulting from a shared infrastructure, such as clusters and networks [12]. For example, McKinsey's report showed that the clustered transport system could significantly decrease transport cost, since fewer large-scale pipelines would be needed to connect them to storage sites; for example, combining two nearby CO_2 emitters using individual 24-in. (61 cm) pipelines into one 36-in. (91 cm) pipeline would reduce estimated CO_2 transport cost by 30 % (compared to point-to-point transport) [66]. Global CCS Institute report also demonstrated that there are potential cost savings through increasing CO_2 flow through a single pipeline by combining CO_2 flows from multiple sources for delivery to a single storage site [96]. For example, the cost of transporting CO_2 from an average size single-user pipeline (flow of about 3–6 Mt/year) is about US$1–2 per ton CO_2. However, combining CO_2 flows from three to four sources (with total flow of 14–18 Mt/year) through a single pipeline would result in a reduction in the CO_2 transport cost down to US$0.5–0.7 per ton CO_2 [96]. The impact of the CO_2 flow on LCOE follows the same trend as CO_2 transportation cost.

The cost analysis of marine CO_2 transport by ships is much more complicated than that of pipeline transport, and since no marine CO_2 transport systems have been implemented on a large scale, the cost of CO_2 transport by ships is still at the level of estimates. Several authors reported that the economics of CO_2 ship transport is more favorable compared to pipeline transport for longer distances (e.g., the break-even distances for transporting six million tons of CO_2 per year by ships and offshore and onshore pipelines are about 1,000 km and 1,600 km, respectively) [25]. But distance is not the only factor determining the competitiveness of marine CO_2 transport; other factors including fuel cost, loading terminals, pipeline shore crossings, seabed stability, security, should also be considered.

7.7.3 Cost of Geological CO_2 Storage

Most experts agree that in near-to-mid term, geological storage is the only economically feasible CO_2 storage option, and within this option, storage in depleted oil and gas fields and deep saline aquifers shows the most promise and potential [66]. The cost of geological CO_2 storage is highly site specific, and it is determined by the type of a storage option (e.g., saline formation, depleted oil/gas reservoir), storage depth, and geographic factors (e.g., the site location, terrain), which results in a high degree of variability in the CO_2 storage cost estimates [99]. The most significant components of the capital cost for CO_2 storage relate to wells (which could run from about US$200,000 for an onshore site to US$25 million for an offshore well), in-field pipelines, facilities, and infrastructure [25]. The major operating costs include manpower, fuel, and maintenance. In most cases, the costs of geophysical and engineering feasibility studies, the reservoir evaluation, and licensing are also included in the cost estimates. Typically, an offshore storage option is more costly compared to onshore one since it involves expensive platforms and a subsea equipment coupled with higher operational costs.

Detailed cost estimates for geological CO_2 storage in the USA, Europe, and Australia have been reported in a number of publications and reports [12, 25]. The cost estimates for CO_2 storage in onshore saline formations, depleted oil field, and disused oil/gas fields are in a rather close range (about US$4–5 per ton CO_2), whereas the cost of offshore CO_2 storage in saline formations and disused oil/gas fields is about twice that of onshore storage (US$8–12 per ton CO_2) [25]. The recent techno-economic analysis of CO_2 storage options identified the significant cost advantages of onshore storage projects over offshore storage, as well as the cost savings of depleted oil/gas reservoirs relative to deep saline formations, particularly, when utilizing reusable legacy wells [12]. The study also demonstrated that despite the relative advantages of certain storage options, the lowest cost storage reservoirs contribute the least to the total available storage capacity. Thus, based on the current understanding of reservoir capacity, especially in Europe, there is more storage capacity offshore (more expensive option) than onshore (less expensive), and there is more storage capacity in deep saline formations (more expensive option) than in DOGR (less expensive) [12].

The cost of geological CO_2 storage may be substantially offset if it is combined with EOR or enhanced coal bed methane (ECBM) recovery. For North American sites, the net cost of onshore CO_2 storage coupled with EOR was estimated at a negative value of—US$14.8 per ton CO_2 stored [100]. The offshore CO_2 storage costs with EOR are somewhat higher compared to onshore storage: from the negative value of –US$10.5 to +US$21.0 per ton CO_2 stored [101]. The potential economic benefits from coupling CO_2 storage with EOR would increase with the increase in oil prices; e.g., the increase in the oil price from US$20 to US$50 per barrel would increase the potential benefit from combining EOR with CO_2 storage from US$16/ton CO_2 to US$30/ton CO_2 [25]. Thus, the economic benefits from oil production may provide incentives for the earlier implementation of EOR for geological CO_2 storage compared to other options.

Although EOR and other beneficial reuse options have a potential to provide early revenues when applied to CCS, one has to take into consideration that the sheer volumes of CO_2 being emitted from even small power stations (500 MW or higher) would place a profound supply–demand mismatch with many EOR projects that would be capable of utilizing no more than 1–2 Mt/year CO_2 and most definitely will not last the 30 year life span of a power plant. Many factors can influence the economics of EOR projects. It was reported, for example, that in Alberta (Canada), during 2000–2010, several provincially funded CO_2-based EOR pilot projects were initiated, but none resulted in a large-scale commercial development [102]. Producers initially blamed the lack of affordable CO_2 supply, but the technology improvements intervened and multi-frac horizontal wells became the EOR technology of choice. Instead of capital-intensive CO_2-based EOR projects with long payback periods, producers could drill long horizontal wells with multiple fracture stimulations that could be put on production immediately. This has caused CO_2-EOR projects in Alberta to practically stall. On the other hand, the development of multi-frac horizontal wells hasn't precluded the use of CO_2 in EOR projects in the USA. Although most of the CO_2 for EOR in the USA is of natural origin, an increasingly significant amount of it also comes from a variety of industrial sources.

Summarizing, CO_2 storage costs are highly site specific and the local geological characteristics of the site will drive the cost. Numerous analytical studies show that when comparing the economics of carbon storage to other parts of the CCS chain, the estimated cost of onshore storage (at more or less favorable conditions, estimated at approximately US$10 per ton CO_2) is a relatively minor contributor to the overall cost of a full-scale CCS project. However, a move to an offshore geological storage would dramatically increase the cost of both storage and transportation compared to onshore storage, especially in the existing relatively tight market for offshore drilling rigs and platforms.

7.7.4 Cost of Ocean CO_2 Disposal

There are a very few studies reported in the literature on the economics of the ocean CO_2 storage. Akai et al. estimated the costs of ocean storage for the following scenarios: (1) CO_2 transport and injection at the depth of 3 km from a floating platform, (2) CO_2 transport and injection at 2.0–2.5 km depth from a moving ship via a towing pipe, and (3) CO_2 injection via pipeline transport at 3.0 km depth [103]. It was assumed in this study that CO_2 was produced by three 600 MW coal-fired power plants and transported 100 or 500 km by a CO_2 tanker ship with the capacity of 80,000 m^3. The estimated costs for the first and second scenarios for the ship transport distance of 500 km are (in US$/ton CO_2 net stored): 13.2 and 15.7, respectively. The pipeline transport scenario was found to be the most expensive: the cost of ocean storage of CO_2 transported by a pipeline 100 km or 500 km and injected at 3 km depth was estimated at US$6.2 and US$31.1 per ton CO_2 net stored, respectively [103]. No literature data are available on the cost of ocean CO_2 storage via the CO_2 "lake" option.

7.7.5 Economics of Integrated CCS System

In general, the cost of an integrated CCS project comprises the cost of the base plant and additional costs due to CO_2 capture, transportation, and storage equipments, plus costs due to losses in overall efficiency and CO_2 leakage from transport, handling, and storage systems. It has been pointed out that although the costs of individual components of the CCS system provide a basis for determining the overall CCS cost, it's not a simple operation of summing up the costs of CO_2 capture, transport, and storage [25]. Due to the difference between the categories of CO_2 captured or avoided during generation of electricity (or an industrial product), the cost of CO_2 mitigation should be expressed either in terms of CO_2 captured or CO_2 avoided. Typically, the cost of CO_2 mitigation in terms of CO_2 avoided is greater than that of CO_2 captured (per ton CO_2). It is recognized that the carbon mitigation cost is best represented by the cost of CO_2 avoided.

Since no large-scale integrated CCS project is currently operational anywhere in the world, early demonstration projects are needed to be built where different technologies are tested along the entire (capture–transport–storage) CCS chain. The cost of these demonstration integrated projects will be markedly higher than that of early commercial and, especially, full-scale commercial projects. McKinsey report provides data on the cost differential between first demonstration (about 2015), early commercial (2020+), and mature commercial (2030+) reference power plants in Europe [66] (the data are summarized in Fig. 7.12).

Although there is a significant spread between individual projects (due to local specifics), the trend is clear: the cost of CO_2 abated drops from €60–90/ton CO_2 for demonstration projects to €35–50/ton CO_2 for early commercial projects (about 45 % drop). Note that transport and storage costs for demonstration and early commercial-scale projects are comparable. The difference between early commercial and mature commercial projects, though, is not that great: €35–50 vs €30–45 per ton CO_2 abated (about 15 % difference). It is very difficult to predict the carbon price in the long term, but some estimates are reported for Phase II of the European Union Emission Trading system ranging from €30 to €48 per ton CO_2 abated, and this value is expected to remain up until 2030. Assuming the median carbon price of

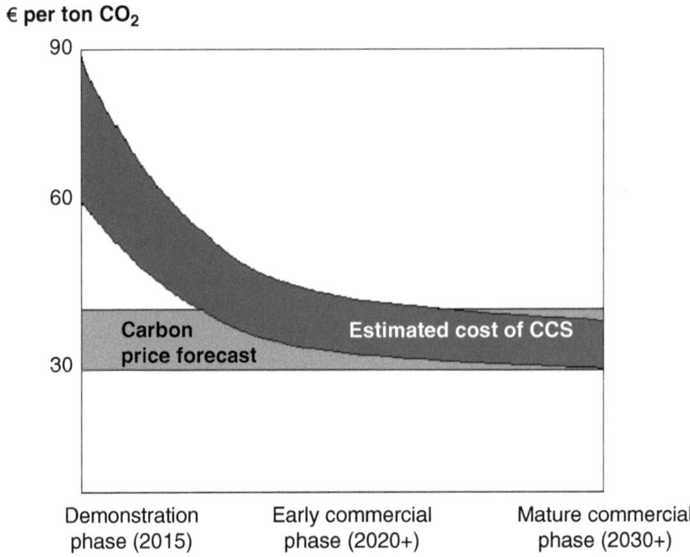

Fig. 7.12 Forecast of CCS development costs and carbon price. *Source* [66]

€35 per ton CO_2, there will be an economic gap of €25–55 for the demonstration projects to be commercially viable. However, the economic gap will significantly shrink as the CCS projects move to an e arly demonstration stage and will practically disappear when the CCS project will reach a mature commercial status with its cost being in the range of the future carbon prices.

The introduction and implementation of new and emerging technologies will have a great potential to reduce the CCS costs as emphasized in a number of reports [12]. It was estimated that (based on the current status of development), the introduction of ITM may reduce plant capital costs by about 10 % due to the combination of decreased plant size and reduced ASU equipment costs [96]. According to McKinsey's Global scenario (which calls for 500–550 CCS projects by 2030), the introduction of new breakthrough technologies (e.g., chemical looping, membranes) could potentially reduce the cost of CCS down to €20–35 per ton CO_2 abated [66].

The available data on the impact of CCS introduction on the incremental production costs in power generation and industrial applications are summarized in Fig. 7.13.

The costs of applying the Post-CCC technology (including capture, compression, transportation, and storage) to blast furnace steel and cement production are in the range of US$49–54/ton CO_2, which are comparable to that of OFC, PC/USC, and IGCC. The costs of CO_2 avoided for NG processing and fertilizer production are significantly lower (US$19 and US$20 per ton CO_2, respectively) than those of Post-CCC in power generation [96]. This can be attributed to the fact that these industrial facilities currently include CO_2 separation/capture process; thus the additional auxiliary load is needed for the compression component only, whereas in the Post-CCC case, both CO_2 capture and compression are required.

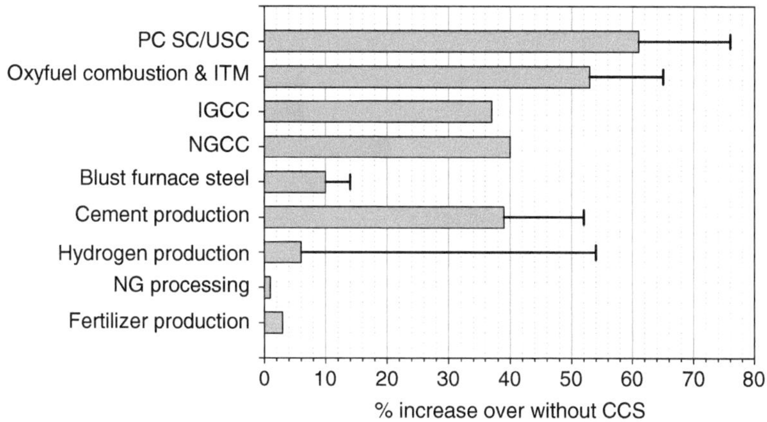

Fig. 7.13 Economic assessment of CCS technologies for industrial applications. Percentage increase of production costs with CCS over without CCS for industrial and power generation applications. FOAK and NOAK are *first of a kind* and *nth of a kind* systems, respectively. *Source* [25, 96]

The general trend in the CCS technology development is to lower its cost and increase a CO_2 storage capacity. Figure 7.14 summarizes the cost–capacity relation for the various CO_2 sequestration options. Terrestrial ecosystems are considered the lowest cost option for CCS followed by storing CO_2 in unmineable coal seams and depleting oil and gas reservoirs. Ocean sequestration is considered the highest cost storage option, although it has an immense storage capacity as a carbon sink.

The economic impact of the large-scale deployment of CCS technology is a subject to major uncertainties. The estimates reported in the literature on the cost of CCS deployment vary from US$350–440 billion [104] to trillions of US dollars [105]. In addition to an uncertain short- and long-term economic impact of CCS technologies, there are factors that are not well known, but could significantly affect the cost of the CCS deployment, e.g., costs related to a possible environmental damage, liability, monitoring cost, and issues related to public acceptance. Due to the enormous potential cost of the CCS deployment, different policy options for providing sufficient funding for massive-scale CO_2 capture and storage are being considered. One study suggested that the World Trade Organization (WTO) should become an active player in the CCS market [106]. According to the study, WTO could jump-start the CCS market by issuing long-term contracts to purchase *bona fide* sequestration-derived CO_2 credits. This measure could bring forth the investment in CCS needed to achieve storage of up to ten billion tons of CO_2 per year by 2025 (including seven billion tons from biological ocean storage and three billion from geologic and terrestrial storage). According to one set of assumptions, the net WTO subsidy would reach US$86 billion by 2022 (which is about 1.01 % of the WTO's tariff on imports and exports). If the policy is successful, WTO would subsidize the CCS deployment in the near to medium term and then recoups its investment and gains large profits over the long term.

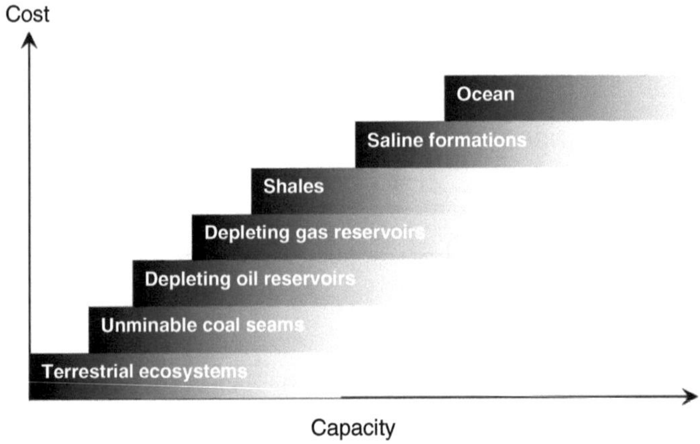

Fig. 7.14 Cost-capacity relation for the various CO_2 sequestration options. *Source* [17]

7.8 Current Status of CCS Projects

The individual component technologies for CO_2 capture, transport, and storage are in different stages of development; generally, most of them are well understood and, in some cases, they have reached commercial maturity. Figure 7.15 summarizes the current (2013) technical readiness levels of main CCS technologies.

Figure 7.15 shows that most of the CCS components/technologies are in the prototype and pilot scale of development, with many of them having reached TRL 9. For example, CO_2 capture from NG (the so-called gas sweetening), geological storage of CO_2 for EOR application, and pipeline transport of CO_2 have been commercially practiced for many years. However, the largest challenge facing the widespread deployment of CCS is the integration of individual component technologies into large-scale demonstration or commercial projects.

According to different (e.g., Global CCS Institute, NETL) databases, there are close to 300 active, planned, and proposed CCS projects across the range of technologies and project types in close to 30 countries around the world. Most of these projects are concentrated in North America (USA and Canada), followed by Europe (Norway, UK, Spain, Netherlands, and others), Australia, and China. Cumulative spending on the CCS large-scale demonstration projects (including component technologies in the CCS chain) between 2007 and 2012 has reached US$10.2 billion, of which US$7.7 billion came from private financing [107]. This is the clear indication of a growing confidence in CCS and a broad commitment to this technology.

7.8.1 Overview of Active and Planned CCS Projects

Much of the increased confidence in CCS technology comes from the experience gained from the operation of *large-scale integrated CCS projects (LSIP)*.

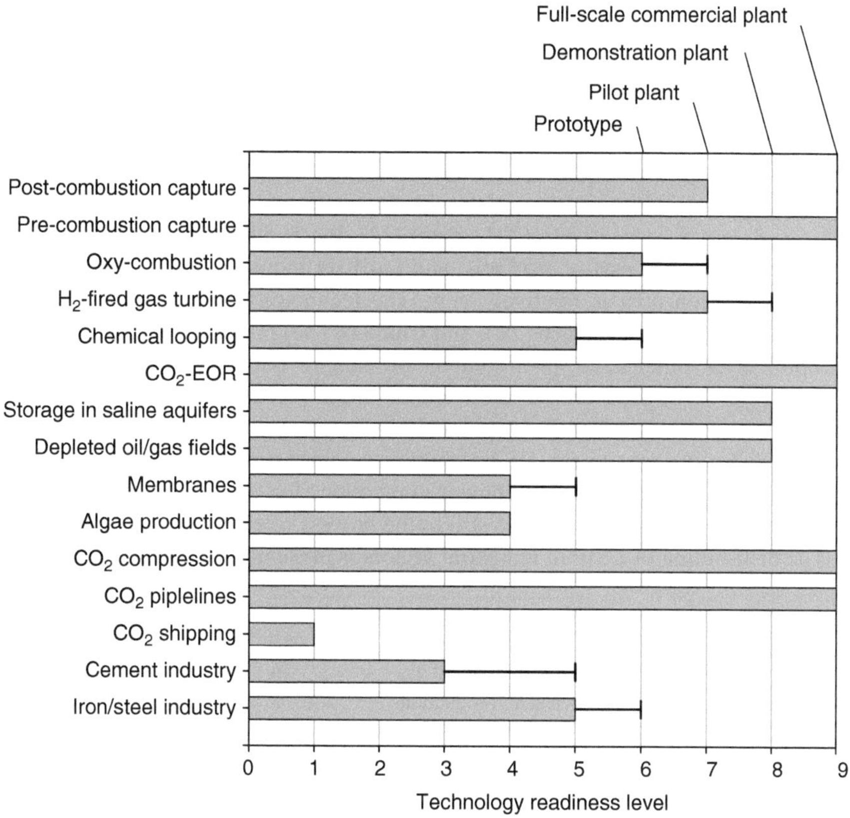

Fig. 7.15 Technical readiness levels of main CCS technologies and sectors. *Source* [63, 66, 107]

The importance of LSIP is that they include all parts of the CCS chain from CO_2 capture to its transport to its permanent storage or utilization (hence, they are called "integrated," as opposed to CCS projects that deal with individual elements of CCS, e.g., only capture or storage). LSIP are defined as the CCS projects that involve all components of the CCS chain at the scale of at least 0.8 Mt/year for coal-fired power plants or 0.4 Mt/year for other industrial point sources, including NG-based power plants [12].

According to the Global CCS Institute database, in 2012, there were 78 LSIP around the world at the various stages of development [108]. The USA has the largest number of LSIP projects (25), followed by Europe (21), Canada (9), and other countries and regions. Within Europe, the UK has the leading position in a number of large-scale CCS projects (7), followed by the Netherlands (4) and Norway (3). It should be noted that as of 2012 there were no LSIP projects in other key emitting countries such as Russia, India, and Japan. Recently, China has emerged as a fast mover in CCS technology, accounting for more than half of all newly identified LSIP around the world in 2012 [109]. In the beginning of 2013, China had 11 LSIP in various stages of planning and more pilot CCS projects than any other country (with most projects in the coal-to-power and coal-to-chemical sectors). CCS has been included in China's 12th Five-Year Plan along with other ambitious policies and actions to reduce carbon emissions.

The analysis of CCS projects indicates that, globally, pre-combustion capture is a predominant decarbonization approach followed by post-combustion, oxyfuel combustion, and industrial separation with the captured CO_2 volumes in Mt/year: 114, 28, 8, and 3, respectively (2011 data) [12]. Pre-combustion capture has been long used in gas processing, synfuels, and fertilizer industries, although its use in power generation area is relatively new. The technology is particularly popular in the USA (88 %), Canada (67 %), and China (83 %) (% of all projects) [12]. In Europe, on the other hand, post-combustion capture is the most widely pursued approach (48 % of all users). The difference in the pattern between the USA and Europe can be explained by the large number of gas processing and substitute natural gas (SNG) projects in the USA (where pre-combustion capture is a preferred option) and the specifics of government grant allocation. Oxyfuel combustion is a relatively new technology, and it is still in the process of entering the market.

The distribution of LSIP in industrial sectors is presented below [12]:

Power generation	42
Gas processing	11
Synthetic natural gas	6
Fertilizer production	5
Coal to liquids	3
Hydrogen production	3
Iron and steel production	2
Oil refining	1
Chemical industry	1

CCS units associated with power generation plants represent the lion's share (57 %) of all existing or planned LSIP (most of them relate to coal-fired power plants), because they represent the major stationary (i.e., most "visible") sources of CO_2 emissions and, hence, they have attracted the largest share of government support. This is followed by gas processing and synthetic fuel facilities, and several energy-intensive industries (fertilizer, hydrogen, metallurgical, etc.).

The distribution of existing and planned LSIP (in % of total) by the storage type is as follows [12]:

Enhanced oil recovery	54
Deep saline formations	31
Depleted oil and gas reservoirs	6
Unspecified	9

In the USA and Canada, about 80 % of all LSIP are predominantly focused on the EOR-type storage of CO_2, since this option provides an excellent economic incentive for the deployment of CCS projects (the same trend is true for Middle East and China).

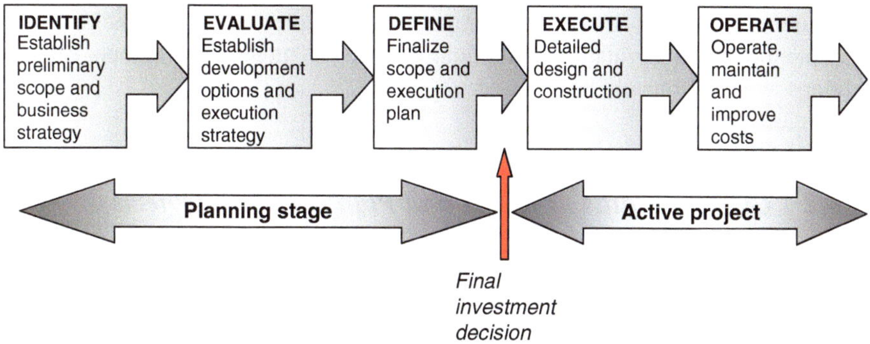

Fig. 7.16 Asset lifecycle model for CCS projects. *Source* [110]

In Europe and Australia, however, CO_2 storage in deep saline formations and DOGR is a more prevalent option [12]. Note that most of the existing and proposed CCS projects in Europe relate to offshore storage (saline formations and DOGR), although they are more expensive compared with the onshore option.

The geological storage and beneficial reuse are the major means of CO_2 storage amounting to about 90 % of all CCS projects. Among beneficial reuse projects, 55 % relate to EOR and 11 % to EGR and ECBM, each, with the remaining projects involving the use of CO_2 in food and chemical industries. Although a complete CCS project integration is easier and less costly to achieve for the EOR-driven projects compared to other geological storage solutions, it is realized that EOR/DOGR option would unlikely have the sufficient storage capacity as a long-term carbon mitigation strategy. The majority of analytical assessments point out that deep saline formations and other forms of geological storage will provide the bulk of carbon storage potential [12].

The *Asset Lifecycle Model (ALM)* widely accepted for the evaluation of the various stages of the project development (i.e., planning, design, construction, and operation) reflects the decision points in the project lifecycle at which developers need to decide whether to continue, delay, or cancel the project for economical, social, or political reasons. Figure 7.16 depicts the schematic diagram of the ALM with regard to the progress of CCS projects [110].

Currently, most of the existing LSIP are in "Evaluate" stage followed by "Define" stage. The carbon capture technologies applied to power generation sector lag far behind those that are part of industrial processes. This can be attributed to the fact that NG processing and chemical industry produce a relatively pure storage-ready CO_2 stream. In contrast, the implementation of CCS technology in the power generation sector would carry substantial additional cost of installing equipment for carbon capture from relatively diluted streams (flue gases, or syngas). The selection of a particular carbon capture option in the power generation sector will be dictated by several factors including the type of power plant (newly built or retrofit), type of fuel (coal, NG, residual oil), geographical location (proximity to onshore or offshore storage sites), specifics of local business, public acceptability, and regulatory conditions. Thus, all three main carbon capture options would ultimately be required.

7.8.2 Current Status of Active and Planned CCS Projects

According to the Global CCS Institute database, at the beginning of 2013, there were 8 operating CCS projects, while the number of "execute"-stage projects reached 9, bringing the total number of projects operating or under construction to 17 and their total capacity to 37 Mt/year [111]. A total of 55 LSIP are in the planning stages of development with projected CO_2 capture capacity of about 104 Mt/year. According to the ALM classification, active projects include CCS projects in "Operate" (i.e., currently commercially operational) and "Execute" (or under construction) stages of development. The following are the highlights of the most significant commercial CCS projects.

The Sleipner Project (Norway) has been successfully operated by the Statoil company since 1996 in the North Sea (about 250 km off the coast of Norway); it is a first commercial-scale project aiming at geological CO_2 storage in saline aquifer formation. In this project, CO_2 is first separated from NG stream with CO_2 content of about 9 vol.% and then compressed and injected into a deep sandstone saline formation located about 0.8–1.0 km below the seabed (above the gas reservoir zone). On an average, about one million ton of CO_2 per year is currently injected into the saline aquifer formation (an additional capacity of 0.7 $MtCO_2$ per year is under construction), and over the lifetime of the project over 20 million tons of CO_2 is expected to be injected and stored underground. The saline formation represents a brine-saturated sandstone beneath the North Sea floor (called Utsira formation) with the top of the formation and the overlaying primary seal consisting of a fairly flat, extensive shale layer 75 m thick. According to estimates, the saline formation has a storage capacity of 1–10 $GtCO_2$. Since the start of the Sleipner Project, the IEA's GHG R&D Program in collaboration with Statoil has been conducting the research activities involving monitoring, baseline data gathering and evaluation, simulation of reservoir geology, geophysical modeling and model verification, and other activities. Of particular importance is the monitoring of the fate and slow movement of CO_2 plume in the saline formation. A seismic time-lapse survey showed that CO_2 plume extends over the area of about 5 km^2, and its migration out of the storage site was effectively prevented by the caprock seal.

In Salah Gas Project (Algeria) located in the Saharan region of Algeria started commercial operation in 2004; it is a joint venture of Sonatrach, BP, and Statoil [112]. In this project, CO_2 is stripped from NG that contains about 10 vol.% CO_2 and injected into a 1.8 km deep sandstone saline formation [1]. CO_2 is injected into the sandstone saline formation via three horizontal wells, each 1.5 km long, at the rate of about 1–1.2 million tons CO_2 per year, and the purified NG is delivered to European markets (over the lifetime of the project, over 17 $MtCO_2$ will be geologically stored) [25]. The onshore-to-onshore-type pipeline transports CO_2 over the distance of 14 km to a storage site [12]. The top seal of the storage formation (caprock) is a layer of mudstones about 950 m thick. The storage site is under continuous surveillance and monitoring with regard to the risk assessment, CO_2 storage integrity, CO_2 migration from the injection wells, etc.

The Snøhvit project (Norway) in the Barents Sea off the coast of northern Norway is a recent (since 2008) addition to the family of the CCS projects utilizing saline formations [113]. In this project, CO_2 is extracted from NG at an onshore liquid NG plant, transported by a 150 km long pipeline, and reinjected into an offshore sandstone saline formation at the depth of 2.6 km beneath the seafloor (the formation is deeper than the gas reservoir, and it has a caprock shale about 30 m thick). Since the beginning of its operation, the project has injected about 0.7 $MtCO_2$ per year into the storage reservoir.

The Weyburn-Midale project (Canada, USA) is a large CCS-EOR operation established in 2000 at the Weyburn Oil Field in Saskatchewan, Canada. With the injection rate of 3 Mt/year CO_2, this project is one of the world's largest carbon storage projects. CO_2 is captured from North Dakota's Great Plains Synfuel plant (based on coal gasification), and a relatively pure stream of CO_2 is delivered in a dense phase form via a 320 km pipeline to Saskatchewan for EOR. Prior to the start of the CCS-EOR project, about 500 million barrels of oil had been recovered, and additional 215 million barrels are expected to be recovered over the life of the project (and over 40 million tons of CO_2 will be stored in the reservoir) [114]. As a result of this measure, the oil field lifetime will be extended by about 25 years. The site is extensively monitored for a potential leakage using high-resolution seismic surveys (as of now, there was no CO_2 leakage to the surface). Surface monitoring also includes the analysis of groundwater and soil.

The Rangely-Weber Sand Unit project (USA) operates since 1986. The project captures CO_2 at ExxonMobil's Shute Creek gas processing plant near LaBarge, Wyoming, and transports it via 283 km long pipeline to the Rangely field (northwestern Colorado), where it is used in EOR operation with the average injection rate of 2.97 Mt/year CO_2 [1, 25]. The Rangely-Weber CO_2-EOR project is projected to produce 129 million barrels of oil via EOR [115].

Salt Creek Project (USA) also relies on CO_2 captured from the Shute Creek gas processing plant. Anadarko Petroleum Corp. transports CO_2 via 200 km pipeline to an old oil field in Natrona County, Wyoming, as well as 53 km pipeline to Monell unit for EOR operation [116]. CO_2 injection began in 2004, increasing the Salt Creek's oil recoverability by 10–15 %. In 2009, the Salt Creek's project reached a milestone of ten million barrel of oil produced as a result of EOR. Anadarko plans to sequester about 50 million tons of CO_2 throughout the life of these projects.

Enid Fertilizer Plant project (USA). In this project, about 680,000 t per year of CO_2 has been captured using pre-combustion technology and transported via a 192 km pipeline from the plant's site in northern Oklahoma for the use in EOR process in Golden Trend and the Sho-Vel-Tum fields, both located south of Oklahoma City [116].

In the *Century Plant project* (USA), Occidental Petroleum jointly with Sundridge Energy is operating (since 2010) a gas processing plant in West Texas, where about five million tons CO_2 are captured annually using pre-combustion technology and transported via 256 km pipeline for use in EOR process. The CO_2 capture capacity is projected to increase to 8.5 Mt/year in 2012.

The Val Verde Natural Gas Plants project (USA) operating in Southwest Texas since 1972 is the first commercial-scale CCS-EOR project in the USA. The project captures up to 1.3 Mt/year of CO_2 from Mitchel, Gray Ranch, Puckett, Pikes Peak, and Terrell gas processing plants, dehydrates and compresses it, and transports it by 132 km pipeline to Sharon Ridge field, where it is used by ExxonMobil for EOR [12].

The Allison Unit Project (USA) of Burlington Resources in New Mexico (USA) was the first commercial ECBM project. As a result of the project, the extent of methane recovery increased from about 77 % to estimated 95 % of the original gas in place. The methane recovery ratio was one volume of methane recovered for every three volumes of CO_2 injected [117]. More details of the project are presented in Sect. 9.2.1.

CCS Projects in "Execute" Stage. The following LSIP projects fit the definition of the "Execute" stage in the ALM classification, which involves detailed design, engineering, and construction of a plant. As of 2012, there were 7 "Execute"-stage LSIP around the world with projected start of the plant operation within 2–3 years:

- The Gordon Project (Australia) [116, 118].
- ADM Illinois Industrial Carbon Capture and Sequestration Project (USA) [119].
- Air Products Steam Methane Reformer EOR Project (USA) [120].
- Argium CO_2 Capture ACTL Project (Canada).
- Boundary Dam Integrated CCS Project (Canada).
- Kemper County IGCC Project (USA) [119].
- Lost Cabin Gas Plant Project (USA) and others.

7.8.3 CCS Industrial Applications

The IEA's 2012 report on the role of CCS in industrial applications projects that CCS has the potential to reduce CO_2 emissions from industrial applications by 4 Gt in 2050 [121]. Such an amount of CO_2 is equal to roughly one-tenth of the total emission cuts needed to reduce emissions by 50 % by the mid-century. The IEA's roadmap focuses on five main industrial applications: high-purity CO_2 sources, biomass conversion, cement, iron and steel manufacturing, and refineries. The Roadmap sets out a vision of CCS in industrial applications up to 2050, including milestones that need to be achieved for technology, financing, policy, and international collaboration. Although most industrial CO_2 capture can be accomplished using pre-, post-, and oxyfuel combustion technologies, in some cases (e.g., iron and steel manufacture, biochemical biomass conversion), the combination of the technologies used in these industries deviate from the conventional carbon capture approaches.

7.8.3.1 Cement Manufacturing

Cement manufacturing industry is the major industrial source of CO_2 emissions: it is responsible for about 5 % of global anthropogenic CO_2 emissions, making the industry an important sector for carbon abatement measures [122]. CO_2 is emitted from different sources: the limestone calcination process, combustion of fuels in the kiln, and from power generation. It is recognized that recent reductions in CO_2 emission rates from cement plants due to technological advancements and improvements in energy efficiency are unlikely to provide the needed drastic carbon reductions without implementation of CCS. The IEA's ETP-2012 report estimates that the cement industry will need to reduce its emissions by around 1.1 $GtCO_2$ per year and that the share of CCS will be between 660 and 940 million tons per year [4]. The IEA also provides interim targets for 2020 that involves 11 CCS plants in the cement sector capturing 13 $MtCO_2$ per year [123].

There have been a number of studies discussing the different aspects of CCS use in the cement industry [122, 123]. The studies indicated that among technologically advanced CO_2 capture technologies, only post-combustion capture and oxyfuel combustion may be applicable to the cement production sector. Pre-combustion capture is not considered suitable for cement plants because it is unlikely to be able to capture the emissions from the conversion of limestone to lime, the key component of cement. These studies also underscored that post-combustion capture can be readily fitted to existing cement plants, although there could be some technical issues related to handling the dust and SO_x emissions from the process and meeting the additional heat requirements for the solvent process (e.g., an auxiliary boiler), which would increase the capital and operational costs. Oxyfuel technology, on the other hand, although potentially beneficial for the cement industry, would require more significant changes to the cement plant operation compared to the post-combustion capture option. The application of oxyfuel technology to cement manufacturing plants would require more R&D efforts, especially, with regard to heat transfer, gas recirculation, etc.

Although no commercial-scale cement plant coupled with CCS exists anywhere in the world, four pilot projects at different stages of technological development (mostly, at TRL of 5–6) should be mentioned [123]:

1. The European Cement Research Academy (ACRA) has proceeded with Phase IV of the ECRA CCS Project [124].
2. In Norcem (Norway), a pilot project at the Brevik cement plant is underway.
3. In Taiwan, the Industrial Technology Research Institute and Taiwan Cement company are collaborating on a calcium looping capture project.
4. In the USA, Skyonic has an operational pilot facility at the Capital Aggregates cement mill. Skyonic focuses on the reuse of CO_2 with production of sodium bicarbonate (baking soda) [125].

Alternative strategies for reducing CO_2 emissions from cement plants include the development of cement formulations that are based on magnesium instead of calcium. These technologies are at early stages of development.

7.8.3.2 Iron and Steel Manufacture

Currently, three process routes are being investigated in the iron/steel industry for coupling with carbon capture at pre-commercial scale:

- Top Gas Recycling Blast Furnace (TGR-BF) is a blast furnace variant, where the top gas of the blast furnace goes through CO_2 capture, but the remaining reducing gas is reinjected at the base of the reactor.
- HIsarna process—a smelting reduction process based on the combination of a hot cyclone and of a bath smelter. The process uses pure oxygen and generates CO_2-rich off-gas which is almost ready for storage.
- ULCORED process—a direct reduction process, which produces Direct Reduced Iron in a shaft furnace, either from natural gas or from coal gasification. Captured CO_2 leaves the plant in a concentrated stream and goes to storage.

7.8.3.3 Biochemical Biomass Conversion

In biochemical biomass conversion processes (e.g., fermentation), living microorganisms break down the feedstock and produce liquid and gaseous fuels. The CO_2-rich off-gases from the fermentation tanks are dried and compressed to facilitate transport and storage. One of the first commercially operated ethanol plants integrated with CCS started operation at the Arkalon bioethanol plant in Kansas (USA) in 2009 [12]. A larger scale Illinois-ICCS project commenced construction in 2011, with operation expected in 2013.

7.9 Environmental Impact of Large-Scale CCS Deployment

Due to the increasing role of CCS technologies in view of carbon abatement and climate policies, the possible environmental impact of large-scale CCS deployment is an active area of experimental and modeling studies.

7.9.1 Environmental Aspects of CO_2 Capture

It is likely that (in most cases) a power plant (or an industrial facility) coupled with a CO_2 capture system will produce a stream of concentrated CO_2 (for storage) along with certain amounts of gaseous, liquid, and solid wastes [92]. Depending on the source of CO_2 and technology utilized for CO_2 capture, the CO_2 stream may contain impurities that would impose some limitations on CO_2 transport and storage, as well as present some safety and environmental concerns. The total concentration of impurities in dried CO_2 from coal- and gas-fired plants with post-combustion capture is targeted at 0.01 vol.% or less, which would be adequate for its use in the majority of industrial applications, including food industry [92]. The pre-combustion CO_2 capture systems utilizing physical solvent-scrubbing processes produce less pure CO_2 streams with typical concentrations of impurities as follows (for the case of coal-fired IGCC plants, in vol.%): H_2 0.8–2.0, CO 0.03–0.4, CH_4 0.01, $N_2/O_2/Ar$ 0.03–0.6, H_2S 0.01–0.6, adding up to the total of 2.1–2.7 [92]. The production of combined CO_2/H_2S streams can potentially reduce the cost of the capture process; however, this option is feasible only if the environmentally acceptable ways of transporting and storing the combined stream are available. In the case of post-combustion capture by amine-based solvents, flue gases may contain traces of solvents and, possibly, ammonia (from the solvent decomposition).

Liquid waste products of the CO_2 capture processes could include degraded solvents, which are typically incinerated in specialized facilities. It has been reported that the post-combustion systems typically produce more liquid wastes than pre-combustion processes [25]. The solid wastes from CO_2 capture systems may contain deactivated reforming and CO-shift catalysts from the pre-combustion capture systems (these wastes are either reprocessed or disposed in an environmentally safe manner) [126].

It was pointed out by several authors (e.g., [5]) that even equipped with CCS, coal plants still produce significant amount of GHG, because CCS technology, generally, does not reduce pollutants aside from CO_2 [5]. For example, in the post-combustion type CCS units, CO_2 after combustion is separated from other gases such as SO_x, NO_x, NH_3 that are typically discharged to the air. Because CCS-equipped coal plants usually consume more fuel to generate a given quantity of electric power, compared to plants with no carbon capture, more non-CO_2 pollutants could be emitted to the air by coal-CCS plants. It was reported that the addition of a CCS unit to an IGCC plant resulted in the increase in fuel usage by 15.7 %, and, consequently, SO_x and NO_x emissions by 17.9 % and 11 %, respectively [25]. According to reported estimates, the addition of CCS equipment to a pulverized coal power plant would increase the fuel usage by 31.3 %, resulting in the substantially increased NO_x and NH_3 emissions of 31 % and 2,200 %, respectively (the addition of an emission-control device decreases SO_x emissions by 99.7 %) [25]. Overall, the life-cycle GHG emissions of coal-CCS power plants are estimated at 255–442 g CO_2-equiv./kWh [5].

7.9.2 *Environmental Impact of CO_2 Storage Systems*

The long-term ecological consequences of CO_2 storage are largely unknown. The assessment of environmental risks is highly specific to the storage method and the techniques used to inject CO_2 into a storage site. The potential negative environmental impact of underground CO_2 storage mainly relates to possible hazards to groundwater and terrestrial and marine ecosystems and to induced seismicity.

7.9.2.1 Hazards to Groundwater

In 2011, the US EPA issued a proposal that would exclude CO_2 injected underground from being treated as a hazardous waste [127]. The main motivation behind this move was the EPA's intention to reduce barriers for the implementation of CO_2 reduction technologies and expand locations where CO_2 can be injected. However, drinking water suppliers voiced serious concerns over the plans to expand CCS projects in some areas arguing that CO_2 injected deep into geologic formations and held there in a supercritical state might come into contact with drinking water wells and slightly acidify it. Although not a problem by itself (CO_2 is not a poisonous gas), CO_2 may result in leaching some naturally occurring toxic minerals (e.g., Pb, Cd, As, Sb) from soil and rocks into drinking water and make it unsafe for the consumers. Indeed, the computer simulation studies conducted at the Lawrence Livermore National Laboratory (USA) indicated that CO_2 can cause leaching of naturally occurring trace elements, such as As, Ba, Cd, Hg, Pb, Sb, Se, Zn and even uranium from geological formations (however, only release of arsenic could really pose a problem due to its very high toxicity) [128]. Moreover, in many cases, CO_2 carries harmful impurities (e.g., H_2S) that may become mobile and further exacerbate the problem. At best, these compounds may alter an odor and taste of water, but in worst cases, this may prevent the use of groundwater for drinking or even irrigation purposes.

7.9.2.2 Hazards to Ecosystems

CO_2 stored in underground formations may have a negative impact on terrestrial and aquatic ecosystems. High concentration of CO_2 and reduced pH may adversely affect microbial populations in deep subsurface and plants and animal life in shallower subsurface and surface. If CO_2 leaks upward to the surface, it would have a detrimental effect on soil causing a vegetation "die-off". Available data indicate that CO_2 concentrations in soil above 5 % are dangerous for plants, and the CO_2 concentrations of 20 % and higher may be fatal for vegetation (normal CO_2 concentration in soil is 0.2–4 %) [25]. Although there are no known records of the negative impact of CO_2 leakage from existing CCS projects on terrestrial ecosystems (which could be due to the relatively short period of exploitation time), some naturally occurring underground seepage of CO_2 may provide some indications of a possible impact. For example, in 1989–2001, CO_2 seepage in a volcanic area near Mammoth Mountain (California, USA) caused an extended plant die-off (CO_2 concentration in soil was 15–90 %).

Potential hazards to ecosystems could be exacerbated if other much more problematic gases such as H_2S, SO_2, and NO_2 are stored along with CO_2, thus, substantially increasing risks to the ecosystems. These gases are routinely formed during combustion or gasification of coal, oil residues and can be present in CO_2 storage reservoirs in relatively high concentrations (e.g., in the Weyburn project, injected CO_2 gas contains 2 % H_2S) [25]. H_2S is a much more toxic gas than CO_2; so, the level of tolerance to the leakage of H_2S-containing gas would be much lower compared to pure CO_2 due to the much higher risk of poisoning. In Colli Albani (Italy), in 1999–2001, a natural seepage of CO_2 containing 2 vol.% H_2S from magmatic activity caused death of 29 cows and 8 sheep from asphyxiation [129]. SO_2 is another problematic impurity in CO_2, which upon dissolution in groundwater would produce much stronger acid compared to CO_2, resulting in the leaching of much higher quantities of toxic metals and minerals in groundwater, thus, significantly increasing the risk of exposure to the hazardous compounds.

7.9.2.3 Induced Seismic Activity

The injection of CO_2 into deep subsurface formations (e.g., porous rock) at significant pressures (exceeding formation pressures) can potentially induce fracturing and fault activation (i.e., movement along faults). This may result in two undesirable consequences: (1) increased fracture permeability (CO_2 migration from a storage site to the surface) and (2) induced earthquakes (potentially, large enough to cause a damage). Although there are no direct evidences of an induced seismic activity related to CO_2 storage projects, it has been suggested that the deep-well injection of waste fluids were, in all likelihood, responsible for the local earthquakes of moderate magnitudes in a number of locations in the USA, e.g., in Denver (1967) with M_L of 5.3 and Ohio (1986, 1987) with M_L of 4.9 [25] (M_L is an earthquake local magnitude scale).

Stanford University researchers published an article casting serious doubts on the overall viability of large-scale geologic carbon storage based on the concerns over possible triggering of earthquakes [130]. By drawing an analogy between recently experienced earthquakes resulting from brine injections in the USA and other countries, the authors point to the impact of CO_2 pressure buildup in the hosting formations and its potential to induce the earthquakes that would result in fractures and faults. The main concern is that even small to moderate triggered earthquakes could be accompanied by fracturing that could propagate upward and compromise the integrity of an overlaying geologic seal of a CO_2 repository.

A report from the US National Research Council suggests that CCS may have the potential for inducing seismic events [131]. The report examines the potential for different energy technologies such as hydraulic fracturing, CCS, geothermal energy production, and conventional oil and gas development to cause earthquakes. The report indicates that CCS may potentially induce seismic events because enormous volumes of fluids are injected underground over long periods of time. On the other hand, insufficient information exists to evaluate the potential of CCS to cause earthquakes, because of a limited number of large-scale continuously operating projects. The committee recommended that continued research would be needed to examine a potential for induced seismicity in large-scale CCS projects.

7.9.2.4 Environmental Impact of Ocean CO_2 Storage

There is very limited knowledge of the possible environmental impact of CO_2 storage on the ocean ecosystem. It is recognized that regardless of the injection method the large-scale deployment of CO_2 storage strategies would result in the production of the tremendous volumes of seawater with an altered chemical balance (e.g., drop in its pH and change in CO_3^{2-} concentration). Most of the studies on the impact of ocean CO_2 storage on ecosystems are at the level of mathematical models; however, making extrapolations from small-scale field experiments, although useful, are unlikely to predict all the ecological consequences of the worldwide implementation of ocean CO_2 storage. It is also unclear how the ocean biota would adapt to sustained elevated CO_2 levels. To make the matters even more complex, the changes in seawater acidity (pH), CO_2, carbonate ion and bicarbonate ion concentrations may have different and specific effects on marine life, e.g., the impact of CO_2 accumulation could be more severe than the effects of pH reduction or an increase in carbonate ion concentration. Due to evolutionary selection, deep-sea organisms are most likely to be more sensitive to environmental disturbance compared to shallow water relatives [25]. Of particular concern is the environmental impact of harmful contaminants in CO_2, especially H_2S. The risk management issues related to ocean carbon storage have not been adequately studied.

7.9.2.5 Environmental Impact of Mineral Sequestration

The environmental impact of mineral sequestration technology is directly associated with the issues related to the large-scale mining, mineral processing, and the disposal of carbonates and solid waste products. Due to the potential displacement of million tons of soil and rock, these operations might adversely impact soil, water, and air, along with land clearing, leaching of heavy metals, and other problems [25]. In particular, mining activities involving blasting, drilling, earth moving would create dust and aerosol matter that may affect respiration and pollute local vegetation. Moreover, due to the weak acidity of by-products, there is the possibility of metals leaching resulting in the contamination of groundwater. To minimize the negative effects of water and soil contamination, mine reclamation programs have to be implemented. On the other hand, after CO_2 is chemically bound into mineral carbonates, this method provides almost 100 % "leak-proof" method for long-term CO_2 storage, since there would be practically no CO_2 leakage from the disposal sites. This would practically eliminate the need for long-term storage site monitoring, which is an important issue with other CO_2 storage options, especially the geological and ocean CO_2 disposal.

7.10 Risk Factors Associated with Large-Scale CCS Deployment

7.10.1 CO_2 Emissions and Leakage Due to CCS Deployment

It is important to emphasize that CCS does not completely eliminate all CO_2 emissions, and CO_2 storage is not necessarily 100 % permanent. Certain fraction of stored CO_2 is likely to escape to the atmosphere via a variety of pathways such as molecular diffusion

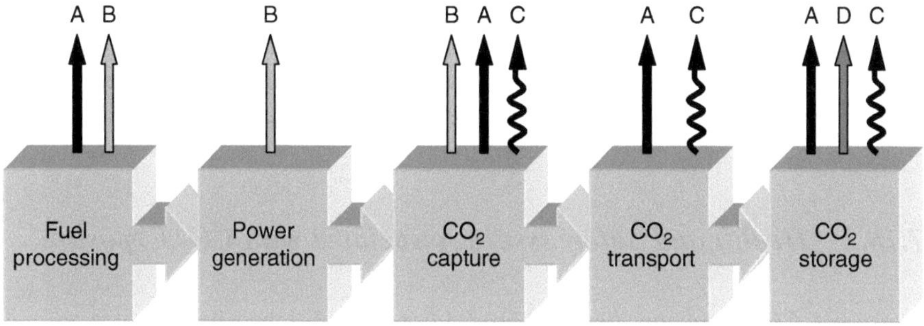

Fig. 7.17 Flow diagram of potential CO_2 emissions and leakage sources during the deployment and operation of the entire CCS chain. (*A*) Emissions from additional energy/fuel usage. (*B*) Emissions from fuel processing, power generation, technological processes. (*C*) Fugitive emissions. (*D*) Leakage from storage

through the caprock, a gradual leakage through faults, fractures, wells, or an abrupt leakage, and an injection well failure. [132]. In the case of ocean storage, the retention time depends on the depth of injection: shallower the injection faster the leakage rates.

Since CCS is a sequence of several steps, CO_2 capture, transportation, and storage, the CO_2 leakage could occur at each stage of this sequence. Figure 7.17 depicts the flow diagram of the potential sources of CO_2 leaks over the entire CCS technological chain.

CO_2 capture processes produce two types of emissions: the emissions due to imperfect CO_2 capture and those resulted from an additional energy use due to CO_2 capture. The capture process can produce GHG other than CO_2, e.g., methane, NO_x, CO, NH_3, SO_2 (with some of these gases resulting from the degradation of the solvents used in the CO_2 capture processes, e.g., amine and potassium carbonate scrubbing). CO_2 transport related emissions result from additional energy requirements for this operation (compressing, pumping) and fugitive CO_2 emissions released during its transport. The pipeline transport of CO_2 consumes electrical energy used in compressors installed at a CO_2 capture site (longer distances would require more compressors resulting in more CO_2 emissions). The emissions related to CO_2 transportation by ships, rails, and trucks come from mobile combustion sources, e.g., vehicles. Fugitive losses during CO_2 transportation are insignificant.

CO_2 emissions generated during the CO_2 storage phase include emissions from an additional energy use for CO_2 injection (underground or the deep ocean), fugitive emissions during the injection operation, and CO_2 leakage from a storage reservoir. The assessment of CO_2 emissions due to additional energy requirements for its injection to some extent is similar to that related to CO_2 emissions due to its capture and transport. The quantification of a physical leakage from geological and ocean storage sites is a focus of numerous studies (e.g., [25]). A physical leakage from storage reservoirs could occur via either gradual long-term release or sudden rapid release due to the disruption or perturbation of a storage reservoir. The numerical simulations of CO_2 permanence in geological storage sites indicated that the probability of CO_2 release is rather insignificant. For example, a simulation study using a probabilistic model concluded that over 5,000 year period a statistical mean release of 0.001 % and a maximum release of 0.14 % of the total stored amount of CO_2 would occur [133]. Although the estimates of the fraction of retained CO_2 in geological storage sites are highly site specific, models show that for well-selected and monitored storage sites the fraction of retained CO_2 is likely to exceed 99 % over the first 1,000 years [25].

7.10.2 Health and Safety Issues Associated with CO_2 Exposure

In general, CO_2 is not considered a harmful or toxic substance, because it is present in the air we breathe and it is a product of human metabolism. The majority of healthy people can tolerate relatively high concentrations of CO_2 (up to 1.5 vol.%) for several hours without health complications; however, an exposure to higher CO_2 concentrations for much longer durations could be harmful to human health. In particular, the exposure to CO_2 at 3 vol.% level or higher in air can cause a hearing loss, visual impairment, headache, breathing difficulties, dizziness; at the levels of 7–10 vol.%, CO_2 acts as an asphyxiant (CO_2 enters bloodstream displacing oxygen). At CO_2 concentrations above 20 vol.%, death could occur in 20–30 min [25]. The potentially harmful effect of high concentrations of CO_2 on human health is reflected in the current USA occupational exposure standard [134] limiting the maximum permissible CO_2 concentration in air to 0.5 vol.% for 8 h of a continuous exposure and 3 vol.% for a short period (15 min) of exposure (the CO_2 level of 5 vol.% is considered immediately dangerous to life and health).

Anthropogenic CO_2 leak sources could originate from a number of sources, e.g., pipelines, storage tanks, industrial units, open pits, and buildings. Since CO_2 is about 1.5 times denser than air, it tends to accumulate in low-lying confined or enclosed spaces; adding to the potential dangers of CO_2 exposure is the fact that it is a colorless and an odorless gas, which cannot be detected by human senses until CO_2 intoxication already occurred. Therefore, CO_2 monitors (or sensors) should be installed where the accumulation of CO_2 in relatively high concentrations is likely to occur, along with adequate ventilation in the areas where CO_2 leak is possible.

7.10.3 Public Acceptance of CCS Risks

Although issues related to the adverse environmental impact of fossil fuels and the implications for climate change are a hot topic of discussion nowadays; in general, there is an insufficient public knowledge and awareness of various carbon mitigation options and, particularly, the role of CCS technologies. There are several studies with regard to the public perceptions of CCS technology and associated risks. For example, the survey of public perception of CCS conducted in the USA showed that on the scale from 1 (very negative) to 7 (very positive), the respondents rated ocean and geological storage at 3.2 and 3.5, respectively; however, after receiving more detailed information on the CCS options, the respondents shifted their ratings to a more negative side of the scale, i.e., 2.4 for ocean and 3.0 for geological storage [135]. In a survey in Japan, on a scale from 1 (negative) to 5 (positive), the respondents rated the CCS options as follows: 2.24–2.47 for ocean storage and 2.57–75 for geological storage [136]. After receiving an additional information, the ratings slightly increased for all categories by about 0.1–0.3 points. In general, the public perception of the ocean storage option was more negative compared to geological storage.

One should also not underestimate the powerful NIMBY factor, which resurfaces every time when local interests collide with a broader public purpose. One example of the NIMBY factor as applied to the CCS technology relates to a 200 MW coal-fired power plant retrofitted with oxyfuel combustion and CO_2 capture technologies. Originally, CO_2 sequestration site was planned in nearby Mattoon (Illinois, USA); however, the officials in Mattoon have announced that they are not interested in having the sequestration site in that location [137]. Due to a strong public opposition (concerned with possible leaks and potential health hazards), Germany's Bundesrat has struck down a law for testing CCS technology [6]. In 2010, Shell's Barendrech CCS project in the Netherlands was canceled by the government, which cited "the complete lack of local support" as the main reason [138]. The project planned to store around ten million tons of CO_2 over a period of 25 years from Shell's Pernis refinery under the town of Barendrecht. Local public opposition to the plan was fueled by fears that the project would endanger the town. However, in many cases, the NIMBYsm stems from the lack of information. For example, the results of a recent Eurobarometer survey of the public awareness of the CCS technology in Europe indicated that 67 % of public have never heard of the technology and just 10 % know what it is about [139].

7.11 Current Trends and Challenges to CCS Technologies

7.11.1 Current Trends in CCS Technologies

At the basic level, all CO_2 capture technologies are technically proven and viable, and some of them (e.g., Pre-CCC) have been commercially practiced for several decades in gas processing industry. However, none of them have been integrated with coal- or gas-fired power plants at a full commercial scale. The use of a particular carbon capture option depends on the type of the application or the process which best fits CO_2 capture technology. Currently, the available carbon capture technologies are still very costly and far from optimal in their performance. The reduction in carbon capture costs is critical for the technology to be broadly deployed beyond initial pilot and demonstration projects.

Further optimization, integration, and technology improvements will be necessary to reduce the energy losses and capital costs associated with carbon capture and to operate commercial-scale facilities with carbon capture. It is recognized that the major reductions in CO_2 emissions from fossil fuel plants can be achieved through an increase in the energy efficiency of basic technologies, such as pulverized coal combustion and gas turbines because this will partially offset the detrimental impact of carbon capture on the plant's performance.

Current trends in the CCS technologies indicate a shift away from deep ocean storage and mineralization options toward geological storage, preferentially in saline formations, oil and gas reservoirs, and unmineable coal seams. One of the main reasons why ocean storage is losing a favor relates to concerns over the potential adverse impact of the enormous quantities of stored CO_2 on marine biota. With regard to the mineralization option, it is realized that the unrealistically large quantities of reactants would be required for the widespread deployment of this technology (e.g., lime production would have to be increased by several orders of magnitude to meet the increased demand for this reagent). Biological CO_2 fixation and storage technologies, particularly, via algal growth are currently gaining momentum. The worldwide interest in this technology is fueled by the realization that algae could be converted into fuels that are similar to conventional petroleum-based transportation fuels.

Due to the high carbon intensity of coal as fuel, so far, the emphasis in carbon capture has mainly been focused on coal-fired power plants. But there is an increasing recognition that CCS technologies also have to be applied to NG, especially, considering that its role in power generation and industrial sectors may substantially strengthen in the coming decades (due to increased supply of unconventional gas, particularly, shale gas) (IEA projects gas to become a predominant fuel by mid-century) [140]. Despite relative cleanness of gas (compared to coal), in order to meet climate policy objectives, gas-fired power plants and industrial units will have to be equipped with CCS.

More small-scale CCS projects are needed in a variety of storage media around the world as precursors to larger-scale projects. The benefits of smaller-scale projects are that they will help clarify regulatory issues and increase the public awareness (and, hopefully, acceptance) of the technology, which otherwise may delay (or even stall) the larger projects. Moreover, these projects will provide valuable information on the storage characteristics of selected geological sites, which will be needed for designing larger demonstration plants and help develop local expertise to implement and operate the larger projects [63].

7.11.2 Challenges Facing Large-Scale Deployment of CCS

Currently, CCS technology developers around the world face a number of challenges beyond "normal" technical hurdles that are part of the development of any new technology. They found themselves in unfavorable political and economic climate: a very high financial barrier without any economic stimuli or regulatory drivers to encourage investments in the expensive carbon mitigation technologies. The lack of public support and, in many cases, poor perception of the technology exacerbate the situation.

One of the key hurdles is that in many countries, governments are reluctant to give an unconditional commitment to the GHG limits, which could potentially lead to an effective penalty-avoidance, regulatory and tax frameworks, which in turn would drive CCS economics [116]. For example, despite the fact that in 2009 the US House of Representatives initially cleared a legislation to require CO_2 emission reduction and a cap-and-trade program that would potentially have put cost on carbon emissions, a year later the bill died in the US Congress, and as of 2012, the majority of the Congress members show no interest in putting a price on carbon [119]. With the demise of this legislative initiative, the primary drivers for CCS—a price on carbon emissions and CO_2-trading mechanism—also expired.

The ongoing financial and regulatory impasses adversely affect large CCS projects in the USA, and in some cases even kill them. For example, in July 2011, the Midwestern utility company American Electric Power (AEP) canceled a project dealing with the construction of a commercial-scale advanced CCS add-on to its 1.3 GW Mountaineer Power Plant in West Virginia [119]. If materialized, the Alstom-built CCS unit would have been the world's largest unit installed at a coal-fired power plant, capturing 1.5 Mt/year CO_2, or 90 % of total CO_2 by 2015. However, after 2 years of planning and construction of a $100 million pilot facility at the site, AEP terminated the project, citing the combination of the US climate policy confusion, the regulatory uncertainties, and weak economy as main reasons for the cancellation.

Similarly, in Europe, some uncertainties are befogging the progress of the CCS technology. Since the vast majority of Europe's most advanced CCS projects are reliant on obtaining substantial government funding to proceed, particularly, from the NER300[1] program, the recent drop in the emission allowance prices represent a significant risk for the deployment of CCS technology in the region over the next decade. It was reported that based on the results of the auctioning of emission allowances by the European Investment Bank, around €1.2 billion will be available to the CCS projects under the NER300 program, and it is likely to be spread across three to five projects, as opposed to the initial target of eight CCS projects [141].

Other major hurdles to the widespread introduction of CCS to the world marketplace are the lack of a general consensus on climate change and, in many cases, a public opposition to CCS projects. Today, a significant fraction of policymakers and even many environmental scientists challenge the scientific basis of global climate change. The CCS opponents are also concerned about its efficacy and long-term ecological uncertainties and that CCS would provide only a temporary relief and make humankind even more dependent on fossil fuels, thus, making the necessary changes later even more difficult. These debates negatively reflect upon the public acceptance of CCS technology. These developments, if further persist, may seriously hinder the dynamics of the market penetration of CCS technology.

A very high financial wall to overcome is another major hurdle to the deployment of large-scale CCS projects. Already high, the cost of CCS technology has

[1] New Entrants Reserve (NER300) is one of the world's largest funding programs for innovative low-carbon energy demonstration projects as part of the EU Emission Trading System.

recently further risen in line with the overall trend in increasing price tags of large-scale energy projects (about 20–30 % higher over last 2–3 years) [116]. The additional cost increase incurred by incorporating CCS into a power plant would amount up to 40–75 %. The cost of CCS infrastructure, including pipelines, gas compression stations, the identification and development of secure CO_2 storage sites, is also very high, and it is continuously rising. This explains why today almost all commercially operating CCS projects utilize CO_2 capture technology (predominantly, of pre-combustion type) as part of an already established industrial process, e.g., they use CO_2 to generate revenue through EOR (that can potentially double the output of oil fields [119]) and, in most cases, they have an access to low-cost storage sites based on existing geologic information. This fact by itself underscores the challenges and hurdles facing the CCS projects that cannot count on the EOR revenues, or do not include CO_2 capture as part of the technological process. No wonder, five out of eight operating commercial CCS projects involve EOR. Therefore, many in industry believe that EOR is a prerequisite for the deployment of the majority of new CCS projects with the potential for substantial incremental oil recovery (only in the USA, EOR provides 281,000 barrels per day of oil) [116]. The increasing prevalence of applying CCS practice to EOR has attracted a range of oil and gas companies, pipeline operators and CO_2 source companies, and end users to forge mutually attractive business opportunities. These trends will continue in future in the USA and other countries as long as CCS-EOR suitable fields will remain available and oil prices are at the levels that encourage such an investment [12].

On the other hand, some energy analysts argue that while EOR has been practiced by oil and gas industry for many decades, and it provides a substantial economic payback, its potential CO_2 storage benefit is far less than that of storing CO_2 in much larger storage reservoirs such as deep saline formations [116]. The same could be true for depleted oil and gas formations that are well characterized by the industry over their history, but have relatively small geologic CO_2 storage capacity compared to deep saline formations.

The pace of CCS progress (at least in the USA) is affected not only by a high economic barrier and climate policy confusion but also by the coal–NG interplay in the power generation arena. Ironically, in order to protect coal and coal-related jobs, the US coal industry and coal-rich state politicians and lobbyists have fought hard and been rather successful in blocking climate legislation, but in doing so, they have killed the primary reason for developing CCS technology [119]. Without CCS technology, it will be increasingly difficult for coal to compete with cleaner energy sources (e.g., NG) in the carbon-constrained world. To avoid many uncertainties (economic, political, environmental) associated with coal, energy companies will increasingly gravitate toward NG for electricity generation since it produces almost half of the carbon emissions against coal. Historically, low prices for NG in the USA will only intensify this trend. Taking into consideration the record levels of production and plummeting NG prices, the growth in NG share is projected to continue (with simultaneous decrease in coal's share of the electricity market in the USA).

In March 2012, the US EPA, in a move to reduce carbon emissions to fight climate change, proposed a new rule that would limit CO_2 emissions from all fossil fuel-based power plants to 454 kg per MWh of electricity generation [119]. Modern NG-fired power plant can easily meet this target without resorting to CCS technology, whereas even the new state-of-the-art coal-fired plants (750–795 kg/MWh) would fall far short of this limit. Thus, to meet the EPA's proposed emission limits, all new coal-fired power plants would have to be equipped with CCS technology. On the other hand, utilities would be very reluctant to build a new coal power plant with CCS in the absence of clear climate policies and without knowing what regulations they must comply with. A classical Chicken-Egg problem. With plentiful and cheap gas, there would be no reason to build a coal plant and there would be no incentives to develop improved CCS technologies [142].

Given the dominant role of fossil fuels in global economy and the current worrisome trends of rising CO_2 emissions, many experts emphasize that the urgency of CCS deployment is only increasing. As the OECD Secretary-General A. Gurria put it: "Without CCS, continued reliance on coal-fired power is a road to disaster" [143]. At the same time, the 2013 IEA's report [107] acknowledged that the CCS deployment is running far below the trajectory required to meet the long-term carbon reduction targets in order to limit the global average temperature increase to 2 °C. The IEA's analysis shows that for CCS to help reach atmospheric CO_2 stabilization targets consistent with the 2DS pathway, the total CO_2 capture and storage rate levels must grow from the tens of megatons of CO_2 captured in 2013 to over 50 Mt/year in 2020, 2 Gt/year in 2030, and 7 Gt/year CO_2 in 2050 [107]. Cumulatively, 120 $GtCO_2$ would need to be captured between 2015 and 2050, across all regions of the world. By mid-century, CCS would need to contribute 14 % of the cumulative CO_2 emission reductions against a business-as-usual scenario [107]. According to IEA, the next 7 years are critical to the accelerated development of CCS for achieving CO_2 stabilization goals in the framework of climate policies.

7.11.3 Knowledge Gaps in CCS Technologies

7.11.3.1 CO_2 Storage

While the extensive experience gained from the existing large-scale commercial CO_2 storage projects gives a confidence in the feasibility of geological storage, it does not by itself promote the global deployment of CCS. There are still many unknowns and gaps in fundamental knowledge as well as the practical demonstration of large-scale CO_2 storage in various underground reservoirs. Although CO_2 storage in the North America's oil reservoirs is well documented, not all oil fields are suitable for using CO_2 for EOR, and not all the suitable sites have the sufficient capacity to store CO_2 in quantities commensurable with climate policy objectives. Therefore, alternative options for geological (and not only geological) storage need to be proven and demonstrated. Furthermore, due to the regional variations in geology, it is crucial to conduct similar tests in other locations worldwide (this would also help build public confidence and acceptance of the technology) [63].

The following is the list of major technology R&D gaps in the CO_2 storage area [63]:

- There is a lack of quantitative data on CO_2 leakage rates from different types of storage sites; there is a need for reliable hydrogeological–geochemical–geomechanical simulation models that can accurately predict long-term storage performance.
- The subsurface impacts of impurities in CO_2 streams on the storage site integrity should be investigated; if CO_2 from both Pre-CCC and Post-CCC sources is stored at the same site, there is a possibility that H_2S and SO_2 would be simultaneously present in the reservoir, which could potentially lead to the formation of elemental sulfur via Claus reaction negatively impacting the site's permeability.
- There are many gaps in capacity estimates at the global, regional, and local levels, in particular, in the areas that are likely to experience the greatest growth in energy consumption, such as China, India, Southeast Asia, Middle East, Brazil, Eastern Europe (including Russia), and South Africa.
- The methods for the quantitative assessment of risks to human health and local environment from CO_2 leakage have to be developed.
- The impact of biological (i.e., microbial) processes on CO_2 storage reservoirs in the deep subsurface are virtually unknown and have to be elucidated.
- There is a lack of methodologies for estimating and dealing with the CO_2 leakage resulting from system failures due to seismic activities, pipeline disruptions, accidents, and other unexpected circumstances.

The long-term operation of the existing commercial CO_2 storage projects and further development and field demonstrations in this area would decrease the uncertainties and knowledge gaps allowing to make a more informed decision on the prospects of CCS as a carbon mitigation solution.

7.11.3.2 CO_2 Transport

Although CO_2 pipeline transport has been practiced for decades in the USA, Europe, and other countries, there are still remaining R&D needs with regard to this technological option, in particular:

- Better define the thermodynamic properties of CO_2 and its mixtures with potential impurities such as Ar, N_2, O_2, CO, NH_3, and H_2S at supercritical conditions [63]
- Investigate the compatibility of non-steel materials (e.g., elastomers and polymers for seals and gaskets) with CO_2 impurities
- Examine the possible impact of impurities from newly developed CO_2 capture technologies on the corrosion rates of carbon steel pipelines
- Investigate possible hydrate formation to avoid operational downtime

For the readers who are interested in more detailed information on the different aspects of CCS technologies, we would recommend recently released books [144–147].

References

1. Global CCS Institute (2009) Strategic analysis of the global status of carbon capture and storage. Final report. http://www/globalccsinstitute.com/downloads/reports/2009/worley/foundation-report-1-rev0.pdf. Accessed 3 Aug 2010
2. Outcomes from UNFCCC conference in Doha (2013) Carbon Capture J 32
3. Global CCS Institute (2013) Status of carbon capture and storage update. Carbon Capture J 32
4. International Energy Agency (2012) Energy technology perspectives. Pathways to a clean energy system. IEA/OECD, Paris, France
5. Jacobson M (2009) Review of solutions to global warming, air pollution and energy security. Energy Environ Sci 2:148–173
6. Quaile I, Gräßler B (2011) Deutsche Welle. http//www.dw-world.de/dw/article/0,15402725,00.html. Accessed 10 Dec 2011
7. Thompson J (2012) The dash for gas—no climate cure without CCS. Global CCS Institute. http://www.globalccsinstitute.com/insights/authors/jyhompson/2012/10/22/das-gas-%E2%80%93-no-climate-cure-without-ccs. Accessed 23 Oct 2012
8. International Energy Agency (2013) Tracking clean energy progress. IEA input to the Clean Energy Ministerial, IEA, Paris. www.iea.org/publications/TCEP_web.pdf. Accessed 10 Sep 2013
9. IEA calls for more funds to support carbon capture technology (2012) Bloomberg, June 11, 2012. http://www.eco-business.com/news/iea-calls-for-more-funds-to-support-carbon-capture-technology. Accessed 10 Jul 2012
10. Mankins J (1995) Technology readiness levels, NASA Office of Space Access and Technology, Advanced Concepts Office. http://www.hq.nasa.gov/office/codeq/trl/trl.pdf. Accessed 20 Oct 2012
11. WorleyParsons, Schlumberger, Baker & McKenzie and EPRI (2009) Strategic analysis of the global status of carbon capture and storage. http://cdn.globalccsinstitute.com/sites/default/files/publications/5751/report-5-synthesis-report.pdf. Accessed 10 Jul 2012
12. Global CCS Institute (2011) The global status of CCS: 2011 update. Canberra, Australia, ISBN 978-0-9871863-0-0
13. He H, Li W, Zhong M et al (2013) Reversible CO_2 capture with porous polymers using the humidity swing. Energy Environ Sci 6:488–493
14. US Department of Energy (2007) Carbon sequestration technology. Roadmap, Washington, DC
15. Rochelle G (2009) Amine scrubbing for CO_2 capture. Science 325:1652–1654
16. Ciferno J, Fout T, Jones A et al. (2009) Capturing carbon from existing coal-fired power plants. Chem Eng Progress:33–47
17. US Department of Energy (2003) Carbon sequestration. Technology roadmap and program plan. US DOE Office of Fossil Energy, National Energy Technology Laboratory, Morgantown, West Virginia
18. Millward A, Yaghi O (2005) Metal-organic frameworks with exceptionally high capacity for storage of carbon dioxide at room temperature. J Am Chem Soc 127:17998–17999
19. Carbon sponge could soak up coal emissions (2013) Carbon Capture J 32:21
20. CO_2 capture in vehicles and home heating systems (2012) Carbon Capture J. http://www.carboncapturejournal.com/displaynews.php?NewsID=1043&PHPSESSID=18dkjsha7qaa6s815mlrslvtq4. Accessed 30 Nov 2012
21. Lin H, Merkel T, Baker R (2007) The membrane solution to global warming. 6th annual conference on carbon capture and sequestration, Pittsburgh, PA
22. Du N, Park H, Robertson G et al (2011) Polymer nanosieve membranes for CO_2-capture applications. Nat Mater 10:372–375
23. US Department of Energy (2006) Carbon Sequestration Technology Roadmap. Office of Fossil Energy, NETL, Morgantown, West Virginia
24. US Department of Energy (2005) Roadmap for the hydrogen economy. Workshop on manufacturing R&D for the hydrogen economy, Washington, DC

25. Intergovernmental Panel on Climate Change (2005) IPCC special report on carbon dioxide capture and storage. Prepared by Working Group III of the Intergovernmental Panel on Climate Change. In: Metz B, Davidson O, de Coninck H et al (eds). Cambridge University Press, Cambridge
26. International Energy Agency (1996) Decarbonization of fossil fuels, IEA Report PH2/2, March 1996, IEA Greenhouse Gas R&D Programme, Cheltenham
27. Muradov N (2009) Production of hydrogen from hydrocarbons. In: Gupta R (ed) Hydrogen fuel. Production, transport and storage. CRC Press, Boca Raton
28. Wilson E, Gerard D (2007) Carbon capture and sequestration. Integrating technology, monitoring and regulation. Blackwell Publishing, Ames, IA
29. Ball M, Weindorf W, Bunger U (2009) Hydrogen production. In: Ball M, Wietschel M (eds) Hydrogen economy. Opportunities and challenges. Cambridge University Press, Cambridge, UK
30. Lin S (2009) Hydrogen production from coal. In: Gupta R (ed) Hydrogen fuel. Production, transport and storage. CRC Press, Boca Raton
31. US Department of Energy (2010) Worldwide gasification database online, NETL. Pittsburgh, PA. www.netl.doe.gov/coalpower/gasification/models/dtbs.pdf. Accessed 12 Aug 2012
32. US Department of Energy (2012) Turbine program. Enabling near-zero emission coal-based power generation, Office of fossil energy, NETL. http://www.netl.doe.gov/technologies/coalpower/turbines/refshelf/brochures/brochure%209-19-05.pdf. Accessed 20 Aug 2012
33. Fukuizumi Y, Shiozaki S, Muyama A et al (2004) Large frame gas turbines. The leading technology of power generation industries, Mitsubishi Heavy Industries, Technical Review, vol.41, No. 5, 1–5 Oct 2004. http://www.mhi.co.jp/technology/review/pdf/e415/e415254.pdf. Accessed 20 Aug 2012
34. Bancalari E, Chan P, Diakunchak I (2007) Advanced hydrogen turbine development, Siemens, DOE contract No. DE-FC26-05NT42644. http://www.energy.siemens.com/co/pool/hq/energy-topics/pdfs/en/igcc/6_advanced_hydrogen.pdf. Accessed 20 Aug 2012
35. Mak S (2003). Where are construction materials headed. http://www.dbce.csiro.au/inno-web/0899/sustainable.htm. Accessed 20 Jan 2008
36. Stern M, Simeon F, Herzog H et al (2013) Post-combustion carbon-dioxide capture using electrochemically mediated amine regeneration. Energy Environ Sci. doi:10.1039/c3ee41165f
37. CO_2 Technology Centre Mongstad (2012) Technologies. http://www.tcmda.com/en/technology/. Accessed 15 Jul 2012
38. Brennecke J, Gurkan B (2010) Ionic liquids for CO_2 capture and emission reduction. J Phys Chem Lett 2010:3459–3464
39. Ritter S (2010) Carbon dioxide's unsettled future. Chem Eng News 88:36–37
40. Trachtenberg M, Cowan R, Smith D et al (2008). Membrane-based enzyme-facilitated efficient CO_2 capture. Proceedings of the 9th international conference on Greenhouse Gas Control Technologies, Washington, DC, p 353–360
41. Tuinier M, Hamers H, van Sing Annaland M (2011) Techno-economic evaluation of cryogenic CO_2 capture—a comparison with absorption and membrane technology. Int J Greenhouse Gas Control 5:1559–1565
42. US Department of Energy (2012) Cryogenic carbon capture. ARPA-E project DE-AR0000101. NETL. http://www.netl.doe.gov/publications/proceedings/12/co2capture/presentations/5-Posters/D%20Frankman-SES-Cryogenic%20Carbon%20Capture.pdf. Accessed 20 Jan 2013
43. Harris S (2013) Material could enable cheaper method of carbon capture. The Engineer, June, 2013. http://www.theengineer.co.uk/sectors/energy-and-environment/news/material-could-enable-cheaper-method-of-carbon-capture/1012898.article. Accessed 14 Sep 2013
44. Zafar Q, Mattisson T, Gevert B (2005) Integrated hydrogen and power production with CO_2 capture using chemical-looping reforming-redox reactivity of particles of CuO, Mn_2O_3, NiO and Fe_2O_3 using SiO_2 as a support. Ind Eng Chem Res 44:3485–3496
45. McGlashan N (2008) Chemical looping combustion—a thermodynamic study. J Mech Eng Sci 222:1005–1019
46. Brandvoll O, Bolland O (2004) Inherent CO_2 capture using chemical looping combustion in a natural gas fired power cycle. ASME J Eng Gas Turb Power 126:316–321. ASME Paper GT-2002–30129

47. Naqvi R, Bolland O (2007) Multi-stage chemical looping combustion (CLC) for combined cycles with CO_2 capture. Int J Greenhouse Gas Control 1:19–30
48. Habib M, Badr H, Ahmed S et al (2011) A review of recent developments in carbon capture utilizing oxy-fuel combustion in conventional and ion transport membrane systems. Int J Energy Res 35:741–764
49. Mattisson T (2007) Chemical-looping combustion using gaseous and solid fuels. 2nd Workshop on international oxy-combustion research network, 25–26 Jan 2007, Windsor, CT
50. Tan X, Wang Z, Lin H et al (2008) Enhancement of oxygen permeation through $La_{0.6}Sr_{0.4}Co$ $0.2Fe_{0.8}O_{3}$-δ hollow fibre membranes by surface modifications. J Membr Sci 324:128–135
51. Park J, Do Par S (2007) Oxygen permeability and structural stability of $La_{0.6}Sr_{0.4}Co_{0.2}Fe_{0.8}O_{3}$-δ membrane. Korean J Chem Eng 24:897–905
52. Esquiro A, Brandon N, Kilner J et al (2004) Electrochemical characterization of $La_{0.6}Sr_{0.4}$Co0.2F $e_{0.8}O_{3}$-δ cathodes for intermediate temperature SOFC. J Electrochem Soc 151:A1847–A1855
53. Wang H, Cong Y, Yang W (2002) Oxygen permeation study in a tubular $Ba_{0.5}Sr_{0.5}Co_{0.8}Fe_{0.2}O_{3}$-δ oxygen permeable membrane. J Membr Sci 210:259–271
54. Zhu X, Cong Y, Yang W (2006) Effects of synthesis methods on oxygen permeability of $BaCe_{0.15}Fe_{0.85}O_{3}$-δ ceramic membranes. J Membr Sci 283:158–163
55. Arnold M, Wang H, Feldhoff A (2007) Influence of CO_2 on the oxygen permeation performance and microstructure of perovskite type $(Ba_{0.5}Sr_{0.5})$ (Co0.$_8Fe_{0.2})O_{3}$-δ membranes. J Membr Sci 293:44–52
56. Yan A, Liu B, Dong Y et al (2008) A temperature programmed desorption investigation on the interaction of $Ba_{0.5}Sr_{0.5}Co_{0.8}Fe_{0.2}O_{3}$-δ perovskite oxides with CO2 in the absence and presence of H_2O and O_2. Appl Catal B 80:24–31
57. Benson S, Waller D, Kilner J (1999) Degradation of $La_{0.6}Sr_{0.4}Co_{0.2}Fe_{0.8}O_{3}$-δ in carbon dioxide and water atmospheres. J Electrochem Soc 146:1305–1309
58. Efimov K, Arnold M, Martynczuk J et al (2009) Crystalline intermediate phases in the sol-gel-based synthesis of La_2NiO_{4+}δ. J Am Ceram Soc 92:876–880
59. Kvamsdal H, Jordal K, Bolland O (2007) A quantitative comparison of gas turbine cycles with CO_2 capture. Energy 32:10–24
60. Sandquist S, Julsrud S, Vigenand B et al (2007) Development and testing of AZEP reactor components. Int J Greenhouse Gas Control 2007:180–187
61. Sundquist S, Griffin T, Thornhaug N (2011) AZEP-development of an integrated air separation membrane-gas turbine. 2nd Nordic minisymposium on carbon dioxide capture and storage. Alstom Power, Gothenburg, Sweden, October 2011
62. US Department of Energy (2012) Praxair: IEP—oxy-combustion CO_2 emissions control, OTM-based oxycombustion for CO_2 recovery, Project No.: FC26-01NT41147 & FC26-07NT43088 http://www.netl.doe.gov/technologies/coalpower/ewr/co2/oxy-combustion/otm-based.html. Accessed 14 Jan 2013
63. Global CCS Institute (2009) Strategic analysis of the global status of carbon capture and storage. Report 5: synthesis report. http://cdn.globalccsinstitute.com/sites/default/files/publications/5751/report-5-synthesis-report.pdf. Accessed 12 Feb 2012
64. Seiersten M (2001) Material selection for separation, transportation and disposal of CO_2. Proceedings of the Corrosion National Association. Corrosion Engineers, paper 1042
65. Guijt W (2004) Analyses of incident data show US, European pipelines becoming safer. Oil Gas J 102:68–73
66. McKinsey and Co (2008) Carbon capture & storage: assessing the economics. McKinsey and Company. http://assets.wwf.ch/downloads/mckinsey2008.pdf. Accessed 12 Feb 2012
67. International Energy Agency (2004) Ship transport of CO_2. IEA Greenhouse Gas R&D Programme, Report PH4/30, Cheltenham
68. Marchetti C (1977) On geoenginering and the CO_2 problem. Clim Chang 1:59–68
69. Kaarstad O (1992) Emission-free fossil energy from Norway. Energy Convers Manag 33:781–786
70. Koide H, Tazaki Y, Noguchi Y et al (1992) Subterranean containment and long-term storage of carbon dioxide in unused aquifers and depleted natural gas reservoirs. Energy Convers Manag 33:619–626

71. Bachu S (2003) Screening and ranking of sedimentary basins for sequestration of CO_2 in geological media. Environ Geol 44:277–289
72. Pershad H (2012) CCS for gas—results of Element Energy study. Carbon Capture J 13–15
73. Doyle A (2013) Norwegian sea can hold 100 years of Norway's CO_2. Scientific American, 25 Jan 2013. http://www.scientificamercian.com/article.cfm?id=Norwegian-sea-can-hold100-years-of-Norways-CO_2. Accessed 10 Jul 2013
74. Johnson J (2009) Huge CO_2 storage potential in China. Chem Eng News 87:27
75. Bachu S, Gunter W, Perkins E (1994) Aquifer disposal of CO_2: hydrodynamic and mineral trapping. Energy Convers Manag 35:269–279
76. Perkins E, Czernichowski-Lauriol I, Azaroual M et al (2004) Long term predictions of CO_2 storage by mineral and solubility trapping in the Weybourn Midale Reservoir. Proceedings of the 7th international conference on Greenhouse Gas Control Technologies, vol II, Vancouver, Canada, p. 2093–2096
77. Casey A (2008) Carbon cemetery. Canadian Geographic Magazine, January/February:61
78. US Department of Energy (2012) NETL. Enhanced oil recovery information. www.netl.doe.gov/KMD/cds/disk44/F-General/EOR_Brochure.pdf. Accessed 20 Dec 2012
79. Holt T, Jensen J, Lindeberg E (1995) Underground of storage of CO_2 in aquifers and oil reservoirs. Energy Convers Manag 36:535–538
80. Ferguson R, Nichols C, van Leeuwen T et al (2009) Storing CO_2 with enhanced oil recovery. Greenhouse Gas Technologies-9, Elsevier Science Direct, Energy Procedia 1: 1989–1996
81. Baviere M (2007) Basic concepts in enhanced oil recovery processes. Elsevier Applied Science, London, ISBN 1-85166-617-6
82. Stevens S, Kuuskraa J, Schraufnagel R (1996) Technology spurs growth of US coal bed methane. Oil Gas J 94:56–63
83. Key R, Kozyr A, Sabine C et al (2004) A global ocean carbon climatology: results from GLODAP. Global Biochem Cy 18:GB4031
84. Giles J (2002) Norway sinks ocean carbon study. Nature 419:6
85. Brewer P, Peltzer E, Walz P et al (2005) Deep ocean experiments with fossil fuel carbon dioxide: creation and sensing of a controlled plume at 4 km depth. J Mar Res 63:9–33
86. Intergovernmental Panel on Climate Change (2007) IPCC 4th assessment report, climate change 2007. The physical science basis. Cambridge University Press, Cambridge
87. O'Connor W, Dahlin D, Rush G et al (2005) Aqueous mineral carbonation, Final Report, DOE/ARC-TR-04-002, 15 March 2005
88. US Department of Energy (1999) Carbon sequestration. State of the Science. US DOE, Office of Science, Office of Fossil Fuels, Working paper on carbon sequestration science and technology. Washington, DC
89. Andersen R (2005) Algal culturing techniques. Elsevier, Amsterdam, pp 189–203
90. Nakamura T (2004) Recovery and sequestration of CO_2 from stationary combustion systems by photosynthesis of microalgae, Technical report to DOE, NETL, No. PSI-1356, December 2004
91. National Laboratory Directors (1997) Technology opportunities to reduce US greenhouse gas emissions. Oak Ridge National Laboratory, Tennessee
92. International Energy Agency (2004) Improvements in power generation with post- combustion capture of CO_2, IEA report PH4/33. IEA Greenhouse gas R&D Programme, Cheltenham, UK
93. Davison J (2005) CO_2 capture and storage and the IEA Greenhouse Gas Programme. Workshop on CO_2 issues, Miffelfart, Denmark, May 2005, IEA Greenhouse Gas Programme, Cheltenham
94. Power from gas+CCS at 54% efficiency? (2013) Carbon Capture J 14–15
95. Tola V, Pettinau A (2014) Power generation plants with carbon capture and storage: a techno-economic comparison between coal combustion and gasification technologies. Appl Energy 113:1461–1474
96. Global CCS Institute (2011) Economic assessment of carbon capture and storage technologies. 2011 update. WorleyParsons, Schlumberger. Canberra, Australia. http://cdn.globalccsinstitute.

com/sites/default/files/publications/12786/economic-assessment-carbon-capture-and-storage-technologies-2011-update.pdf. Accessed 12 May 2012
97. Thayer A (2009) Chemicals to help coal come clean. Chem Eng News 87:18–20
98. McCoy S, Rubin E (2008) An engineering economic model of pipeline transport of CO2 with application to carbon capture and storage. Int J Greenhouse Gas Control 2:219–229
99. National Energy Technology Laboratory (2012) Carbon storage. Geologic storage focus. http://www.netl.doe.gov/technologies/carbon_seq/corerd/storage.html. Accessed 12 Oct 2012
100. Bock B, Rhudy R, Herzog H et al. (2003) Economic evaluation of CO_2 storage and sink options. DOE research report, DE-FC26–00NT40937, Washington, DC
101. Hendriks C, Graus W, van Bergen F (2002) Global carbon dioxide storage potential and costs. Report Ecofys & The Netherlands Institute of Applied Geoscience TNO, Ecofys Report EEP, p 63
102. Roche P (2012) Widespread CCS unlikely unless CO_2 capture cost reduced, conference hears. Daily Oil Bull. http://airwaterland.ca/issues/printer.asp?article=dob%2F121017%2Fdob2012_oh0032.html. Accessed 23 Oct 2012
103. Akai M, Nishio N, Iijima M et al. (2004) Performance and economic evaluation of CO2 capture and sequestration technologies. Proceedings of the 7th international conference on Greenhouse Gas Control Technologies, Elsevier, Oxford
104. World Energy Investment Outlook (2003) OECD/IEA, 75775 Paris. ISBN: 92-64-01906-5
105. Edmonds J, Feund P, Dooley J (2000) The role of carbon management technologies in addressing atmospheric stabilization of greenhouse gases. Proceedings of the 5th international conference on Greenhouse Gas Control Technologies. Sponsored by the IEA Greenhouse gas R&D Programme, Cairns
106. Kithil P (2007) A policy option to provide sufficient funding for massive-scale sequestration of CO_2, the Smithsonian/NASA Astrophysics Data System. http://adsabs.harvard.edu/abs/2007AGUFM.U43C1410K. Accessed 14 Jan 2010
107. International Energy Agency (2013) Technology roadmap. Carbon capture and storage, 2013 edn. http://www.iea.org/publications/freepublications/publication/TechnologyRoadmapCarbonCaptureandStorage.pdf. Accessed 5 Sep 2013
108. Status of CCS project database (2012) Carbon Capture J:24–25
109. China becoming a global leader in CCS projects (2013) Carbon Capture J. http://www.carboncapturejournal.com/displaynes.php?NewsID=1126&PHPSESSID=18dkjsha7qaa6s815mlrslvtq4. Accessed 15 Jun 2013
110. Global CCS Institute (2011) The global status of CCS: 2010, Canberra, Australia, ISSN 1838-9481. http://www.globalccsinsitute.com/resources/publications/global-status-ccs-2010. Accessed 25 May 2011
111. Global CCS Institute status of CCS update (2013) Carbon Capture J. http://www.carboncapturejournal.com/. Accessed 20 Jul 2013
112. Michael K, Allison G, Golab A (2009) CO_2 storage in saline aquifers II- experience from existing storage operations. Proceedings of the greenhouse gas technologies-9 conference. Energy procedia, vol 1, p. 1973–1980
113. Statoil (2009) Development solution. http://www.statoil.com/statoilcom/snohvit/svg02699.nsf?opendatabase&lang=en. Accessed 25 Sep 2009
114. Petroleum Technology Research Center (2007) Weyburn-Midale CO_2 monitoring & storage project. Proceedings of the 6th annual conference on carbon capture & sequestration. Pittsburg, PA, 7 May 2007
115. Wackowski R (2007) Rangely Weber Sand unit CO_2 flooding case study, a long history of CO_2 injection. Proceedings of the CO_2 capture and storage conference, The Canadian Institute, Calgary, AB, 7 Feb 2007
116. Beckwith R (2011) Carbon capture and storage: a mixed review. JPT:42–45
117. Reeves S, Davis D, Oudinot A (2004) A technical and economic sensitivity study of enhanced coalbed methane recovery and carbon sequestration in coal. DOE topical report, Washington, DC

118. Flett M, Beacher G, Brantjes J et al (2008) Gordon Project: subsurface evaluation of carbon dioxide disposal under Barrow Island. Society of Petroleum Engineering, SPE 116372
119. Johnson J (2012) Stumbling on the path to "clean coal". Chem Eng News 90:37–39
120. Reisch M (2011) Air products inks carbon-capture deals. Chem Eng News 89:24
121. International Energy Agency (2012) Carbon capture and storage in industrial applications. doi:10.1787/9789264130661-en. http://www.oecd-ilibrary.org/energy/carbon-capture-and-storage-in-industrial-applications_9789264130661-en. Accessed 18 Dec 2012
122. Worrell E, Price L, Martin N et al (2001) Carbon dioxide emissions from the global cement industry. Annu Rev Energy Environ 26:303–329
123. Van Puyvelde D (2013) An update on CO_2 capture from cement production. Global CCS Institute.http://www.globalccsinstitute.com/insights/authors/dennisvanpuyvelde/2013/02/20/update-co2-capture-cement-production. Accessed 20 Aug 2013
124. European Cement Research Academy (2012) Carbon capture technology ECRA's approach towards CCS. ECRA Commun Bull. http://www.ecra-online.org/fileadmin/redaktion/files/pdf/ECRA_CCS_Communication_Bulletin_29.08.12neu.pdf. Accessed 20 Feb 2013
125. Skyonic (2013) Profitable, scalable, solid carbon capture. www.skyonic.com. Accessed 12 Mar 2013
126. Velzy C, Grillo L (2007) Waste-to-energy combustion. In: Kreith F, Goswami Y (eds) Handbook of energy efficiency and renewable energy. CRC Press, Boca Raton, USA
127. Johnson J (2011) EPA proposes easing CO_2 controls. Chem Eng News 89:25
128. Johnson J (2009) Water and CO_2 shouldn't mix. Chem Eng News 87:32
129. Carapeza M, Badalamenti B, Cavarra L et al (2003) Gas hazard assessment in a densely inhabited area of Colli Albani Volcano. J Volcanol Geothermal Res 123:81–94
130. Zoback M, Gorelick S (2012) Earthquake triggering and large-scale geologic storage of carbon dioxide. Proc Natl Acad Sci U S A. http://www.pnas.org/cgi/doi/10.1073/pnas.1202473109. Accessed 10 Mar 2013
131. CCS may have potential to induce seismic events (2012) Carbon Capture J. http://www.carboncapturejournal.com/displaynews.php?NewsID=959&PHPSESSID=9sd636g2pot2m8t5t1e8n2gsc3. Accessed 20 Mar 2012
132. Stone E, Lowe J, Shine K (2009) The impact of carbon capture and storage on climate. Energy Environ Sci 2:81–91
133. Zhou W, Stenhouse M, Arthur R et al (2005) The IEA Weybourn CO_2 monitoring and storage project—modeling of long-term migration of CO_2 from Weybourn. Proceedings of the 7th international conference on Greenhouse Gas Control Technologies, vol 1, Vancouver, Canada, p 721–730, Elsevier, UK
134. National Institute for Occupational Safety and Health (1997). Washington, DC
135. Palmgren C, Granger Morgan M, Bruine de Bruin W et al (2004) Initial public perception of deep geological and oceanic disposal of CO_2. Environ Sci Technol 38:6441–6450
136. Itoaka K, Saito A, Akai M (2004) Public acceptance of CO_2 capture and storage technology: a survey of public opinion to explore influential factors. Proceedings of the 7th international conference on Greenhouse Gas Control Technologies, 2004, Vancouver, Canada
137. Johnson J (2010) Gasification plant funds shifted by DOE. Chem Eng News 88:34
138. Shell Barendrecht project cancelled (2010) Carbon Capture J. http://www.carboncapturejournal.com/displaynews.php?NewsID=676. Accessed 20 Jul 2012
139. CCS awareness higher in Saskatchewan than Europe Projects (2011) Carbon Capture J. http://www.carboncapturejournal.com/displaynews.php?NewsID=873. Accessed 15 Jan 2012
140. International Energy Agency (2011) World Energy Outlook. Are we entering a golden age of gas? IEA, Paris, France
141. Global CCS Institute (2012) Global status of large scale integrated projects. 2012 update. http://cdn.globalccsinstitute.com/sites/default/files/publications/41146/globalstatusoflargescaleintegratedprojectsjune2012update.pdf. Accessed 20 Jan 2013
142. Johnson J (2013) EPA tries capping power plant carbon. Chem Eng News 91:42–43

143. Gurria A (2013) Carbon price vital to address climate change. Responding to climate change (RTCC), 9 Oct 2013. http://www.rtcc.org/2013/10/09/oecd-chief-carbon-price-vital-to-address-climate-change/. Accessed 14 Oct 2013
144. Wilcox J (2012) Carbon capture. Springer, New York, p 321
145. Cook P (2012) Clean energy, climate and carbon. CSIRO Publishing, Collingwood, p 214
146. Mills R (2011) Capturing carbon. The new weapon in the war against climate change. Columbia University Press, New York, p 465
147. Jackson F (2009) Conquering carbon. Carbon emissions, carbon markets and the consumers. New Holland, London, UK, p 256

Chapter 8
Transition to Low- and Zero-Carbon Energy and Fuels

Abstract Switching from high-carbon to low- and zero-carbon energy sources and fuels is considered Holy Grail of the decarbonization policy. The evolutionary model of the substitution of primary energy sources predicts that methane followed by hydrogen will take over the energy market during the current century. The interplay of three energy systems based on methane, electricity, and hydrogen (dubbed *Decarbonization Triangle*) can greatly facilitate and expand the decarbonization of global economy. Many challenges hindering the expansion of intermittent renewable energy sources (solar and wind) could potentially be addressed by means of interconnected electricity, methane, and hydrogen grids that form a large integrated low-carbon energy network. Due to the complimentary and synergistic nature of the basic elements of the networks, in combination, they can provide more energy services per unit of primary energy with associated economic and environmental benefits. The main strategies and pathways to transitioning to low-to-zero carbon energy systems and the prerequisites for building Methane and Hydrogen Economies are analyzed in this chapter.

8.1 Pathways to Low- and Zero-Carbon Energy and Fuels

8.1.1 The Decarbonization Triangle Concept

The evolutionary model of the substitution of primary energy sources (see Chap. 5) implies that methane (the major component of NG) followed by hydrogen will take over the energy market during the current century. Besides methane and hydrogen, non-carbon electricity is projected to be a critical component of the progressive primary energy substitution process in near-to-mid term future. All three players, methane, low-to-zero carbon electricity, and hydrogen, are well positioned to catalyze decarbonization of the global energy system. Muradov recently analyzed this three-way integrated energy system dubbed "*Decarbonization Triangle*" (DT) [1].

N. Muradov, *Liberating Energy from Carbon: Introduction to Decarbonization*, Lecture Notes in Energy 22, DOI 10.1007/978-1-4939-0545-4_8,

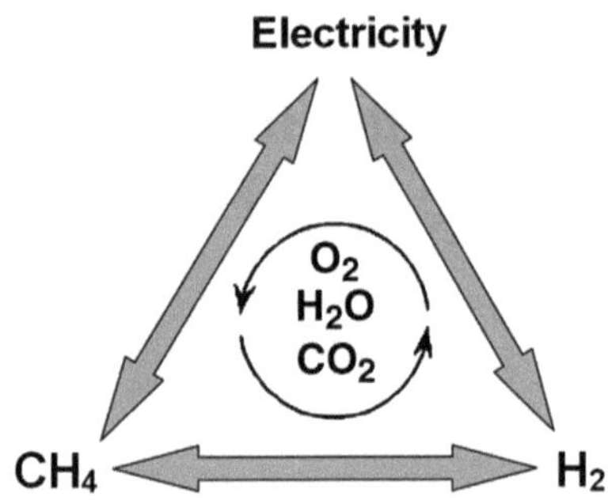

Fig. 8.1 Schematic diagram of Decarbonization Triangle concept. *Source* [1]

Figure 8.1 shows the scheme of DT energy system consisting of methane ↔ electricity, electricity ↔ H_2, and methane ↔ H_2 pairs.

The main advantage of the DT system is in its multifaceted functionality and flexibility: all three "pillars" of the system are interchangeable, synergistic, and complimentary to each other; the Nature's building blocks, water, oxygen, and CO_2, are regenerable agents in this closed loop energy system. The synergistic aspects of the DT particularly evident from the analysis of its right arm (electricity ↔ H_2 pair): besides being interconvertible (with high energy conversion efficiencies, via use of FC and electrolyzers), in combination, electricity and H_2 can offer the solutions that are nonexistent for either one acting alone. For example, in contrast to H_2, electricity cannot be stored in very large quantities; on the other hand, in contrast to electricity, H_2 cannot transfer information or generate light; in combination, they can do both [2].

Table 8.1 summarizes the technologies and processes that form the basis of the DT concept. In particular, the table includes the data on operational temperatures, energy efficiencies of the state-of-the-art devices, and the current status of the technological development.

Table 8.1 shows that except for the electricity → CH_4 transformation (which is still in an early R&D stage), all other processes are either being commercially practiced or approaching the commercialization stage.

Potentially, the DT-based energy systems will be able to accelerate the emergence of new technologies that hold the promise of higher productivity and environmental compatibility. In particular, the implementation of decarbonized electricity and hydrogen technologies could potentially lead to the decrease in demand for many energy-intensive materials and processes (*dematerialization*). For example, the utilization of hydrogen, which has the lowest atomic mass of all the elements, would dramatically reduce the total mass flow of products related to energy activities with the associated reduction in emissions. Decarbonized electricity is practically free of any (direct) material emissions. Within the framework of the DT concept, it would be possible to capture and convert atmospheric CO_2 to CH_4 (and further to other value-added products, if necessary), thus, providing a viable solution to stabilizing or reducing atmospheric CO_2 levels.

Table 8.1 Technologies and processes forming the basis of the Decarbonization Triangle concept

No.	Reaction	Technology	Temperature (°C)	Efficiency[a] (%)	Status
1	$H_2O + \varepsilon_{DC} \rightarrow H_2 + 1/2O_2$	Water electrolysis			
		Alkaline	50–100	75–80 [3]	Commercial
		PEM	80–100	80–90 [4]	Commercial
		Solid oxide membrane	800–1,000	80–90 [4]	Pilot
2	$H_2 + 1/2O_2 \rightarrow H_2O + \varepsilon_{DC}$	Hydrogen fuel cells			
		Alkaline FC	90–100	50–60 [5]	Commercial
		PEM FC	50–100	45–60 [5]	Commercial
		Solid oxide FC	700–1,000	50–60 [5]	Pilot
		Molten carbonate FC	600–700	45–50 [5]	Pilot
3	$CH_4 + 2H_2O \rightarrow 4H_2 + CO_2$	Steam reforming	700–900	80–89 [6]	Commercial
4	$4H_2 + CO_2 \rightarrow CH_4 + 2H_2O$	Methanation process	350–400	65–70 [7]	Commercial
5	$CH_4 + 2O_2 \rightarrow CO_2 + 2H_2O + \varepsilon_{LC}$	Gas turbine combined cycle	1,250–1,400	55–60 [8]	Commercial
		Solid oxide FC	800–1,000	50–60 [5]	Pilot
6	$CO_2 + 2H_2O + \varepsilon_{DC} \rightarrow CH_4 + 2O_2$	Electrochemical reduction	25–80	60–70[b]	R&D [9]

ε_{DC} and ε_{LC} are electricity generated from zero-carbon (solar, wind, geo, nuclear, biomass), and low-carbon (NG, NG/CCS, coal/CCS, etc.) sources, respectively

[a]State-of-the-art or projected energy efficiencies

[b]An estimate

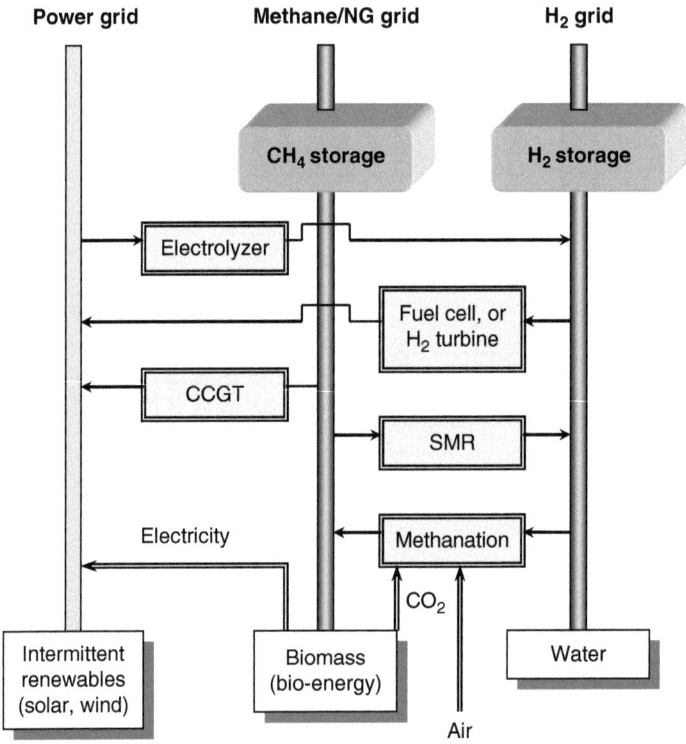

Fig. 8.2 Schematic diagram of integrated electricity–methane–hydrogen networks. *CCGT* combined cycle gas turbine, *SMR* steam methane reforming

8.1.2 Interplay of Electricity, Methane, and Hydrogen Networks

In practical terms, the realization of the DT concept would greatly catalyze the penetration of low- and zero-carbon[1] energy sources and fuels to the marketplace. Currently, among major factors hindering the expansion of intermittent renewable energy sources (solar and wind), large-scale energy storage and long-distance transmission are particularly challenging. These challenges could potentially be addressed through the implementation of interconnected electricity, methane/NG, and hydrogen grids that form a large integrated low-carbon energy network, as shown in Fig. 8.2.

Zero-carbon renewable (solar, wind, geothermal, biomass) and nuclear sources form the basis of the electrical power grid, whereas NG, biogas (biomethane), and synthetic methane form the basis of the methane grid. The main source of hydrogen in this integrated system is water, although certain portion of H_2 is also produced

[1] In this chapter and elsewhere, "zero-carbon" energy sources relate to non-carbon sources (e.g., solar, wind, geo), as well as, zero net CO_2 sources, such as biomass and biofuels.

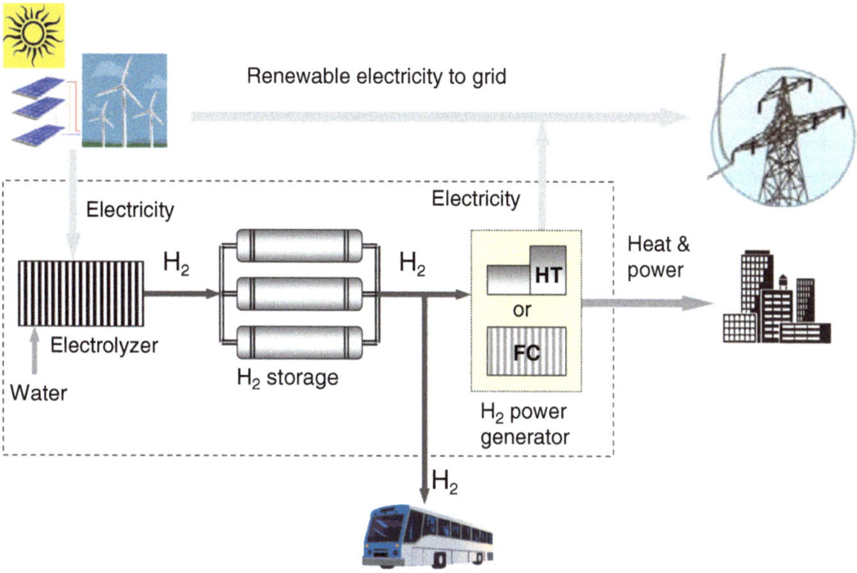

Fig. 8.3 Schematic diagram of an of integrated electricity–H_2 system for large-scale intermittent energy storage. *HT* hydrogen turbine, *FC* fuel cell

from methane. The analysis of an interplay between different branches of the integrated methane–electricity–hydrogen network follows.

8.1.2.1 Interplay of Electricity and Hydrogen Networks

One of the most promising options for large-scale storage of intermittent renewable electricity (RE) (e.g., from solar and wind farms) is through hydrogen production/storage. In this system, hydrogen is produced via water electrolysis using surplus electricity, stored in a suitable storage system, and on-demand converted back to electricity via FC or H_2 turbines with the estimated "round-trip" efficiency of about 35–40 %. Figure 8.3 depicts one example of the intermittent energy storage system involving production, storage, and conversion of H_2 to electricity.

Several projects around the world are taking advantage of this approach. For example, in the HyUnder project launched in June 2012, the potential for the large-scale seasonal storage of intermittent renewable (solar, wind) electricity using underground H_2 storage is being explored [10]. The project coordinated by Spain's Fundacion H_2 Aragon is budgeted at €1.766 million and is to run through 2014; it involves a 12-member consortium from seven European countries and companies: Shell, Solvay, France's Atomic Energy Agency, and others.

In a "smart community" project set in the Higashida area of Japan, hydrogen is being used to store surplus electric energy [11]. In this project, a 100 kW solar facility and a small wind turbine are combined with a 400 kW fuel cell to supply power (the equivalent of 10 % of the total electricity supply) to the area. The test field involves 230 regular homes, along with corporate offices, hospitals, and commercial facilities. The amount of electricity supplied by the facilities and consumed by the users is monitored by a regional energy monitoring system called the Cluster Energy Management System (CEMS). Based on the supply–demand pattern, CEMS is able to forecast demand for electricity, which allows to control the amount of electricity generated, and thus, conserve electricity. As part of the "smart community" project, the validation of "Hydrogen Town" concept is also being conducted in this region. Hydrogen is being used not only for storing electricity, but also for a back-up power supply. As a byproduct of nearby steel plants, hydrogen is delivered to the "Town" by a pipeline and is directed to hydrogen-fueling stations and FC-based power cogeneration plant.

In 2011, ENERTRAG AG (the German subsidiary of French oil and gas provider TOTAL) announced putting into an operation the world's first renewable energy hydrogen hybrid power plant that produces both electricity and transportation fuel in Prenzlau, Germany [12]. The €21 million project combines the production of electric energy from wind and biogas to produce "green" hydrogen to store energy. The stored hydrogen is combusted in a CHP plant at times of low or no wind and is also shipped to car-refueling stations in Hamburg and Berlin (under the existing Clean Energy Partnership). Three wind generators (2 MW each) are linked directly to a 500 kW pressurized electrolyzer and through a transformer to the 220 kV high-voltage grid operated by Vattenfall Europe Transmission GmbH. Hydrogen produced in the electrolyzer at the rate of 120 Nm^3/h is compressed to 31 bar and stored in five high-pressure tanks with a total capacity of 1,350 kg. Also included in the energy system are two 350 kW CHP plants, which operate on a variable mixture of hydrogen and biogas with maximum H_2 content of 70 % and minimum biogas content of 30 %, depending on the demand (if needed, the plant can operate on 100 % biogas). About 225 MW_{th} of heat can be pipelined to the Prenzlau municipal heating grid. The entire system is controlled by a computer, which produces 8-h forecasts and adjusts operations of the various components in order to optimally match the plant's output to the demand and expected wind conditions.

In combination, hydrogen and electricity provide more electrical and thermal energy services (via the use of FC) per unit of primary energy compared to conventional systems through the exergization of energy systems. Along with carbon-free electricity, FC can provide thermal energy to an end-user in a wide range of temperatures varying from 80 °C (e.g., PEMFC, for home heating) to 500–900 °C (e.g., MCFC and SOFC, for industry). Thus, the combination of hydrogen and electricity would drastically increase the amount of useful energy obtained from a primary energy source via complementing the prevailing thermodynamics of the Carnot-limited thermal systems with electrochemical systems such as FC and batteries [2]. In transportation, the implementation of DT would boost the use of electric-drive vehicles. Besides being an energy storage medium, hydrogen could also play a role of an enabler of peak-load electricity. For example, off-peak electricity can be used to produce H_2 and O_2 that will be stored and then used in H_2–O_2 steam turbine cycles or FC to produce peak electricity. The advantage of this system is in its capacity to convert the peak-load electricity demand into base-load demand.

8.1.2.2 Interplay of Electricity and Methane Networks

The interplay between electricity and methane/NG networks is rather complex and, besides economic factors, is determined by the local specifics and end-user requirements. Despite a general trend toward electrification of industrial and residential sectors (see Sect. 8.3), there are areas where NG is more efficient than electricity, e.g., for heat supply and other thermal applications in homes, businesses, and some industrial processes. For example, when electricity is used for cooking or heating industrial units almost two-thirds of energy in NG is lost during generation of electricity. On the other hand, if gas is used directly to cook or heat water, only about 8 % of the calorific value of NG is lost [13].

In addition to heating homes, NG can cool houses through gas-powered air conditioning (GAC). The technology is not new: it provided most of the air conditioning (AC) requirements in the mid of last century. Due to recent technological advancements and the increase in efficiency, GAC is returning back to the market. Although GAC units are more expensive to purchase compared to electric AC units with a comparable cooling capacity, they are considerably more efficient and require less maintenance (expected working life is about 20 years), and, in the long run, could be more cost effective. But many homes and businesses, even in developed countries (e.g., the USA), are not plumbed for gas; thus, there is an enormous potential to save fuel by switching from electricity to NG (with associated environmental benefits in countries where the share of coal in energy mix is high). This example illustrates the advantages of the integration of electricity and NG networks since it would provide an opportunity to increase the overall end-use energy efficiency and optimally manage energy consumption through fine-tuning of load distribution between the networks depending on which of the two players has an advantage over another one in a particular application area.

Use of NG as a back-up and storage medium for renewable electricity has recently been a hot topic among developers of clean energy technologies. In a back-up mode, methane will be taken from the grid and through the use of high-efficiency combined cycle gas turbine (CCGT) power generators will compensate for the shortfall in intermittent solar/wind energy sources to achieve the around-the-clock supply of low-carbon electricity into the power grid. In the storage mode, methane will be produced via methanation (or Sabatier) reaction (see Table 8.1, line 4) utilizing H_2 from RE-powered water electrolysis and CO_2 from flue gas of bioenergy plants, biogas, or the atmosphere (the additional amount of biomethane will be recovered from biogas by gas separation techniques).

The concept of using methane for storing intermittent renewable energy (solar, wind) has recently been promoted by German researchers and clean energy developers. The Zentrum für Sonnenenergie-und Wasserstoff-Forschung Baden-Württemberg (ZSW, the Center for Solar Energy and Hydrogen Research) in collaboration with Fraunhofer Institute for Wind Energy and Energy System Technology (IWES), SolarFuel Company have developed the "*Power-to-Gas*" concept [14]. According to the concept, at times when more renewable energy from solar and wind farms is generated that could be used or transported across the power network, the excess energy is used to split water into H_2 and O_2. Hydrogen along with CO_2 are converted through the Sabatier process to methane-rich gas that is fed into NG network. During shortages in renewable power generation, the stored gas can be fed back to gas-fired power plants to help to bridge this shortage for the periods up to 2 months. This option could enormously expand available capacities for transporting and storing renewable energy in Germany, according to estimates, roughly 3,000 times the country's pumped storage capacity [14]. Renewable methane could be produced at the cost of US$150–200 per barrel (equivalent).

Currently, the Power-to-Gas Strategy Platform is managed by the Deutsche Energie Agentur and is part of Germany's Energiewende program (which represents the long-term transformation and restructuring of Germany's energy system with main emphasis on renewables and energy efficiency). The Platform includes 31 partners that represent leading German research institutes and high-tech companies such as ITM Power, Bayerngas, Volkswagen AG, 50 Hz, and others [15]. The first large Power-to-Gas demonstration project with the capacity of 250 kW built by ZSW started operation in October 2012 in Stuttgart, Germany [16]. The plant consisting of an alkaline pressurized electrolyzer, a methanation unit, and a process control system demonstrated good flexibility to fluctuating power supply from the wind and sun units and an adequate response to sudden interruptions. The next step would be to scale-up the Power-to-Gas plants to the 1–20 MW range relevant to future large-scale industrial plants. SolarFuel company is already constructing a 6 MW Power-to-Gas plant for the automaker Audi in Werlte, Lower Saxony, Germany [17]. In this project, power from four 3.6 MW offshore wind turbines will be used to produce methane fuel for 1,500 turbo-compressed Audi A3 vehicles for a year. Audi is planning to begin serial production in 2014.

Gas turbines are near-perfect technology to supplement and compensate for renewable energy intermittency. They have quick start-up time that enables frequent starts and stops to support the variable generation mode. Recently, aeroderivative gas turbines (AGT) have been introduced to the market, that have faster start-up time compared to conventional frame-type turbines, which allows for a much better load management. Advancements in a combustor design and catalytic reduction technology resulted in significantly reduced NO_x emissions from the AGT systems. General Electric and other turbine manufacturers have recently expanded their AGT fleet to specifically tailor to the requirements of wind and solar industry. For example, General Electric LM6000 turbines have been successfully deployed as back-up or peaking units to support wind farms [18].

It can be expected that the combination of renewables with NG back-up will experience further growth and expansion in the energy market, since it is considered one of the most cost effective near-term clean energy options. More discussion on this topic can be found in Sect. 8.2.7.

8.1.2.3 Interplay of Methane and Hydrogen Networks

Methane and hydrogen are interconvertible through two commercially significant processes: the SMR process can be used for CH_4 to H_2 conversion and methanation (or Sabatier) reaction for H_2 to CH_4 conversion. SMR is the most economical and widely used process for the large-scale industrial manufacturing of H_2 (see Sect. 7.2.2). The Sabatier process, although being commercially practiced for many decades (mostly, for CO_2 removal from gaseous streams), is not currently used for the large-scale production of methane from CO_2 (albeit, the related syngas methanation process is widely used for commercial manufacturing of substitute NG in the USA, China, and other countries).

In principle, CO_2 feedstock for the Sabatier process could originate from a number of sources, such as flue gases from fossil fuel- and biomass-fired power plants, cement manufacturing and SMR plants, and even from the atmosphere. Hydrogen, on the other hand, is to come only from water and non-fossil (e.g., renewables, nuclear) sources. Thus, this approach would require very large volumes of economical "green" H_2—the challenge that is yet to be fully overcome. There are numerous efforts underway to develop the cost-effective ways of producing H_2 from water, among which electrolysis of water shows most promise in the near-to-mid term future. For example, the ZSW research center in Germany is developing a process coupling the Sabatier process with water electrolysis in the framework of the Power-to-Gas project discussed in the previous section.

As fuels, methane and hydrogen are compatible, and, in many cases, they or their suitable mixtures can be used in the same or similar devices, e.g., in gas turbines and internal (spark) combustion engines. Hydrogen-enriched mixture with methane containing 20 vol.% H_2 and lower can be distributed by the mid-pressure NG network without substantial technical modification [19]. (Note that the mixture of 20 % H_2–80 % CH_4, by volume, is a registered trademark of Hythane®.) The H_2–CH_4 mix can be stored under pressure using the conventional methane storage technologies. In some respect, the fuel characteristics of hydrogen–methane mixtures could be superior to those of the components used separately (hence, another example of synergism in the DT concept); for example, the H_2–CH_4 mix has greater volumetric energy density than does pure H_2 (which is essential for onboard storage), and it features improved combustion characteristics and reduced GHG (CO_2 and NO_x) emissions compared to methane used alone (with environmental benefits). It has been reported that the best balance between a spark ignition engine efficiency and pollutants (CO_2, NO_x) emissions reduction could be reached through the use of a fuel blend composed of 30 vol.% H_2 and 70 vol.% CH_4 [20].

Advantages and synergies in production, storage, distribution, and combustion of H_2–CH_4 mixtures in ICE have been analyzed by Klell et al. [21]. Flex-fuel vehicles for the operation with NG, H_2, and any H_2–CH_4 mixture is of particular practical interest. Since ICE vehicles are likely to remain on the road in the near-to-medium term perspective, the use of multifuel vehicles opens the opportunity to apply H_2 and NG as clean fuels. The EU legislation is accounting for this option by adopting a regulation for fuel consumption and pollutants emission related to H_2 and H_2–NG vehicles [21]. The feasibility of using H_2–CH_4 mixtures (sometimes also called *hydromethane*) in road vehicles has been tested and proved by several research groups worldwide. NREL (USA) tested several buses on the road in 2002–2004 using Hythane® fuel. In Malmo (Sweden), two vehicles of the local public transport were fueled with different H_2–NG mixtures (8–25 vol.% H_2). The tests showed a reduction in the fuel consumption and pollutants emissions [22]. ENEA researchers have also conducted on-road tests using two buses fueled with different hydromethane blends (ENEA is Italian Agency for Energy and New Technologies and Environment). The optimal blend was reported to contain 15 vol.% of H_2 [22]. The researchers demonstrated significant energy savings and reductions in pollutant emissions when using the hydromethane blends.

At the same time, from the viewpoint of the efficient and safe exploitation of separate methane/NG and H_2 networks (shown in Fig. 8.3), the direct use of CH_4–H_2 blends in either grid should be avoided (or limited to very small percentages). For instance, the H_2 content of gas entering the NG grid is strictly regulated in many countries (e.g., in Germany, not higher than 3 vol.% of H_2 is permitted in the gas mix [17]). Similarly, the presence of methane in the H_2 grid in appreciable quantities might complicate some hydrogen applications, e.g., H_2 liquefaction, production of metal hydrides, onboard storage in hydrides, etc.

8.1.3 Decarbonization Potential of Electricity–Methane–Hydrogen Network

The integrated electricity–methane–hydrogen energy system that forms the basis of the DT concept has a significant decarbonizing potential. Figure 8.4 provides a quantitative assessment of the carbon reduction potential of the DT-based energy system. In particular, it depicts the reduction in CO_2 emissions (in %) from different DT-based electricity generation systems against the baseline CO_2 emissions from electricity produced by the world energy mix (in 2009, it was 500 kg CO_2 per MWh [23]). For the comparison, CO_2 emissions from conventional coal and advanced coal IGCC power plants are also included in the diagram.

The data in Fig. 8.4 show that if all fossil fuel-based power generation (globally, about 67 % of total) would be converted to NGCC technology, this would result in 54 % reduction in CO_2 emissions compared to the current world average (which includes non-carbon sources such as hydropower and nuclear). Switching power plants from coal to NG would benefit both energy and carbon intensity factors in the Kaya equation as follows: (1) NG burns cleaner than coal (per kWh of electricity generated, NG releases 330–370 g CO_2, compared to 830–940 g CO_2 for coal, the reduction of 60–65 %) [23], and (2) NGCC (nearly 60 % efficient) is significantly more efficient than state-of-the-art USC coal-fired plants (47 % efficient) [24].

It is recognized, however, that, despite the relative cleanness of NG as fuel, its increased usage will eventually necessitate equipping NG-fired power plants with CCS systems in order to fully appreciate its carbon reduction potential. According to estimates, the NGCC plants equipped with CCS systems capturing 90 % of CO_2 byproduct would reduce CO_2 emissions by 95 % compared to the current world energy mix. The largest carbon reductions (close to 98–99 %) can be achieved through the implementation of various combinations of carbon-free electricity

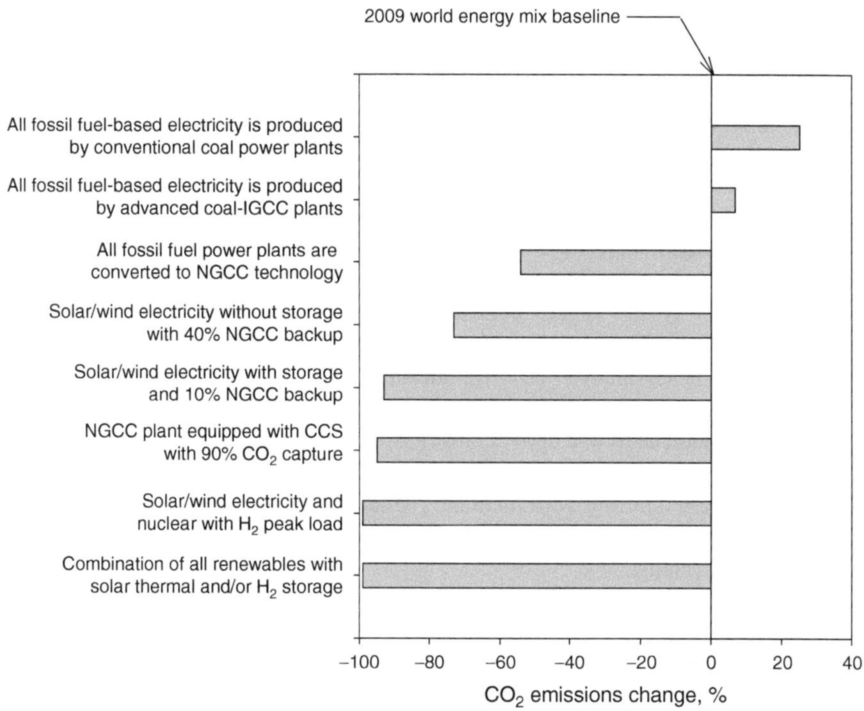

Fig. 8.4 The percentage of CO_2 emission reduction from electricity generation if main elements of the Decarbonization Triangle concept would be implemented. 2009 world energy mix baseline is 500 kg CO_2 per MWh [23]. *Note*: only direct CO_2 emissions are presented in this diagram, without accounting for the lifecycle emissions. *Source* [1]

sources (e.g., solar, wind, geo, nuclear) with energy storage and H_2-based peak-load electricity [1]. Budischak et al. have modeled many combinations of intermittent renewable electricity sources (onshore and offshore wind, solar PV) with electrochemical storage (batteries and H_2-FC) incorporated into a large (72 GW) grid system [25]. The model screened over 28 billion combinations of renewables and energy storage systems, each tested over the period of 4 years of load and weather data. It was concluded that assuming the year 2030 technology costs, the electric system could be powered 90–99.9 % of the time entirely on renewable electricity at the costs comparable to today's costs. Similar conclusions have been drawn by other investigators, that it would be possible to provide 99.8 % of electricity through the combination and efficient management of renewable energy sources in places like California (see discussion in Chap. 6).

In summary, the DT concept, which is based on the interplay of three networks, electricity, methane, and hydrogen, could greatly facilitate and expand decarbonization of the global energy system. The concept allows efficiently integrating these networks and taking the advantage of complimentary and synergistic nature of the basic elements of the networks. Together, electricity, methane, and hydrogen (and their associated networks) are capable of accomplishing things that would not be possible when they act alone (e.g., transferring information, or storing large amounts of energy, or backing up each other), and in combination, they can provide more energy services per unit of primary energy (i.e., at higher exergetic efficiency), with associated economic and environmental benefits. Most importantly, the integration of the electricity, methane, and hydrogen networks would help to overcome many energy shortage problems on a global scale, because the basic elements of the system could be realized in any country (which has important geopolitical implications). Although many elements of the DT system already exist and function on a large scale (see Table 8.1), and, to some extent, the integration of the networks is already underway (e.g., Power-to-Gas project in Germany), the full-scale integration of electricity, methane, and hydrogen networks will most likely accelerate after 2020 with the accent gradually shifting from methane to non-carbon electricity and hydrogen.

In the following sections, the role of all three components of the DT system is analyzed in detail, starting with methane (which is already used as a major decarbonization tool in the USA and other countries), followed by zero-carbon electricity and hydrogen systems.

8.2 An Advent of Methane Economy

Since methane/NG is the "cleanest" of fossil fuels (in terms of CO_2 emissions per unit of thermal or electric energy produced), many energy analysts consider replacing coal with NG in a variety of applications to be the most cost-effective near-term decarbonization option. Some authors see a 30-year window, in which carbon emissions would remain below previous peak levels, as gas continues to replace coal and other higher-carbon energy sources [13]. Besides the obvious environmental benefits, the major advantages of methane-based energy system are as follows:

- Abundant conventional and unconventional resources (larger than previously thought; according to some estimates, sustaining over 250 years at the current rate; see Chap. 1) geographically well spread across the world.
- Existing delivery and distribution infrastructure (e.g., transport via gas pipelines, liquid NG tankers).
- Flexibility of use (it is widely used in power generation, industry, transportation, and commercial/residential sectors).
- In many applications, methane is safer than gasoline and other liquid petroleum-based fuels since it is lighter than air and quickly disperses (e.g., in the event of a spill or leak).
- No need for the expensive onsite storage of fuel (in contrast to gasoline, diesel fuel, heating oil).
- In most cases, lower capital cost, easier maintenance, smaller physical footprint of NG-based units (compared to coal, heating oil).
- Enhanced energy security (via fuel diversification).

Back in 1980s, IIASA's energy analysts Nakićenović, Grübler, and Marchetti predicted the coming of the *Methane Economy* as a natural progression of the world energy system evolution [26]. In accord with this amazing foresight, in its 2011 World Energy Outlook report, IEA predicted a "Golden Age" for NG in coming decades [24]. Among the major boosters to establish the Methane Economy, is a recent boom in the supply of relatively cheap gas, which is often called "*Shale Revolution.*"

8.2.1 Technology Behind Shale Revolution

8.2.1.1 Introduction to Hydraulic Fracturing

The Shale Revolution was made possible through the development and large-scale implementation of two transformative drilling and gas-extracting technologies. The *hydraulic fracturing* (HF) process (also known as "hydrofracking" or, simply, "fracking") allows producing additional gas from the older wells previously thought depleted and to bring about new production in the areas once estimated as non-perspective to develop. According to estimates, without HF as much as 80 % of gas production from shale formations would be practically impossible [27]. The HF technology has an enormous impact on domestic energy production in the USA and other regions around the world; it offers the opportunity for many countries to reduce their reliance on energy imports and potentially reach energy independence.

The HF is not a new technology. The first commercial applications of HF can be traced to the mid-1940s when the technology was used to stimulate the production of oil and gas in the Hugoton field in Kansas and near Duncan in Oklahoma (USA). Since then, HF has been frequently used in the completion of vertical gas wells. In the earlier days of the technology, because of the impermeability of shale formations and their layered structure, the gas industry was not able to economically extract shale gas. HF alone would not have enabled the exploitation of the shale gas resources because the amount of recovered gas from the vertical wells would not be sufficient to achieve the economical gas production. For this to happen, breakthroughs and technological advances in *horizontal drilling* technology were necessary.

Drilling of horizontal wellbores at relatively low cost became a standard industry practice in the 1980s. Whereas vertical wells allow an access to tens-to-hundreds of meters (typically in the range of 20–100 m) across a flat-lying formation, in the horizontal wells, the drill-hole extends parallel to the formation layer for thousands of meters (e.g., up to 3 km in Bakken formation in North Dakota). This allows for a much greater exposure to a gas formation compared to the vertical wellbores and enables gas extraction from the areas where vertical wells are not feasible. Another noteworthy advantage of the horizontal drilling is that it reduces surface disturbance/disruption since it requires fewer wells than vertical drilling (per unit of fuel produced), which has significant environmental and esthetic implications. As a result of direct investments by the US DOE and other agencies in research, development and demonstration in the directional drilling, microseismic three-dimensional imaging and other innovative techniques, the company Mitchell Energy achieved first economical recovery of gas from a shale gas well in 1998 [28].

Up to date, the HF technology has been used in over 1 million producing wells, and, as it continues to develop and improve, as many as 35,000 wells of all types (vertical and horizontal, oil and natural gas) are being fractured each year [27]. In the USA, besides already mentioned Bakken formation, the HF technology is used in such shale gas formations as Barnett (Texas), Montney, Haynesville and Marcellus (Pennsylvania). Although the general idea of the HF concept is simple and easy to understand, the technology and physical/chemical processes involved are extremely complex.

8.2.1.2 Mechanics of Hydraulic Fracturing

From a mechanistic viewpoint, HF is the development and propagation of fractures in a rock in response to the action of a highly pressurized fluid. (Note that HF can also be a naturally occurring process causing such fluids as gas and oil to migrate from source rocks to reservoir formations via a system of veins and dykes in response to seismic activity.) [29]. The currently used HF technique refers to "*induced hydrolic fracturing*," by means of which engineers greatly accelerate the natural fracturing process by significantly increasing the rate at which fluids (gas and oil) are recovered from such subterranean reservoirs as limestones, shale rocks, coal beds, dolomites, sandstones, etc. The HF technique enables the recovery of NG from the reservoirs located at the depths of 1.5–6 km and deeper.

The process of HF involves pumping of the fracturing fluid into a wellbore at a certain rate (dictated by the specifics of the rock formation: depth, permeability, porosity, etc.). This results in the rock fracturing (cracking) with the fluid penetrating deeper into the formation and causing the crack to extend farther. Figure 8.5 depicts the schematic diagram of a typical HF process.

The water-based fluid (typically, a slurry) used in the fracturing operations contains chemicals and so-called *proppants* [30]. The solid proppants are used to keep a fracture (or a crack) open after the injection of the fracturing fluid is ceased. The typical proppants include a sieved silica sand, resin-coated sand, ceramics, depending on the rock's permeability, hardness, and other mechanical properties. The propping of the fracture increases permeability and enables the formation fluids (the mixture of water, gas, oil, and fracturing fluids) to flow to the well.

HF equipment typically consists of a slurry blender, a set of powerful high-pressure high-volume triplex and quintiplex pumps, fracturing tanks, storage and handling of the proppant, a chemical additive unit (to precisely meter the addition of chemicals to the mix), flexible hoses, a monitoring unit, and a variety of gauges and meters to measure fluid flow rate and density, treating pressure, etc. The fracturing equipment operates under extreme conditions rarely found in other industries: pressure and injection rate could reach 1,000 atm and 265 L/s, respectively.

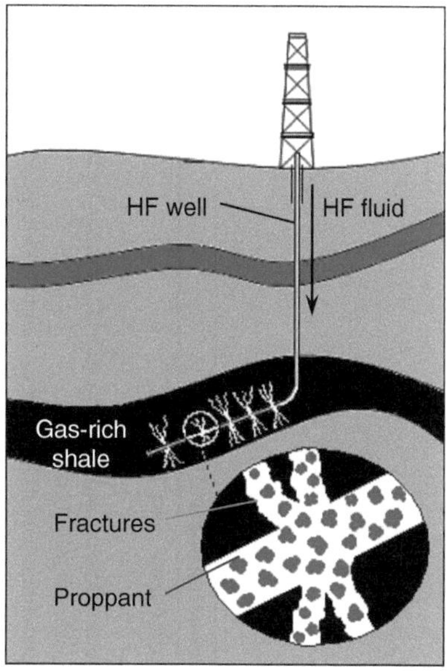

Fig. 8.5 Schematic diagram of a typical HF process. The drawing is not to the scale

8.2.1.3 Chemistry of Hydraulic Fracturing

The fracturing fluid is an extremely complex mixture (slurry) of water, chemical additives, compressed gases (air, N_2, CO_2) and proppants; in most cases, it consists of 98–99.5 % of water, depending on the well type [30]. Fresh water is a primary carrier fluid in the HF process, thus, its availability in reasonable quantities and its quality are of a major concern (the poor quality of water can adversely affect the efficiency of the chemical additives used in the fracking fluid). The high usage of water is one of the main controversies surrounding the HF technology. Depending on the permeability of reservoirs, the volume of the fracturing fluid could vary between 20,000 and 80,000 gallons per a well (low-volume HF operation) to 2–3 million gallons of the fluid per a well (high-volume HF) [31] (one gallon is equal to 3.79 L). Typically, water comes from surface water sources such as rivers, lakes, and municipal supplies. In some areas, groundwater can also be used to supplement surface water supplies if present in sufficient quantities.

Chemical additives are essential part of the fracturing fluid: their composition not only has to be taylored to a specific rock formation (from a geochemical viewpoint), but also has to be designed to protect the well, improve its operation, and facilitate its post-treatment. Under the pressure of environmental groups, shale gas companies are now required to disclose the chemical composition of the fracturing fluids. Table 8.2 summarizes the set of chemicals most frequently used in the fracturing fluids.

It should be noted that not all of the above-listed chemicals are utilized in every HF operation; the composition of the fracking "cocktail" depends on the geological, mechanical, and chemical specifics of the well such as its depth, thickness, chemical nature of the rock formation, etc. In many cases, the composition of the fracturing fluid changes in the course of the fracturing operation.

Table 8.2 Chemicals most often used in hydraulic fracturing processes

Product function	Chemical/physical action	Chemical name and formula
Acidification	Clears cement debris, dissolves carbonate and other minerals and initiate cracks in the rocks	Hydrochloric acid, HCl
Biocide	Prevents growth (and kills) bacteria in water that may interfere with HF operation (e.g., by producing corrosive byproducts)	Glutaraldehyde, ammonium chloride, tetrakis hydroxy-methyl phosphonium sulfate
Breaker	Delays break-down of the gel	Ammonium persulfate, magnesium peroxide, magnesium oxide
	Stabilizes product	Sodium chloride, NaCl; Calcium chloride, $CaCl_2$
Clay stabilizer	Prevents clays from swelling or shifting	Choline chloride, tetramethylammonium chloride, sodium chloride
Corrosion inhibitor	Product stabilizing and/or winterizing agent	Isopropanol, methanol
	Prevents the degradation of pipes and the steel well casing	Formic acid, acetaldehyde, *N-N-dimethyl* formamide, ammonium bisulfite
Crosslinker	Enhances the ability of the gelling agent to transport the proppant (sand); maintains fluid viscosity as temperature increases	Potassium metaborate, triethanolamine zirconate, sodium tetraborate, boric acid
	Product stabilizing and/or winterizing agent	Ethylene, glycol, methanol
Friction reducer	Reduces tubular friction and, hence, the pressure needed to pump HF fluid into wellbore (by "slicking" water)	Polyacrylamide, potassium chloride
	Carrier fluid for polyacrylamide friction reducer	Petroleum distillate
Gelling agent	Thickens water in order to suspend the sand and facilitate its transport	Guar gum, polysaccharide blend
	Carrier fluid for guar gum in liquid gels	Petroleum distillate
	Product stabilizing and/or winterizing agent	Methanol, ethylene glycol
Iron control/ stabilizing agent	Prevents precipitation of iron and other metal oxides by keeping them in a soluble form	Citric acid, acetic acid, hydrochloric acid, thioglycolic acid, sodium erythorbate

Non-emulsifier	Prevents the formation of emulsions in the fracture fluid	Lauryl sulfate
	Product stabilizing and/or winterizing agent	Isopropanol, ethylene glycol
pH adjusting agent	Adjusts pH of fluid to maintain the effectiveness of other components, such as crosslinkers	Sodium hydroxide, potassium hydroxide, acetic acid, sodium carbonate, potassium carbonate
Scale inhibitor	Prevents scale deposits in the pipe and controls the precipitation of certain carbonate and sulfate minerals	Copolymer of acrylamide and sodium acrylate, sodium carboxylate, phosphinic acid salt
Oxygen scavenger	Removes oxygen from fluid to reduce pipe corrosion	Ammonium bisulfite
Proppant	Keep fractures open	Silica quartz sand
Surfactant	Increases the viscosity of the fracture fluid	Lauryl sulfate
	Product stabilizing and/or winterizing agent	Ethanol, methanol, isopropyl alcohol, 2-butoxyethanol
	Carrier fluid for the active surfactant ingredients	Naphthalene

8.2.1.4 Commercial Practice of Hydraulic Fracturing

Typically, the HF operation consists of four stages. It starts with pumping tens of thousands liters of diluted hydrochloric acid (HCl) into the well to clear the passage for other HF fluids, to score the perforations and initiate cracks and fractures near the wellbore (e.g., by dissolving carbonates and other minerals). In the next so-called pad stage, the wellbore is filled with about 400,000 L of friction-reducing "slickwater," which facilitates the flow of the proppant material. In the third stage (which may consist of many substages), high-pressure HF fluid containing a proppant is pumped into the wellbore with the objective of propping the created fractures (i.e., keeping them open) after the pressure is diminished. Besides the proppant, the fracking fluid contains a gelling agent, which facilitates the delivery of the proppant into the formation. As the HF operation proceeds, other ingredients are also added to the fracking fluid (see Table 8.2). During this stage, the size of the proppant particles may increase from fine powder to coarser grains. At some point, viscosity-reducing agents such as oxidizers and enzyme breakers are also added to the fluid to diminish the effect of the gelling agents and facilitate flowback [31]. This is the most water-intensive stage of the whole HF operation that can consume many hundred thousand liters of water. In the final "flushing" stage of the process, fresh water (typically containing a friction reducing component) is pumped under pressure into the wellbore to flush the excess proppant and other components of the HF fluid from the well.

The measurements of the HF fluid pressure and injection rate (and, in some cases, microseismic monitoring and injection of radioactive tracers) along with the available data on the site geology provide the means of monitoring the growth and development of the hydraulic fractures within the formation [32]. After the HF operation, the recovered fracking fluid can be handled by different techniques: (a) treatment and discharge, (b) recycling, (c) storage in pits, and (d) temporary storage in containers [33]. Currently, new treatment technologies are being developed to more efficiently process wastewater and improve the HF fluid reusability.

HF is commercially practiced in many countries: Australia, Canada, Ireland, New Zealand, South Africa, UK, USA (although, some provinces and states in these countries have a moratorium in place on the HF practice). In the USA, HF is slowly gaining political acceptance, because it invites the opportunity for companies to create thousands of high-paid jobs, generates a sustainable profit, lowers domestic energy costs, and contributes to the goals of energy security and independence [34]. Intensive confidence-building policies are underway at regional and national levels to assure public that the fracking operation can be done safely, without harming environment and communities.

The fracking technology itself is still evolving, and more advanced and productive modifications of HF technology are being introduced for unlocking deep shale formation around the world. In particular, some energy companies are increasing the productivity of the shale gas wells by creating more cracks and reaching the deeper layers of the rock in a process dubbed "*superfracking*" [35]. Halliburton company (USA) is developing a technological approach called "*Frack of the Future*" aimed at substantially speeding up gas production and cutting down on materials, while Baker Hughes company has unveiled the "*Disintegrating Ball*" technology to plug the wells. Schlumberger Ltd. company has introduced "*HiWay*" technology that allows to make the cracks bigger to recover more gas; the technology has already been reduced to practice at the Barnett Shale [35]. Although the developers of new technology claim that from the environmental viewpoint it is no worse than a normal HF operation, many of these new processes are still at the initial stages of commercial practice and environmental implications are yet to be established.

8.2.1.5 Environmental Aspects of Hydraulic Fracturing

With the explosive growth of HF technology, the voices of environmental groups and concerned citizens uneasy about its health effects and long-term ecological impact are becoming increasingly louder. HF has become a contentious and even political issue in many countries. Currently, several countries either completely ban (e.g., France, Bulgaria) or have a moratorium in place (e.g., New South Wales, Australia; Karoo basin, South Africa; Quebec, Canada; several states in the USA) on the HF practice. The major environmental concerns over the HF practice include the potential risks of

- Groundwater contamination
- Air pollution
- Migration of gases and chemicals to the surface
- Induced earthquakes
- Mishandling of wastes
- Health effects
- Impact of methane leakage on climate
- Radioactive contamination
- Shortage of sand
- Shortage of water

Some of these concerns are more serious than others, and some proved to be baseless. The following discussion will sort out myths from reality with regard to the environmental and social impacts of the HF technology.

Groundwater Contamination

Groundwater contamination is one of the most publicized and contentious environmental impacts of the HF technology, and there are many reported incidents where HF is suspected of drinking water contamination. The main problem is that once chemicals contaminate underground drinking-water sources, it is very difficult and costly to remove those contaminants. Understandably, oil and gas industry executives have maintained that HF has nothing to do with the groundwater contamination. They argue that HF occurs thousands of meters below drinking-water aquifers and because of that the fracking fluids pose no risk [36]. This positive attitude toward the HF technology ("*Hydraulic fracturing is a safe production technique*" [37]) was often emphasized by the US EPA administrators.

The US EPA have been looking into the cases where the HF operation was a suspect in the groundwater contamination since late 1980s (e.g., a case of the suspected leak of fracking fluid in a water well in Jackson County, West Virginia) [36]. In May 2011, the EPA Administrator Lisa Jackson during her testimony in a Senate Hearing Committee stated "*I'm not aware of any proven case where the HF process itself has affected water*" [38]. Recent studies and reports, however, paint a quite different picture: there were reports on groundwater contamination in Clark and Pavillion (Wyoming) and Dimock (Pennsylvania) [39]. The authors of the 2013 study provide a direct evidence of a link between HF and methane levels in drinking water; in particular, they analyzed 141 drinking water wells across the Appalachian Plateaus (Pennsylvania, USA), aiming at correlating methane concentrations with their proximity to HF wells [40]. The results indicated that closer a well is to a HF site, higher is the methane concentration in drinking water. In particular, methane was detected in 82 % of water samples tested, with the average concentration six times higher for the homes within 1 km from the HF wells compared to homes located farther away [41]. The same was true for ethane content in water with even more dramatic difference: 23 times.

Currently, eight US states require that all chemicals (excluding proprietary ones) must be published online, whereas, in other states, many companies are already voluntarily disclosing this information [42]. However, many of the chemicals included in the fracking fluid are not currently regulated by the US Safe Drinking Water Act, raising public concerns about the possible contamination of drinking water supply. According to available data, from 2005 to 2009, about 750 chemicals were used in HF process ranging from benign components (e.g., coffee grounds and walnut hulls), to 29 components that are hazardous and present health-related problems if leaked into the drinking water supply [42] (see Table 8.2).

In response to reported evidences of a link between HF operation and the contamination of groundwater with methane, the industry supporters make a case that there is absolutely no reason for concern: even if methane is present in water, it is not toxic, and because of its low solubility in water, its concentration in most of the wells is far below the "action level" recommended by the government. With regard to chemical contamination of groundwater, the HF supporters argue that in many cases, chemicals may not necessarily come directly from the injected HF fluid, but from poorly constructed wastewater evaporation ponds and pipelines that allow chemicals making their way down to the groundwater systems below. To prove that the fears about toxicity of the fracking fluid are exagerrated, some "volunteers" claimed that the fluid is even "drinkable" [43]. The HF industry officials also argue that the contamination of groundwater with fracking liquids is highly improbable (if the wells are designed properly) based on physical principles (e.g., Darcy's law) that govern flow of fluids through a porous medium, such as a rock.

Thus, as of now, in the case of the direct contamination of groundwater by fracking fluids the jury is still out. The case will be closed when all alternative explanations will be exhausted. The improved understanding of the fate and transport of potentially harmful contaminants and wide-spread long-term monitoring and data dissemination will help effectively manage water quality risks associated with unconventional gas industry today and in the future [42].

Air Pollution

Air emissions from HF operations include hydrocarbons present in NG (methane, ethane, propane) and volatile organic compounds (VOC) originating from the chemicals present in the fracking fluids. Of particular concern are aromatic hydrocarbons, such as benzene, toluene, xylene, and others, that could cause neurological problems, birth defects, and cancer [44]. Elevated levels of VOC have been detected in air in some counties in the state of Wyoming (USA); on the other hand, the examination of air quality around NG sites near Fort Worth, Texas, did not show any significant health threats [45]. This indicates that, in many cases, air pollution could be significantly minimized through enhancing the VOC control on sites and elevating safety standards.

Wastes from HF

Although, in general, the concentrations of the chemicals in the HF fluids are relatively low (typically, 1–2 %) and many ingredients are environmentally benign, the process' wastewater can be hazardous due to the presence of hydrocarbons and other substances washed out from the formations during the HF operation. The US EPA plans to develop national standards for the disposal of wastewater discharged from HF and other shale gas drilling activities [46]. As part of a Clean Water Act rule, the EPA will establish technology-based standards for the pre-treatment of water from shale gas wells going to municipal treatment plants for the final processing before discharge. Environmental groups and law firms insist on this pre-treatment because some of the chemicals in the fracking fluids are carcinogens (or suspected carcinogens) and could get to rivers, streams, and drinking water. But NG-producing companies, such as America's Natural Gas Alliance, argue that shale gas wastewater disposal is already actively regulated at the state level. Meanwhile, the Marcellus Shale Coalition and other shale gas producers have agreed to disclose the composition of the chemicals used in the drilling process on the online database at the FracFocus website (www.FracFocus.org). It has been reported that in some China's shale-gas operations, 10–90 % of fracking fluids are returned to the surface [47]. The inadequate treatment of these wastewater streams introduces heavy metals, acids, pesticides, and other hazardous materials to soil and aquatic systems, exacurbating China's polluted water and environment problem.

Radioactive Contamination

The problem of radioactive contamination linked to the HF operations has recently been brought to a public attention by an article in *The New York Times* [48].

The sources of the radioactive contamination are naturally occurring radioactive minerals washed out by fracking fluids and gamma-emitting tracer isotopes (e.g., Iodine-131, typically carried by sand particles) used to determine the injection profile and the location of fractures during HF operation. The problem is particularly severe in Pennsylvania, which has seen a drastic increase in drilling, with roughly 71,000 active gas wells in operation. Drillers trucked at least half of this waste to public sewage treatment plants in Pennsylvania in 2008 and 2009, according to state officials [48]. Philadelphia Water Department posted a notice that Iodine-131 was found in the water supply [49]. Pennsylvania state officials, while acknowledging that water treatment plants are neither equipped to remove the radioactive material from the wastewater nor even required to test for it, affirmed that municipal drinking water is safe and meets the required standards. To some extent, the problem stems from the fact that safe drinking water standards are yet to be set for the radioactivity levels of substances involved in the HF process [48].

Earthquakes

Recently, there have been reports linking small earthquakes and earth tremors to shale gas operations. For example, several small earthquakes in northeastern Ohio and north Texas (USA) are suspected to be caused by the injection of wastewater used in HF process to disposal wells [50, 51]. A recent UK report also linked some earth tremors in Lancashire to a shale gas operation [52]. One tremor of magnitude 2.3 (1 April 2011) was followed by a second of magnitude 1.4 (27 May 2011). A study by The British Geological Survey placed the epicenter for each quake about 500 m away from the Preese Hall-1 well, near Blackpool, UK. As a result of an ensuing public outcry, the energy company Cuadrilla suspended its shale gas test drilling over fears of the links to the earthquakes. However, the following geomechanical study carried out by independent experts indicated that the combination of the geological factors that caused the quakes was rather rare and would be unlikely to occur together again at future well sites.

According to another study published in *Science* magazine, micro-earthquakes (with magnitudes below 2) commonly accompany large HF operations, so, the process appears to pose a low risk of inducing destructive earthquakes [53]. More than hundred thousands wells have been subjected to HF in recent years, and the largest induced earthquake was of the magnitude of 3.6—weak enough to produce a serious damage. On the other hand, wastewater disposal into very deep wells poses the significantly higher risk of inducing larger earthquakes. It is suspected that several recent relatively large earthquakes in the USA have been triggered by nearby deep disposal wells (the largest among them had a magnitude of 5.6 and occurred in Oklahoma destroying 14 homes). Researchers have found that some of the largest quakes induced by the deep injection are preceded by a warning sign—the distinctive swarm of smaller tremors [54].

Sand Shortage

There are warnings that a sand shortage in the USA could be a serious problem caused by HF operations. The amount of industrial sand used in HF has quadrupled from 2000 to 2009, according to the US Geological Survey [55]. It was estimated that about 6.5 million metric tons of sand were used in hydraulic fracturing in 2009 and, most likely, this number substantially increased since then (the data are not yet available). As the HF technology becomes more complex and wells keep getting bigger, the need for sand is expected to increase in the USA and other countries.

Shortage of Water

The access to clean water resources is projected to be one of the biggest challenges facing HF technology. Currently, a public resistance grows to the use of potable water in HF operations, especially in the areas where it is already in shortage. This may lead to the investments in infrastructure to make use of recycled wastewater, processing of brackish water, and the growth in desalination projects [56]. As the HF infrastructure expands, wastewater storage and distribution systems are likely to become more efficient and reliable. The drilling and HF operation of a single horizontal well in the Marcellus Shale requires between 7.6 million and 26.6 million liters of water; the projected water consumption for gas production in the region will reach about 71 million liters per day in 2013 [57]. Although this constitutes only 0.2 % of the total annual water withdrawals in Pennsylvania (water withdrawals are similarly low in other areas), some temporary problems at the local level could be expected due to flow variability, especially during drought periods [42].

China has ambitious plans to produce 6.5 BCM of shale gas by 2015, with 13 provinces selected as priority areas [58]. However, a recent report shows that these plans might be threatened by severe water shortages in many provinces [47]. In particular, more than half of the designated provinces are already plagued by water shortages, with less than 2,000 m^3 per person, which is less than one-quarter of the world average. Due to the complex geological conditions, most of China's shale-gas wells consume 10,000–24,000 cubic meters of water per each well [47]. In Sichuan province, the targeted production of 1.5 BCM of gas would require 171 million cubic meters of water, which is equal to 10.5 % of the province's water demand. Thus, HF extraction of gas in many China's provinces is expected to compete for limited water resources with agricultural, industrial, and domestic sectors.

8.2.1.6 Concluding Remarks on Hydraulic Fracturing

Although NG recovered from shale formations has enormous economic potential, the drilling technology itself may present a variety of environmental risks that cannot be underestimated, especially with regard to the contamination of groundwater. The 2011 report issued by US DOE advisory committee called for the full disclosure of the chemical composition of the fluids used in HF. The report recommends the creation of a national database on shale-derived gas to make the information on the drilling technology more accessible to the public. It also calls for the establishment of a shale gas industry group to implement best practices and improve drilling standards [59]. However, the industry trade association—American Petroleum Institute—adamantly opposes the federal regulation of the hydraulic drilling technology arguing that the shale gas and oil development is safe and is already regulated by the states where it is practiced. Most companies in the USA began making the disclosures voluntarily [60]. The full support by the US administration of NG development indicates that the outright ban on fracking some environmentalists demand is unrealistic.

In 2011, the US EPA proposed new regulations aimed at limiting emissions from oil and NG drilling operations, including hydraulic fracturing [61]. The proposed measures would reduce atmospheric emissions of smog-forming VOC by about 25 %; for new and updated gas wells utilizing fracking technique, the reduction in these emissions would reach up to 95 %. The proposed EPA regulations were in response to the threat of a lawsuit from environmentalist groups that pointed to rapidly deteriorating ecological situation (increased smog pollution) in the rural areas where gas drilling by the hydraulic fracturing method is booming (e.g., in eastern Wyoming). According to EPA estimates, the new rules would trim nationwide emissions of VOC and methane by 25 % and 26 %, respectively.

8.2.2 Trends in Methane Demand

Due to the flexibility, versatility, and relative "cleanness" of NG as fuel, a broad increase in NG demand has taken place across practically all sectors: power generation, buildings, industry, transportation, etc. Particularly, gas-fired power capacity has risen dramatically over the last two decades as high-efficiency combined cycle gas turbine plants started competing with traditional base-load power plants. Between 1990 and 2011, 930 GW of gas-fired capacity was built—an increase in 170 % over last two decades [62].

Coal-to-gas switching in the energy sector is a rather complex mechanism greatly influenced by regional market dynamics. In the USA, shale gas revolution has substantially lowered the gas prices, increasing the share of gas-fired electricity at the expense of coal (in April 2012, gas prices in the USA fell below US$2/GJ [62], from near US$16 in early 2006). From 2006 to 2011, gas-fired power generation rose by 24 % to more than 1,000 TWh, whereas, coal-fired generation dropped by 12 % to less than 1,900 TWh (against a backdrop of stagnating power demand) [62]. In contrast, in Europe, over the same period, gas-fired electricity generation had faced difficulties competing against coal-fired generation mostly due to high gas prices (in 2012, European gas prices were between US$8/GJ and US$10/GJ [62]). As a result, between 2010 and 2011, gas-fired generation in OECD-Europe decreased by 9 % (with its share of total generation falling from 23.5 to 21.6 %), while coal-fired generation rose 4 % [62]. This trend deepened in 2012, and gas share in the European power sector is expected to further decline by 13 % between 2011 and 2017 [63]. A different trend could be observed in China, where USC coal plants produce electricity at the cost of $42/MWh, which is about two times cheaper than that from NGCC at NG price of US$(9–11)/GJ [24]. Thus, the ongoing increase in gas-fired power in China is largely driven not by economical but government policies concerned with local air pollution and diversification of the energy mix.

IEA projects that power generation will remain the largest source of NG demand; it will grow by about 0.8TCM between now and 2035 (exceeding 2 TCM by 2035) [24]. Switching from coal-fired to efficient NGCC power plants will play a major role in transition to low-carbon electricity before 2020. According to the IEA's 2DS pathway, global gas-fired electricity generation will increase to over 5,800 TWh in 2020 (up 23 % from 2010 level) and will contribute carbon emissions reductions of about 1,200 Mt CO_2 relative to 4DS pathway [62]. About 80 % of these reductions would result from coal-to-gas switching and the remainder 20 % from plant efficiency improvements. After 2025, the role of NG in the 2DS pathway will be altered, with the emphasis gradually shifting to energy systems with a lower carbon footprint than gas-fired power (which will peak in 2030 at around 6,800 TWh). By mid-century, gas-fired electricity generation will be decreased to 4,800 TWh, with roughly a third of generated power coming from the gas-fired plants equipped with CCS [62]. NG will continue to provide back-up capacity to balance variability from intermittent renewable energy sources (solar, wind).

8.2.3 Methane in Transportation

The low prices and abundant resources are making methane very attractive and increasingly competitive with conventional fuels (e.g., gasoline) in the land transportation market. Other incentives for switching to NG in the transportation sector have to do with the enhanced fuel security (in the USA, for example, 98 % of all NG supply is local or from Canada) and drastically reduced local air pollution. IEA projects an increase in the NG consumption in road transport to 0.155 TCM in 2035, mostly, at the expense of oil, which will drop by 60 Mtoe (or 1.2 million barrels per day) [24]. Currently, the lack of adequate refueling infrastructure hinders the growth in the share of NG vehicles (NGV) sales to private consumers (this factor plays a lesser role for the commercial fleets, such as buses, municipal vehicles, public transportation, taxis, delivery trucks, that have an access to central refueling depots). IEA projections put about 186 million NGV on the road by 2035 [24]. As a result, CO_2 emissions from road-transportation sector will drop by 165 Mt in 2035.

Selling points for NGV are two-fold: not only is NG cheaper fuel than oil right now (e.g., in the USA and other countries), but NGV emissions are much cleaner than that of gasoline or diesel vehicles (EPA has reported that NGV emit 25 % less GHG emissions than equivalent gasoline-fueled vehicles [64]). The advantages of running vehicles on NG were recently emphasized by Burns (retired vice president of General Motors for R&D), who underscored that NG converted to electricity or hydrogen would yield nearly twice the driving range with half as much CO_2 emissions (i.e., through battery-electric or FC vehicles) [65]. Burns estimated that the investment for supporting first-generation commercial FC-based vehicles (FCV) would be on the order of 1/20th to 1/10th of vehicle investments by five major automakers or about $1–2 million per station for 100–200 stations [65].

8.2.3.1 Challenges to NGV

Companies operating big fleets are switching to NGV to save on fuel costs. NG is already used in a big way in the commercial-truck market. But the much more daunting task is to get light-duty vehicles and passenger cars running on NG, which is associated with three main challenges: cost, infrastructure, and psychology [66]. These challenges are significant, but the pay-off is high: according to MIT's Prof. Knittel, with the right policy incentives, NGV could increase the nation's energy security, decrease the susceptibility of the US economy to recessions caused by oil-price shocks, and reduce greenhouse-gas emissions and other pollutants [64]. These three challenges are discussed below.

Cost of NGV

NGV are still more expensive (in terms of upfront cost) than regular gasoline-fueled models and would require more frequent refueling. The biggest issue with NGV is the fuel tank. Since most of current NGV use compressed NG (CNG) stored in cylindrical containers at high pressure of 200–300 atm, the tanks must be stronger, and, consequently larger and heavier, driving the price up. The only passenger NGV sold in the USA, the Honda Civic GX, was sold (in 2012) at about $5,200 more than a comparable gasoline vehicle and $3,600 more than the gasoline/electric hybrid Civic [66].

The developers of NGV argue that a big part of the upfront-cost problem is perception, because people do not realize how much they can save over the life of the vehicle. For example, in many parts of the USA, driving on NG fuel could cost motorists about half of that of gasoline cost considering comparable vehicles and 2012 fuel prices. But the pay off time could be long: assuming NGV costs $5,500 more than its gasoline counterpart, and assuming $1.40/gallon price advantage for CNG, it could take more than 9 years before the car owner breaks even [66].

Researchers are working on new materials and tank designs to cut the size, weight, and costs of NG storage. Recent developments in adsorbed NG (ANG) technology may lead to significant reductions in the storage tank pressure and volume. For example, MOF with extremely high surface area have emerged as promising methane storage materials [67]. The University of Missouri (USA) researchers have developed a smaller tank that allows to store NG at much lower pressure by adsorbing it in the charcoal briquettes made out of corncobs [66]. The main driving force behind improvements in the ANG system is to reduce onboard NG storage pressure to 35 atm, which would allow using an existing NG pipeline system for refueling. No external compression equipment would be necessary, so home refueling could become feasible. Additionally, low pressure would make it possible to use lighter materials for the storage tank, and, thus, reduce car weight and improve fuel efficiency.

Among other noteworthy efforts to bring down the NGV tank cost, 3 M Corp. (USA) is developing the fuel tanks that use plastic linings wrapped in carbon-composite materials. The tanks could potentially be 10–20 % lighter with 10–20 % more volumetric capacity than current tanks [66]. Gas Technology Institute (USA) in collaboration with the Cummins Westpoint Inc. are developing and readying for the commercialization the new heavy-duty engine that operates exclusively on CNG or L-NG [68]. This advanced NG-fueled engine targets regional trucking, vocational and refuse markets in North America. Medium-duty (6.7 L) ultra-low emission variation of the NG engine will be ideal for school buses, package delivery, and class 5–7 trucks.

Infrastructure

The lack of adequate refueling infrastructure is currently viewed as another major bottleneck to the wide-spread use of NGV in the private transportation sector. Currently, NG refueling stations are few and far between, and building new ones is expensive. Of the 1,500 NG refueling stations available in the USA, only about half are accessible to the public (the rest are reserved for fleet vehicles): this is less than 1 % of the 118,000 public gasoline stations spread across the country [66]. A major barrier to the widespread deployment of NG fueling stations is their high cost. The average cost for building a gasoline station (coupled with a convenience store) in the USA was about $2.3 million in 2010; adding a compressor and storage tanks needed for NG dispensing could increase the price by as much as $0.5 million, which is quite a big investment considering few people are driving NGV [66].

One potential alternative to expensive NG stations would be in-home refueling appliances that tap into homes' existing NG line and could refill the fuel tank. One such NG refueling appliance is already on the market, and it is called "Phill." It is about the size of a large upright vacuum cleaner, and can be installed in a garage with an access to a 240-V electrical outlet and NG pipeline. A flexible hose fills a car in about 6 h [66]. Even though NGV are not a hot item on the passenger vehicles market now, some gas retailers are optimistic. Kwik Trip Inc., the operator of gas stations and convenience stores, has opened in 2012 its first NG fueling station servicing passenger-car drivers in La Crosse, Wisconsin (USA), with plans to build several more. Love's Travel Stops & Country Stores, of Oklahoma City, plans to open 10 retail outlets with compressed NG pumps.

Consumer Psychology

Besides economic and infrastructure-related challenges of introducing NGV to the passenger vehicles market, there are some psychological barriers. Many consumers are not even aware of NG as alternative fuel. Drivers have to be convinced that it makes sense and is beneficial for them to put NG in their cars. Will low emissions and fuel produced in the USA be strong enough motivations to persuade them? What about safety fears (high-pressure fuel tank)? As one form of the motivation to buy the gas-fueled vehicles, in California freeways, NGV can drive in high-occupancy-vehicle lanes.

8.2.3.2 Current Status of NGV

Currently, there are about 13 million NGV worldwide with the largest number in Asia (Pakistan, Iran, India) and Latin America (Argentina, Brazil). In some countries, governments have mandated a switch to NGV, regardless of the higher cost of vehicles. For example, in Pakistan and Iran, the governments had to make the change because of the lack of sufficient gasoline-refining capacities (these countries now have about 2.7 million and 1.9 million NGV, respectively) [66]. That kind of mandate is unimaginable in the USA, where NGV still remain a niche market. However, the share of NGV in the USA is expected to significantly grow due to the abundant supplies of shale gas and low prices. In 2012, the US administration proposed new tax credits for NGV, and it is working to develop up to five highway NG corridors [60, 69]. More stringent standards on GHG emissions and air quality regulations could also boost the deployment of NGV. It should be noted that as a clean alternative to gasoline vehicles, NGV will compete with electric-drive vehicles, e.g., battery-electric, plug-in hybrids, and the local specifics will determine the winner.

8.2.4 Environmental Aspects of Methane Economy

Over the past few years, there has been a major progress in our understanding of the potential impact of methane on the global climate system (methane as GHG is discussed in Chap. 2). Although combustion of NG generates much less CO_2 emissions than oil and coal, its GWP is significantly greater than that of CO_2. As a potent GHG, methane leaking from NG systems could be a significant contributor to overall anthropogenic GHG emissions, which would greatly diminish its advantages as relatively clean fuel. For example, in the USA, in 2011, total methane emissions amounted for 551 Tg CO_2-equiv., which is about 8.2 % of overall GHG emissions (including CO_2, NH_3, N_2O and halocarbons) from all sources (equal to 6,708 Tg CO_2-equiv.) [70]. Of these total methane emissions, about a third (33.9 %) came from fugitive emissions due to the exploitation and operation of NG systems, primarily from NG transmission and distribution and, to a lesser extent, well completions (approximately two-thirds of methane emissions came from other human-related activities such as livestock, agriculture and landfills, coal mining, etc.) [71].

Emissions from the US' NG system in 2009 (both conventional and unconventional gas) are estimated at 2.6 % of methane produced from NG wells over their production lifetime, with 1.7 % from upstream and midstream emissions (i.e., leaks at the well site and in gas processing plant, respectively), and 0.9 % from downstream emissions (i.e., storage systems, and transmission and distribution pipelines) [72]. Estimates for the combined upstream and midstream methane emissions from unconventional gas obtained from shale and tight-sand vary from 0.6 to 4 % methane over the lifecycle of a well (with most estimates at about 3 % level, over the lifecycle of a well) [73]. A major difference between methane emissions from conventional and unconventional gas operations relates to the emissions during the flow-back of fracking fluids, which occurs during the 1–2 week period following the HF operation. An average conventional gas well in the USA releases 1.04×10^3 m^3 of methane during a well completion [74]. In totality, the fugitive methane emissions associated with the development of NG from conventional wells were estimated at the range of 1.7–6.0 % of methane produced over the lifetime of a well [75].

The fact that fugitive methane emissions from unconventional gas production are higher than those from conventional wells is often used by the opponents of the shale gas development. In a number of publications, Howarth and coauthors evaluated the GHG footprint of NG obtained by high-volume HF operations with a special emphasis on methane emissions [73, 76]. According to the authors, 3.6–7.9 % of methane from HF operations will escape into the atmosphere as a result of venting and leaks over the lifetime of a shale gas well. The largest share of the fugitive methane emissions associated with a shale gas operation will come from gas transport, storage, and distribution (1.4–3.6 %), followed by a well completion (1.9 %) and venting and equipment leaks (0.3–1.9 %) [73]. These emissions are at least 30 % greater than those from conventional gas production (since the latter does not involve such operations as flow-back, drill out, and others). The study implies that over short timescales (20 years) the GHG footprint of shale gas is at least 20 and 50 % greater than that of coal and oil, respectively (per unit energy released during combustion).

However, the validity and main conclusions of this analysis were questioned by several researchers [77, 78]. The critics of the Howarth's work pointed to serious flaws of the analysis, e.g., the overestimation of the leakage rate of fugitive methane emissions and some wrong assumptions with regard to time interval over which climate impact of gas and coal are computed, and the comparison between gas and oil based on heat production rather than electricity, etc. [79]. The authors of the latter study concluded that if those considerations are factored in the GHG footprint of shale gas is about half if not a third that of coal.

The US DOE projects that the major use of shale gas over the next 23 years will focus on replacing conventional reserves of NG, as they become depleted [80]. Considering that the rates of methane emissions associated with shale and tight gas are greater than those for conventional gas, this will likely increase methane emissions from the US' NG industry by estimated 40–60 % in methane emissions [80]. This potential increase in methane emission from unconventional gas production is especially concerning in light of the recent upgrading of methane GWP to 33 and 105 for the 100- and 20-year integrated time frame, respectively [81]. Using these values, Howarth et al. calculated that methane would contribute 19 % and 44 % of the warming impact of the entire US GHG inventory (from human activities) over the next 100 and 20 years, respectively [72]. At the 20-year time scale, the methane emissions from NG systems would contribute 17 % of the entire US GHG inventory [80].

Could the fugitive methane emissions be reduced? In principle, methane emissions from NG systems could be properly addressed provided adequate incentives would be in place for NG facilities to invest in preventing these leakages from the NG infrastructure. The easiest approach to curbing methane emissions would be to target the emissions from well completions, which typically occur at the well-head during flow-back stage of HF operation. Practically all of these emissions could be avoided by implementing existing technologies and best management practices. There are several technological approaches to addressing the leakage at well-heads, e.g., through minimizing the number of wells drilled, better understanding of flow physics, subsurface properties, etc.

The advancement and further improvements in drilling and gas processing and handling technologies can substantially reduce the quantity of the fugitive methane emissions produced by the gas industry. According to the US General Accountability Office (GAO) and industry reports, up to 99 % reduction in venting methane emissions have been achieved in the San Juan basin through the use of smart-automated plunger lifts [82]. Other technological improvements such as the use of flash-tank separators or vapor recovery units can reduce methane emissions from dehydrator operations by as much as 90 % [83].

The relatively large GHG footprint of NG (especially, shale gas) systems undercuts the logic of its use as a "bridging" fuel over coming decades in the context of climate mitigation policies. Some studies went as far as claiming that shale gas may be worse than coal in terms of a climate change impact [84]. The study conducted at Cornell University (USA) concluded that compared to coal, the carbon footprint of shale gas is at least 20 % greater on the 20-year horizon, and is comparable over 100 years, mainly due to the methane leaks while the well is being exploited. Many energy analysts warn that NG is not a "panacea" to solve climate change; it is likely to make up about one-quarter of the world's energy supply by 2035, but that would lead the world to a 3.5 °C temperature rise [85]. Nobuo Tanaka, executive director of the IEA emphasized that "*While natural gas is the cleanest fossil fuel, it is still a fossil fuel. Its increased use could muscle out low-carbon fuels such as renewables and nuclear, particularly in the wake of the Fukushima. An expansion of gas use alone is no panacea for climate change.*" IEA is concerned that, as gas prices continue to go down, this will put governments under pressure to reduce support for low-carbon energy and opt for gas instead, as oil and gas companies have been urging, which could imperil the fight against climate change.

Along with a surge in supply and associated boom in industrial applications, gas is also benefiting from a temporary turmoil surrounding the alternatives: coal suffers from high carbon emissions, renewables are still expensive, and there are safety fears over nuclear, especially, after the Fukushima. As the IEA's chief scientist Birol put it: "*Gas is a fortunate fuel because all its competitors have some problems*" [85]. At the same time, Birol warned that the concerns around development of unconventional gas resources should lead companies to adopt more stringent safety and environmental measures.

8.2.5 Coupling Methane with CCS

Due to the appreciable carbon footprint of NG, especially, unconventional gas, there is a recognition among many experts that without coupling with CCS, NG will be unable to achieve needed carbon reductions from the power generation sector (some even argued that "coal *with* CCS is better than gas *without* CCS") [86]. Without CCS, the best NG can do is to cut CO_2 in half compared to coal; to make further reductions, the gas-fired plants have to be equipped with CCS (which will capture at least 90 % of CO_2 emissions). Critics of the "Dash-for-Gas" policies are concerned that the recent developments in gas industry will simply allow gas to be used as fuel unabated for decades to come and with no serious commitment to capturing CO_2 emissions [86]. This would be a serious setback for the development and commercial deployment of CCS technology, which will require substantial lead time. There are concerns that, in the long run, this policy will delay rather than speed-up the GHG reductions from the power generation sector.

In principle, post-combustion capture technology could be readily applied to gas-fired power plants, including NGCC plants. In fact, the technology faces fewer technical hurdles for its practical implementation compared to coal-fired plants, in part, because the emissions from gas combustion contain fewer harmful contaminants and no ash particles or mercury. At the same time, new carbon capture technologies are being developed that could significantly reduce NG-CCS costs and facilitate its deployment. For example, CO_2 Technology Centre Mongstad (Norway) has been developing and testing novel technology for capturing 85 % of CO_2 from the flue gases of nearby NGCC heat and power plant and refinery cracker (CO_2 contents are about 3.5 % and 13 %, respectively) [87]. The developers of the technology believe that it might be possible to generate electricity from gas with CCS at 54 % efficiency [88].

Governments around the world are warming up to the idea of coupling NG with CCS. For example, in UK, the newly formed Department of Energy and Climate Change emphasized that "*…in the longer term we see an important role for gas with CCS*" [86]. The Department set out plans for using gas-fired power plants fitted with CCS after 2030. Studies on integrating a full carbon capture scheme onto NGCC electric power generation hub for offshore operations are being conducted in Norway [89]. The European Commission recently published an Energy Roadmap to achieve deep CO_2 cuts by 2050, which identifies a substantial increase in the capacity for gas-fired plants with CCS (Gas-CCS) by the 2030s [90].

However, it is realized that there are multiple hurdles to the NG-CCS deployment, such as, the late or weak application of capture readiness requirements, the low levels of "bankable" storage capacity, restrictions on onshore storage, the absence of integrated CO_2 transport networks, and others. Modeling studies and proposed scenarios indicate that it would be practical for NG-CCS to play an increasingly important role in the supply of low-carbon electricity in 2030 and beyond in the countries with the largest predicted gas-fired power generation, particularly, in UK, Germany, France, Spain, and Italy [90].

At the times when many believe the progress on climate change drastically slowed down and even seemingly stalled because CO_2 emissions continue to grow globally at a high rate, NG-CCS could create a new low-carbon low-cost option that would help to drive down global CO_2 emissions while accommodating energy demand growth through the mid century. IEA in its 2DS pathway projects that roughly a third of power generation would come from gas-fired plants equipped with CCS [62].

8.2.6 Methane Dissociation as an Alternative Decarbonization Strategy

Due to high cost and long-term ecological uncertainties associated with the capture and storage of large amounts of CO_2 produced in the SMR process (which is currently a primary H_2 manufacturing process), there have been proposals to alternatively produce hydrogen by dissociation (or decomposition) of methane [91–94]:

$$CH_4 \rightarrow C + 2H_2, \qquad \Delta H^\circ = 75.6\,kJ/mol \tag{8.1}$$

The advantages of this approach are twofold: not only the process does not produce CO_2 byproduct, but it generates a value-added product—clean carbon, which could enhance the economic competitiveness of the process in carbon-constrained world.

The required heat input to this process is much less than that to SMR (37.8 kJ/mol H_2 vs. 63.3 kJ/mol H_2 for SMR). Although, according to the reaction stoichiometry, the methane decomposition produces half the amount of hydrogen produced by SMR (2 mol vs. 4 mol H_2 per mole of CH_4, see (7.2)), if one takes into account the amount of NG to fuel the strongly endothermic SMR process (i.e., almost a third of the total amount of NG) and the energy loses due to CO_2 capture and sequestration, the net H_2 yields and energy efficiencies of the SMR-CCS and methane decomposition processes would become comparable [93]. One should also remember that in the latter option, the chemical energy of carbon product is not lost, but is safely stored for the possible future use (if ecological situation would permit).

There are several technological options for methane decomposition to hydrogen and carbon, including thermal, thermocatalytic, and plasma (thermal and non-thermal) decomposition. Thermal decomposition of NG has been commercially practiced since 1930s until recently (*Thermal Black* process) for the production of carbon black (with hydrogen byproduct being a supplemental fuel for the process). Kværner Company of Norway has developed and operated a commercial thermal plasma process for decomposition of methane to hydrogen and carbon black [95]. Among recent technological developments, one should also mention plans by BASF, the Linde Group and ThyssenKrupp (Germany) to develop a two-stage environmentally friendly process involving methane decomposition process [96]. In the first step, NG is split into hydrogen and carbon through an high-temperature technology, and in the second step, H_2 produced reacts with CO_2 to produce syngas, which will be used in a variety of industrial applications (fuels, chemicals). The German Federal Ministry of Education and Research is subsidizing the project, which started in 1 July 2013.

Thermocatalytic decomposition (TCD) is an active R&D area. Two types of catalysts have been developed: metal- and carbon-based catalysts for the use in the TCD of methane. Metal catalysts such as Ni-, Fe-, Co-based catalysts have been most commonly used for the TCD process; however, they proved prone to rapid deactivation due to carbon build up on the catalyst surface. At optimized operational conditions, high-value multi-wall carbon nanotubes could also be produced as a byproduct of the metal-catalyzed methane decomposition process [97]. The use of carbon-based catalysts offers certain advantages over metal catalysts due to their durability, sulfur resistance, and low cost. Muradov et al. screened a variety of carbon materials and demonstrated that efficient catalytic methane decomposition could be accomplished over high-surface area disordered carbons [98]. Recently, a series of novel carbon-based catalysts with improved stability have been reported [99–101].

Techno-economic evaluation of the NG-TCD process indicates that the hydrogen production cost is the function of carbon selling price [102]. In order to become competitive with SMR, TCD-produced carbon should be sold at about $350/ton carbon. As a high quality substitute for petroleum coke it could potentially be sold for $310–460/ton C (for manufacturing of electrodes in aluminum and ferroalloy industries) [103]. The TCD process could potentially become competitive with the SMR even without carbon credit if the cost of CO_2 capture and storage (about 30–50 % of the H_2 product cost) or carbon tax on CO_2 emissions would be added to the cost of hydrogen manufactured by SMR. In order for the TCD process to become a major decarbonization tool in terms of global carbon mitigation objectives, sufficiently large markets for its carbon products would need to be found (in the order of tens to hundreds million tons) [104]. The perspectives of using solid carbon products generated by decomposition of NG as structural materials capable of replacing steel and concrete were discussed by Muradov [94, 102, 104] and Halloran [105].

8.2.7 Methane as a "Bridge" to Renewable Energy

Many energy analysts predict NG to play a crucial role in ushering the greater penetration of renewable energy to the marketplace (see discussion in Sect. 8.1.2). Current trends indicate that for renewable energy to become a viable source of energy, the integration of their variable output with the NG-fueled gas turbines matching (i.e., peaking and balancing) load will be a necessary step toward the growth of renewable electricity share. Combined with the relative "cleanness" of NG as fuel, this makes methane decarbonizing "*bridge fuel*" that could facilitate and accelerate the transition from "dirty" (coal) to "clean" (renewables) energy and fuels. NG is the only utility-scale source of energy with high flexibility and relatively low carbon footprint that can be efficiently adjusted to match fluctuations in the grid load (This is in contrast to coal plants, which produce significantly more NO_x and CO_x emissions during cycling up or down operations to meet wind-energy supply fluctuations) [106]. Three main prerequisites for NG to become a "bridge" to the renewable energy future are:

- Steady supply of NG at competitive rates.
- Improvement in fast-cycle CCGT technology.
- Optimizing match between renewable energy and CCGT plants.

Although, historically, NG prices fluctuated significantly, recent developments in fracking and other gas-extracting technologies point to plentiful supply of affordable NG for the near future. Most importantly, the NG sources are widely geographically distributed around the globe, so many countries and regions have an access to these supplies. Proximity to the supply of NG also easies the risk of power loss due to interruptions in long-distance gas pipelines or electrical transmissions. At the NG price of about $2/GJ (mid of 2012, in USA), electricity could be produced at $30/MWh; based on the NG futures prices forecasted at below $6/GJ through 2016, the cost of NG-generated electricity is projected to be $55/MWh [106].

In the recent years, there has been a significant improvement in fast-start/rapid-response gas turbine technology that are much better suited to integration with intermittent renewable energy sources (e.g., solar and wind) compared to conventional turbines. For example, CCGT is a good match with a PV plant that generates less power in the late afternoon or early evening when power consumption peaks and power prices are highest. The development of power plants that combine a solar concentrator and a gas turbine (and, optionally, a fuel cell) is considered an attractive technological option with potential to achieve efficiencies up to 70 % [62]. There are several opportunities for a technological integration between CCGT and renewable sources, e.g.

- A concentrating solar power facility can be coupled with CCGT, and during the day, the solar-powered facility can supplement the steam supply to the gas turbine; this solution is being practiced in the *Archemedes* solar plant in Italy [107] (see the scheme of Archemedes project in Fig. 8.6).
- If CCGT and wind or solar installations are located close to each other, they could conveniently share interconnection facilities, which would also facilitate the regulation of the power output.
- If CCGT is located along the energy transmission corridor, it could regulate and optimize the power supply from several renewable energy generation sites to end-users.
- CCGT plant can be designed to meet specific market requirements to balance expected conditions and efficiencies, e.g., small (100 MW) vs. large (1,000 MW) plant capacities [106].

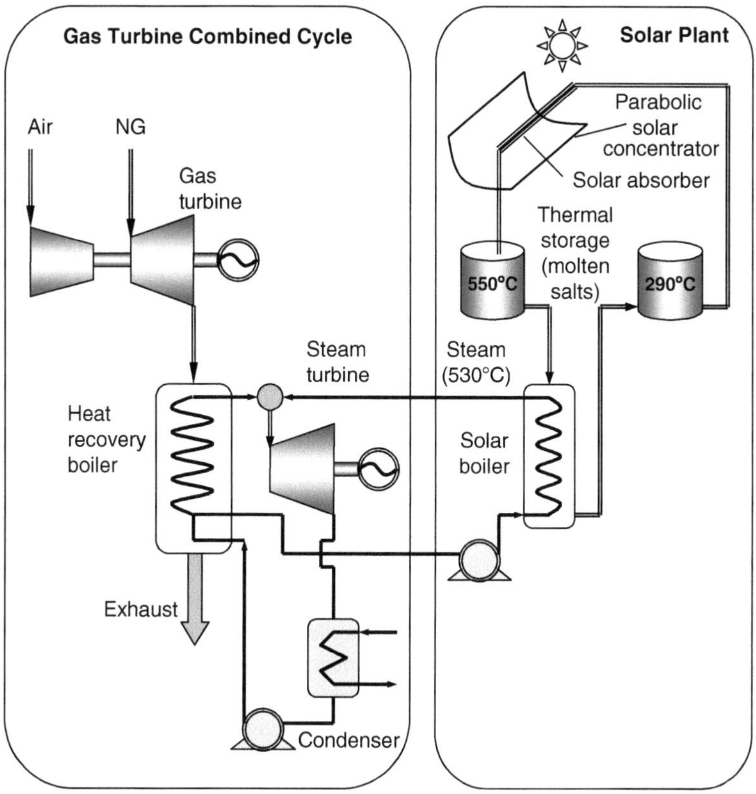

Fig. 8.6 Scheme of the Archemedes project in Italy. *Source* [107]

An alternative design of a hybrid solar-gas power plant is under development at the Pacific Northwest National Laboratory (PNNL) (USA) [108]. The system uses solar heat to convert NG into syngas (H_2+CO), which is fed to a gas-fired power plant. PNNL uses a parabolic dish concentrator that focuses solar beam onto a microchannel-type catalytic methane reforming reactor. Because syngas has about 20 % higher energy content than methane fuel, a power plant can consume about 20 % less NG still giving the same electrical output. This also substantially lowers the power plant's carbon footprint at a cost competitive with traditional fossil power, since significant part of the energy input to the system comes from carbon-free solar source. When the sun sets or during cloudy days, the solar-booster component of the system is bypassed and NG is directly burned in the turbines without interruptions in power generation. The concept is especially attractive for power plants located in sunshine-drenched areas (e.g., the US Southwest). A 500 MW gas-fired power plant would require about 3,000 solar dishes equipped with methane reformers. Integrated solar hybrid gas-fired combined cycle power plants are under construction or in planning stages in Algeria, Egypt, India, Iran, Italy, Mexico, Tunisia, and the USA [62].

Due to high power prices during the day, the integration of CCGT with wind farms is particularly economically attractive. The capacity factor for wind power is in the range of 13–40 % (typically higher at night) depending on the particular location [106]. Modern commercial CCGT plants operate at the capacity factor of 40–60 %, and, typically, they run during the day when demand and electricity prices are highest. Thus, the capacity factors of CCGT and a wind farm can potentially balance each other resulting in a higher overall capacity factor and increased power output.

The quite unexpected aspect of NG role as a "bridge-to-renewables" fuel has emerged recently: whether NG will be a collaborator or a competitor to renewable energy. On the one hand, NG has the potential and ability to almost instantaneously supplement renewable power generation ensuring an uninterrupted power delivery in the event of falling of solar or wind supply. On the other hand, many energy analysts are concerned that the excessive push for NG could simply "crowd out" renewable energy sources. There is a notion that "*buying time with natural gas is not very useful if you don't use the time*" to get renewable technologies to the marketplace on a big scale, and this is the critical decade to accomplish that [13]. The OECD Secretary-General A. Gurria clearly expressed his concerns with regard to the transitional role of NG: "*The question then becomes: how do you ensure that gas is a transitional step towards an eventual goal of zero emissions? If we invest too much in dedicated pipelines and other infrastructure, the transition risks becoming a new and permanent dependency. Any new fossil resources brought to market—conventional or unconventional—risk taking us further away from the trajectory we need to be on…*" [109]. It is realized that building an expansive and expensive new infrastructure for NG while developing new zero-carbon technologies to eliminate the use of gas may prove very challenging (because of powerful gas supporters in industry and government, among other factors).

8.3 Electrification as an Efficient Decarbonization Strategy

Electricity as a versatile energy carrier is increasingly substituting fossil fuels in many areas, e.g., industrial processes (e.g., resistive heating vs. fuel combustion), transportation (battery-electric, plug-in hybrid, and FC-electric vehicles), household appliances, and others. Presently, nuclear sector followed by hydroelectric plants and renewable sources are the major producers of non-carbon electricity worldwide (11.0, 2.1, and 1.2 % of total, respectively) [110]. Most near-to-mid term future energy scenarios project that this trend will intensify with an emphasis shifting from high-carbon to low-carbon to zero-carbon electricity. Reported estimates indicate that as a result of electricity-driven decarbonization measures, CO_2 emissions from electricity generation will drop from 500 g CO_2 per kWh (2009 worldwide average) to about 25 g CO_2 per kWh by 2050 [23].

Analytical studies show that switching from the direct uses of fossil fuels to electricity has a tremendous potential to reduce the carbon intensity of economy. For example, the recent analysis of measures aiming at drastically reducing CO_2 emissions in

California (USA) indicated that the share of electricity as an end-use energy would need to grow from current 15 to 55 % in 2050 in order to meet the target of reducing emissions by 80 % by 2050 (against the 1990 level) [111]. According to the study, the largest share of GHG emission reductions would need to come from the electrification of transport, where 70 % of vehicle-miles traveled will be powered by electricity (e.g., via the use of PHEV and BEV), 20 % powered by biofuels, and remaining 10 % by conventional fossil fuels. In addition, 65 % of non-heating/cooling fuel use in buildings and 50 % of industrial fuel use would also need to be electrified.

IEA projects that by mid-century, the share of PHEV, BEV, and FC vehicles in the light-duty transportation market will reach 80 % [112, 113]. In the USA, electric vehicles are slowly gaining ground: in March 2013, there were 34,000 BEV on the road [114]. The US consumers have now 14 different BEV models to choose from, a figure that is likely to triple in the next couple model years. However, for BEV to overtake their ICE counterparts, their price will have to come down, battery life to go up, and the charging station infrastructure will need to expand to make longer trips easier and convenient.

The remarkable decarbonization effect of the end-use electrification measures could be attributed to the fact that electric/electrochemical processes are much more efficient than thermal/thermochemical processes constrained by the Carnot limitations. FC-based systems are particularly advantageous because they can generate simultaneously electricity and heat with very high exergetic efficiency (see Chap. 5). The detailed discussion of the exergization aspects of electricity-driven systems can be found in publications by Winter (e.g., [2]).

8.4 Transition to Hydrogen Economy

The evolutionary fuel progression trend manifesting itself in further increase in H/C ratio predicts the gradual transition to hydrogen fuel (essentially, with no further limits to the increase in the H/C ratio). This historical trend would signal the unrelenting penetration of hydrogen into the energy market, cleansing it from carbon, and establishing (according to Ausubel [115]) "*hydrogen metabolism*," more commonly known as "*Hydrogen Economy*."

Similar to methane, hydrogen has a capacity to reduce both energy and carbon intensity factors in the Kaya Identity equation, but with a much greater impact. In contrast to methane, however, hydrogen does not exist in nature in a free form, thus, it has to be produced from (primarily) water with an energy input from (preferably) non-carbon primary energy sources (e.g., nuclear and renewables) with inevitable energy losses/penalties. In this regard, hydrogen is an *energy carrier* (or a *vector*, or "*currency*"), similar to electricity.

The global commercial production of hydrogen is about 65 million tons per year [116]; however, most of it is used in non-energy application areas, ammonia production, oil refineries, food industry, metallurgy, etc, which is not exactly in the mainstream of the carbon mitigation policies. Currently, steam reformation of NG is the primary process for the commercial manufacturing of hydrogen (48 % of total), with 30 % and 18 % of hydrogen produced from petroleum and coal, respectively, and the remaining 4 % produced by water electrolysis [116].

Despite remaining technological challenges, it appears that hydrogen energy is no longer a "dream fuel" but, rather, a realistic clean energy solution, and markets are gradually warming up to embrace the potential of hydrogen. Practically all industrialized and many developing countries have ongoing hydrogen-related R&D programs and commercial activities. In the USA, over 9 million metric tons of H_2 are produced annually and close to 2,000 km of H_2 pipelines exist [117]. Hydrogen buses are in operation in Austin (Texas), Birmingham (Alabama), Cleveland (Ohio), and several other US cities [118]. The US DOE coordinates hydrogen and FC activities in the country; it addresses technical barriers through the wide range of fundamental research, R&D, technology validation, and demonstration projects. As of mid-2013, there were 60 hydrogen fueling stations in the USA. Research activities enabled a doubling of FC durability and reducing the cost of automotive FC systems down to $47/kW (or 80 % since 2002), approaching the target of $30/kW, which will put the FC-powered systems in cost-parity with ICE (assuming high volume cost) [117].

In Germany, the *H_2 Mobility* initiative is targeting installing a network of about 400 hydrogen-fueling stations across the country to supply H_2-powered vehicles [119]. The first mass market hydrogen FC cars are expected to hit the country's roads in 2015. This initiative has a support from gas supply firms (Air Liquide, Linde), automaker (Daimler) and petroleum companies (Shell, Total and OMV). According to reported estimates, the new infrastructure will cost about US$475 million and will take about a decade to complete [119]. There's also a large hydrogen bus fleet in Reykjavik (Iceland), where there's a vast supply of renewable energy (mostly, geothermal) to cheaply produce hydrogen fuel [118]. Energy analysts are already predicting an energy market transformation with a planned roll-out of hydrogen-FC vehicles starting in 2014 from Mercedes Benz, BMW, Toyota, GM, Nissan, Honda, and Ford.

Contrary to a widespread perception that the transportation application of hydrogen is limited to FC vehicles only, a significant progress has recently been achieved in the development of hydrogen internal combustion engines (HICE). A team of engineers from Argonne National Laboratory (USA) have demonstrated one-cylinder research engine with direct injection of H_2 attaining peak brake thermal efficiency (PBTE) of 45.5 %, which exceeds the US DOE's 2010 efficiency goal of 45 % PBTE (PBTE measures the amount of fuel converted into useful power during combustion) [120]. These improvements are due to both optimized engine geometry and an upgraded injection system. If this hydrogen engine were used in a mid-size sedan with a conventional power train and 5-speed automatic transmission, it would achieve the fuel economy of 32.4 mile per gallon (mpg) in city driving and 51.5 mpg on the highway (with the combined average of 38.9 mpg (1 mpg is equal to 235.2 L gasoline per 100 km). This research shows that HICE holds promise as a bridging technology toward a large-scale hydrogen transportation infrastructure.

Despite the rapid progress and major advances in the development of hydrogen technologies, significant challenges must be overcome on the path to the mature Hydrogen Economy. Among the most important barriers are the high cost of renewable hydrogen

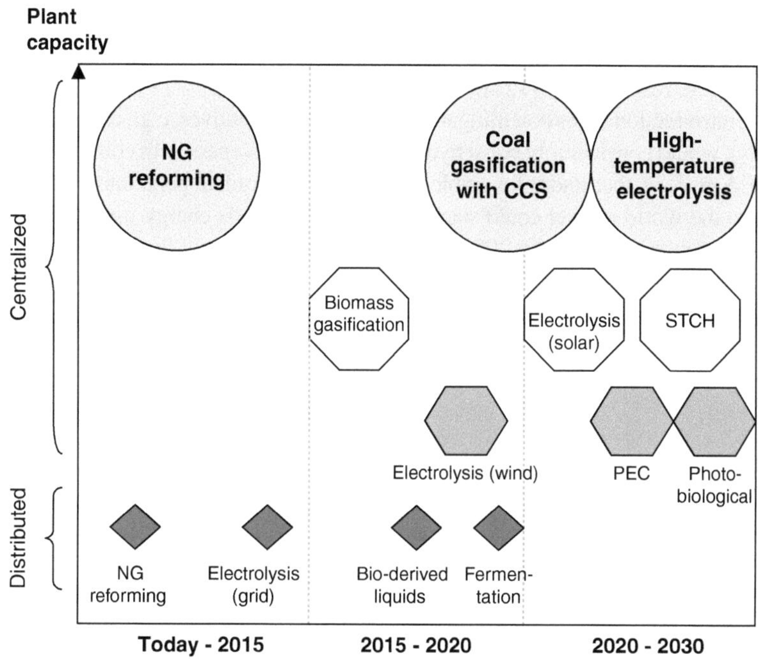

Fig. 8.7 US DOE scenario for hydrogen production pathways as a function of estimated plant capacity and anticipated technology readiness. Estimated plant capacities (in tons H_2 per day) are shown by the symbols: *diamonds*—up to 1.5, *hexagons*—50, *octagons*—100, *circles*—more than 500. *PEC* photoelectrochemical systems, *STCH* solar thermochemical cycles for hydrogen production. *Source* [121]

and onboard H_2 storage systems. As of mid-2013, the cost of renewable hydrogen is more than $10 per gallon gasoline equivalent (gge) [117]. In order to be competitive with gasoline fuel, the hydrogen cost should drop to $2–4/gge. Advanced hydrogen storage systems are needed for the widespread commercialization of FCEV across all vehicle platforms. The existing on-board hydrogen storage systems are still far from the US DOE's targets in storage density and cost. Most of the present-day hydrogen-vehicle projects involve compressed hydrogen in light-weight composite tanks. Recently, Linde Company (Germany) announced that it has developed lighter composite storage materials withstanding higher pressure, which resulted in more than doubling the amount of H_2 stored compared to existing tanks [119].

Due to the enormous cost of building a new hydrogen infrastructure, the initial major deployment of hydrogen systems in the marketplace will most likely occur through distributed (or decentralized) hydrogen production units based on SMR and renewables/electrolysis technologies (e.g., wind-electrolysis, solar PV-electrolysis). The next step would be a transition to centralized hydrogen production based on such low- and zero-carbon technologies as NG/CCS, coal/CCS, nuclear, and renewable sources. Figure 8.7 depicts the US DOE vision of the market penetration of hydrogen production technologies starting with reforming of domestically abundant NG and focusing on biomass and solar pathways for mid- and long-term production of hydrogen, respectively [121].

The introduction of carbon mitigation policies and incentives, e.g., carbon price (carbon tax), would considerably boost hydrogen's competitiveness with conventional fuels. The IEA projects that under favorable assumptions, a gradual penetration of hydrogen fuel into the world market could start around 2020, with its energy use reaching 12.5–22 EJ (exajoules) per year by 2050 [116]. The establishment of the matured Hydrogen Economy will complete the evolutionary decarbonization process.

To the readers interested in more detailed discussion of the different aspects of Hydrogen Economy, we would recommend the following sourcebooks [122–124].

References

1. Muradov N (2013) Decarbonization at crossroads: the cessation of the positive historical trend or a temporary detour? Energy Environ Sci 6:1060–1073
2. Winter C-J (2012) Hydrogen energy—abundant, efficient, clean: a debate over the energy-system-of-change. In: Muradov N, Veziroglu N (eds) Carbon-neutral fuels and energy carriers. CRC, Boca Raton, FL, pp 115–202
3. Kruse B, Grinna S, Buch C (2002) Hydrogen. The Bellona Foundation Publ., Oslo, Norway. ISBN 82-92318-05-4
4. Tamburic B, Ademovic A, Garnha E et al (2011) Residential fueling with hydrogen, Imperial College London, Design report. http://www.hydrogencontest.com/pdf/2011/Imperial%20College%20London%20Entry%202010_11.pdf. Accessed 5 Dec 2012
5. U.S. Department of Energy (2010) Fuel cells, fuel cell technologies program, Office of Energy Efficiency and Renewable Energy. November 2010. http://www1.eere.energy.gov/hydrogenandfuelcells/pdfs/fct_h2_fuelcell_factsheet.pdf. Accessed 10 Aug 2011
6. Spath P, Mann M (2000) Life cycle assessment of hydrogen production via natural gas steam reforming. National Renewable Energy Laboratory Report. NREL/TP-570-27637, Golden, Colorado
7. Cheb L, Nolan R, Avadhany S (2009) Thermodynamic analysis of coal-to-SNG process. Massachusetts Institute of Technology, Boston, MA. Report. http://www.mit.edu/people/leichen/Thermodynamic_analysis_of_Coal_to_SNG_process(final).pdf. Accessed 12 Apr 2012
8. US Department of Energy (2012) DOE technology successes "Breakthrough" gas turbines, Office of Fossil Energy. http://www.fe.doe.gov/programs/powersystems/turbines/turbines_successes.html. Accessed 12 Aug 2012
9. Kaneco S, Katsumata H, Suzuki T et al (2006) Energy Fuel 20:409–414
10. Hoffmann P (2013) HyUnder workshops on underground H_2 gas storage launch 2-year effort. Hydrogen Fuel Cell Lett 28:2
11. Radnedge S (2013) Hydrogen used in storing electricity. Gas World. July 16. http://www.gasworld.com/news/regions/north-pacific/hydrogen-used-in-storing-electricity/2002435.article?goback=.gde_77156_member_258618138. Accessed 12 Aug 2013
12. Hoffmann P (2011) World's first renewable hydrogen hybrid power plant starts production. Hydrogen Fuel Cell Lett 26(11):1–2
13. Johnson J (2013) The natural gas advantage. Chem Eng News 91:29
14. Fraunhofer Institute (2012) Research highlights. "Power-to-Gas"—renewable gas connects energy networks. http://www.iwes.fraunhofer.de/en/highlights20112012/highlights2010 2011/_power-to-gas_renewablegasconnectsenergynetworks.html. Accessed 15 Jun 2013

15. ITM Power (2013) ITM power joins German energy Agency's Power to Gas strategy platform. 20 June 2013. http://www.itm-power.com/news-item/itm-power-joins-german-energy-agencys-power-to-gas-strategy-platform. Accessed 10 Aug 2013
16. Distributed Energy (2012) World's largest power-to-gas plant for generating methane enters operation. http://www.distributedenergy.com/DE/articles/19724.aspx?format=2. Accessed 20 Jun 2013
17. Wind Power-to-Gas (P2G) Technology (2013) Energy Storage J. March 2013. http://energystoragejournal.com/wind-instrument-power-to-gas-technology. Accessed 20 Jun 2013
18. Rathi C (2012) Will natural gas compete or coexist with renewable power? Power Eng 116(11) http://www.power-eng.com/articles/print/volume-116/issue-11/departments/view-on-renewables/will-natural-gas-compete-or-coexist-with-renewable-power.html. Accessed 20 Jun 2013
19. De Falco M, Basile A (2012) Editorial. Int J Hydrogen Energy 37:11487
20. Moreno F, Munoz M, Arroyo J et al (2012) Efficiency and emissions in a vehicle spark ignition engine fueled with hydrogen and methane blends. Int J Hydrogen Energy 37: 11495–11503
21. Klell M, Eichlseder H, Sartory M (2012) Mixtures of hydrogen and methane in the internal combustion engine—synergies, potential and regulations. Int J Hydrogen Energy 37: 11531–11540
22. Villante C, Genovese A (2012) Hydromethane: a bridge towards the hydrogen economy or an unsustainable promise? Int J Hydrogen Energy 37:11541–11548
23. International Energy Agency (2011) CO_2 emissions from fuel combustion, IEA report, 2011 ed. Paris, France
24. International Energy Agency (2011) World Energy Outlook. Are we entering a golden age of gas? IEA, Paris, France
25. Budischak C, Sewell D, Thomson H et al (2013) Cost-minimized combinations of wind power, solar power and electrochemical storage, powering the grid up to 99.9 % of the time. J Power Sources 225:60–74
26. Ausubel J, Grubler A, Nakićenović N (1988) Carbon dioxide emissions in a methane economy. Clim Change 12:245–263
27. FracFocus (2012) Chemical disclosure registry. http://fracfocus.org/hydraulic-fracturing-how-it-works/site-setup. Accessed 12 Dec 2012
28. Shellenberger M, Nordhaus T (2011) A boom in shale gas? Credit the feds. Washington Post, 16 Dec 2011 http://www.washingtonpost.com/opinions/a-boom-shale-gas-credit-the-feds/2011/12/07/giqa-ecfizo_story.html. Accessed 10 Jan 2012
29. Laubach S, Reed R, Olson J et al (2004) Co-evolution of crack-seal texture and fracture porosity in sedimentary rocks: cathodoluminescence observations of regional fractures. J Struct Geol 26:967–982
30. US Department of Energy (2009) Hydraulic fracturing fluids—composition and additives. Modern shale gas development in the United States, US DOE. http://geology.com/energy/hydrolic-fracturing-fluids/. Accessed 15 Jan 2011
31. Andrews A, Folger P, Humphries M et al (2011) Unconventional gas shales: development, technology and policy issues. http://www.fas.org/sgp/crs/misc/R40894.pdf. Accessed 20 Jan 2012
32. Scott G (1997) Method for monitoring the hydraulic fracturing of a subterranean formation, US Patent 5,635,712. US PTO, Washington, DC
33. US Department of Energy (2009) Modern shale gas development in the United States: a primer. http://www.netl.doe.gov/technologies/oil-gas/publications/ereports/shale_gas_primer_2009.pdf. Accessed 12 Feb 2012
34. Stepniak B (2012) Fracking gains political acceptance. Wealth Wire, 7 Feb 2012. http://www.wealthwire.com/news/energy/2650. Accessed 20 Feb 2012
35. Panzica B (2012) Super fracking: The "Holy Grail" of natural gas, Wealth Wire, 20 Jan 2012 http://www.wealthwire.com/news/energy/2551. Accessed 20 Feb 2012

36. Urbina I (2011) A tainted water well and concern there maybe more. The New York Times, 3 Aug 2011. http://www.nytimes.com/2011/08/04/us/04natgas.html?pagewanted=all. Accessed 26 Feb 2012
37. US Senate Committee on Environment and Public Works (2009) Obama administration: no documented cases of hydraulic fracturing contamination. 8 Dec 2009. http://epw.senate.gov/public/index.cfm?FuseAction=PressRoom.PressReleases&ContentRecord_id=70289BE8-802A-23AD-479D-CA2D6F6B36CD. Accessed 26 Feb 2012
38. US Senate (2011) Pathways to energy independence: hydraulic fracturing and other new technologies, 6 May 2011. http://epw.senate.gov/public/index.cfm?FuseAction=PressRoom.PressReleases&ContentRecord_id=23eb85dd-802a-23ad-43f9-da281b2cd287. Accessed 12 Feb 2012
39. Osborn S, Vengosh A, Warner N et al (2011) Methane contamination of drinking water accompanying gas-well drilling and hydraulic fracturing. Proc Natl Acad Sci U S A 108:8172–8176
40. Jackson R, Vengosh A, Sarrahet T et al (2013) Increased stray gas abundance in a subset of drinking water wells near Marcellus shale gas extraction. Proc Natl Acad Sci U S A. doi:10.1073/pnas.1221635110
41. Johnson J (2013) Methane found in drinking water. Chem Eng News 91:15
42. Vidic R, Brantley S, Vandenbossche J et al (2013) Impact of shale gas development on regional water quality. Science 340:826. doi:10.1126/science.1235009
43. Wolfgang B (2013) I drank fracking fluid, says Colorado Gov. John Hickenlooper. Washington Times. 12 Feb 2013. http://www.washingtontimes.com/blog/inside-politics/2013/feb/12/colorado-gov-hickenlooper-i-drank-fracking-fluid/. Accessed 12 Jun 2013
44. US Environmental Protection Agency (2001) Toxicity and human health effect following exposure to oxygenated or reformulated gasoline, May 2001. http://www.epa.gov/region04/foiapgs/readingroom/hercules_inc/toxicological_and_human_health_effects_following_exposure_3v.pdf. Accessed 10 Jan 2012
45. City of Fort Worth (2011) Natural gas air quality study. 13 July 2011. http://forthworthtexas.gov/uploadedfiles/gas_wells?aitqualitystudy_final.pdf. Accessed 10 Jan 2012
46. Hess G (2011) EPA to regulate shale gas waste. Chem Eng News 89:28
47. Yang H, Flower R, Thompson J (2013) Shale-gas plans threaten China's water resources. Science 340:1288
48. Urbina I (2011) Regulation lax as gas wells' tainted water hits rivers, New York Times. http://www.nytimes.com/2011/02/27/us/27gas.html?_r=1. Accessed 5 Mar 2012
49. Philadelphia Water Department (2011) Iodine-131 found in Philadelphia's drinking water. http://www.phila.gov/water/pdfs/qa_iodine131.pdf. Accessed 5 Mar 2012
50. Lamont-Doherty Earth Observatory Institute (2012) Ohio quakes probably triggered by waste disposal well, say seismologists, Columbia University. http://www.ldeo.columbia.edu/news-events/seismologists-link-ohio-earthquakes-waste-disposal-wells. 12 Oct 2012
51. Jenkins J (2013) Natural gas boom rewrites the energy rules. Discovery. January: 31
52. BBC News (2011) Fracking tests near Blackpool "likely cause" of tremors. http://www.bbc.co.uk/news/uk-england-lancashire-15550458. Accessed 3 Apr 2012
53. Ellsworth W (2013) Injection-induced earthquakes. Science 341:142
54. Van der Elst N, Savage H, Keranen K et al (2013) Enhanced remote earthquake triggering at fluid-injection sites in the Midwestern United States. Science 341:164
55. Driver A (2011) Critics of energy 'fracking' raise new concern: sand. Reuters. 21 Sept 2011. http://www.msnbc.msn.com/id/44612454/ns/us_news-environment/t/critics-energy-fracking-raise-new-concern-sand/. Accessed 10 Jan 2012
56. Wealth Wire (2012) How the US shale boom will change the world. http://www.wealthwire.com/news/energy/2712. Accessed 5 Nov 2012
57. Rogers M, Dillenbeck R, Eid R (2004) Society of petroleum engineers. Paper 90829
58. China National Development and Reform Commission (2012) China's shale gas development plan (2011–2015). www.sdpc.gov.cn/zcfbtz/2012tz/t20120316_467518.htm. Accessed 15 Jan 2013

59. Hess G (2011) Energy panel sees risks of gas drilling. Chem Eng News 89:25
60. Hargreaves S (2012) Obama's energy plan: the winners and winners, CNN Money. http://money.cnn.com/2012/01/30/news/companies/obama_energy/index.htm?hpt=hp_t3. Accessed 2 Apr 2012
61. Hess G (2011) EPA takes aim at fracking emissions. Chem Eng News 89:26
62. International Energy Agency (2013) Tracking clean energy progress. http://www.iea.org/publications/TCEP_web.pdf. Accessed 20 Aug 2013
63. International Energy Agency (2012) Medium term gas market report. OECD/IEA, Paris
64. Knittel C (2012) Leveling the playing field for natural gas in transportation, Brookings. http://www.brookings.edu/~/media/research/files/papers/2012/6/13%20transportation%20knittel/06_transportation_knittel.pdf. Accessed 20 Oct 2012
65. Hoffmann P (2011) Cheap natural gas as fuel cell enabler. Hydrogen Fuel Cell Lett 26:6–7
66. Fowler T (2012) America, start your natural gas engines. Wall Street J. 18 June 2012. The journal report: innovations in energy. http://online.wsj.com/article/SB10001424052702304192704577406431047638416.html. Accessed 12 Dec 2012
67. Mason J, Veenstra M, Long J (2014) Evaluating metal–organic frameworks for natural gas storage. Chem Sci. doi:10.1039/C3SC52633J
68. Gas Technology Institute (2012) Unlocking the global potential of natural gas annual report. GTI-13/0008.7-13 ABS2m. 2013. Des Plains, IL, USA
69. Silverleitb A, Cohen T (2012) Obama unveils plan to control college cost. CNN News. 27 Jan 2012. http://www.cnn.com/2012/01/27/politics/obama-trip/index.html?hpt=hp_t3. Accessed 27 Jan 2012
70. US Environmental Protection Agency (2012) Inventory of U.S. Greenhouse gas emissions and sinks: 1990–2011. US-GHG-Inventory-2011-Chapter-1-Introduction.pdf. Accessed 18 Dec 2012
71. US Environmental Protection Agency (2010) GHG emissions. Technical note on the 1990–2009 inventory estimates for natural gas systems, Washington, DC
72. Howarth R, Santoro R, Ingraffea A (2012) Venting and leakage of methane from shale gas development: reply to Cathles et al. Clim Change. doi:10.1007/s10584-012-0401-0
73. Howarth R, Santoro R, Ingraffea A (2011) Methane and greenhouse gas footprint of natural gas from shale formations. Clim Change. doi:10.1007/s10584-011-0061-5
74. US Environmental Protection Agency (2010) Greenhouse gas emissions reporting from the petroleum and natural gas industry. Background technical support document. http://www.epa.gov/climatechange/emissions/downloads10/subpart-W_TSD.pdf. Accessed 20 Jan 2011
75. US Environmental Information Administration (2010) Natural gas navigator. Natural gas gross withdrawals and production. http://www.eia.gov/dnav/ng/ng_prod_sum_dcu_NUS_m.htm. Accessed 20 Jan 2012
76. Howarth R, Ingrafea A (2011) Should fracking stop? Extracting gas from shale increases the availability of this resource, but the health and environmental risks may be too high. Nature 477:271–275
77. Jiang M, Griffin M, Hendrickson C et al (2011) Life cycle greenhouse gas emissions of Marcellus shale gas. Environ Res Lett 6: paper 034014. doi: 10.1088/1748-9326/6/3/034014
78. Hultman N, Rebois D, Scholten M et al (2011) The greenhouse impact of unconventional gas for electricity generation. Environ Res Lett 6: paper 044008. doi: 10.1088/1748-9326/6/4/044008
79. Cathles L, Brown L, Taam M et al (2012) A commentary on "The greenhouse-gas footprint of natural gas in shale formations" by R.W. Howarth, R. Santoro, and Anthony Ingraffea. Clim Change 113:525–535. doi:10.1007/s10584-011-0333-0
80. Howarth R, Shindell D, Santoro R et al (2012) Methane emission from natural gas systems. Background paper prepared for the National Climate Assessment. Reference No: 2011–0003. 25 Feb 2012
81. Shindell D, Faluvegi G, Koch D et al (2009) Improved attribution of climate forcing to emissions. Science 326:716–718
82. US General Accountability Office (2010) Federal oil and gas leases: opportunities exist to capture vented and flared natural gas. Which would increase royalty payments and reduce

greenhouse gases. GAO-11-34 US General Accountability Office, Washington, DC. http://www.gao.gov/new.items/d1134.pdf. Accessed 16 Apr 2012
83. Fernandez R, Petrusak R, Robinson D et al (2005) Cost effective methane emissions reductions for small and midsize natural gas producers. Reprint from J Petrol Technol. http://www.icfi.com/markets/environment/doc_fiels/methane-emissions.pdf. Accessed 16 Apr 2012
84. Black R (2011) Shale gas 'worse than coal' for climate. BBC News. 12 Apr 2011. http://www.bbc.co.uk/news/science-environment-13053040. Accessed 16 Apr 2012
85. Harvey F (2012) Natural gas is no climate change 'panacea', warns IEA. The Guardian. 6 June 2012. http://www.guardian.co.uk/environment/2011/jun/06/natural-gas-climate-change-no-panacea. Accessed 10 Oct 2012
86. Thompson J (2012) The dash for gas- no climate cure without CCS. Global CCS Institute. http://www.globalccsinstitute.com/insights/authors/jyhompson/2012/10/22/das-gas-%E2%80 %93-no-climate-cure-without-ccs. Accessed 23 Oct 2012
87. CO_2 Technology Centre Mongstad Technologies (2012) http://www.tcmda.com/en/Technology/. Accessed 20 Feb 2013
88. Power from gas+CCS at 54 % efficiency? (2013) Carbon Capture J. 14–15 March–April. http://www.carboncapturejournal.com/news/power-from-gas--ccs-at-54-efficiency/3350.aspx. Accessed 23 Aug 2013
89. Hetland J, Kvamsdal H, Haugen G et al (2009) Integrating a full carbon capture scheme onto a 450 MW_e NGCC electric power generation hub for offshore operations: presenting the Sevan GTW concept. Appl Energy 86:2298–2307
90. Pershad H (2012) CCS for gas—results of Element Energy study. Carbon Capture J. 13–15 November–December
91. Muradov N (1993) How to produce hydrogen from fossil fuels without CO_2 emission. Int J Hydrogen Energy 18:211–215
92. Choudhary T, Sivadinarayana C, Chusuei C et al (2001) Hydrogen production via catalytic decomposition of methane. J Catal 199:9–18
93. Steinberg M (1999) Fossil fuel decarbonization technology for mitigating global warming. Int J Hydrogen Energy 24:771–777
94. Muradov N (2001) Hydrogen via methane decomposition: an application to decarbonization of fossil fuels. Int J Hydrogen Energy 26:1165–1175
95. Lynum S, Hildrum R, Hox K et al (1998) Kværner based technologies for environmentally friendly energy and hydrogen production. Proc. 12th World Hydrogen Energy Conf. Buenos Aires, Argentina. p 697
96. Holzhäuser V (2013) Research cooperation develops innovative technology for environmentally sustainable syngas production from carbon dioxide and hydrogen. BASF. Press Release. 2 July 2013. http://www.basf.com/group/pressrelease/P-13-351. Accessed 20 Aug 2013
97. Mordkovich V, Dolgova E, Karaeva A et al (2007) Synthesis of carbon nanotubes by catalytic conversion of methane: competition between active components of catalyst. Carbon 45: 62–69
98. Muradov N, Smith F, T-Raissi A (2005) Catalytic activity of carbons for methane decomposition reaction. Catal Today 102(103):225–233
99. Serrano D, Botas J, Pizarro P, Guil-Lopez R, Gomez G (2008) Ordered mesoporous carbons as highly active catalysts for hydrogen production by CH_4 decomposition. Chem Commun 6585–6587
100. Suelves I, Pinilla J, Lazaro M, Moliner R (2008) Carbonaceous materials as catalysts for decomposition of methane. Chem Eng J 140:432–438
101. Muradov N, Choi P, Smith F et al (2010) Integration of direct carbon and hydrogen fuel cells for highly efficient power generation from hydrocarbon fuels. J Power Sources 195: 1112–1121
102. Muradov N, Veziroglu N (2008) "Green" path from fossil to hydrogen economy. An overview of carbon-neutral technologies. Int J Hydrogen Energy 33:6804–6839
103. Gaudernack B, Lynum S (1996) Hydrogen from natural gas without release of CO_2 to the atmosphere. Proc. 11th World Hydrogen Energy Conference, Stuttgart, Germany

104. Muradov N, Veziroglu N (2005) From hydrocarbon to hydrogen-carbon to hydrogen economy. Int J Hydrogen Energy 30:225–237
105. Halloran J (2008) Extraction of hydrogen from fossil fuels with production of solid carbon materials. Int J Hydrogen Energy 33:2218–2224
106. Ng Y (2012) Natural gas key to a renewable energy future. Power Eng 116:84–86
107. Lorenzini G, Biserni C, Flacco G (2010) Solar thermal and biomass energy. WIT, Southampton
108. Solar booster shot for natural gas power plants (2013) Science Daily. 13 April 2013. http://www.sciencedaily.com/releases/2013/04/130411152332.htm?goback=%2Edge_77156_member_232872827. Accessed 20 Aug 2013
109. Gurria A (2013) Carbon price vital to address climate change. Responding to climate change (RTCC). 9 Oct 2013. http://www.rtcc.org/2013/10/09/oecd-chief-carbon-price-vital-to-address-climate-change/. Accessed 14 Oct 2013
110. International Energy Agency, *2011* Key world energy statistics, Paris, France, 2011. http://www.iea.org/textbase/nppdf/free/2011/key_world_energy_stats.pdf. Accessed 20 Apr 2012
111. Williams J, DeBenedictis A, Ghanadan R et al (2012) The technology path to deep greenhouse gas emissions cuts by 2050: the pivotal role of electricity. Science 335:53–59
112. International Energy Agency (2010) Energy technology perspectives 2010—scenarios and strategies to 2050. ISBN 978-92-64-08597-8, 2010
113. IEA- progress towards clean energy stalled (2013) Carbon Capture J. 17 April 2013. http://www.carboncapturejournal.com/displaynews.php?newsid=1125&phpsessid=18dkjsha7qaa6s815mlrslvtq4. Accessed 25 Aug 2013
114. Niiler E (2013) Electric vehicles are gaining ground. Discovery. April 2013. http://news.discovery.com/autos/fuel-and-alternative-fuel-technologies/electric-vehicles-gaining-ground-130416.htm. Accessed 25 Aug 2013
115. Ausubel J (1996) Can technology spare the Earth? Evolving efficiencies in our use of resources suggest that technology can restore the environment even as population grows. Am Sci Mag 84:166–178
116. International Energy Agency (2007) Hydrogen production and distribution: IEA energy technology essentials, Paris. http://www.iea.org/techno/essentials5.pdf. Accessed 20 Nov 2011
117. Satyapal S (2013) Commentary from the US DOE director of fuel cell technologies. IAHE Newsletter, v.5, Summer 2013. http://www.IAHEnewsletter_summer2013-1.pdf. Accessed 28 Sep 2013
118. Sterling S (2012) Hydrogen on the rise. A renewable energy based storage solution, North American Clean energy. September/October, 2012, p. 100, 101, nacleanenergy.com. Accessed 5 Feb 2013
119. Bomgardner M (2013) Germany to get H_2 fueling network. Chem Eng News 91:24
120. Hoffmann P (2011) Argonne builds experimental high-efficiency hydrogen I.C. engine. Hydrogen Fuel Cell Lett 26:7
121. US Department of Energy (2013) Hydrogen production research and development. Office of Energy Efficiency and Renewable Energy. Funding opportunity announcement: DE-FOA-0000826. https://eere-Exchange.energy.gov. Accessed 26 Nov 2013
122. Ball M, Wietschel M (eds) (2009) Hydrogen economy: opportunities and challenges. Cambridge University Press, Cambridge, UK
123. Sigfusson T (2008) Planet hydrogen—the taming of the proton. Coxmoor Publ, Oxford, UK, p 218
124. Gupta R (ed) (2009) Hydrogen fuel. Production, transport and storage. CRC, Boca Raton, USA

Chapter 9
Industrial Utilization of CO_2: A Win–Win Solution

Abstract Carbon capture and utilization (CCU) is an attractive carbon abatement strategy because of its potential for not only preventing CO_2 emissions to the atmosphere but also converting CO_2 to value-added products: a win–win solution. This approach can potentially make the carbon capture process more profitable and substantially reduce the investment needs for a rather expensive CO_2 storage infrastructure. Over the last few years, interest in CCU has grown significantly, and many innovative technological approaches to the industrial CO_2 utilization are under development, such as CO_2 conversion to construction materials, plastics, fertilizers, fuels, etc. At the same time, the analysis of the CO_2 utilization market shows that all existing industrial CO_2 applications consume relatively small quantities of CO_2, thus for the CCU to present a practical interest as a sink for anthropogenic CO_2 emissions, the markets for the CO_2-derived products would need to be increased by orders of magnitude. In this chapter, existing and emerging CO_2 utilization technologies are analyzed in terms of their technological maturity, market size, permanence of CO_2 storage, environmental impact, potential revenue generation, and carbon mitigation potential. The current status and outlook for CO_2-to-fuel conversion technologies and CO_2 utilization in algal systems are highlighted in this chapter.

9.1 Introduction

The economic challenges of CCS have caused major delays in its commercial deployment; therefore, the capture and industrial use of CO_2 is considered a promising approach to overcome these challenges. *Carbon capture and utilization* (*CCU*) presents a very attractive alternative to conventional CCS because of its potential for not only preventing CO_2 emissions to the atmosphere but also converting CO_2 to value-added products, which is seen by many as a win–win solution. The support for CCU has been steadily growing worldwide and intensified during the last couple decades. Currently, a number of commercial technologies are turning CO_2 into

N. Muradov, *Liberating Energy from Carbon: Introduction to Decarbonization*,
Lecture Notes in Energy 22, DOI 10.1007/978-1-4939-0545-4_9,

building materials, plastics, chemicals, fertilizers, and several projects converting CO_2 into fuels are now approaching a demonstration stage.

Although CO_2 has been utilized in a variety of industrial applications for decades (e.g., in food/beverage and chemical industries), the current CO_2 utilization efforts focus on novel pathways and approaches to reducing CO_2 emissions through its beneficial uses (i.e., *reuses*), particularly in areas where other means of CO_2 storage (e.g., geological storage) may not provide an optimal solution. The advantages of the CCU option are threefold:

- The industrial use of CO_2 provides a carbon sink, the extent of which will be determined by the size of the market for CO_2-derived products and lifetime of these products
- CO_2 could be converted to the products generating significant revenues that could potentially offset part of the CO_2 capture and transport costs and make the entire process more profitable and potentially viable without the government subsidies
- CCU reduces the need for CO_2 storage site locations and associated infrastructure to deposit the gas underground at the risk of its leaking out (e.g., due to natural or induced seismic activity).

On the other hand, the utilization of CO_2 as a feedstock for technological processes may face the following complicating factors:

- The products produced from CO_2 have different lifecycle characteristics depending on their manufacturing process, use and disposal, and, therefore, they can function as carbon sinks for the periods of time from weeks to months to centuries (i.e., until CO_2 is released back to the atmosphere via oxidation or decay processes). Evidently, only products with sufficiently long lifetime (e.g., centuries) can be considered as a means of long-term carbon storage.
- CO_2 is a chemically (relatively) inert compound, and, as such, it features certain disadvantages and technological challenges as a reactant and feedstock. Due to its inertness, the processes that utilize CO_2, in general, have greater energy input requirements (e.g., higher temperatures) compared to alternative chemical routes, which could translate into potentially higher CO_2 emissions (e.g., from heat or electricity input sources). From this viewpoint, every existing and emerging industrial process utilizing CO_2 should be carefully examined and verified that in a long run the amount of CO_2 emitted does not exceed that of CO_2 utilized.

CO_2 utilization technologies can be categorized by a number of criteria:

- CO_2 content of the feedstock (i.e., concentrated vs. diluted CO_2 streams).
- Permanence of CO_2 storage (nonpermanent, semipermanent, and permanent).
- Maturity (the stage of technological development and commercialization).
- Additional CO_2 emissions from the reuse.
- Potential revenue generation.

Not all the CO_2 reuse technologies require concentrated stream of CO_2, and not all of them provide permanent carbon storage. Figure 9.1 provides a summary of existing CO_2 utilization technologies categorized according to CO_2 content requirement and storage permanence.

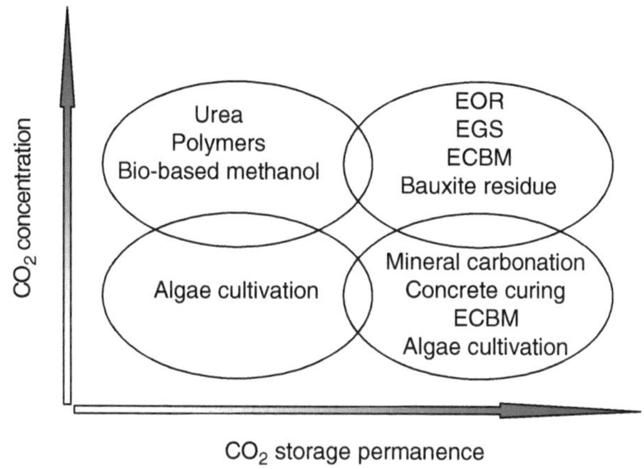

Fig. 9.1 Classification of existing CO_2 utilization technologies by CO_2 content and storage permanence. *Source*: [1]

As seen from Fig. 9.1, the CO_2 reuse technologies fall into four main categories, namely, the technologies:

- Requiring concentrated CO_2 and permanently storing CO_2
- Requiring concentrated CO_2 and not permanently storing CO_2
- Not requiring concentrated CO_2 and permanently storing CO_2
- Not requiring concentrated CO_2 and not permanently storing CO_2

The first two categories of technologies are based on the sources that provide a concentrated CO_2 stream either as a process byproduct (e.g., from NG processing, ethanol production), or through the addition of a CO_2 capture unit that concentrates dilute CO_2 emissions (e.g., from power, cement, and steel manufacturing plants). The CO_2 reuse technologies that utilize dilute sources (e.g., flue gases, off-gases) directly do not require a capture plant (though, they may require some form of gas clean-up) thus, they might provide a lower cost option for CO_2 utilization, and, even, provide some form of revenue. For this reason, these technologies have a potential to act as a transitional option to conventional CCS, especially in the cases where integrated CCS projects are delayed due to problems related to timely accessing viable storage sites [1].

The ability of reuse technologies to permanently store CO_2 is an important attribute, which determines the viability of the technology as a CO_2 abatement option. Some reuse technologies could result in practically permanent storage of CO_2 (e.g., for thousands of years, as in a carbonate mineralization process). Therefore, such technologies as EOR, ECBM, mineralization, concrete curing, bauxite residue carbonation that practically permanently store CO_2 are considered alternative forms of CCS. On the other hand, such CO_2-derived products as urea-based fertilizer, polymers, and others may start breaking down and releasing CO_2 to the atmosphere in as short as 6 months after their use, and some fuels produced from CO_2 will release CO_2 once they are combusted in engines. As evident from Fig. 9.1, algae belongs to both permanent and nonpermanent storage categories, because the permanence of the algae-based products would depend on the end product; e.g., biofuels would not permanently store CO_2, but some products could provide semipermanent and permanent storage, e.g., pigments, bioplastics. Generally, the CO_2 reuse technologies that do not provide permanent storage are facing some ambiguity due to the uncertainties around the carbon price liability, regulation, and climate policies.

9.2 Existing Industrial CO_2 Utilization Processes

Currently, CO_2 is used in a great variety of industrial applications. Figure 9.2 summarizes the majority of the current and potential industrial uses of CO_2.

The existing industrial uses of CO_2 include a number of important processes and commercial applications such as production of urea, methanol, "dry" ice, the use in refrigeration systems, food and beverage industries, fire extinguishers, horticulture, enhanced oil recovery, water treatment, solvents (supercritical CO_2), and many others. However, many of these uses are relatively small and typically emit CO_2 to the atmosphere during or after their use (e.g., fire extinguishers, solvents, wastewater treatment, etc.), and, as such, they do not reduce overall CO_2 emissions to the atmosphere.

The current global demand for CO_2 is estimated at 80 Mt/year CO_2 [1]. Of that amount, 25 Mt/year is in a liquid and solid form and the remainder in gaseous and supercritical form. The distribution of major industrial CO_2 users (in % of total) is shown below (Note that only non-captive[1] uses of CO_2 are presented) [1]:

Enhanced oil recovery	67.5
Food industry	12.1
L-CO_2 applications	8.1
Beverage carbonation	6.7
Precipitated $CaCO_3$	3.1
Oil and gas industry (other than EOR)	1.4
Others	1.1

Of the total CO_2 demand, at least 50 Mt/year is utilized in EOR (most of it, in USA and Canada). The remaining portion of CO_2 use relates to all other users, predominantly, in food and beverage industry. The future potential demand for CO_2 (for 2020) is estimated at 140 Mt/year (based on the predicted growth of current technologies such as EOR and other uses) [1].

The global CO_2 demand is currently met by natural geological CO_2 reservoirs and a number of industrial processes that produce CO_2 as a byproduct (e.g., SMR, NG processing, ethanol production, etc.). In particular, over 80 % of CO_2 used in EOR in the USA is originated from natural wells, and there is a potential to replace natural CO_2 with man-made CO_2 in the EOR applications. The total potential supply of CO_2 from large point sources is estimated at 18 Gt/year CO_2 [1]. The CO_2 capture

[1] Non-captive CO_2 use refers to the processes where CO_2 is supplied from an external source, as opposed to captive use, wherein CO_2 is only an intermediate product, and it is ultimately consumed in a later stage of the process.

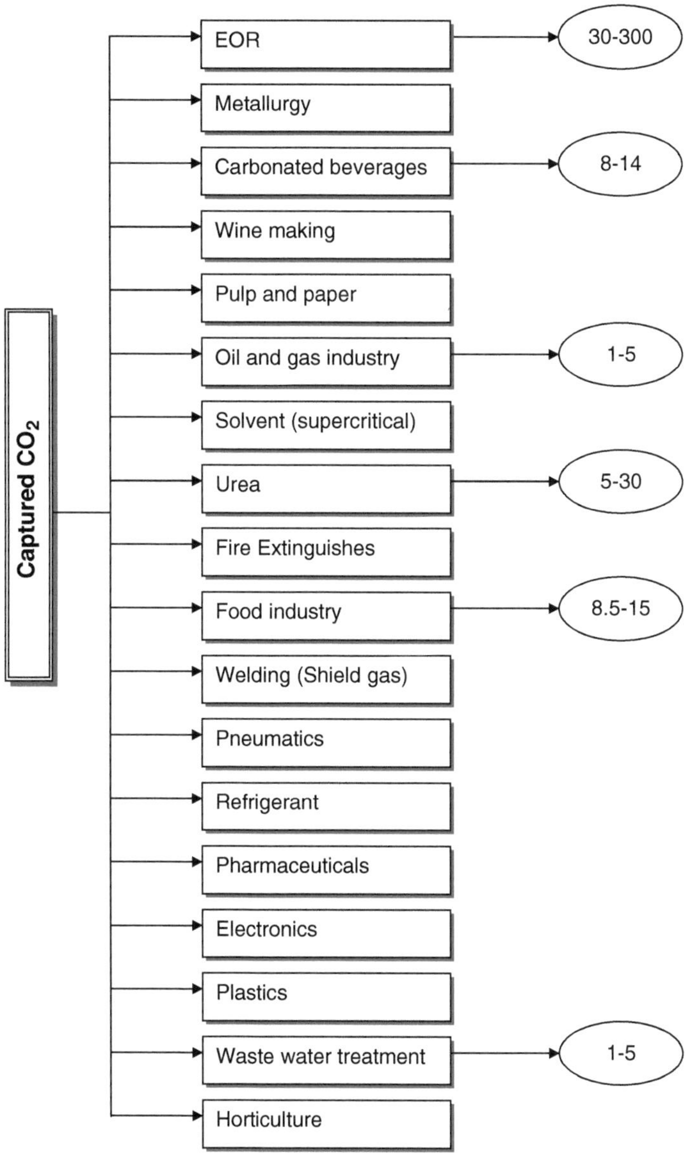

Fig. 9.2 Existing CO_2 industrial uses and current and potential CO_2 demand. The range of current and future CO_2 demand for a particular application is shown in ovals in Mt/year CO_2. In cases where ovals are not shown, the current and future demand is less than 0.1 Mt/year CO_2. *Source*: [1]

cost depends on the source of CO_2: the most expensive option is to capture it from power plants (which accounts for 70 % of the total CO_2 supply), while from industrial sources such as NG processing and ammonia and ethanol production it could be relatively cheap. It was estimated that the lowest and low-to-moderate (i.e., less than US$35 t^{-1} CO_2 avoided) cost sources could provide 0.5 Gt/year and 2.5 Gt/year CO_2, respectively. Considering future CO_2 demand of "only" 0.14 Gt/year, one can see a very large mismatch between demand and potential supply. Moreover, considering anticipated restrictions on CO_2 emissions in the near-to-mid term future the CO_2 supply surplus is likely to grow even further.

9.2.1 *Current CO_2 Prices*

The price of bulk[2] CO_2 from industrial sources varies in a wide range depending on the nature of the source and its location (it is typically negotiated between parties and, in most cases, is not available to public). The following are examples of CO_2 prices from available literature sources [1, 2]:

- The US ammonia producers sold CO_2 at the price range of US$3–15 per metric ton of gaseous/supercritical CO_2 (depending on location)
- The Great Plains Synfuels Plant (Dakota Gasification Company) sold US$453.2 million worth CO_2 at a price of US$19 per metric ton (which incorporates the cost of CO_2 transport)
- Cardinal Ethanol LLC sold 40,000 t of CO_2 at a price of US$5 t^{-1} (the recipient paid for CO_2 transport)
- The price for pipelined CO_2 has historically been in the range of US$9–25 t^{-1} (including the cost of the pipeline infrastructure).

Considering the substantial supply surplus and the prospects of regulatory constraints on CO_2 emissions, large-scale facilities (e.g., power, steel, and cement plants) that will install CO_2 capture technology are likely to be price-takers in the CO_2 market. Future prices for bulk gaseous/supercritical CO_2 are expected not to exceed current prices since the current surplus is likely to increase.

9.2.2 *Industrial CO_2 Utilization Markets*

9.2.2.1 EOR

CO_2 utilization in EOR operation is discussed in detail in the Chap. 7.

9.2.2.2 Urea

Urea is a major nitrogen-based fertilizer (almost 50 % of the world's nitrogen fertilizer production) and a chemical intermediate. In 2009, 154.9 Mt of urea was produced globally, which consumed 116.2 Mt of (captive) CO_2. It is produced from

[2] Bulk CO_2 is defined as unprocessed gaseous stream with CO_2 content of more than 95 vol.%.

CO_2 by reacting it with ammonia at high pressure and temperature via intermediate formation of ammonium carbamate:

$$CO_2 + NH_3 \rightarrow H_2NCOOH \quad (9.1)$$

$$H_2NCOOH + NH_3 \rightarrow H_2NCOO^- NH_4^+ \rightarrow CO(NH_2)_2 + H_2O \quad (9.2)$$

Typically, CO_2 used in the process is recovered from waste streams of ammonia manufacturing plants. From this viewpoint, urea production can be considered a "captive" use of CO_2, since it is produced and consumed in the same process (0.735–0.75 t CO_2 is consumed for every ton of urea produced [3]). However, when NG is used as a feedstock for the urea production (as often is the case), there is about 5–10 % surplus of ammonia, which could be reacted with an external (non-captive) source of CO_2 to produce additional urea. This non-captive part of CO_2 presents interest from the CCU perspective. Currently, an increasing number of urea production facilities are equipped with units for CO_2 capture from flue gas (e.g., Mitsubishi Heavy Industries), thus, this technology (referred to as "urea yield boosting," *UYB*) can be considered mature. However, in the context of carbon storage, it is not a permanent solution: once urea granules are applied to agricultural land, urea reacts with water releasing ammonia and CO_2, which ends up in the atmosphere. Besides, the level of CO_2 emissions in the process of the CO_2 reuse is rather high: 2.27 t CO_2-equiv./t reused CO_2 [1]. Due to these shortcomings, the utilization of CO_2 in urea production will unlikely present a practical interest as a carbon storage option.

9.2.2.3 Other CO_2 Application Areas

Other industrial CO_2 application areas include:

Food and beverage. CO_2 is widely used in food and beverage industries in three main areas: beverage carbonation, foodstuff packaging and, in the form of dry ice in chilling and freezing operations. As an example of the project involving industrial CO_2 use in food/beverage market, Prosint methanol plant in Brazil captures 32,850 t of CO_2 per year and uses it in soft drink manufacturing process [4]. In the USA, Big Brown SkyMine captures 112,500 t of CO_2 per year and supplies it to different food processing facilities. CO_2 is also used for controlled atmosphere packaging of food products to extend their shelf life (CO_2 inhibits growth of bacteria).

Metallurgical industry. CO_2 is used for fume suppression during charging of furnaces. It is also used in Cu and Ni production processes during tapping of electric arc furnaces [5] and in some basic oxygen furnaces as a bottom stirring agent, and for dust suppression.

Pharmaceuticals. CO_2 is used in supercritical fluid extraction procedures, chemical synthesis, as inert gas, and as a cooling agent for storage and transportation of temperature-sensitive substances.

Pulp and paper. CO_2 is used for reducing pH of recycled chemical pulps after an alkaline bleaching, and for enhancing the performance of paper manufacturing equipment.

Electronics. CO_2 is used as a supercritical fluid for removing photoresists from wafers, as a cooling medium during environmental testing of electronic devices, and for abrasive cleaning of residues on wafers (in the form of snow).

Wastewater treatment. Addition of CO_2 to waste streams enables adjusting and controlling the pH of aqueous effluents.

Fire extinguishers. Solid CO_2 in the form of snow is used in fire extinguishers. It provides a heavy blanket of gas that reduces the oxygen flow suppressing combustion.

Supercritical solvent. CO_2 is used for high-pressure extraction of targeted compounds, such as fragrances and flavors. It is now attracting interest in dry-cleaning industry.

As shown in Fig. 9.2, except for EOR, UYB and food/beverage areas, the existing industrial CO_2 applications require relatively small quantities of CO_2 (less than 0.1 Mt/year) with a little chance of any dramatic increase in the future demand. To present a practical interest as a potential sink for anthropogenic CO_2, industrial CO_2 utilization technologies will need to have very large markets, in orders of magnitude larger than existing ones.

9.3 Emerging Industrial CO_2 Utilization Processes

The development of new large markets for anthropogenic CO_2 is a focus of a great number of R&D and demonstration projects worldwide. In this section, the main emphasis is placed on CO_2 reuse technologies, where the "reuse" is defined as any practical application of the captured CO_2 that adds value (e.g., revenue generation) or provides tangible environmental, societal, or other benefits. The beneficial reuse of CO_2 should be distinguished from a broad definition of CO_2 industrial applications, where CO_2 could originate, e.g., from natural CO_2 sources, since in the latter case, CO_2 will be used with a minimal (or without any) environmental benefit.

The reuse technologies that permanently (at least, for several hundreds of years) store CO_2 are considered to be an alternative form of CCS and sometimes referred to as "alternative CCS" [1]. For CO_2 reuse technologies to contribute to an accelerated CCS deployment at a scale commensurate with climate change policies, they must show a potential to absorb very large quantities of CO_2 comparable to those released by power generation and other large industrial sources.

The list of emerging CO_2 reuse technologies presenting an interest from a carbon abatement viewpoint is shown in Fig. 9.3.

According to the Global CCS Institute data, several emerging industrial CO_2 applications (e.g., ECBM, enhanced geothermal systems, algae cultivation, CO_2 concrete curing, and others) could potentially exceed the global CO_2 reuse potential of 5 Mt/year CO_2 (which was suggested by the Global CCS Institute as CO_2 utilization potential commensurate with future carbon capture requirements from large industrial sources [1]).

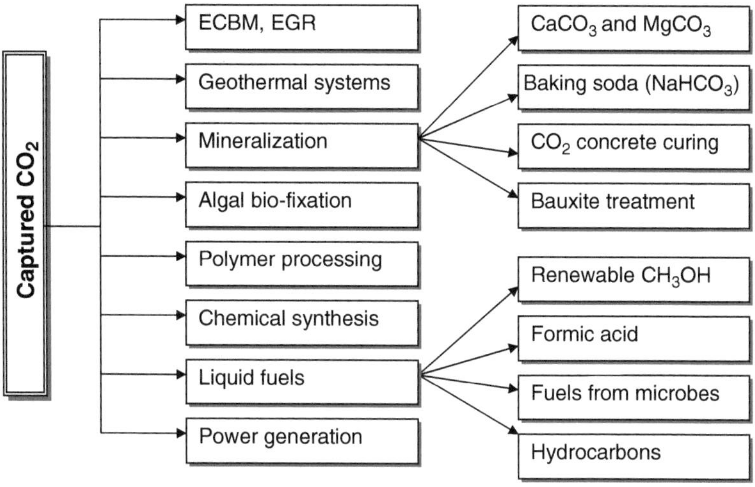

Fig. 9.3 Emerging CO_2 reuse technologies. *Source*: [1]

9.3.1 Enhanced Coal Bed Methane Recovery

Most coal seams contain naturally occurring methane (along with trace quantities of light hydrocarbons, N_2, and CO_2), which is formed during the geological time-scale transformation from organic matter to peat to anthracite coal in underground coal seams. Methane is adsorbed to the surface of coal micropores with its content typically increasing with the coal bed depth, permeability, coal rank, and pressure in the coal bed. The amount of methane trapped within coal seams could be in excess of 100 m^3/t of coal [1].

The ECBM technology is based on the capacity of coal to have much higher affinity to CO_2 than to methane (the ratio of CO_2:CH_4 adsorption capacities over coal surface varies in the range of 1–10, depending on the type of coal); as a result, when CO_2 is injected into coal seams, it displaces methane [5]. Advantageously, the ECBM process can utilize CO_2 in either dense or gaseous forms depending on the coal seam depth. The ECBM operation can potentially increase the amount of recovered methane to about 90 % of the total gas, compared to only 50 % achievable by a conventional pressure depletion recovery method [6]. CO_2 utilization rate using ECBM technology depends on the nature of the coal bed/seam, the pressure of the seam, and the CO_2/methane storage ratio (this ratio typically varies from 2:1 to 13:1) [1].

9.3.1.1 ECBM Technology Status

ECBM technology is still in the development phase, mostly, involving pilot and demonstration projects. As of 2012, there were no ECBM commercial projects, primarily, due to the lack of a commercial incentive for the process, rather than to technical challenges. (Note that commercial coal bed methane production is currently limited to conventional extraction technology, without the use of CO_2.) Several research projects are underway in the USA, Canada, Australia, Switzerland, the UK, the Netherlands, and China [1]. Most of the research efforts are focused on understanding of the many unknown variables in the ECBM process and modeling of the multicomponent adsorption–desorption characteristics of different coal types.

The Allison Unit Project (operated by Advanced Resources International and Burlington Resources) in San Juan County in southern New Mexico (USA) was the first large-scale demonstration ECBM project. Injection of CO_2 for methane recovery at the Allison unit started in 1995 and lasted 6 years during which a total of 322,000–335,000 t of natural CO_2 was pumped into the reservoir at pressure of 10.4 MPa (of that amount 277,000 t remained in the storage reservoir, with the rest produced with methane), with incremental methane recovery of about 30,000 t [1, 4, 7]. The US DOE is currently sponsoring several CO_2-ECBM pilot scale projects. In one of them (SECARB project), about 1,000 t of gaseous CO_2 was injected into two unmineable coal seams 490–670 m deep in Virginia (USA) in 2009 [8].

A multinational CO_2-ECBM demonstration project was announced in 2010 with the participation of CSIRO (Australia), United Coal Bed Methane Corp. (China), and JCOAL (Japan). The project plans to inject 2,000 t of CO_2 into the Liulin Gas Block, Shanxi Province, at a depth of approximately 500 m [1]. Another multinational ECBM project is ongoing in Alberta, Canada, where abundant coal-bed methane resources are located. Five injection wells are installed in this area, and tests are being performed on the adsorption of CO_2 into a coal seam at a depth of 500 m [1]. RECOPOL ECBMR project was launched in 2001 in Poland with the objective of evaluating the technical and economic feasibility of storing CO_2 in coal seams and producing methane. Further research activities in ECBM technology are expected to expand, especially in countries with large coal reserves.

9.3.1.2 CO_2 Utilization Potential of ECBM

Globally, CO_2 utilization potential of ECBM is immense. It was estimated that the implementation of ECBM on a global scale would increase coal-bed methane production by 18 TCM, while simultaneously sequestering 345 Gt of CO_2 [1]. Simulation studies show that an injection of CO_2 in an 80 acre area containing low-rank coal would result in storage of 1.27–2.25 BCF of CO_2 and recovery of 0.62–1.10 BCF of coal-bed methane (at the injection depth of 1.9 km) [9] (BCF is billion cubic feet; 1 BCF is equal to 28 million cubic meters). According to reports, the CO_2 storage potential of 29 coal-bed sites in China has been estimated at about 143 Gt with simultaneous production of methane from ECBM reaching 3.4–3.8 TCM [10]. This could sequester CO_2 emissions from major industrial sources for about 50 years (based on the 2000 emission level), and would produce methane in quantity equivalent to 218 years of its production (at China's 2002 production rate). In the Netherlands, four potential ECBM areas were evaluated with estimated combined capacities to store between 54 Mt and 9 Gt of CO_2 [11].

9.3.1.3 Market Potential of ECBM

The main market drivers for ECBM technology are NG prices and potential carbon incentives and regulations (carbon price, carbon trading, etc.). Besides the above factors, the ECBM potential for revenue generation will also depend on value of produced methane, cost of processing, cost of transportation, and others. ECBM technology is region specific since it requires suitable coal seams located in proximity of CO_2 point sources. Favorable market opportunities for the technology exist in North America (the USA, Canada), Europe (Poland, Italy, the UK, Russia, Ukraine), Australia, and New Zealand. There is a great potential for the ECBM technology in developing countries (China, India, Indonesia, and other countries with large coal deposits), where energy demand and storage regulations could be favorable to ECBM application.

9.3.1.4 Challenges and Barriers for ECBM

Potential challenges and barriers for ECBM technology include

- The technology is at a relatively early stage of technological development. The potential detrimental effect of the permeability deterioration due to CO_2 injection on the extent of methane recovery from coal seams has to be further evaluated.
- Further research is required to understand the fundamental issues related to ECBM, particularly, multicomponent adsorption/desorption behavior of different types of coal.
- The economic viability of the technology depends on NG prices; current low NG prices does not favor the wide-scale implementation of the technology in many regions.
- Potentially high cost barriers due to the region-specific nature of ECBM [1].

9.3.2 *CO_2 as Working Fluid for Enhanced Geothermal Systems*

The use of CO_2 as a working fluid for enhanced geothermal systems (EGS) is an emerging geothermal technology whereby supercritical CO_2 is circulated as the heat exchange fluid (instead of water or brine, as in a conventional process) to recover geothermal heat from the reservoir. Supercritical CO_2 (SC–CO_2) can be used as a working fluid of the power cycle in SC–CO_2 turbine. In this technology (which is also called "hot fractured rocks" or "hot dry rocks"), CO_2 is pressurized and injected to the depths of 4–5 km [1]. The force generated by CO_2 injection creates fissures in the hot dry rock, through which CO_2 flows generating hot (200–300 °C) geothermal fluid that expands back to the surface (some CO_2 remains sequestered into the geological formation). At the surface, the hot SC fluid expands through a gas turbine, and cooled SC–CO_2 is compressed and reinjected into the reservoir.

CO_2 sequestration potential of the EGS technology depends on the geological and chemical specifics of the hot rock formation. Long-term studies on fluid losses during fluid recirculation in pressurized reservoirs suggest that approximately 5 % of the injected CO_2 gets trapped, which implies that this method could continuously sequester CO_2 through its diffusion into the rock mass surrounding the reservoir [1]. It was estimated that the potential for geological storage is about 24 t CO_2/day/MW of EGS, which is an equivalent of achieving geological storage of CO_2 emitted from 3 GW coal-fired power plant (or 1 t CO_2 per second per 1 GW of electricity generated) [1]. The process has the potential to sequester CO_2 permanently, with the extent of permanence depending on the geological characteristics of the site (e.g., the presence of suitable capping formation).

9.3.2.1 Present Technological Status of EGS

EGS is still at an early development stage; the fundamentals of the process are yet to be understood. Currently, no large-scale commercial units operate anywhere in the world. Although there are several EGS units operating in the USA, Europe, Australia, and Japan, all of them use water (not SC–CO_2) as a heat transfer medium. Notably, two demonstration units with 1.5 MW and 3 MW electrical output operate in France and Germany, respectively, and one 25 MW plant is proposed in the Cooper Basin in Australia [1].

Several companies are working on the development of SC–CO_2 EGS technology: Heat Mining Company, Symmex Technologies, and GreenFire Energy (all in the USA), and Geodynamics (Australia). Heat Mining Company LLC is developing CO_2 plume geothermal (CPG) technology for using CO_2 as the subsurface working fluid in geothermal power generation at typical CCS or EOR sites [12]. The CPG systems are expected to operate at the electricity-production efficiency of 0.5–4 times that of conventional water-based geothermal systems. For example, a CPG system using 2.5 km deep reservoir at 100 °C, operates at 11.8 % geothermal-to-electricity energy conversion efficiency, compared to 3.4 % efficiency for conventional binary water/brine power system. Preliminary estimates indicate that direct CO_2 geothermal systems may cost less than half of what conventional water-based binary systems cost. CPG is particularly advantageous for harvesting geothermal energy from areas not viable for conventional geotechnologies [13].

In summary, the SC–CO_2 EGS technology is expected to be a cost-effective way to sequester CO_2 from existing fossil-based power plants and generate new base load power round the clock. The estimated EGS resource base in the USA is more than 13 million EJ (the extractable portion is over 200,000 EJ) [1]. The main challenges facing the practical implementation of the SC–CO_2 EGS technology include [1] (1) limited knowledge on the geochemistry of SC–CO_2, (2) the long-term CO_2 retention, (3) possible seismic triggers and events causing CO_2 leakage, and (4) the proximity of a CO_2 source or electric grid to EGS site.

9.3.3 CO_2 as Feedstock for Polymer Processing

Recently, new approaches to producing a number of important polymers (plastics) with the use of CO_2 as a chemical feedstock have been developed. In particular, in the polycarbonates and polyurethane synthesis processes, CO_2 substitutes traditionally used feedstocks. (Note that an initial motivation for the use of CO_2 in these processes was not carbon mitigation considerations, but the attempts to solve the toxicity problem via replacing the toxic compound—phosgene, $COCl_2$—with CO_2.) In the new approach, CO_2 is combined with traditional feedstocks to synthesize industrially important polymers—polycarbonates. For example, polyethylene and polypropylene carbonates (PEC and PPC, respectively) can be produced by copolymerization of CO_2 and appropriate epoxy compounds in the presence of $Zn(CH_2CH_3)_2$ catalyst in benzene solution at room temperature and elevated pressure of 60 atm, as follows [14]:

$$H_2C(O)-CHR+CO_2 \rightarrow 1/n(-H_2C-CHR-O-CO-O-)_n \quad (9.3)$$

As seen from the equation, such polymers (e.g., PEC) can contain up to 50 wt% of CO_2; therefore, the replacement of traditional materials by these polymers would significantly reduce the carbon footprint and create a substantial demand for CO_2 (for each ton of the plastic manufactured, up to 0.5 t of CO_2 is sequestered) [1].

The polymers produced from CO_2 could potentially replace traditional petroleum-based plastics such as polyethylene, polypropylene, polystyrene, and polyvinylchloride in many application areas (where properties of plastics are comparable), thus, greatly reducing a demand for finite oil-based feedstocks. Potential markets for CO_2-derived polymers are as follows:

- Coatings and laminates: protective finishes for wood, metal, furniture, flooring, appliances, automotive parts, linings for food products, and medical components.
- Packaging: plastic bags, film wraps, bottles; the polymers can be processed using common manufacturing techniques such as injection moulding, extrusion (film and sheet), blow moulding, etc. The barrier protection of CO_2-derived polymers exceeds that of petroleum-based plastics by two orders of magnitude [15].
- Surfactants for EOR application; PPC-derived surfactants improve the solubility of SC–CO_2 in oil, thus, increasing oil recovery and creating permanent storage for CO_2 (in the form of surfactant).

Storage permanence characteristics of this method depend on the end use of the plastic products. In a pure form, CO_2-derived polymers can degrade in about 6 months in favorable conditions (releasing CO_2 into the atmosphere). However, they could last much longer if the polymers are protected or embedded in relatively stable long-life products (e.g., composites). For example, as a finished product, PPC could have a very long life cycle, thus, extending the permanence of CO_2 storage.

9.3.3.1 Current Technological Status

Although the CO_2–epoxide copolymerization to produce polycarbonates was discovered in 1969, the process is still in a pilot scale of development. Since 2009, Novomer Ltd is operating a pilot scale (a 1,500 L batch reactor) production of CO_2-based polycarbonates in Rochester (USA). The developers of the technology claim to come up with novel proprietary Zn-based catalysts that allow conducting the process at low temperature and pressure. In 2010, Novomer partnered with Praxair to scale-up the process [15]. According to estimates, the CO_2 utilization potential of the polymer processing approach is about 22.5 Mt CO_2; for comparison, current global markets for polyethylene and polypropylene are approximately 80 Mt and 45 Mt, respectively, thus, creating drivers for entering the polymer market [1]. A start-up company Newlight Technologies (USA) is using CO_2 and methane as the input for its polyhydroxyalkanoate (PHA) resins [16]. The company announced that it has more than 50 t of the product annual capacity at their fermentation-based demonstration facility in Irvine, California.

9.3.3.2 Challenges

The wide-scale deployment of the technology as a carbon abatement option faces the following challenges:

- In general, the permanence of CO_2 storage is not great and depends on a particular material and application; e.g., under the right compost conditions, degradation of polycarbonates could occur in about 6 months. Thus, increasing chemical resistance of the materials to degradation is an important objective.
- Due to great chemical stability of CO_2 molecule, the CO_2-conversion processes, in general, are energy intensive and not selective; thus, more research is necessary to develop efficient catalysts that could reduce the energy activation of the reactions and improve their selectivity.
- Depending on the CO_2 source, additional purification and polishing may be required to meet the process feedstock specifications, which will add to the cost of the polymer product.
- The technology is still at relatively early (pilot) stage. To date, there are no commercial incentives for this technology since it is not likely that CO_2-derived plastics would be sold at a lower price than existing plastics, nor it is likely that consumers would prefer the polycarbonate polymers (as being superior) over existing petroleum-based products. The reason for that is that today, the main market target for these materials is the packaging industry, which is a low-end application where the product acceptance will be entirely driven by cost.

9.3.4 *Mineral Carbonation*

Mineral carbonation (also called "carbonate mineralization" or "mineral sequestration," or "enhanced weathering") encompasses a diverse range of processes and applications involving the use of exhaust gas CO_2 to transform naturally occurring minerals and industrial wastes into value-added products and materials. Since CO_2 is consumed in the process, mineral carbonation enables CO_2 sequestration without costly capital and energy-intensive CO_2 capture and separation processes; it also avoids the cost and challenges of supercritical CO_2 pipeline transport and storage infrastructure [17]. The energy and materials intensive industries such as iron and steel, glass, cement manufacturing, and minerals mining could particularly profit from mineral carbonation; it can also provide low-cost CO_2 capture materials as a scalable solution for a wide range of industrial and power generation CCS markets.

The mineral carbonation process variants can range in scale and complexity from simply sprinkling crushed rock particles onto large areas to soak up atmospheric CO_2, to rather sophisticated chemical processes designed to produce pure metals, ceramics, silica, calcium and magnesium oxides, and carbonates [17]. This relatively new approach is based on the reaction of CO_2 with metal oxides or complex metal oxides (e.g., silicates of alkaline-earth metals) to produce corresponding insoluble solid carbonates and silica. Of a particular practical interest are oxides and silicates of Mg and Ca, since they are widely occurring in nature and produce insoluble stable carbonates. The mineral carbonation approach is considered by many as a prospective beneficial CO_2 reuse technology because carbonates produced are stable and ecologically safe over long time scales and can be used in a variety of applications such as construction, mine reclamation, etc. Alternatively, these materials can be disposed of without need for long-term monitoring or concerns over possible CO_2 leakage, that would pose health, safety, or environmental risks [1]. Figure 9.4 provides a sketch of the general concept of mineral carbonation, where carbonate minerals produced are used in a variety of industrial applications.

9.3.4.1 Technology Status

Currently, the mineral carbonation technology is in pilot and demonstration stages of the development. Due to slow kinetics of carbonation reactions, current R&D activities are mostly focused on increasing the reaction rates commensurate with the rate of CO_2 supply from industrial sources, and increasing the energy efficiency of the mineralization process. Two main variations of the mineral carbonation technology involve the use of natural rock silicates and industrial mineral waste (e.g., fly ash from power plants, wastewater/brine).

The first approach involves the utilization of abundant magnesium silicates (serpentine, olivine), and it is still in the research phase of development. In particular, a gas/solid mineral carbonation system utilizing a fluidized bed reactor is being researched in Finland [1]. Absorption of 1 t of CO_2 using this method would require 3 t of serpentine [18]. The main technological challenges are concerned with the efficient extraction of MgO (which is the reactive component) from the mineral, and improvements in the carbonation reaction kinetics. Other limiting factors include additional energy intensive operations of mining, crushing, milling and transporting the minerals to a processing plant. These factors may hinder the commercial realization of the silicate mineral carbonation approach.

The carbonate mineralization of industrial wastes is a much more technologically advanced option (compared to natural silicates), and is currently practiced at a pilot scale by at least two companies: Calera and Skyonic Corp. (both in USA).

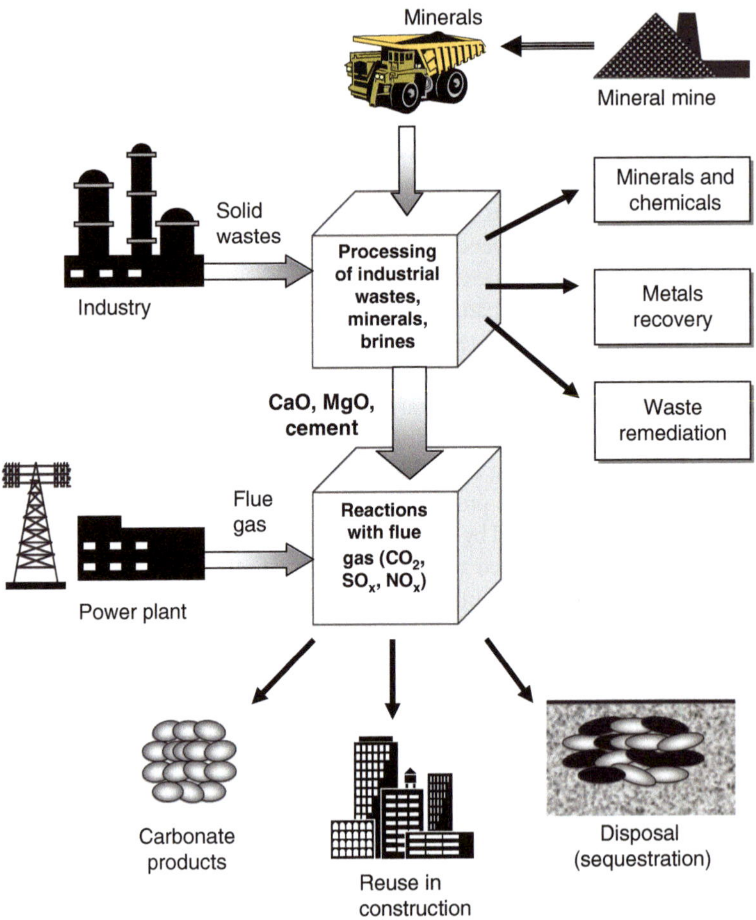

Fig. 9.4 Scheme of the mineral carbonation concept. *Source*: [17]

In the Calera's SMAP (Carbonate Mineralization by Aqueous Precipitation) process, CO_2 emissions from a power plant (flue gas) are chemically combined with water/brine and fly ash (as the source of alkalinity) to produce of cementitious materials, aggregates, and other building and construction materials. Absorption of 1 t of CO_2 using this process would require about 1 t of brine or alkalinity source (NaOH) and fly ash. Thus, each ton of cement or other carbonated product contains half ton of mineralized CO_2. Currently, Calera Corp. is running a continuously operating pilot-scale plant in Moss Landing (California) with the capacity of 5 t/day of supplementary cementitious material [1]. A demonstration plant is currently under construction, which will use a slipstream from the 1.5 GW Dynergy Moss Landing gas-fired power plant as a CO_2 source. Recently, Calera Corp. and Bechtel Power Corp. have formed a strategic alliance to deploy Calera's technology worldwide.

Skyonic Corp. (Texas) is developing SkyMine® carbon mineralization process, which removes CO_2 from industrial waste streams through cogeneration of carbonate and/or bicarbonate (baking soda) materials; the latter will be sold in the market along with other byproducts of the process. In collaboration with ConocoPhilips, BP, and others, the company has initiated Capitol-SkyMine demonstration facility at Capitol Aggregates, Ltd, cement plant in San Antonio, Texas, targeting capture of 75,000 t of CO_2 from flue gas and converting it to 143,000 t of baking soda ($NaHCO_3$) (which will be tested in several industrial applications, including feedstock for algae-based fuels) [1]. The SkyMine® technology is effective with low CO_2 concentrations in the flue gas, making it suitable for natural gas plants [19]. It efficiently scrubs SO_x, NO_x, mercury, and metals from the flue gas of coal-fired utilities with the cleanup efficiency of 99 %.

The *Olivine Foundation* (Netherlands) advocates a relatively inexpensive approach copied from nature: they utilize olivine to take up CO_2 from atmosphere by contacting the crushed mineral with water and air [20]:

$$Mg_2SiO_4 + 4CO_2 + 4H_2O \rightarrow 2Mg^{2+} + 4HCO_3^- + H_4SiO_4 \tag{9.4}$$

One ton of olivine captures 1.25 t of CO_2. The developers of the technology (referred to as *Smart Stones* technology) propose to spread the olivine particles as a replacement for sand in large areas (sport fields, along motorways, dikes, beaches, sandpits, etc.), or for soil enrichment, road construction, roofing materials, etc.

Cambridge Carbon Capture Ltd. (UK) is developing a low-cost chemical digestion process to extract amorphous silica and pure calcium and magnesium oxides from olivines and steel slugs [17]. The obtained CaO and MgO can be used as chemical sorbents for CO_2 capture from exhaust gases and zero- or low-carbon substitutes for the conventional CO_2-intensive lime in cement industry. Carbonate products of sufficient purity may find high volume applications as pigments, paper fillers, while silica can be used in glass, electronics, construction, and plastic industries. *Carbon Sciences* of Santa Barbara (California) uses flue gas and Ca, Mg-rich wastewater from mining operations to produce cement.

9.3.4.2 Markets

Potential markets for the products produced by mineral carbonation are (1) mine reclamation, (2) building and construction materials—aggregates, and (3) the supplant portion of cement. The cement and aggregates produced via mineral carbonation could be marketed as alternatives to traditional Portland cement and building aggregates. According to Calera Corp. estimates, their products can be sold competitively in the current market and replace 1.5 billion tons of Portland cement and another 30 billion tons of aggregates used in concrete, asphalt, and road construction [1].

9.3.4.3 Challenges and Barriers

Mineral carbonation as CO_2 reuse technology is facing several limitations and barriers as follows [1]:

- The immense quantities of natural rocks have to be handled (the mass of silicate rocks to store CO_2 will be over eight times larger than the mass of coal used to produce that amount of CO_2). The mining operation will be of a magnitude similar to that of coal industry [21].
- Environmental consequences of large-scale mining operations and disposal of carbonates are mostly unknown.
- High energy intensity of the mining operation and carbonation process itself.
- The resources of some silicate ores (e.g., Mg-containing ores) that can be technically exploited may be limited.
- Geographical constraints to the technology, since the carbonation process would need to be arranged in a close proximity to a mine.

The challenges facing the Calera and similar reuse technologies utilizing industrial wastes are:

- Currently, there are no significant commercial benefits from the realization of the technology since it is unlikely to produce the products that will be superior to existing products in the market (e.g., Portland cement), and, for this reason, it may be rejected by the cement industry. Therefore, the commercialization of the technology will be largely driven by environmental incentives (carbon price/tax, carbon trading, carbon regulations, etc.).
- The implementation of the technology depends on the availability of suitable subsurface brine in very large volumes to provide the requisite minerals content and alkalinity. In lieu of natural alkalinity, it has to be provided by manufactured alkalinity-enhancing agents (e.g., NaOH), the economical viability of which still needs to be demonstrated for large-scale systems.

9.3.5 Use of CO_2 for Concrete Curing

Cement manufacturing is one the most carbon-intensive industrial processes. Several emerging technologies focus on reducing CO_2 emissions from conventional Portland cement manufacturing. For example, such companies as Novacem (UK), TecEco, and Calix (Australia) are working on new technologies that could eliminate or significantly reduce CO_2 emissions (that would otherwise be emitted during conventional Portland cement production process), and/or absorb CO_2 from the atmosphere during the cement curing process [1].

Carbon Sense Solutions (CSS) from Halifax (Nova Scotia, Canada) is aiming at using CO_2 to limit the need for heat and steam curing of precast concrete products. In this process, CO_2 from combustion sources (e.g., flue gases) cures concrete products, delivering the same performance as the traditional energy-intensive steam curing process. The process sequesters up to 120 kg of CO_2/t of precast concrete.

Currently, the technology is tested at full-scale demonstration plant. The main market drivers for this technology is the price and demand for concrete and the existence of environmental incentives (carbon price, trade, etc.). Similar to mineral carbonation, the concrete curing process presents permanent storage of CO_2 for centuries in the form of concrete products.

9.3.6 CO_2 Use in Bauxite Residue Carbonation

The extraction of alumina from bauxite ore according to the Bayer process produces a highly alkaline (pH 13–13.5) residue slurry, referred to as "red mud." The alkaline bauxite residue can capture CO_2 from concentrated streams, thus, locking CO_2 in carbonate products and reducing the pH of the slurry (via a neutralization reaction) to a less hazardous level (pH ~ 10). This technology has been operated by the Alcoa company at its Kwinana plant for the last several years [1]. The plant uses CO_2 pipelined from a nearby ammonia plant at the rate of 70,000 t/year CO_2; as a result, the entire Kwinana's residue byproduct (2–2.5 Mt/year) is converted to carbonated products. The theoretical CO_2-locking capacity of the red mud (typically treated with seawater) is very high—750 kg CO_2/t of red mud; however, from practical viewpoint, Alcoa limits it to 30–35 kg CO_2/t of red mud.

There is a large market for implementing the technology in aluminum plants around the world. Worldwide, over 70 million tons of bauxite residues (dry basis) are generated, and more that 200 million tons have already accumulated and mostly stored in tailing ponds. The potential environmental benefits include neutralization of the existing excess stores of highly alkaline bauxite residue and reduction in red mud dust (currently, environmental hazard). The neutralized bauxite residue could potentially be used as road base, building materials, or soil amendment on acidic soils. In all likelihood, the treated bauxite residue will have a limited commercial value (and even could be offered free), although economic benefit will arise from the costs savings due to negating need for handling the unwanted hazardous residue.

Technical challenges facing the technology include (1) a requirement for relatively high purity and high-pressure CO_2, (2) a need for local source of CO_2 (to make the process commercially viable), (3) no or very small revenues from byproduct, and (4) relatively low CO_2 storage capacity. The permanence of CO_2 storage depends on the final product: if CO_2 is converted to carbonates, it is permanently sequestered; if bicarbonates are formed, CO_2 will be partially released to the atmosphere (due to its conversion to carbonate).

9.3.7 CO_2 Conversion to Fuels

Conversion of CO_2 to fuels (especially, liquid transportation-ready fuels) is an extremely attractive idea that has been actively pursued by researchers for many years with efforts dramatically intensified during last couple decades.

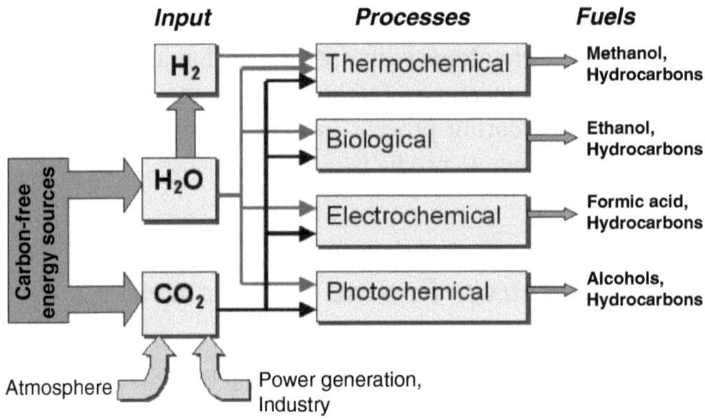

Fig. 9.5 Technological routes to CO_2 conversion to alternative fuels

The utilization of CO_2 as a feedstock for alternative fuels production is a rather broad category of the CO_2 reuse technologies that include the production of such fuels as methane, methanol, ethanol, dimethyl ether, petroleum-equivalent fuels (PEF, i.e., liquid hydrocarbons in gasoline-diesel range), formic acid, and others.

From a carbon mitigation viewpoint, the primary energy input to the CO_2-to-fuels conversion processes has to be from zero- or low-carbon energy sources such as nuclear, renewables, or NG-CCS, otherwise, the large quantities of CO_2 may be produced in the process, thus, defeating the purpose. This is a very important requirement, because due to relative chemical inertness of CO_2 these processes, in general, are energy intensive, and, in many cases, have relatively low-energy conversion efficiency (i.e., the ratio of chemical energy of fuel to the total energy input). Hydrogen (if it is required) for fuels production also has to come from non-fossil sources (e.g., water electrolysis powered by nuclear, solar, or wind sources). If this approach is successful, CO_2 (including atmospheric CO_2) would become a practically infinite source for the production of alternative fuels. Figure 9.5 shows a conceptual diagram of CO_2 conversion to fuels via thermochemical, electrochemical, biological, and photocatalytic routes.

Despite daunting technical challenges, certain scientific and technical advancements have been achieved in the conversion of CO_2 to fuels. Several technological routes to conversion of CO_2 to liquid fuels are currently under consideration, including

- Catalytic conversion of CO_2 to methanol and dimethyl ether
- Electrocatalytic reduction of CO_2 to methane, alcohols, formic acid, and other products
- Hydrocarbon production by hydrocarbon-excreting microorganisms (helioculture)
- Bioorgano-catalytic direct CO_2 conversion to hydrocarbons
- High-temperature thermochemical metal oxide-mediated CO_2–H_2O conversion to liquid fuels

- Photocatalytic reduction of CO_2 to fuels in aqueous solutions
- Reduction of CO_2 to CO followed by catalytic hydrogenation of CO to fuels
- Direct catalytic hydrogenation of CO_2 to liquid hydrocarbons
- Artificial photosynthesis

While some of the above technologies are moving from laboratory toward pilot units most of them are still in an early R&D stage.

9.3.7.1 Conversion of CO_2 to Methanol

Methanol is a major product of chemical industry: 32 million tons of methanol is produced worldwide. Currently, methanol is manufactured commercially by passing H_2–CO–CO_2 mixture over alumina-supported Cu–Zn catalyst at 250–300 °C and elevated pressure of 50–100 atm. Typically, the H_2–CO–CO_2 mixture is produced by reforming of NG or gasification of coal, and, as such, the process cannot be considered CO_2 abatement technology (because CO_2 is produced from the same source as H_2 and CO). In order to function as a CO_2 sink, methanol has to be produced from an external source of CO_2 (e.g., power plants or the atmosphere) and H_2 has to produced from water and carbon-free energy source (e.g., via water electrolysis using carbon-free electricity):

$$CO_2 + 3H_2 \rightarrow CH_3OH + H_2O, \quad \Delta H^\circ = -90.7\,\text{kJ/mol} \tag{9.5}$$

The idea of converting atmospheric CO_2 to methanol as a carbon mitigation strategy has been floating in literature for many decades. In his 1980 book "Energy Options," Bockris outlined the concept of the *Methanol Economy*, in which methanol fuel would play a major role between fossil and carbon-free (nuclear, solar) eras [22]. According to the concept, methanol is to be produced from CO_2 captured from the atmosphere and hydrogen will be generated from water via nuclear- or solar-powered electrolysis. The 2006 book by George Olah (Nobel laureate in chemistry) on the prospects of the Methanol Economy also emphasized the importance of producing methanol from atmospheric CO_2 and carbon-free hydrogen [23].

Technological advancements in this area vary widely from fundamental research to pilot scale systems to, at least one, demonstration plant. Icelandic-American company *Carbon Recycling International* (*CRI*) has developed the renewable methanol production process and is in the process of its commercialization. In this process, hydrogen is produced by water electrolysis powered by a geothermal power station. The concentrated stream of CO_2 (captured by a conventional amine-based capture method) is compressed to 5 MPa and introduced (together with H_2) into a reactor at 225 °C to produce methanol [1]. The overall thermal efficiency of the catalytic process is estimated at 75 %.

Currently, CRI is constructing a commercial demonstration plant with the capacity of 4.2 million liters of methanol in Iceland. The plant will be located in a proximity of 76.5 MW Svartsengi Geothermal Power Station, which will be the source of both power and CO_2 for the methanol production process. Note that the carbon emission intensity of Iceland's grid is very low −310 kg CO_2/MWh (compared to world average of 500 kg CO_2/MWh), and if power for renewable methanol production will be solely supplied by the geothermal plant, the emissions intensity would further drop to 171 kg CO_2/MWh. Methanol will be blended with conventional unleaded gasoline and sold at Olis gas stations in the greater Reykjavik area. The major motivation for building a methanol plant in Iceland is that in this country gasoline/electricity price ratio is one of the highest in the world. The major engineering challenge for making CO_2-derived fuels a commercial reality is scaling up the process while making it cost-effective. Methanol could be further converted to dimethyl ether (DME) or gasoline according a commercial methanol-to-gasoline Mobile-process.

9.3.7.2 Conversion of CO_2 to Methane

Conversion of CO_2 to methane via catalytic hydrogenation reaction is a well-known process called *methanation* or *Sabatier process* (see also Chap. 6). The Sabatier process involves the reaction of H_2 with CO_2 at the preferred temperature range of 300–400 °C and elevated pressures in the presence of transition metal catalysts to produce methane and water:

$$CO_2 + 4H_2 \rightarrow CH_4 + 2H_2O_{liq}, \quad \Delta H^\circ = -253.1\ \text{kJ/mol} \tag{9.6}$$

As seen from the above equation, this reaction presents a reverse steam methane reforming reaction, and it is highly exothermic. Among the transition metals catalyzing the Sabatier reaction, Ni and Ru are the most catalytically active metals (Ru shows higher activity than Ni, but it is more expensive than Ni catalyst, which makes the latter a preferred catalyst for this reaction). The methanation process is commonly used in the purification of hydrogen streams of CO_x impurities.

Currently, no large-scale industrial facilities solely dedicated to conversion of CO_2 to methane via the Sabatier process exist anywhere in the world. (Note that there are several commercial applications of the methanation reaction, but they involve methanation of CO–CO_2 mixture to synthetic NG, e.g., in the Great Plains Synfuels Plant in South Dakota, USA [24].) There are several R&D and demonstration projects pursuing the applications of the Sabatier process to conversion of CO_2 emissions from power plants to methane as a carbon mitigation approach. This approach, however, requires large quantities of cost-effectively produced "green" (i.e., non-fossil fuel based) hydrogen, which is still a challenging problem.

Considerable R&D work has been conducted by the RCO2 company in Langhus (Norway) and by the Desert Research Institute (DRI) in Reno (Nevada, USA) on the improvement of the Sabatier process efficiency and moving the process toward commercialization [25]. Pilot-scale units have been constructed and tested using a simulated flue gas from NG combustion. Both RCO2 and DRI projects are designed to use H_2 produced by water electrolysis using non-carbon electricity: hydropower and solar/wind, respectively. CO_2 contained in the flue gas was converted to methane at 350 °C and space velocity of more than 15,000 h^{-1} with the conversion efficiency of 98 % [25]. The objective of both projects is to develop a commercial process in which methane will be recycled back to the inlet of gas turbines (where it will be mixed with the make-up gas) of the power plant, thus, addressing GHG emissions problem. In another development (so-called, Power-to-Gas Platform), German researchers utilize Sabatier reaction as part of the scheme for storing intermittent renewable energy in the form of methane that is injected into natural gas grid (see details in Sect. 8.1.2.2).

9.3.7.3 Conversion of CO_2 to Liquid Hydrocarbons

Direct conversion of CO_2 to liquid hydrocarbons is an attractive approach to production of synthetic alternatives to traditional petroleum-based fuels (e.g., gasoline, jet, and diesel fuels) because of their high energy density and existing infrastructure.

$$2nCO_2 + (m + 4n)H_2 \rightarrow 2C_nH_m + 4nH_2O \tag{9.7}$$

where C_nH_m is a liquid hydrocarbon ($n > 5$).

This technological route, however, faces technical challenges associated with unfavorable thermodynamics of the reaction and, consequently, low products yields; it would require more fundamental research, especially, in the catalyst development area [26]. It has been reported that direct hydrogenation of CO_2 to hydrocarbons over Co–Pt/Al_2O_3 yielded liquid hydrocarbons [27]. Direct catalytic hydrogenation of CO_2 to liquid hydrocarbons (C5+) over K-promoted iron catalysts (Fe–K/Al_2O_3) in fluidized bed and slurry reactors was studied by Kim et al. [26].

Indirect conversion of CO_2 to liquid hydrocarbons consists of two stages: reduction of CO_2 to CO followed by catalytic hydrogenation of CO to fuels via Fischer-Tropsch synthesis:

$$CO_2 + 2H_2O \rightarrow CO + 2H_2 + 1.5O_2 \tag{9.8}$$

$$CO + 2H_2 \rightarrow 1/n(CH_2)_n + H_2O \tag{9.9}$$

where $(CH_2)_n$ represents liquid hydrocarbon

Researchers at Sandia National Laboratory (New Mexico, USA) and ETH university in Zurich (Switzerland) use a solar concentrator that provides high-temperature heat to break down CO_2 and water with production of syngas, which could be further converted to "green" gasoline via reaction (9.9) [28, 29]. Photothermal dissociation of CO_2 to CO and O_2 using a concentrated solar radiation source has been demonstrated [30].

9.3.7.4 CO_2 Conversion Through Artificial Photosynthesis

For many years, researchers have been striving for mimicking Mother Nature and developing "artificial photosynthesis" or bio-inspired systems, which involve solar-driven catalytic reduction of CO_2 and water to a variety of compounds and fuels such as alcohols, organic acids, hydrocarbons, which are usually derived from petroleum. One example of a such reaction (producing formic acid) is shown below:

$$CO_2 + H_2O + \text{solar photons} \rightarrow HCO_2\,H + 0.5O_2 \qquad (9.10)$$

Princeton University spin-off company Liquid Light claims it can produce more than 20 different fuel products from CO_2 using this approach [29]. In particular, the researchers showed the production of range of C1–C4 compounds including formic and oxalic acids and synthesis gas [31]. Solar light-induced photoelectrochemical reduction of CO_2 to CO and other products have been demonstrated recently by a number of research groups [32, 33]. Anpo reported on photoelectrochemical reduction of CO_2 in presence of water to CH_4 and methanol on dispersed TiO_2 photocatalyst as a model of artificial photosynthesis, according to the following generic chemical equation [34]. The work on the artificial photosynthesis systems is still in an early R&D stage.

9.3.7.5 Electroreduction of CO_2 to Fuels

Since all the known systems for the direct conversion of CO_2 to fuels using solar radiation suffer from very low solar conversion efficiencies, there have been attempts to bypass some of the constraints imposed by the extremely complex natural photosynthetic process by turning to electrochemical and electrocatalytic systems. Stanford University researchers have developed electrochemical cell consisting of copper electrodes immersed in an electrolyte solution. When stream of CO_2 flowed through the cell under the applied voltage, a broad range of simple hydrocarbons and oxygenated compounds was formed: total of 16 products were detected, including methane, ethylene, formate, ethanol, CO, ethylene glycole, glycolaldehyde, hydroxyacetone, and glyoxal [35]. At the University of Delaware (USA), researchers used Bi-modified electrodes for electroreduction of CO_2 to CO with relatively high efficiency [36]. Researchers at University of Messina (Italy) have developed gas-phase electrochemical system based on carbon nanotubes-supported iron electrode. The catalytic cell was optimized to produce isopropyl alcohol, but it could also convert CO_2 to C1–C8 hydrocarbons [31]. The researchers hope that using iron, instead of traditionally used Pt catalyst, would help to develop a low-cost system for CO_2 conversion to value-added products.

University of Illinois (USA) researchers in collaboration with researchers at Dioxide Materials Co. have developed novel catalytic system based on an ionic liquid medium that they claim reduces the energy requirements (due to a lower overpotential for CO_2 electroreduction) [37]. DNV company of Høvik (Norway) has recently developed a trailer-sized pilot electrochemical reactor for converting CO_2 to formic acid and related compounds [31]. To make the technology useful for commercial applications, the system throughput and CO_2 conversion rate should be dramatically increased.

More detailed discussion of catalytic, photoelectrochemical, electrocatalytic, and bio-mimicry options for CO_2 conversion to different fuels could be found in the following review papers and book chapters [28, 38–44].

9.3.8 CO_2 Conversion to Chemicals and Value-Added Products

There has been a practical interest in using CO_2 as a building block in synthesizing various oxygen-rich chemical compounds and value-added products [40, 41]. Besides the applications where CO_2 is utilized as a monomeric building block for production of polymers (see Sect. 9.2.2), there are many examples of using CO_2 in synthesis of a variety of practically important oxygenated compounds via selective carboxylation reactions at relatively mild conditions. For example, it was shown that the members of propionic acid family could be synthesized from CO_2 and terminal alkynes via the carboxylation reaction in the presence of Cu-based catalysts [43]. Cyclic carbonates were synthesized using CO_2 as a feedstock with very high selectivity (almost 100 %) at room temperature [44]. A practically important example of CO_2 utilization include the production of arylcarboxylic acids by carboxylation of aromatic compounds with CO_2 at mild conditions (20–80 °C) [45].

The reaction of CO_2 with methane to produce acetic acid is another example of a synthesis of an important chemical product from CO_2, because it is an intermediate in the production of a variety of pharmaceuticals, polymers, etc.:

$$CH_4 + CO_2 \rightarrow CH_3COOH \tag{9.11}$$

LanzaTech, a producer of low-carbon fuels and chemicals from waste gases, and PETRONAS, the national oil company of Malaysia, are working together to accelerate the development and commercialization of technologies to produce sustainable chemicals (e.g., acetic acid) from carbon dioxide and NG. The driving force behind the collaboration is to create technology that not only complies with the emissions reduction requirements, but also generates revenue along the way [46]. The electrochemical reduction of CO_2 to ethylene has been reported by Ogura [47]. For the readers interested in production of chemicals and value-added products from CO_2 we would recommend the following reviews and books (e.g., [40, 41, 48]).

9.4 CO_2 Use in Algal Systems

The unique capacity of a photoautotrophic organism, algae, to capture CO_2 from both the atmosphere and industrial sources and transform it to biomass that can be further processed into transportation fuels, foodstuff, pharmaceuticals, and chemical products recently attracted an enormous interest. The R&D and commercial efforts in this area are receiving an ever increasing financial support from governments and entrepreneurs worldwide. Of particular interest is the potential of algae to produce petroleum-equivalent products at a price that is competitive with those produced from crude oil. The growth and processing of algae as a CO_2 utilization approach could potentially reduce CO_2 emissions by up to 95 % versus fossil fuels [1]. The following is a discussion of CO_2 utilization in algal systems consisting of two major steps: CO_2 conversion to algae followed by the algae conversion to fuels and other value-added products.

9.4.1 Status of CO_2-to-Algae Technology

Three types of light-harvesting algal species present interest from the viewpoint of CO_2 utilization: cyanobacteria, microalgae, and macroalgae.

Cyanobacteria are prolific carbohydrate and secondary metabolite producers (can double in yield in less than 10 h), but, generally, they do not accumulate storage lipids. Many strains of cyanobacteria can produce hydrogen, and some could be genetically manipulated making them attractive for biofuels production [49]. Despite recent progress in cyanobacteria research, a comprehensive understanding of their metabolism and regulation is not yet available, which makes it difficult to apply them to biofuel production systems.

Microalgae represent highly diverse unicellular eukaryotic species—the product of over three billion years of evolution [49, 50]. *Chlamydomonas reinhardtii*, *Chlorella*, *Dunaliella salina*, and diatoms are well-studied microalgae species. Current research efforts are concerned with the identification of species that are adept at making fuel precursors and have high productivity under various environmental conditions.

Macroalgae, or seaweeds, represent the broad and diverse group of eukaryotic photosynthetic aquatic organisms that are abundant in the world's oceans and coastal waters. In contrast to microalgae, they are multicellular and feature plant-like structural characteristics. Historically, macroalgae are classified as *Phaeophyta* (brown algae), *Chlorophyta* (green algae), and *Rhodophyta* (red algae). Typically, macroalgae have low lipid content but are high in carbohydrates. As such, they can be efficiently converted to various fuels.

Most of the discussion in this chapter will primarily be concerned with microalgae, as they show the most promise from the viewpoint of biofuels production, and, currently, it is the most technologically advanced area of algae-related activities worldwide.

9.4.1.1 Algae as a Carbon Sink

Algae are one of the most abundant and highly adapted forms of life on the Earth forming the foundation of food chains and playing an important role in the carbon cycle. Many algae are photosynthetic organisms capable of harvesting sunlight and converting CO_2 and water to oxygen and macromolecules such as carbohydrates, lipids, and proteins. Certain microalgal species naturally accumulate large amounts of triacylglycerols (up to 60 % of dry weight), whereas, macroalgae and cyanobacteria mostly accumulate carbohydrates, with lipid content being typically less than 5 % (on dry weight basis), although some species demonstrated lipid concentrations up to 20 % [51, 52].

The efficiency of the photosynthesis process is a critical parameter related to the productivity of algae, since it would ultimately affect the yields of CO_2 capture and biofuels production. The theoretical algal biomass productivity yield was estimated at about 200 g/m^2/day [53]. (However, currently, there is no consensus on the true maximum productivity of algae.) Many reviews cover basics of algal photosynthetic processes (e.g., [54]).

Despite the recent technological progress in maximizing light exposure of the algae cultures, light utilization efficiency still remains one of the limiting factors in the commercial development of algae utilization systems. In large-scale high cell density systems, the substantial portion of incident light is not efficiently utilized because the cells nearer to the light source tend to absorb practically all the incoming light, thus, depriving more distant cells of light [53]. On the other hand, when an algal culture is exposed to high-intensity light, the algal photosystem has built-in mechanism to prevent the over-absorption of light energy (which could potentially lead to an oxidative damage) by dissipating most of the absorbed incident light in the form of heat (thus, this portion of light could be considered as wasted). Photoinhibition due to "light damage" could occur under a variety of light regimes, which could result in the reduction in photosynthesis efficiency [49]. There are several strategies to overcome this problem and increase the efficiency of light utilization, and reducing the size of the chlorophyll antenna is one of them [55]. The better understanding of the dynamics and regulation of the algal photosynthetic apparatus is important from the viewpoint of better engineering systems for the efficient utilization of light energy and algae biomass production.

In contrast to terrestrial plants that directly uptake CO_2 from air, microalgae can utilize CO_2 from several sources: atmospheric CO_2, industrial streams (flue gases), and soluble carbonate solutions [56]. Upon dissolution in water, CO_2 exists in aqueous solutions in the form of dissolved CO_2 in equilibrium with bicarbonate HCO_3^- and carbonate CO_3^{2-} ions. This dynamic equilibrium could be affected by the changes in the nutrient concentrations, pH of the aquatic system, and other factors that influence the function of *Rubisco* (*ribulose-1,5-biphosphate carboxylases/oxygenase*), which is a key enzyme participating in CO_2 fixation.

Numerous types of microalgal strains have biomass productivity superior to that of terrestrial plants (trees, grasses, etc.) by at least one order of magnitude. Moreover, the utilization of microalgae allows avoiding many challenges associated with terrestrial biomass, in particular, through the use of non-productive land area and non-potable water for cultivation. To achieve high CO_2 uptake rates (and, hence, biomass production rates) microalgae has to be grown under the optimal conditions of light, temperature, pH, nutrient, and CO_2 concentrations. The photosynthesis-based cellular growth occurs according to the following generic equation [57]:

Table 9.1 CO_2 utilization capacity of selected microalgal strains using different CO_2 sources at near ambient conditions

Microalgae strains	CO_2 concentration in gas stream (vol.%)	CO_2 utilization capacity (gCO_2/L/day)
Haematococcus pluvialis	16–34	0.143
Scenedesmus obliquus	air	0.031
Scenedesmus obliquus	18	0.260
Chlorella vulgaris	15	0.624
Spirulina sp.	12	0.413

Source: [56]

$$106\,CO_2 + 16\,NO_3^- + H_2PO_4^- + 122H_2O + 17H^+ \rightarrow (C_{106}H_{263}O_{110}N_{16}P) + 138O_2 \quad (9.12)$$

The stoichiometry of reaction (9.12) indicates that for every 106 mol of CO_2, 16 mol of nitrate and 1 mol of phosphate are consumed (as nutrients).

During the evolution, microalgae adapted to the low concentration of atmospheric CO_2 (about 0.03 vol.%), therefore, the productivity of algal systems could be significantly below the values needed for their application in the industrial CO_2 capture and sequestration systems. From this viewpoint, novel and advanced microalgae species have to be developed (e.g., through genetic engineering) to accommodate to much higher concentrations of CO_2 typical of power plant emissions (e.g., CO_2 concentrations of 5 vol.% or even higher).

The selection of suitable microalgae strains for CO_2 biomitigation could have a significant impact on the process efficacy and economics. Among desired attributes for algae as a carbon sink are (a) high growth rate, (b) ease of harvesting and downstream processing, (c) tolerance to changes in temperature, (d) high tolerance to impurities in CO_2 streams (e.g., SO_x and NO_x in flue gases), (e) the possibility of producing multiple value-added products, (f) possibility to use the strains in conjunction with wastewater treatment, etc. [56]. No single algal strain is known to satisfy all the above requirements; however, many experimental observations point to the potential of microalgae for carbon biofixation under various conditions. The CO_2 utilization capacities of selected microalgal strains using different CO_2 sources are shown in Table 9.1.

The reported findings indicate that *Chlorella vulgaris* grown on wastewater discharged from a steel plant and using gaseous stream with 15 vol.% CO_2 sequestered 0.624 g CO_2/L/day [56]. Velea et al. have screened 35 different microalgal strains for their growth rate and the potential for the biofixation of CO_2 from emissions of a typical coal-fired power plant [58]. The authors demonstrated that the selected microalgal strains *Chlorella* sp, *Scenedesmus* sp, and *Chlorobotrus* sp had the productivity of about 1 kg/day of *Chlorella* algal biomass obtained from 1.5 kg of CO_2 consumed by the algal strain. It was concluded that the studied algal system could be a basis for the development of a CO_2 biofixation system with the production of value-added products: horticultural oils, proteins, and carbohydrates.

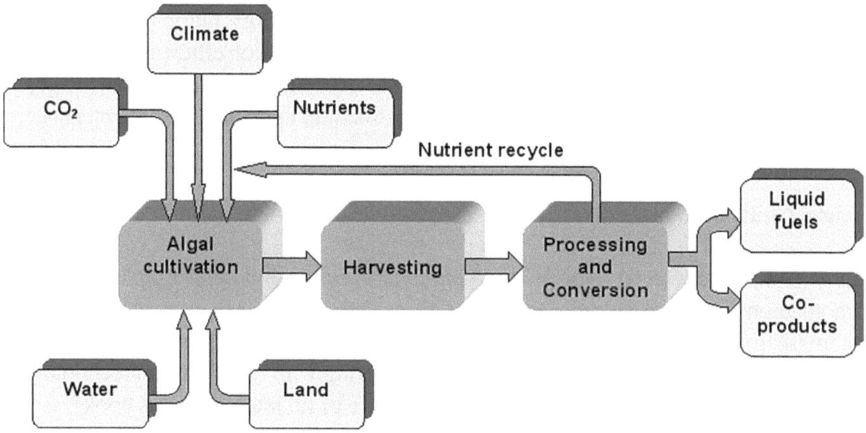

Fig. 9.6 Main resource elements for algal biofuel production. *Source*: [49]

The technological development and scaling-up of algal biofuels production involves several stages:

- Selection of suitable microalgae species with high productivity and tolerance to environmental variations (temperature, pressure, pH, CO_2 concentration) and impurities in CO_2 stream
- Development of an efficient photo-bioreactor and a light source (if using artificial light)
- Cultivation, collection, and handling of biomass products
- Conversion algae biomass to desirable products (fuels, food, chemicals).

9.4.1.2 Algae Cultivation and Processing

Critical requirements for the large-scale production of algal biomass include the following factors:

- Suitable climate (solar radiation, temperature, precipitation, evaporation)
- Suitable land and location
- Available water resources
- CO_2 and nutrients supply
- Possible ecological impact
- Local socioeconomic factors and environmental regulations

These requirements must be appropriately aligned in terms of their geo-location, characteristics, availability, and affordability, as well as a variety of socioeconomic, public perception, and environmental factors [49]. Figure 9.6 provides a simplified high-level overview and interplay of the major resource and environmental parameters affecting algae growth and biofuels production.

As seen in Fig. 9.6, the inputs of climate, water, CO_2, energy, nutrients, and land are critical from the viewpoint of siting, facilities design, production efficiency, and process economics. Additional resources include capital equipment, materials and reagents, labor, and other inputs associated with the infrastructure, operation, and maintenance.

Algae Cultivation

In general, algae can be cultivated via *photoautotrophic* and *heterotrophic* methods. Photoautotrophic algae require light and CO_2 as carbon source to grow and create new biomass. In contrast, heterotrophic algae can grow without light, and are typically fed sugars (not CO_2) as a carbon source, thus, competing with other biofuel technologies for the feedstocks (as such, heterotrophic algae are of no interest to this book).

Depending on the specific type of algae and technological approach used, algae can be cultivated in either "open" or "closed" systems. Open systems are technologically simple and inexpensive, because, in most cases, they represent shallow ponds where algae, water, and nutrients circulate using a large paddlewheel (Fig. 9.7a).

The shallow design of the pond is dictated by the need to keep algae exposed to sunlight (besides, evaporative cooling maintains temperature). Typically, several ponds are combined in an "algae farm," the productivity of which is measured in terms of biomass weight produced per day per unit of available surface (or land) area. (Note that in this case, the productivity is a function of the surface area, as opposed to volume of the system, since the surface area is critical to capturing sunlight.) The major challenges facing the open system are that they are

- prone to bacterial contamination or overtaking by "foreign" algae (e.g., carried in the wind) (i.e., it is very difficult to maintain monocultures),
- subject to daily and seasonal changes in temperature and humidity, and
- prone to excessive loss of water, especially, in dry hot climates.

Closed systems typically represent the rows of transparent tubes or plastic bags known as "photo-bioreactors" through which water-born algae and CO_2 are pumped (Fig. 9.7b). The closed systems are not subject to contamination with alien organisms (i.e., they have superior long-term culture maintenance), they are easy to control (less loss of water), and they enjoy relatively high biomass yields. However, closed systems face the following challenges:

- Compared to open systems, the photoreactor-based closed systems are complex and capital intensive
- They require periodic cleaning due to biofilm formation on the photoreactor walls (the film reduces algae exposure to light)
- Temperature maintenance may be an issue (since they do not have evaporative cooling)
- There may be a need to maximize the light exposure of algae
- There could be photoreactor scalability problems when moving to commercial units.

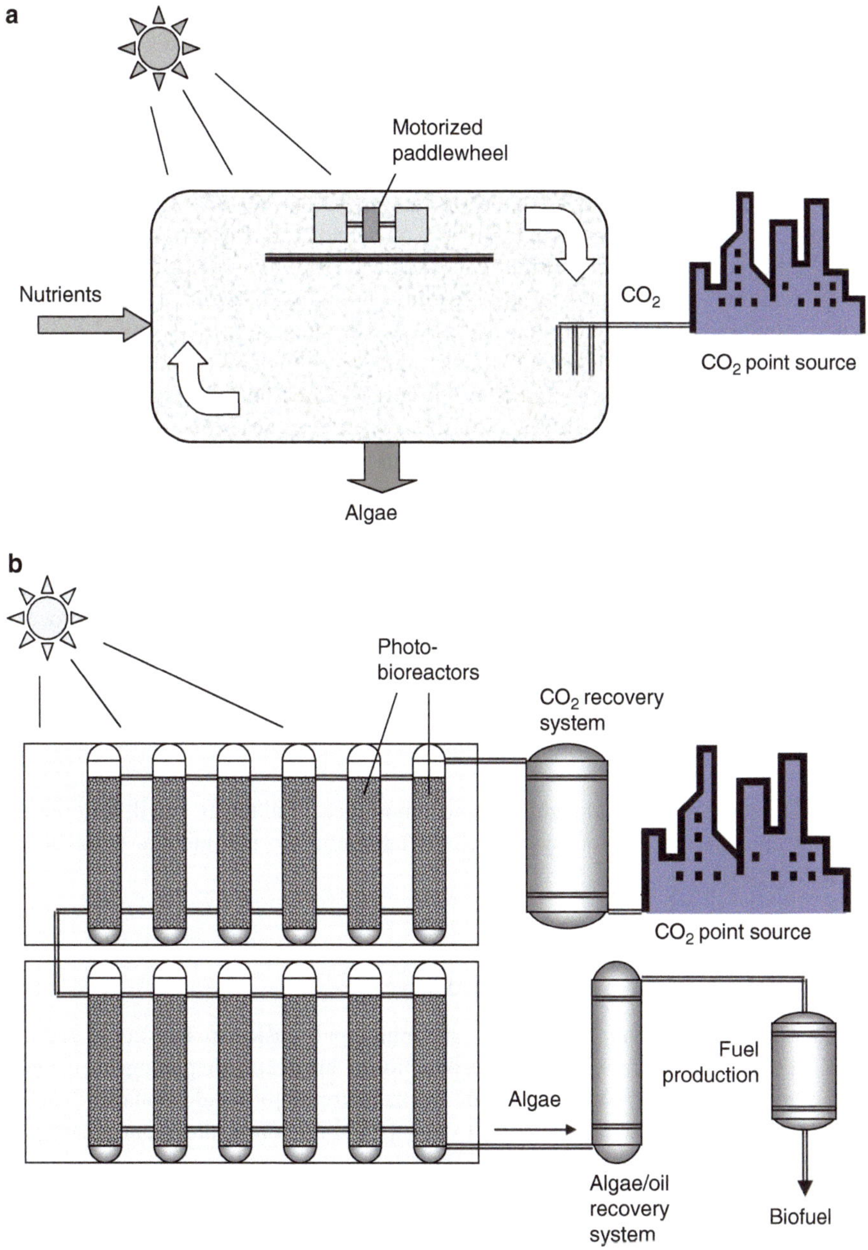

Fig. 9.7 Schematic of reactors for algae cultivation. (**a**) an open (pond) type reactor, (**b**) a closed reactor (a set of photo-bioreactors)

Closed photobioreactors have shown much higher algal biomass growth rates compared to open systems, e.g., 3.8 g/L/day for *Chlorella* grown in a photoreactor, compared to typical growth rates of 0.05–0.32 g/L/day in open pond systems [56]. The use of advanced materials and the photoreactor engineering design optimization could potentially reduce the cost of the closed systems, which is a focus of intensive R&D efforts. Some advanced systems utilize optical fiber-based photoreactor systems, which could dramatically reduce the surface area requirements (at comparable algae production rates). In both open and closed systems, it is critical to supply the required amounts of nutrients and ensure optimum exposure to sunlight.

In addition to open and closed cultivation systems, there are hybrid systems, where photobioreactors play a critical role as breeders or feeders linked to open ponds, providing high cell density inocula to enhance the production capacity of the ponds [59]. Currently, it is premature to project whether algal cultivation in closed, open, or hybrid systems will prevail in the industry, because this to a large extent will also depend on the economics of upstream and downstream processing.

Macroalgae (photosynthetic aquatic plants, seaweeds) require unique cultivation strategies. They can be cultivated offshore, near-shore or in open pond facilities. As CO_2 remediation approach or a contributor to biofuels market, macroalgae technology is not as advanced as microalgal systems; significant advances in strain selection, scale-up activities, and major technological improvements in the process efficiency and cost reduction are needed.

Challenges to Algae Cultivation Systems

Currently, there are several technical and economical challenges and difficulties of scaling up algal cultivation systems from laboratory to commercial-scale units, including [49]

- Culture stability
- Nutrient source sustainability and scaling-up
- Water conservation, management and recycling

Tapping into existing agricultural and municipal waste streams could reduce nutrient costs, but carries the risk of introduction of unacceptable pathogens, chemical compounds, or heavy metals into the biomass stream [60]. Additionally, in large cultivation systems, it will be difficult to maintain algal monocultures, and it is little known about artificial pond ecology or pathology.

Large-scale algae cultivating systems could also face resource requirement and environmental challenges (e.g., water, land, CO_2 source, etc.).

Water. Although algal culture could grow on non-fresh water sources such as brackish and saline ground water, there may be some challenges with regard to availability of water resources, especially for open pond systems located in arid environments with high evaporation rates. Water utilization for algal biomass growth and downstream production of biofuels warrants close attention and has to be a key element of life cycle analysis for algal biofuels production [49].

Carbon dioxide. Several major factors will affect the technical and economic feasibility of coupling a stationary CO_2 source with an algae production facility (assuming the suitability of climate):

- The logistics of CO_2 capture and delivery to the algae facility
- Availability of land for algae cultivation within the reasonable proximity of a CO_2 source
- Availability of sustainable water resources at the site
- Matching of CO_2 supply with the algal cultivation facility demand for CO_2.

Land. Due to relatively low light-to-chemical energy efficiency of algal photosynthesis, large-scale algae production in either open or closed systems would require very large areas of land. Even assuming an aggressive target of annual average algal biomass production of 30–60 g/m^2/day with 30–50 % lipid content (dry weight basis), such systems would require roughly 800–2,600 acres of land to produce ten million gallons of oil feedstock [49]. In addition to coastal and inland microalgae production, offshore marine environment concepts are getting increasingly attractive (as they completely do away with land constraints).

The earlier assessments of resource requirements for large-scale commercial photoautotropic microalgal cultivation indicated that sufficient land, water, and CO_2 resources are available to support the production of billions of gallons of algal biofuels at affordable cost [49]. More research work is needed to understand and manage the algal communities, and to develop large-scale cultivation risk mitigation and remediation strategies.

Algae Downstream Processing

Downstream processing of algae biomass cultivated in ponds or photo-bioreactors or offshore systems consists of the following three main steps: (1) harvesting, (2) dewatering/drying, and (3) the extraction of fuel precursors, e.g., lipids or carbohydrates. (Note that the last step could be eliminated if whole algae is directly processed to fuel.) These are very energy-intensive processes that could markedly affect the economic feasibility of the algae-to-fuel process. Typical algal cultures contain as low as 0.02–0.07 wt% algae, and they have to be concentrated to slurries containing at least 1 wt% algae to make their recovery practicable. The final slurry content will depend on two main factors: (1) the extraction methods to be employed (which will also impact the required energy input) and (2) plant location (due to associated transportation, water quality and recycling issues) [49].

In the well-managed high growth rate cultures, microalgae remain in suspension due to their microscopic dimensions (from about 1 to 30 μm). The small size of the algal species facilitates transport and circulation of microalgae within the cultivation medium and their exposure to light, but, on the other hand, this makes harvesting very difficult. Several methods are currently practiced to separate algae from water: flocculation, sedimentation, filtration, centrifugation [61]. Preliminary estimates have been reported on the energy balance of harvesting, dewatering, and

Table 9.2 Composition of 17 most common algae species

	Composition, wt%	
Algae components	Overall range	Most typical range
Carbohydrates	4–64	9–27
Proteins	8–71	43–51
Lipids	3–40	10–15
Nucleic acid[a]	1–6	3–5

[a]Data are based on six algae species

drying processes. The energy content of most algae strains is in the order of about 5 Wh/g, which includes typical energy content and percentage of lipids, carbohydrates, and proteins [62]. The energy penalty associated with the drying stage is significant, which could adversely affect energy balance of the systems since very large volumes of water are to be evaporated. The development of innovative dewatering technologies and integrated systems could potentially lower energy intensity of algae drying process.

Once harvested, algal biomass can be processed in a variety of ways depending on the nature of the algal species and the desired range of products to be extracted. The chemical composition of algae varies in a very broad range, depending on the particular species. The summary of available data on the composition of 17 most common algae species is shown in Table 9.2.

A particular technological approach to algae processing will be dictated by the choice of the targeted component, in most cases, lipid or protein. From CO_2 abatement viewpoint, the extraction of lipids (natural oils) would be a preferred route since it could lead to production of biodiesel. While considerable technical information exists on large-scale extraction of oils/lipids from oily plant biomass, most of the information on the extraction of algal lipids is limited to laboratory and pilot scale systems. In general, a suitable extracting solvent should be able to penetrate through the matrix enveloping lipid, contact and solvate it. Current practices for lipid extraction from microalgal biomass include a large number of techniques, such as

- Cell rupture (mechanical and non-mechanical disruption)
- Organic cosolvent mixtures (the combination of cosolvents)
- Accelerated solvent extraction [63]
- Selective extraction
- Subcritical and supercritical fluid extraction [64, 65]

The main challenge with the extraction processes relate to the presence of large amounts of water (water can promote the formation of stable emulsions or participate in side hydrolysis reactions). The energy balance of the system could be an issue, especially in energy-intensive processes. The following benchmark was proposed by the US DOE for sustainable biofuels production (based on algae energy content 5 Wh/g): the extraction process should consume (per day) no more than 10 % of total energy produced (per day) [49]. Many extraction techniques require highly concentrated substrates (which sharply increases energy penalty), thus, some algae-to-biofuels processes attempt to bypass the extraction step by either processing whole algal biomass or by engineering algal system such that they secrete desired products directly into the growth medium.

9.4.2 *Algae-to-Fuel Conversion Technologies*

It is widely recognized that algae-derived biofuels have a great potential to contribute to meeting transportation fuel needs, displacing petroleum-based products and at the same time, CO_2 mitigating efforts. Currently, all conventional transportation fuels (gasoline, jet, and diesel fuels) are produced at refineries from crude oil. Although this *modus operandi* will continue for the foreseeable future, the oil industry is targeting a broader range of feedstocks and, recently, its attention turned to biomass-based feedstocks, particularly, algae. A major difference between petroleum and biomass-based feedstocks is that the latter have a high oxygen content (along with other ingredients), which creates two major problems. First, without substantial chemical transformation, biomass-derived fuels will not be able to meet a multitude of engine fuel performance specifications such as energy content, volatility, initial and final boiling points, autoignition, flash point, and many others. Second, it would be practically impossible to directly "inject" these biomass-derived feedstocks into existing refinery chain due to their drastic differences in physical and chemical properties with petroleum feedstocks.

9.4.2.1 Algae-to-Fuels Technological Routes

A wide range of potentially viable fuels can be produced from algal biomass ranging from gaseous (hydrogen and methane) to liquid (alcohols, biodiesel, hydrocarbons) fuels. Algae are considered important *sustainable*[3] bioenergy feedstock sources. Figure 9.8 summarizes existing technological routes to the conversion of algal biomass to biofuels.

As can be seen from Fig. 9.8, technological pathways to the conversion of algae to biofuels fall into the following three main categories:

- Direct production of biofuels from algae without the need for extraction or any other manipulation (e.g., such fuels as hydrogen, methane, ethanol, and alkanes)
- Processing of whole algal biomass to biofuels (through pyrolysis, gasification, etc.)
- Processing algal extracts (e.g., lipids, carbohydrates) to biofuel

[3] According to the US DOE definition, *sustainable feedstocks* are the feedstocks that (1) are managed to reduce required inputs of water and nutrients, (2) can potentially improve soil health and water quality, (3) may provide additional ecosystem services, and (4) the feedstock itself is not considered an invasive species where it will be grown [66].

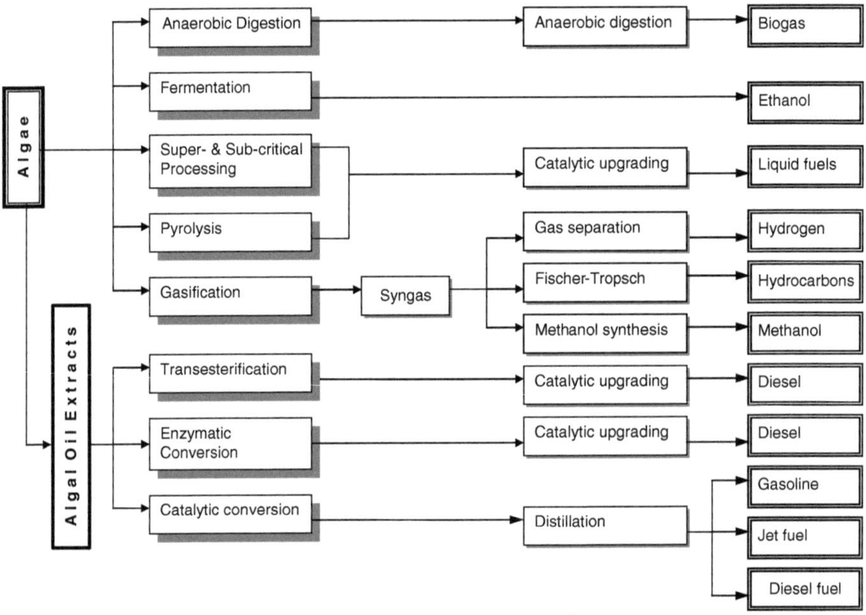

Fig. 9.8 Technological approaches to the conversion of algal biomass to biofuels. *Source*: [49]

Although wide range of biofuels can be produced from algae, of particular importance are liquid biofuels that are widely considered the best-value targets, because they are compatible with existing transportation, petroleum refining, and fuel distribution infrastructures.

9.4.2.2 Direct Production of Biofuels from Algae

The direct production of biofuels from algal biomass has advantages of potentially low cost production of biofuels due to the elimination of several processing steps, e.g., oil extraction, water separation, or chemical/biological processing. Several liquid biofuels (alcohols and hydrocarbons) can be directly produced from algae.

Alcohols

It was demonstrated that such algae species as *Chlorella vulgaris* and *Chlomydomonas perigranulata* were capable of producing ethanol and others alcohols [67]. The production of alcohols can be accomplished through photosynthetic generation and storage of starch within the algae followed by anaerobic dark fermentation of starch. These alcohols could potentially be extracted from an algal culture media, practically eliminating the need for separating algae from water and extracting oils, thus, drastically reducing capital and energy intensity of the process.

The technology for the direct algae-mediated production of alcohols from CO_2 as a carbon source is still under development. There have been feasibility studies and reports of preliminary engineered systems that involved tubular photobioreactors and supply of CO_2 (potentially, from power plants) [68]. It was estimated that more than 90 % of CO_2 could be photosynthetically converted to sugars, portion of which is further converted to ethanol that is secreted into the culture medium and collected. Theoretically, with this technology, 1 t of CO_2 could be converted into about 60–70 gal (228–266 L) of ethanol [49]. In practice, this technology is estimated to yield 4,000–6,000 gal of ethanol per acre per year (potentially, upon further development, up to 10,000 gal ethanol/acre/year). To remain cost competitive, the process would require the cost of captured CO_2 not to exceed US\$10 t^{-1} CO_2 [49].

Algenol company (USA) is actively pursuing this technological approach [69]. Their patented DIRECT TO ETHANOL® technology employs enhanced blue–green algae or cyanobacteria to photosynthetically convert CO_2 into ethanol in salt water with intermediate production of sugar (pyrovate) according to the following (overall) equation:

$$2CO_2 + 3H_2O + \text{sunlight} \rightarrow C_2H_5OH + 3O_2 \tag{9.13}$$

The Algenol's process can produce 166 gal (639 L) of ethanol per each metric ton of CO_2 consumed. The technology is well suited to capturing CO_2 from industrial sources such as power plants and cement plants (more details on the technology can be found in the Sect. 9.3.6).

Alkanes

Several strains of algae can directly produce hydrocarbons (alkanes) via heterotrophic metabolic pathways. The available technical information related to this technology is limited to heterotrophic algal organisms fed by sugars under dark conditions (e.g., [70]). This approach does not present interest from CO_2 mitigation viewpoint.

Hydrogen

Biological production of hydrogen (or biohydrogen) as promising environmentally clean energy carrier using algal systems has been active area of research since early 1970s. It was discovered that some microalgae and cyanobacteria can produce hydrogen according to the following reaction:

$$2H_2O + \text{light energy} \rightarrow O_2 + \left[4H^+ + 4e^-\right] \rightarrow O_2 + 2H_2 \tag{9.14}$$

Under normal photosynthesis conditions, CO_2 is fixed using electrons from water, but in H_2-producing systems, an electron acceptor is switched at the level of ferredoxin from CO_2 to protons under the conditions favorable for hydrogen generation. The topic of algae-mediated production of hydrogen only indirectly relates to the CO_2 abatement problem discussed in this chapter (i.e., as a source of renewable hydrogen for thermochemical CO_2 conversion to fuels), and, therefore, is not discussed here. The topic of biological hydrogen production is covered in detail in excellent reviews by King [71] and Das [72].

9.4.2.3 Processing of Whole Algae to Fuels

As in the case of direct production of biofuels from algae, the direct processing of whole algae into fuels could benefit from reduced costs associated with the elimination of a number of steps, such as, products extraction and their post-processing (although, at some level, dewatering/drying may still be required). An additional advantage of this approach is that it is amenable to processing of a broad and diverse range of algae strains, which can substantially expand the feedstock supply. Figure 9.8 (upper section) summarizes technological pathways to the conversion of whole algae to biofuels. The whole algae processing technologies fall under two major categories: thermochemical and biological (or fermentative) processes. Thermochemical conversion methods include pyrolysis, gasification, hydrothermal liquefaction, and supercritical processing.

As shown earlier, microalgae primarily consist of proteins, lipids, and carbohydrates—together making up to 80 % of the dry algae mass basis. This is in contract to terrestrial plants that are composed of cellulose, hemicelluloses, and lignin that make up more than 95 % of the plant mass on dry basis. The differences between the chemical composition of aquatic and terrestrial biomass profoundly affect the choice of thermochemical conversion method. For example, most aquatic species including algal biomass do not contain lignin—a macromolecule that provides a structural support and protection for the lignocellulosic plants. Extraction of lignin from lignocellulosic biomass is an energy intensive process.

Another important issue to consider is the content of nitrogenous compounds in the algal biomass that mostly originate from the protein fraction of the biomass feedstock. The presence of the N-containing compounds in the biooil produced by liquefaction of biomass is detrimental to hydrogenation catalysts used in the refinery operations. Furthermore, some oxygenated compounds such as higher alcohols present additional problems during upgrading processes within the refinery due to excessive water formation.

Pyrolysis

Pyrolysis is anaerobic chemical transformation of biomass at elevated temperatures. In general, biomass pyrolysis results in three major products: pyrolysis gas, pyrolysis oil (or biooil), and char (or bio-char). In most cases, biooil is a target product of pyrolysis because it could be further processed to liquid hydrocarbon product similar to conventional diesel fuel. The data on the typical elemental composition of biooil from microalgae pyrolysis and petroleum are presented in Table 9.3.

Table 9.3 Comparison of the typical elemental composition of biooil from microalgae pyrolysis and petroleum

	Composition, wt%	
Elemental composition	Microalgal biooil	Petroleum
Carbon (C)	62.1	83–87
Hydrogen (H)	8.8	10–14
Oxygen (O)	11.2	0.1–1.5
Nitrogen (N)	9.7	0.01–0.7
Other	balance	balance

Source: [56]

It is evident from Table 9.3 that biooil is rich with oxygen, therefore, in most cases, the first step in its post-treatment involves hydrogenation or hydrodeoxygenation reactions, where large fraction or entire oxygen content is removed from the biooil at elevated temperatures (280–350 °C) in the presence of hydrogen and suitable catalyst (typically supported Ni, Co, Ru).

While plant biomass pyrolysis to biooil is relatively mature technology (several pilot plants have been built in the USA, Germany, and Brazil), pyrolytic conversion of microalgae and macroalgae to biooil and, consequently, to biofuels, is relatively new area of development, and few reported studies deal with laboratory-scale units [71, 72]. Algae as a feedstock for the pyrolysis process have significant differences with plant biomass, as follows: on the one hand, it exists in the form of small particles and does not have fiber tissue to deal with (which is particularly beneficial for fast pyrolysis), but on the other hand, it has a very high moisture content, which might require energy-intensive upstream dehydration step for the process to operate efficiently. The commercial viability of the pyrolysis approach will depend on technological advancements in the entire process chain, including algal feedstock pretreatment (e.g., dewatering, extraction), pyrolysis, posttreatment (e.g., oil stabilization), and conversion of biooil to biofuels.

Gasification

Conversion of algal biomass to biofuels via gasification route provides several important advantages over other methods:

- It is a very flexible way of producing a wide variety of fuels with acceptable and known properties from practically any type of microalgae or macroalgae
- The approach is based on well-established technologies
- Algae gasification can be easily integrated into existing thermochemical infrastructure, e.g., by cofeeding algae into a coal gasification plant in order to reduce the capital investment and improve process efficiency through economy of scale
- The thermal integration of the process would allow using waste heat for drying algae during a dewatering stage using regenerative heat exchangers.

Conceptually, the process includes two main stages: gasification and Fischer-Tropsch synthesis processes. In the first stage, algal biomass is gasified in the presence of steam or steam/oxygen (or steam/air) mixture to produce syngas (or bio-syngas). Syngas is purified of N-, P-, and S-containing compounds, scrubbed of excessive CO_2, conditioned (via water-gas shift reaction) to adjust for desirable H_2/CO ratio, and directed to the FT reactor. In the FT reactor, synthesis of liquid hydrocarbons takes place (reaction 9.9). The raw hydrocarbon product is distilled to desirable factions of gasoline, jet, and diesel fuels.

Despite recent progress in this technology as applied to plant biomass (e.g., [73]), very little information is available on conversion of algal biomass to fuels through the gasification-FT route. It is assumed that once water content of the algal feedstock is adjusted to desirable levels, the processing of algae would be similar to that of woody biomass. Potential key challenges to applying the gasification-FT technology to the algal feedstock at large scale are expected to relate to syngas clean-up and tar removal (since FT catalyst is very sensitive to impurities in syngas).

Hydrothermal Liquefaction

In hydrothermal liquefaction (HTL) process, algal biomass is exposed to subcritical water (i.e., at temperatures above 100 °C and elevated pressure) with the production of so-called bio-crude. The main advantage of the HTL approach to algae processing is that the technology can be applied directly to wet biomass, thus, bypassing the need for excessive dewatering and drying [74]. The subcritical conditions and presence of liquid (activated) water are adequate to decompose algal feedstock into smaller energy-rich molecules (that constitute bio-crude), but not severe enough to degrade algae to permanent gases: H_2, CO, CO_2, and CH_4 (as in the case of gasification). Bio-crude yield typically accounts for about 45 % of the biomass feedstock (on a dry ash-free basis), and has relatively high content of oxygen; it could be further upgraded to liquid fuels similar to petroleum-based fuels via hydrodeoxygenation route. Thermal efficiency of the biomass HTL process could be as high as 75 % [75].

There are a few studies reported in the literature on HTL of algal biomass. For example, algae *Botryococcus braunii* was processed by HTL at temperature of 300 °C in the presence of Na_2CO_3 catalyst with the bio-crude yield of 64 % [76]. In general, HTL is considered a promising technological approach because it allows processing wet algal biomass feedstock, but more research is needed to prove its commercial viability.

Supercritical Processing

The advantages of supercritical processing approach are that it allows to use whole algae without dewatering (thereby increasing the efficiency of the process) and it is capable of simultaneously extracting and converting oils to biofuels. In general, SC fluid extraction of algal oil is more efficient than conventional solvent extraction techniques [77], and it could be combined with a transesterification reaction enabling a "one pot" approach to biofuels production. Although this approach so far has only been demonstrated for vegetable oils, it is expected that the similar approach can be applied to algal oil [49, 78]. Currently, practically no information is available on the SC processing of algal biomass. Thus, further research work is needed to demonstrate applicability of SC technology either to whole algae or oil extract at the scale allowing to estimate potential commercial viability of the process.

Anaerobic Digestion of Algae

Anaerobic conversion of algae to biogas (sometimes called "renewable methane," because it contains methane along with CO_2 and minor ingredients) is an interesting approach to production of gaseous biofuels. The advantage of this technology is that it could eliminate several costly steps associated with algae processing, including drying, extraction, and conversion to fuels, and, as such, it could present a cost-effective approach. Several studies reported the production of biogas by anaerobic digestion of macroalgae and marine algae. For example, the use of macroalgae *Laminaria hyperbore* and *Laminaria saccharina* for biogas production has been reported by Hanssen [49]. In another study, biogas (with methane concentration of 65 vol.%) was produced with the yield of 180.4 mL/g/day utilizing two-stage anaerobic digestion process with different strains of algae [79]. In principle, this approach can be modified for microalgae utilization, and it could be particularly advantageous for the use in integrated wastewater treatment facilities, where the algae strains not optimized for lipid production can be efficiently used [49].

9.4.2.4 Conversion of Algal Extracts to Biofuels

Currently, most of the technological development and commercial activities in the algae-to-fuel area are concerned with the conversion of algal oil (or lipids, triacylglycerides) to biofuels. There are several chemical and biochemical approaches to conversion of algal oil extracts to biofuels (see Fig. 9.8, lower part).

Chemical Transesterification

The transesterification (TE) of algal oil involves the reaction of triacylglycerides (TAG) extracted from algae with alcohols (typically, methanol or ethanol) to fatty acid alkyl esters, e.g., fatty acid methyl ester (FAME), as shown in the simplified scheme:

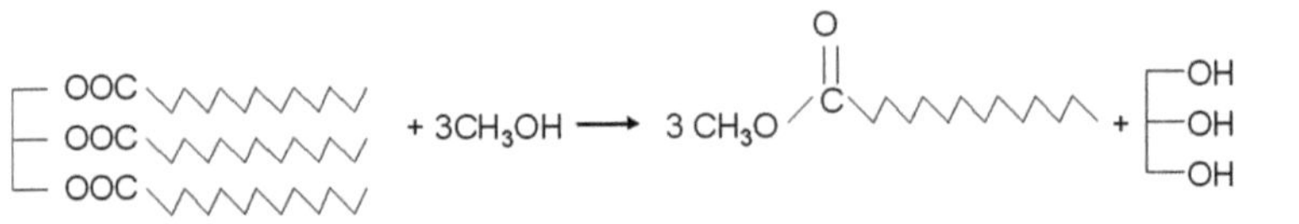

9.15

TAG FAME Glycerol

where  represents long-chain alkyl radical (all three radicals in TAG are not necessarily similar)

Although TE of vegetable oils to biodiesel is considered a mature technology (large-scale commercial installations exist in many countries), the use of algal oil in this process is relatively new area of activities [80]. Similar to TE of vegetable oils, the reaction involving algae-derived TAG can be carried out via non-catalytic and catalytic routes (with the latter being preferable). It has been reported that TE of algal oil can be achieved via conventional base-catalyzed methanol approach, as well as by using ethanol as esterification agent and sodium ethanolate as catalyst. Acid-catalyzed TE of algal oil is less studied, compared to base-catalyzed systems that, in general, are faster and conducted at lower temperatures.

Challenges facing large-scale deployment of TE of algal oil are similar to those of other oils, namely: (1) high energy intensity of the process, (2) production of the large quantities of glycerol byproduct, and (3) difficulty in removing glycerol and catalysts from the system. Alternative approaches to the TE process involving microwave and ultrasonic technologies [81] can potentially be applied to algal oil, but practically no research work has been conducted in this area.

Biochemical Transesterification

Use of biocatalysts (enzymes, lipases) in TE of algae-derived TAG for biodiesel production could potentially address the forementioned challenges facing chemical TE systems and offer some environmental advantages over traditional processes [82]. Although the technology is promising and increasingly attractive to researchers, the process is still expensive (mainly, due to high price of lipase), and it has not been demonstrated at sufficiently large scale. Besides, enzyme's operational life is relatively short due to adverse effects of large quantities of methanol and the coproduct glycerol [49]. Among other challenges are limited solvent and temperature tolerance of enzymes catalyzing the TE reaction. All these problems must be addressed before moving to a commercialization stage.

Heterogeneous Catalytic Esterification

The use of heterogeneous metal-based catalysts in TE can help addressing the product separation and purification problem. The examples of the heterogeneous catalysts include Lewis acids ($AlCl_3$, $ZnCl_2$), Fe–Zn cyanide, MgO, CaO, Al_2O_3, $TiVO_4$, and other catalysts [49, 83]. For example, TE of vegetable oil in the presence of $HTiNbO_3$ catalysts in a fixed-bed reactor at 200 °C and 35 atm produced FAME and glycerol with

91 % yield [49]. Although no information available on the processing of algal oil using metal catalyst-based TE systems, the results similar to vegetable oils could be expected. The main challenge here would be to find heterogeneous catalysts with comparable activity to conventional alkaline catalysts and operating at much lower temperatures and pressures that are currently practiced (200–240 °C and 30–60 atm). Additionally, the catalysts must be prone to poisoning and leaching of metals into the reaction medium.

Catalytic Conversion of Algal Oil to Hydrocarbon Fuels

Conversion of algal oil to biofuels similar to petroleum-based fuels (often called *Third Generation biofuels*, or "green gasoline," "renewable diesel," see Sect. 6.2.3) largely boils down to elimination (as much as possible) of the oxygen content of bio-oil feedstocks to increase their energy density, enhance their chemical stability, and make them compatible with conventional refinery operations. From the refinery's operation viewpoint, the most desirable approach would be to produce the biooil feedstock that can be inserted at any point after vacuum or atmospheric distillation for additional processing, most preferably, in hydrotreaters, fluid catalytic crackers, cokers, or catalytic hydrocrackers. If that can be accomplished, the feedstock would be considered fungible with petroleum and can be utilized for the production of traditional transportation fuels without disruptive changes in infrastructure.

Several catalytic processes can be used for converting algal oils to desirable range of petroleum-like fuels, most of them involve hydrogenation step to remove oxygen, e.g., hydrotreatment, hydrocracking, hydrodeoxygenation, etc. Straight chain alkane moieties present in algal oil are not good starting materials for gasoline production, because they result in low octane number (although, they are desirable for diesel production). Thus, additional structural isomerization step would be necessary to arrive at high-octane blendstocks.

The primary challenge to processing algal oils to renewable hydrocarbon fuels is the development of active and stable catalysts. Crude algal oil may contain high levels of phosphorus (from extracted phospholipids), nitrogen (from extracted proteins), and metals (particularly, Mg from extracted chlorophyll). Presence of the P- and N-containing molecules in the biooil could be detrimental to traditional hydrogenation or hydrocracking catalysts. MgO deposits could also deactivate catalysts. Thus, in order to efficiently and cost-effectively process algal oil, it will be necessary to optimize both the level of purification of algal oils as well as the tolerance of catalysts to contaminants [49].

9.4.3 Algae-Based Biorefineries

Although biofuels production is considered the primary objective of algae processing, the generation of other coproducts could significantly enhance the economics of the process. Indeed, algae is a unique resource in its capacity to address CO_2 mitigation, wastewater treatment (nutrients removal), and deliver a wide range of

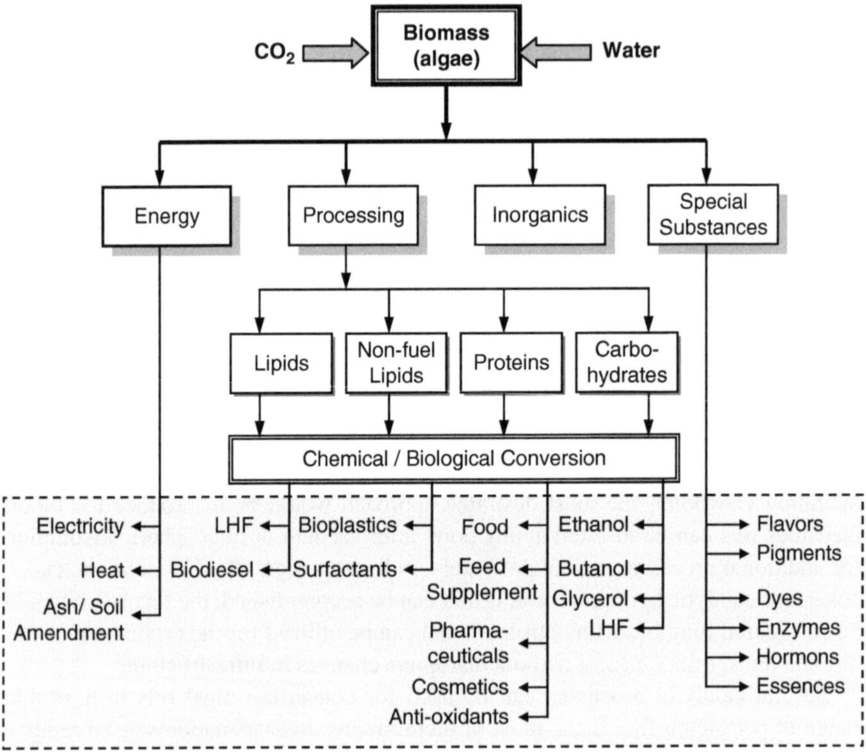

Fig. 9.9 Schematic diagram of algae-based biorefinery. *Source*: LHF is liquid hydrocarbon fuel [49]

valuable products. Using appropriate chemical, enzymatic, or microbial conversion approaches, all primary components of algal biomass, lipids (oils), carbohydrates, proteins, and inorganic compounds, can be converted into a variety of products. The nature of the end products will be determined by the process economics, and they may vary from region to region according to the cost of raw materials [49].

The concept of "biorefinery" has been recently promoted for utilization of all components of a biomass raw material as a means of enhancing the process economics. Under the biorefinery concept, the production of industrial, high-value, and high-volume chemicals from lipids, amino acids, glycerol, and N-containing components of algal biomass could become economically feasible [84]. Figure 9.9 depicts the schematic diagram of an algae-based biorefinery.

A biorefinery is a large integrated system (similar to oil refinery complexes), where multiple products are produced from biologically derived feedstocks. The biorefinery scheme enables serving a variety of market opportunities as they change or emerge. An algae-based biorefinery complex could be located near power stations, preferably, on marginal land not otherwise useful for other forms of agricultural output. A large number of different commercial products have already been produced from microalgae and cyanobacteria (most of them are listed in Fig. 9.9).

Here are some highlights on high-volume commercial products produced from microalgae and the size of markets [49]:

- *Human food supplement*. The use of microalgae as a human food supplement is currently limited to a few species: *Spirulina*, *Chlorella*, *Dunalliella*, and others. The market is about US$2.5 billion/year, and is expected to grow.
- *Animal feed additive*. Microalgae (mainly, *Spirulina*, *Chlorella*) has been used with good results as a food additive for cows, horses, pigs, poultry, etc. The market is estimated at US$300 million/year and is quickly growing.
- *Polyunsaturated fatty acids* (e.g., arachidonic, docohexaenoic acids) are health-promoting additives to human food and animal feed that are produced from microalgae. Market: US$1.5 billion/year.
- *Antioxidants* are sold on a health-food market. The major antioxidant product is β-carotene that is produced from *Dunaliella salina* and sold for up to US$3,000 kg^{-1} [85]. Market: US$380–430 million/year.
- *Coloring agents and pigments* produced from microalgae are used as natural dyes for food, cosmetics, and as pigments in animal feed (e.g., salmon feed to give the fish meat a pink color). Market: US$160 million/year.
- *Fertilizers and soil conditioners*. Currently, microalgae (and macroalgae-seaweeds) are used as plant fertilizers and plant growth regulators, in particular, to improve water-binding capacity and mineral composition of depleted soils.
- *Specialty products*. Bioflocculants, biopolymers and biodegradable plastics, cosmetics, pharmaceuticals and bioactive compounds, polysaccharides, and stable isotopes for research. Due to specialized applications, the market for these products is relatively small.

Macroalgae have high content of structural polysaccharides that are currently extracted for their commercial value. Products from macroalgae are as follows (market in brackets in US$ million) [49]:

- Human food (5,000)
- Agar (for pharmaceuticals, food ingredients) (132)
- Alginate (for pharmaceuticals, food additives, medical, textiles) (213)
- Carrageenen (food additives, tooth paste) (240)
- Fertilizers and soil conditioners (5)
- Animal feed (5)
- Macroalgal biofuels (<1)

The total market for macroalgal products is estimated at US$5.5–6 billion [49].

Several biorefinery concepts have been published in literature. For example, in [88], the authors discussed the feasibility of the integration of microalgal biorefinery with oil refinery. The liquid and gaseous waste streams from oil refining operations were considered for cultivation of *Aphanothece microscopica Nägeli* cyanobacterium in a bubble column photo-bioreactor. Substantial CO_2 conversion rates into bio-products (including algal biodiesel) have been achieved, thus, demonstrating that the process developed could be potentially present a promising biorefinery platform.

9.4.3.1 Integration with Water Treatment Facilities

Municipal wastewater treatment facilities represent a good match with algae production; both operations could be advantageously colocated because nutrient-rich (N, P, K elements) wastewater can be used for algae growth, and algae can help remove potentially harmful organic matter from wastewater. The potential benefits of the integration of algae production with wastewater treatment can be summarized as follows [49]:

- Algae can treat and purify waters polluted with organic matter (including N-, P-containing compounds and synthetic organics), heavy metals, and endocrine disrupting compounds
- A wide variety of wastewater sources can be treated by algae: municipal, organic industrial (e.g., food processing), agricultural (e.g., confined animal facilities), agricultural drainage, and others
- Wastewater treatment revenue can potentially offset algae production costs (by off-setting the cost of commercial N, P, K fertilizers needed for algae growth)
- The ability of algal system to provide natural disinfections and remove trace contaminants
- Lower energy intensity can be achieved compared to conventional wastewater treatment (with the reduction in overall GHG emissions)
- Lower capital and O&M costs compared to conventional wastewater treatment

Although algae-based water treatment would require larger land areas compared to mechanical treatment, in suitable climates, it could be economically viable: the algae-based treatment facilities are typically less expensive to build and operate than conventional mechanical facilities. It was estimated that the total cost of high-productivity algae-cultivating ponds is about 70 % less than that of activated sludge, which is the leading wastewater treatment technology in the USA and other countries [49]. The significant cost savings combined with ever-increasing needs for improved wastewater treatment around the world provide tremendous incentives to install algae production facilities integrated with wastewater treatment plants.

9.4.4 Integration of Algae Production with Stationary CO_2 Sources

The proximity of industrial sources of CO_2-rich flue gases with algae-producing facilities has important technological and economic implications. The advantages of colocation and integration of algae production with stationary industrial sources are as follows [49]:

- There are abundant quantities of CO_2 available from a wide range of stationary industrial source (power plants, cement, ethanol, gas processing plants, etc.).

- Excess heat and power from a power plant may be available to provide heating or cooling to improve thermal management of algae cultivation systems, thus, allowing to operate under a broader range of climate conditions and geographic locations on a near year-round basis.
- Quantities of wastewater or cooling water may be available (especially, near power plants), which would help to overcome a primary resource challenge for large-scale production of algae.

The cost savings from the integration of power plants and algae production operations will be site specific and depend on nature of fossil fuel, CO_2 capture and purification methods, and distance of CO_2 transport from the source to the algae cultivation sites. In principle, coal-fired power plants are a convenient source of CO_2 for algae production; however, they have higher CO_2 emissions per unit energy produced than NG-fired plants, which would necessitate proportionally larger algae cultivation systems and higher costs per unit energy produced. On the other hand, flue gas from coal-fired plants have twice as greater CO_2 concentration (10–15 vol.% vs. 5–6 vol.%) as NG plants, which could bring some advantages in terms of CO_2 capture efficiency and its transport from a power plant to an algae production site. Furthermore, an intermittency factor in CO_2 supply and its impact on algae production should also be taken into consideration: NG plants mostly operate as peaking plants (rather than base-load plants), thus, introducing intermittency in CO_2 delivery. There have been analyses of the combined operation of power plants with algae growth facilities and biorefineries. For example, a reported analytical study indicates that the full utilization of CO_2 in the flue gas emitted by a 50 MW_{el} NG-fired power plant would require about 2,200 acres of algae cultivation area [86].

9.4.4.1 Challenges

There are several challenges facing the integration of algae production with industrial CO_2 sources. Since photoautotropic algae can absorb CO_2 generated by the power plant only during sunlight hours, the CO_2 emissions offset will be limited to an estimated 20–30 % of the total plant emissions due to CO_2 off-gassing during dark (non-sunlight) hours and inevitable parasitic loses associated with algae production [86]. Large coal-fired power base load plants with the output of 1–2.5 GW_{el} would require many tens of thousands of acres of land and immense volumes of water for algae production with similar effective offset of 20–30 % of the CO_2 emitted [49]. Another potential challenge stemming from the photosynthetic nature of the algae relates to the difficulty of matching of CO_2 availability to CO_2 utilization by an algae-cultivating system (especially complicated due to inactive night hours). Moreover, the rate of CO_2 uptake could vary in a wide range with algae species and the specifics of an algae growth system and light conditions, making matching even more difficult.

Transportation of CO_2 from a point source to an algae production facility might become a limiting factor if CO_2 content in gas is low (e.g., as in flue gas) and gas-pumping distance is relatively long. This may require capture and concentration of CO_2 before pipelining to reduce the energy penalties associated with CO_2 transport to algae cultivation facilities. Some CO_2 point sources (e.g., hydrogen and ammonia production plants, NG treatment plants) provide CO_2 in a concentrated form avoiding the need for its concentration [87].

Among other challenges of the integration of algae growth facilities with CO_2 sources are

- Need for nutrient sources: N-, P-, K-based nutrients have to brought in from other (possibly, distant) locations. Ideally, algae production would be colocated with a power plant and a wastewater treatment facility.
- Capital and O&M costs: There will be a need to reduce capital cost and parasitic operational loses and costs for infrastructure and power required to capture and deliver industrial CO_2 to algae ponds. According to reported estimates, approximately 20–30 % of power plant's GHG emissions can be offset by algae biofuel and protein production [86].
- Possible effect of chemicallyaggressive impurities (NO_x, SO_x) in flue gases on algae production and coproducts quality.
- CO_2 is likely to be emitted to the atmosphere during dark hours (unless it is captured and sequestered by other means) due to a mismatch with base-load coal-fired plants [89].

R&D and policy development efforts will be required for successful commercialization of algal cultivation facilities colocated and integrated with large industrial CO_2 emitters and/or wastewater treatment facilities (e.g., there might be a resistance from electric utilities to cooperate with algae production facilities [49]). Policies that encourage partnering between public utilities (and other industrial stationary CO_2 sources) and algal technology and fuel companies should be evaluated and promoted.

9.4.5 Carbon Abatement Potential of Algae

CO_2 utilization potential of algae is very high: typically 1.8–2 t of CO_2 is utilized to produce 1 t of algal biomass (on dry basis), though, this number could be higher or lower depending on the cultivation conditions and algal species used [1]. According to US DOE report, on the biooil productivity basis, algae can potentially produce about an order of magnitude more oil per acre than other oil-producing biomass species [49]. However, recently reported data indicate that from the carbon mitigation viewpoint (i.e., maximizing CO_2 capture), the use of high oil-producing algae may not necessarily be a preferred option as it could compromise the overall productivity of the system [1].

As a CO_2 abatement measure, algae cultivation and processing, in general, is not considered a permanent carbon storage solution. Algae production could provide excellent opportunities for the utilization of fossil-derived CO_2 and complement geological sequestration. However, algae growth does not actually sequester fossil carbon, but rather it provides the means of reusing it in the form of fuels and other products. The extent of permanence depends on the particular technological option. If the targeted product is biodiesel, the storage is temporary because CO_2 is re-released as soon as fuel is burnt in a diesel engine. Algal products could be converted to plastics, which have longer live than fuels, and can be considered a semipermanent storage. In general, carbon abatement credits would come from the displacement of fossil fuels by algal fuels or reduction in fossil fuels consumption due to substitution with algae-derived products.

The CO_2 abatement potential of algae in any carbon-credit or cap-and-trade framework is an important policy question to consider, because CO_2 will be re-released to the atmosphere when algal-derived fuels or products are combusted or completely degraded. While production and utilization of algal biofuels will result in a net reduction in overall GHG emissions, the process of capturing CO_2 from flue gas and converting it into transportation fuels may not strictly be considered carbon sequestration. The climate policy and regulatory implications of this issue will need to be addressed before utilities and fuel companies are likely to widely adapt technologies for CO_2 capture by algae from industrial CO_2 sources [49]. Integration of algae production with power plants or other CO_2-emitting sources could potentially provide carbon credits for utilities, if policy will be established on carbon absorption and reuse as transportation fuel in lieu of permanent sequestration.

9.4.6 Commercial Status of Algae-Based Technologies

The idea of utilizing algae to extract economic value in the form of value-added products is not new (starting from 1978, the US DOE funded a program called "Aquatic Species Program" aiming at developing renewable transportation fuels from algae [90]). Commercial large-scale production of microalgae (*Chlorella*) for food additives applications started in early 1960s in Japan [56]. During last decade, there has been an explosion in algae-related R&D work, with many universities, national research centers and companies focusing on commercial opportunities in converting algal oil to liquid transportation fuels. Along with the countries that have a long track record in algae-related research (the USA, Japan, Israel, and China), other countries (Australia, Spain, Italy, Netherlands, Austria, Portugal, France, New Zealand, and others) are emerging with several large government- and industry-supported programs and collaborations. Numerous pilot and commercial demonstration projects are currently underway around the world (in excess of 200 worldwide). The following provides highlights of recent algae-related activities by some companies in the commercialization of algal fuel technology.

9.4.6.1 USA

Algenol company is developing DIRECT TO ETHANOL® process (see Sect. 9.3.2). It has licensed the technology to BioFields S.A.P.I. de C.V. in Mexico, which has an access to over 55,000 acres of non-arable land in the Sonoran desert in Mexico near a Comisión Federal de Electricidad (CFE) power plant. The area has access to navigable waters for transportation to the United States and growing ethanol markets in South America. Other US-based companies involved in the algae-to-fuel technology are (as of 2012)

- SGI, Solix Biofuels, Sapphire Energy, Algasol (extraction of oils)
- Solazyme (heterotrophic systems) [91]
- Green Car (oil plus ethanol)
- BioFuel Systems SL (pyrolysis of whole algae to biocrude)
- Sapphire Energy (production of jet fuel)
- HR Biopetroleum (algal biorefinety: biofuels and other products)

9.4.6.2 Europe

European Algae Biomass Association (EABA) and European Bioenergy Industrial Initiative (EIBI) are aiming at promoting development, demonstration and commercialization of algae-to-fuel technologies. The list of European companies working in the area of algal biofuels include [92]

- Subitec (Algae photobioreactors) (Germany)
- SEE Algae Technology (biofuels from seaweed-macroalgae) (Austria)
- Abengoa, ECOALGA, Enalg S.p.a., BioSerentia, Bio Fuel Systems (Spain)
- FeyeCon D&I BV, Algae Biotech SA, Clean Algae SA, AlgaeLink N.V. (Netherlands)
- BIOMARA (UK, Ireland), and others

European Commission announced that it will contribute €20.5M to support three projects in the framework of Biofuels from Algae program: BIOFAT, ALL Gas, and InteSusAl (which will form Algae Cluster) [92]. The BIOFAT demonstration project aims at integrating the entire value chain in the production of ethanol and biodiesel from algae that includes strain selection, biological optimization of the culture media, monitored algae cultivation, low-energy harvesting, and technology integration. The developers plan to scale-up the process to a 10-hectare demonstration plant targeting the production of 900 t of algae annually. The project is coordinated by Abengoa Bioenergia Nuevas Tecnologias and includes universities and companies from Italy, Portugal, Israel, Netherlands, France, and Belgium. Besides small companies, such industrial giants as BP, Chevron, ExxonMobil, Dow Chemical, Conoco Philips, Royal Dutch Shell, and others have heavily invested in algae-related research and feasibility studies, and are currently actively participating in demonstration and pre-commercial trials.

However, despite long history of algae-related research and a recent surge in R&D and demonstration activities, bringing the technology to a full-scale commercial stage and entering the competitive marketplace proved very difficult, due to technical and economic reasons, in particular, difficulties with controlling the biomass growth and processing variables at competitive cost, and inability to compete with vastly cheaper supplies of conventional fossil fuels. As a result, by mid 2012, there were no known large-scale commercial facilities producing high volumes of algal biomass year round with the necessary yields for meaningful energy and/or materials production. (Note that some microalgae species for pharmaceutical applications are currently produced at a commercial scale, but the market is rather limited and prices are very high.) Government mandates and incentive programs (e.g., tax credits, loan guarantees for the construction of commercial facilities) could significantly accelerate the initial commercialization of the technology and catalyze its rapid deployment in many regions of the world.

9.4.7 *Markets for Algae-Derived Products*

Existing and emerging market drivers for algae utilization are threefold:

- Energy security (particularly, in the area of liquid transportation fuels)
- Environmental and sustainability issues, CO_2 abatement
- Value-added products (fuels, pharmaceuticals, chemicals, food supplies)

Currently, the main markets for algae-derived products are as follows: advanced biofuels (biodiesel, jet fuel, bio-crude), “green” chemicals (surfactants, lubricants, oleo-chemicals), health care products, pharmaceuticals and nutraceuticals (vitamins), bioplastics, human and animal nutrition (edible oils, food ingredients, animal feed, Omega-3 oils, alternative protein feed source), and others (see also Sect. 9.3.3 and Fig. 9.9). Commercial potential for algae-based products represents a practically untapped resource. Several major existing and emerging commercial applications of algae are listed below.

9.4.7.1 Nutrition, Cosmetics, Aquaculture

Most of the existing commercial applications of algae and algae-derived products relate to the following markets: human and animal nutrition, cosmetics, aquaculture, β-carotene, phycobiliproteins, astaxantin, with the prices ranging from few tens of euros/kg for nutrition products to few thousands euros/kg for β-carotene and astaxantin, to ten thousands euros/g for phycobiliproteins [56]. More details on commercial products produced from algae are in the Sect. 9.3.3.

9.4.7.2 Drop-in Fuels

Because economic value can be extracted from the entire algal biomass, the production process can be efficiently and expediently tailored to adjust to changing market demands. Of particular importance is the potential role of algae cultivation systems in the production of so-called drop-in fuels and the development of bio-refineries. Broadly defining the term, drop-in fuel means the fuel making use of at least some of the existing petroleum infrastructure (in a narrower definition, it is fuel made from a variety of biomass feedstocks that can be blended with petroleum products such as gasoline or diesel fuel) [93]. Algae is considered one of the most promising precursors for the production of drop-in fuels. Algae oil either could be directly blended with diesel fuel, or after a pretreatment (hydrogenation) it could be injected into existing crude oil refineries. Algal biofuels are predicted to be the largest market among algae-derived products. It is projected that by 2022, algae biofuels would reach the capacity of 40 billion gallons accounting for about 37 % of the entire biofuels production capacity of about 109 billions gallons (it will be the largest biofuel category) [1].

9.4.7.3 Jet and Diesel Fuels

Among algae-derived biofuels, the current focus is aimed at possible replacements of jet and diesel fuels, since corresponding power vehicles are not good candidates for electrification (see Chap. 8). Rising jet fuel prices combined with the potential of imposing carbon prices on fossil-derived fuels are putting pressure on airline industry to find clean alternatives to jet fuels to create incentives for biofuels, and particularly, algal fuels. The first commercial flight boosted by biofuel took place in 2008, when a Virgin Atlantic B747 jet flew from London Heathrow to Amsterdam Schiphol aided by fuel derived from Brazilian babassu nuts and coconuts (note that the flight was powered by a blend that included 20 % of biofuel in only one of its four engines, the rest was conventional kerosene) [94]. CO_2 emissions from biodiesel are reduced by about 75 % compared to petrodiesel (and they do not contain SO_x). Biodiesel is the only alternative fuel to have successfully completed the US EPA-required Tier I and Tier II health effect testing under the Clean Air Act [95].

9.4.8 Barriers and Challenges to Deployment of Algae-Based Systems

Barriers and challenges to large-scale deployment of the algal farming systems are as follows:

- Algae farms are currently very expensive: the estimates of capital and operating costs are US$138,000 per hectare and US$43,800 per hectare per annum, respectively [1]. The addition of the cost of CO_2 capture and transport would further increase the cost of the overall system.
- Availability of land. Some studies point to a large land requirement of the algae farming process: about 40 acres of land per each MW_{el} produced by a coal-fired

power plant [4]. High land requirements may limit the commercial viability of the technology in areas with high land prices.

- There are still significant remaining technical issues to overcome, in particular, the reliability of the systems must be proven for year-round operation to ensure supply.
- Algal farming requires large amounts of nutrients similar to existing CO_2-intensive agricultural systems, which would increase lifecycle CO_2 emissions (though, in a captive system, this problem can be managed through CO_2 recycling).
- The algae cultivation technology is best suited to regions with high solar radiation resource and large areas of marginal (inexpensive) land surrounding large CO_2 point sources (e.g., a coal-fired power plant); this will limit the deployment of the technology to certain favorable areas, and discourage its implementation in many other regions.

As of 2013, there has been no known large-scale commercial production of algae for energy, fuels, or materials applications, mainly, because of technological difficulties to control algal growth and downstream processing at competitive costs. On the other hand, many of the recent developments are shrouded in commercial secrecy, so the precise status of the technology is difficult to ascertain.

9.4.9 Carbon Mitigation Potential of Industrial CO_2 Utilization

Recently, a support for industrial CO_2 utilization has been growing for the obvious reason: turning CO_2 into a value-added product could potentially make the process of CO_2 capturing more profitable. This approach also alleviates the need for developing a very expensive infrastructure for depositing CO_2 underground with potential risk of leakage. Due to a generous industrial and government support, there have been significant technological and commercial developments in the CCU area. Figure 9.10 shows the development status and timeline of CO_2 reuse technologies.

Figure 9.10 shows that most CCU technologies will be approaching the commercial stage between 2015 and 2020. As of 2012, none of CCU approaches had proven to cost-effectively produce a CO_2-derived product that could be a sustainable alternative to fossil fuel-derived products. The main challenge to the broad-scale deployment of CCU technologies relates to a rather limited market demand for the CO_2-derived products and high capital costs for plant construction. But that does not mean that the prospects for CCU could not be improved in the near future.

The carbon mitigation potential of industrial CO_2 utilization is determined by two main factors: (1) the overall volume of CO_2-derived products (i.e., its worldwide production and inventory), and (2) the carbon-storing properties of the product (i.e., how long it would take for the product to degrade back to CO_2). If the worldwide production of polycarbonate and polyurethane products were switched to CO_2-based processes, the industrial CO_2 consumption would

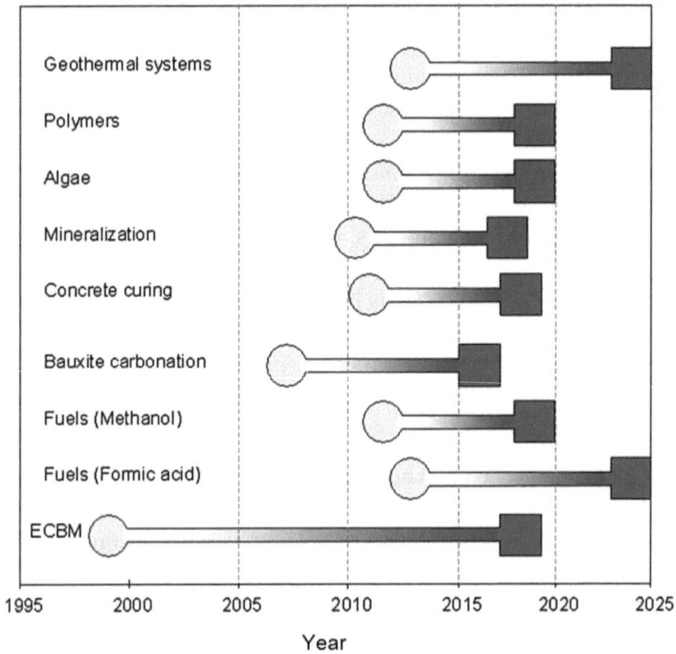

Fig. 9.10 CO_2 reuse technology development timeline. *Light circles* represent the technology at demonstration scale; *dark squares* represent full-scale commercial operation (based on the claims from the developers of the technology and estimates by the Global CCS Institute). *Source*: [1]

increase by about 3.3 Mt/year CO_2. As shown in this chapter, currently, the industrial use of CO_2 can potentially take up a very small portion of man-made CO_2 emissions and no significant increase in the commercial production of traditional CO_2-derived products is expected in the near future (most of the new developments aim mostly at replacing toxic feedstocks by CO_2). According to estimates, realistically, only about 10 % of global carbon emissions could be converted into CO_2-derived synthetic fuels [96].

Another important factor to consider relates to the long-term carbon-storing capacity of CO_2-based products (i.e., storage permanence). The duration of CO_2 storage in the traditional CO_2-derived products varies in a wide range from a few hours-to-days for carbonated beverages and fuel-related applications to a few months-to-years for fertilizers and pesticides to several decades-to-centuries for construction materials (e.g., plastics and laminates). This factor might diminish the capacity of the CO_2-derived products with relatively low storage permanence characteristics to serve as a carbon sink even if they would enjoy a very large market. Many experts believe that despite the enormous appeal and rapid technological development of industrial CO_2 utilization processes, in all likelihood, at least, in near-to-mid term, CCU will be seen as a complementary technology but not as an alternative to CCS [96].

References

1. Global CCS Institute (2011) Accelerating the uptake of CCS: industrial use of captured carbon dioxide. http://www.globalccsinstitute.com/resources/publications/accelerating-uptake-ccs-industrial-use-captured-carbon-dioxide. Accessed 25 May 2011
2. SRI Consulting (2010) Chemical economics handbook. Menlo Park, California
3. Global CCS Institute (2011) The global status of CCS: 2011. Canberra, Australia. ISBN 978-0-9871863-0-0
4. Global CCS Institute (2009) Strategic analysis of the global status of carbon capture and storage. Final report. http://www/globalccsinstitute.com/downloads/rcports/2009/worley/foundation-report-1-rev0.pdf. Accessed 3 Aug 2010
5. Intergovernmental Panel on Climate Change (2005) Carbon dioxide capture and storage. Prepared by Working Group III of the Intergovernmental Panel on Climate Change. In: Metz B, Davidson O, de Coninck H et al (eds). Cambridge University Press, Cambridge and NY, USA
6. Stevens S, Kuuskraa J, Schraufnagel R (1996) Technology spurs growth of US coal bed methane. Oil Gas J 94:56–63
7. Reeves S, Davis D, Oudinot A (2004) A technical and economic sensitivity study of enhanced coalbed methane recovery and carbon sequestration in coal. DOE topical report, March 2004. Washington DC
8. Karmis M (2009) SECARB initiatives in central and southern Appalachia—progress and future opportunities. ECC annual meeting, 12 May 2009, Kingsport, Tennessee
9. Hernandez G, Bello R, McVay D et al (2006) Evaluation of the technical and economic feasibility of CO2 sequestration and enhanced coal-bed-methane recovery in Texas Low-Rank Coals. Soc Petrol Eng J. doi: 10.2118/100584-MS. ISBN 978-1-55563-233-5
10. Hongguan Y, Guangzhu Z, Weitang F et al (2007) Predicted CO2 enhanced coal-bed methane recovery and CO2 sequestration in China. Int J Coal Geology 71:345–357
11. Hamelinck C, Faaij A, Turkenburg C et al (2002) CO2 enhanced coal-bed methane production in the Netherlands. Energy 27:647–674
12. CO_2 storage with geothermal energy production (2012) Carbon Capture J, July 8, 2012. http://www.carboncapturejournal.com/displaynews.php?NewsID=973&PHPSESSID=up5ge1lr524iq0ita9tgcok406. Accessed 8 Aug 2012
13. Randolph J, Saar M (2011) Combining geothermal energy capture with geologic carbon dioxide sequestration. Geophys Res Lett 38, L10401. doi:10.1029/2011GL047265
14. Inoue S, Koinuma H, Tsuruta T (1969) Copolymerization of carbon dioxide and epoxide. J Polymer Sci B Polymer Lett 7:287–292
15. Novomer (2010) http://www.novomer.com. Accessed 12 Dec 2010
16. Bomgardner M (2012) CO_2 pursued as feedstock. Chem Eng News 90:22
17. Priestnall M (2012) Making money from mineralization of CO_2. Carbon Capture Journal. November–December 7–9
18. Hunwick R (2009) A new, integrated, approach to mineralisation-based CCS. Mod Power Syst 29:25–28
19. Hall C (2012) Carbon capture and mineralization: coal's new savior. Carbon Capture Journal, Aug 8, 2012. http://www.energydigital.com/global_mining/carbon-capture-and-mineralization-coals-new-savior. Accessed 10 Oct 2012
20. Smart stones (2012) http://smartstones.nl/index.php/en/challenges/ideeen. Accessed 10 Dec 2012
21. Herzog H (2002) Carbon sequestration via mineral carbonation: overview and assessment. MIT Laboratory for Energy and the Environment, Cambridge, MA
22. Bockris JO'M (1980) Energy options. Australia and New Zealand Book Company, Sydney
23. Olah G, Goeppert A, Prakash S (2006) Beyond oil and gas: the methanol economy. Wiley, Germany
24. Dakota Gasifiication Company (2012) http://www.dakotagas.com/Gasification/index.html. Accessed 10 Dec 2912

25. Ralston R (2010) The Sabatier reaction, possible solution to CO2 emissions. PennEnergy. http://www.pennenergy.com/articles/pennenergy/2010/03/the-sabatier-reaction.html. Accessed 4 Mar 2010
26. Kim J, Lee S, Lee S et al (2006) Performance of catalytic reactors for the hydrogenation of CO_2 to hydrocarbons. Catal Today 115:228–234
27. Dorner R, Hardy D, Williams F et al (2009) Influence of gas feed composition and pressure on the catalytic conversion of CO_2 to hydrocarbons using a traditional cobalt-based Fischer-Tropsch catalyst. Energy Fuel 23:4190–4195
28. Stechel E, Miller J (2013) Re-energizing CO_2 to fuels with the sun: issues of efficiency, scale, and economics. J CO_2 Util 1:28–36
29. Harris S (2012) Chemical potential: turning carbon dioxide into fuel. http://www.theengineer.co.uk/sectors/energy-and-environment/in-depth/chemical-potential-turning-carbon-dioxide-into-fuel/1013459.article. Accessed 9 Aug 2012
30. Traynor A, Jensen R (2002) Direct solar reduction of CO2 to fuel: first prototype results. Ind Eng Chem Res 41:1935–1939
31. Jacoby M (2013) The hidden value of carbon dioxide. Chem Eng News 91:21–22
32. Kumar B, Smieja J, Kubiak C (2010) Photoreduction of CO_2 on p-type silicon using Re(bipy-But)$(CO)_3$Cl: photovoltages exceeding 600 mV for the selective reduction of CO_2 to CO. J Phys Chem C 114:14220–14223
33. Kaneco S, Katsumara H, Suzuki T et al (2006) Photoelectrocatalytic reduction of CO2 in LiOH/methanol at metal-modified p-InP electrodes. Appl Catal B Environ 64:139–145
34. Anpo M (2013) Photocatalytic reduction of CO2 on highly dispersed Ti-oxide catalysts as a model of artificial photosynthesis. J CO_2 Util 1:8–17
35. Kuhl K, Cave E, Abram D (2012) New insights into the electrochemical reduction of carbon dioxide on metallic copper surfaces. Energy Environ Sci 5:7050–7059
36. DiMeglio J, Rosenthal J (2013) Selective conversion of CO_2 to CO with high efficiency using an inexpensive bismuth-based electrocatalyst. J Am Chem Soc 135:8798–8801
37. Ionic liquid catalyst could help turn emissions into fuel (2011) The Engineer, October 7, 2011. http://www.theengineer.co.uk/sectors/energy-and-environment/news/ionic-liquid-catalyst-could-help-turn-emissions-into-fuel/1010535.article. Accessed 5 Feb 2012
38. Lvov S (2012) Hydrogen via electrolysis: a key to utilization of renewable energy resources. In: Muradov N, Veziroglu N (eds) Carbon-neutral fuels and energy carriers. CRC Press, Boca Raton
39. Centi G, Perathoner S (2012) Solar production of fuels from water and CO_2. In: Muradov N, Veziroglu N (eds) Carbon-neutral fuels and energy carriers. CRC Press, Boca Raton
40. Hu B, Guild C, Suib S (2013) Thermal, electrochemical and photochemical conversion of CO2 to fuels and value-added products. J CO_2 Util 1:18–27
41. Centi G, Quadrelli E, Perathoner S (2013) Catalysis for CO_2 conversion: a key technology for rapid introduction of renewable energy in the value chain of chemical industries. Energy Environ Sci 6:1711–1731
42. Jiang Z, Xiao T, Kuznetsov V et al (2010) Turning carbon dioxide into fuel. Phil Trans R Soc A 368:3343–3364
43. Yu D, Zhang Y (2010) Copper- and copper–*N*-heterocyclic carbene-catalyzed C—H activating carboxylation of terminal alkynes with CO_2 at ambient conditions. Proc Natl Acad Sci U S A 107:20184–20189
44. Yang H, Gu Y, Deng Y et al (2002) Electrochemical activation of carbon dioxide in ionic liquid: synthesis of cyclic carbonates at mild reaction conditions. Chem Commun 274–275
45. Olah J, Toeroek B, Joschek J et al (2012) Efficient chemoselective carboxylation of aromatics to arylcarboxylic acids with a superelectrophilically activated carbon dioxide–Al2Cl6/Al system. J Am Chem Soc 124:11379–11391
46. Petronas and Lanzatech to recycle CO_2 into chemicals (2012) Carbon Capture Journal, October 15, 2012. http://www.carboncapturejournal.com/displaynews.php?NewsID=1033&PHPSESSID=l8dkjsha7qaa6s815mlrslvtq4. Accessed 5 Dec 2012
47. Ogura K (2013) Electrochemical reduction of carbon dioxide to ethylene: mechanistic approach. J CO_2 Util 1:43–49

48. Hallman M, Steinberg M (1999) Greenhouse gas CO_2 mitigation. CRC Press, Boca Raton
49. U.S. Department of Energy (2010) National algal biofuels technology roadmap. Office of energy efficiency and renewable energy, biomass program. algal_biofuels_roadmap.pdf., http://biomass.energy.gov. Accessed 19 Dec 2010
50. Falkowski P, Katz M, Knoll A et al (2004) The evolution of modern eukaryotic phytopnkton. Science 305:354–360
51. Hu Q, Sommerfeld M, Jarvis E et al (2008) Microalgal triacylglycerols as feedstocks for biofuel production: perspectives and advances. Plant J 54:621–639
52. Chu W, Norazmi M, Phang S (2003) Fatty acid composition of some malaysian seaweeds. Malays J Sci 22:21–27
53. Christi Y (2007) Biodiesel from microalgae. Biotechnol Adv 25:294–306
54. Eberhard S, Finazzi G, Wollman F (2008) The dynamics of photosynthesis. Annu Rev Genet 42:463–515
55. Polle J, Kanakagiri S, Jin E et al (2002) Truncated chlorophyll antenna size of the photosystems—a practical method to improve microalgal productivity and hydrogen production in mass culture. Int J Hydrogen Energy 27:1257–1264
56. Brennan L, Owende P (2009) Biofuels from microalgae—a review of techniques for production, processing, and extractions of biofuels and co-products. Renew Sust Energ Rev. doi:10.1016/j.rser.2009.10.009
57. Nakamura T (2004) Recovery and sequestration of CO2 from stationary combustion systems by photosynthesis of microalgae, Technical report to DOE, NETL, No. PSI-1356, December 2004. Morgantown, WV
58. Velea S, Dragos N, Serban S et al (2008) Biological sequestration of carbon dioxide from thermal power plant emissions by absorption in microalgal culture media. Romanian Biotechnol Lett 14:4485–4500
59. Benson B, Gutierrez-Wing T, Rusch A (2007) The development of a mechanistic model to investigate the impacts of the light dynamics on algal productivity in a hydraulically integrated serial turbidostat algal reactor. Aquaculture Eng 36:198–211
60. Wilson W, Van Etten J, Allen M (2009) The phycodnaviridae: the story of how tiny giants rule the world. Lesser known large dsDNA viruses, vol 328. Springer, Berlin, pp 1–42
61. Molina Grima E, Belarbi E, Acién Fernández F et al (2003) Recovery of microalgal biomass and metabolites: process options and economics. Biotechnol Adv 20:491–515
62. Illman A, Scragg A, Shales S (2000) Increase in chlorella strains calorific values when grown in low nitrogen media. Enzym Microb Tech 27:631–635
63. Denery J, Dragull K, Tang C et al (2004) Pressurized fluid extraction of carotenoids from *Haematococcus pluvialis* and *Dunaliella salina* and kavalactones from *Piper methysticum*. Anal Chim Acta 501:175–181
64. Herrero M, Cifuentes A, Ibanez E (2006) Sub-and supercritical fluid extraction of functional ingredients from different natural sources: plants, food-by-products, algae and microalgae: a review. Food Chem 98:136–148
65. Metzger P, Largeau C (2005) *Botryococcus braunii*: a rich source for hydrocarbons and related ether lipids. Appl Microbiol Biotechnol 66:486–496
66. U.S. Department of Energy (2012) Energy efficiency and renewable energy, golden field office. Bio-oil stabilization and commoditization: DE-FOA-0000686, https://eere-exchange.energy.gov/. Accessed 20 Nov 2012
67. Hon-Nami K (2006) A unique feature of hydrogen recovery in endogenous starch-to-alcohol fermentation of the marine microalga, *Chlamydomonas perigranulata*. Appl Biochem Biotechnol 131:808–828
68. Hirano A, Ueda R, Hirayama S et al (1997) CO_{22} fixation and ethanol production with microalgal photosynthesis and intracellular anaerobic fermentation. Energy-Oxford 22:137–142
69. ALGENOL Biofuels (2012) Harnessing the Sun to Fuel the World. http://www.algenolbiofuels.com/direct-to-ethanol/direct-to-ethanol. Accessed 5 Jul 2012
70. Xu H, Miao X, Wu Q (2006) High quality biodiesel production from a microalga *Chlorella protothecoides* by heterotrophic growth in fermenters. J Biotechnol 126:499–507

71. Bridgewater A (2004) Biomass fast pyrolysis. Therm Sci 8:21–50
72. Muradov N, Fidalgo B, Gujar A et al (2010) Pyrolysis of fast-growing aquatic biomass—*Lemna minor* (duckweed): characterization of pyrolysis products. Bioresour Technol 101:8424–8428
73. Okabe K, Murata K, Nakanishi M et al (2009) Fischer-Tropsch synthesis over Ru catalysts by using syngas derived from woody biomass. Catal Lett 128:171–176
74. Patil V, Tran K, Giselroed H (2008) Towards sustainable production of biofuels from microalgae. Int J Molecular Sciences 9:1188–1195
75. Goudriaan F, Van de Beld B, Boerefijn F et al (2000) Thermal efficiency of the HTU® process for biomass liquefaction. In: Bridgwater A (ed) Proceedings of progress in thermochemical biomass conversion, Tyrol, Austria, Sep 17–22, 2000. Blackwell Science, Oxford, p 1312–1325
76. Sawayama S, Inoue S, Dote Y et al (1995) CO_2 fixation and oil production through microalga. Energy Convers Manag 36:729–731
77. Mendes R (2007) Supercritical fluid extraction of active compounds from algae. In: Martinez J (ed) Supercritical fluid extraction of nutraceuticals and bioactive compounds. Taylor and Francis, Boca Raton, pp 189–213
78. Hawash S, Kamal N, Zaher F et al (2009) Biodiesel fuel from Jatropha oil via non-catalytic supercritical methanol transesterification. Fuel 88:579–582
79. Vergara-Fernandez A, Vargas G, Alarcon N et al (2008) Evaluation of marine algae as a source of biogas in a two-stage anaerobic reactor system. Biomass Bioenergy 32:338–344
80. Hossain A, Salleh A, Boyce A et al (2008) Biodiesel fuel production from algae as renewable energy. Am J Biochem Biotechnol 4:250–254
81. Kalva A, Sivasankar T, Moholkar V (2008) Physical mechanism of ultrasound-assisted synthesis of biodiesel. Ind Eng Chem Res 48:534–544
82. Svensson J, Adlercreutz P (2008) Identification of triacylglycerols in the enzymatic transesterification of rapeseed and butter oil. Eur J Lipid Sci Technol 110:1007–1013
83. Soriano N, Venditti R, Argyropoulos D (2009) Biodiesel synthesis via homogeneous Lewis acid-catalyzed transesterification. Fuel 88:560–565
84. Mooibroek H, Oosterhuis N, Giuseppin M et al (2007) Assessment of technological options and economical feasibility for cyanophycin biopolymer and high-value amino acid production. Appl Microbiol Biotechnol 77:257–267
85. Spolaore P, Joannis-Cassan C, Duran E et al (2006) Commercial applications of microalgae. J Biosci Bioeng 101:87–96
86. Brune D, Lundquist T, Benemann J (2009) Microalgal biomass for greenhouse gas reductions: potential for replacement of fossil-fuels and animal feeds. J Environ Eng 135:1136–1144
87. Campbell W, Naucler T, Ruijs J (2008) Carbon capture & storage: assessing the economics. McKinsey Climate Change Initiative. http://www.mckinsey.com/clientservice/ccsi. Accessed 12 Oct 2012
88. Jacob-Lopes E, Teixera Franco T (2013) From oil refinery to microalgal biorefinery. Journal of CO2 Utilization 2:1–7. doi:10.1016/j.jcou.2013.06.001
89. Rubin E, Meyer L, de Coninck H (2005) Technical summary. In: Metz B, Davidson O, de Connick H (eds) Carbon dioxide capture and storage. Cambridge University Press, Cambridge, pp 17–49
90. US Department of Energy (1998) A look back at the U.S. department of energy's aquatic species program: biodiesel from algae. NREL/TP-580–24190. Technical report under, contract No. DE-AC36–83CH10093, 1998
91. Solazyme (2012) How the Solazyme biotechnology platform works. www.solazyme.com. Accessed 10 Dec 2012
92. European Biofuels Technology Platform (2012) Algae for production of biofuels. http://www.biofuelstp.eu/algae.html. Accessed 12 Nov 2012
93. Department of Energy (2012) Alternative fuels data center: drop-in biofuels, http://www.afdc.energy.gov/fuels/emerging_dropin_biofuels.html. Accessed 15 Oct 2012

94. Algae Biodiesel (2012) Flight path to a cleaner future. http://algaebiodiesel.com/flight-path-to-a-cleaner-future. Accessed 30 Jun 2012
95. Cellana (2012) Products overview: biofuels. http://cellana.com/products-overview/biofuels/. Accessed 20 Sep 2012
96. Styring P, de Coninck H, Reith H et al (2011) Carbon capture and utilisation in the green economy. Publisher: The Centre for Low Carbon Futures. Report no. 501

Chapter 10
Carbon-Negative Options

Abstract Considering the slow progress of current carbon mitigation policies, there is a growing recognition that the low-risk levels of atmospheric CO_2 cannot be achieved without a significant carbon-negative component. Among the proposed carbon-negative solutions, bioenergy coupled with carbon capture and storage (Bio-CCS) is the most technologically advanced option. The conversion of biomass to biochar associated with negative CO_2 emissions is another promising approach in the context of carbon abatement policies. Removal of CO_2 from atmosphere (air capture) by chemical systems as a carbon-negative option is still in the early stage of technological development and would require the increased industrial and government support for pilot and demonstration-scale projects to drive its costs down. The current scientific and technological status, economic and environmental aspects, as well as opportunities for Bio-CCS, biochar, air capture as carbon-negative solutions are analyzed in this chapter.

10.1 Introduction

The notion of a "*carbon-negative*" energy system is a relatively new concept, mostly driven by the concerns that the likelihood of achieving the required "safe" atmospheric CO_2 stabilization level (to keep global temperature rise below the 2 °C threshold) is increasingly diminishing with every passing year. There is a growing realization that despite decades-long international efforts to curb CO_2 emissions they continue rising, and all the proposed solutions based on current climate change mitigation policies, unfortunately, are not producing the necessary action. Some justifiable concerns exist that even if GHG emissions are somehow dramatically cut at unprecedented rates, we may still face the accumulated impact of past emissions (almost half of CO_2 released today will be in the atmosphere a century from now). All these concerns are exacerbated by uncertainties in the climate response to increased CO_2 levels due to possible (not well-understood) feedback mechanisms.

N. Muradov, *Liberating Energy from Carbon: Introduction to Decarbonization*, Lecture Notes in Energy 22, DOI 10.1007/978-1-4939-0545-4_10,

Thus, it appears that there will be an urgent need for the deployment of radical "catch-up" technologies that would not only reduce carbon emissions, but, actually, start removing some CO_2 from the atmosphere.

Many experts believe that the role of carbon-negative processes will become increasingly important in the context of carbon abatement policy for they provide the only option to fill the gap between required CO_2 emission reductions (in order to meet carbon mitigation targets) and global CO_2 emission trends. Potentially, the carbon-negative systems could lead to a new paradigm in ways to produce and use energy: the production of more energy would lead to not more but, essentially, less CO_2 in the atmosphere. Most recent carbon emission models (e.g., models AIM 6.0, MiniCAM 4.5, IMAGE 2.6) include the technology scenarios involving net negative CO_2 emissions and project that these technologies will be an important component of carbon stabilization policies by the middle of twenty-first century [1].

Currently, several approaches to the development of carbon-negative systems are under consideration:

- Biological systems, e.g.,
  - Bioenergy with CCS
  - Biochar production
- Chemical systems, e.g.,
  - Capture of atmospheric CO_2 coupled with its storage
  - Conversion of captured atmospheric CO_2 to stable carbonaceous products

10.2 Bioenergy with CCS (Bio-CCS)

The extraordinary ability of biological systems to capture carbon from atmosphere, and the potential to use these systems as the source of energy with negative carbon emissions at acceptable cost, have been brought recently to the forefront of carbon mitigation efforts [2]. Biomass conversion to energy and fuels in combination with CCS technologies is one of few options that could actually reduce atmospheric CO_2 concentration and, according to many experts, it is likely to be required to meet stringent climate policy targets. (This technology is referred to in the literature in an abbreviated form as *Bio-CCS*, or *Bioenergy CCS*, or *BECCS*; hereafter, the acronym *Bio-CCS* will be used in this chapter.) Similar to traditional CCS applications, the Bio-CCS approach involves capturing CO_2 from biomass point emission sources, transporting and storing it in geological formations or other means of long-term CO_2 storage [3].

The proponents of the Bio-CCS approach point out that billions of years ago, the Earth's atmosphere was saturated with CO_2, and it was removed by algae (cyanobacteria) that fixed that primordial CO_2 and rendered our atmosphere breathable [4]. In a similar fashion (but on much shorter timescale), the combination of biosystems with CCS would help to drastically reduce atmospheric CO_2 concentrations if efficient commercial Bio-CCS technologies were developed and implemented.

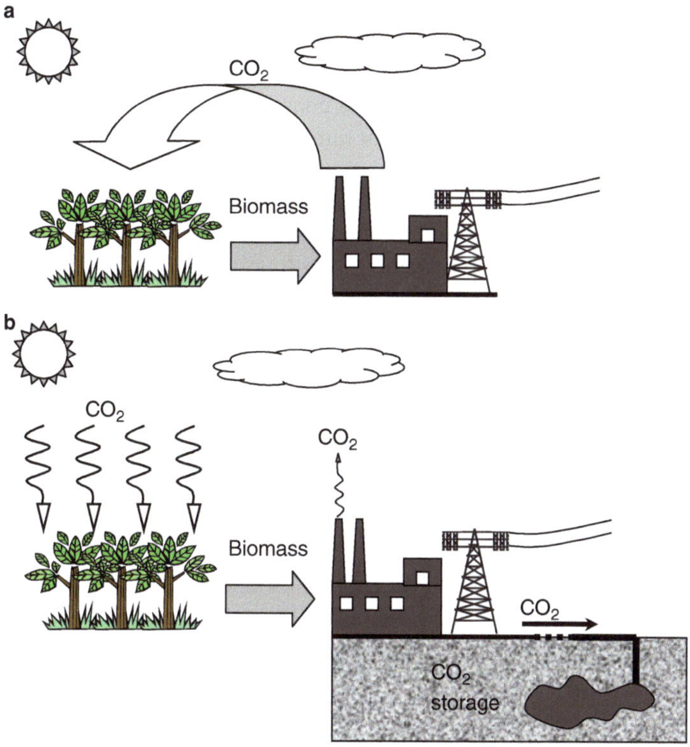

Fig. 10.1 Schematic representation of biomass-based carbon-neutral (**a**) and carbon-negative (**b**) systems

Biomass is widely referred to in the literature as a *carbon-neutral* energy source; therefore, it would be useful to clearly differentiate between biomass-based carbon-neutral and carbon-negative systems. Figure 10.1 depicts a sketch illustrating the difference between biomass-based carbon-neutral and carbon-negative systems.

As part of the natural carbon cycle, during plants growth atmospheric CO_2 is captured and stored in biomass (via *bio-sequestration*), but, as a carbon sink, such plants (or any other primary producer that binds CO_2 into biomass) are not permanent. The carbon sink of this type simply moves carbon from the atmosphere or hydrosphere to the biosphere. This process could be reversed in nature, e.g., by wildfires or degradation, and in industry—by combustion or gasification or thermal conversion. Since during this cycle the amount of released CO_2 approximately equals to that consumed during the plants' growth, the system is considered "*carbon-neutral*." Bio-CCS enables removing CO_2 from this natural cycle and locking it in a sink, thus making possible a net removal of CO_2 from the atmosphere (i.e., acting as a "*carbon-negative system*").

A term "*negative carbon* (*or CO_2*) *emission*" could also be found in the literature. The difference between these two interrelated terms is that the term "carbon-negative system" implies that the amount of atmospheric CO_2 absorbed and sequestered by the system is greater than the CO_2 amount that would ultimately be released back to the atmosphere (e.g., via oxidation, a leakage, diffusion, degradation, and other processes). The "negative carbon emission" is considered the direct opposite of carbon emission, and it results in the permanent removal of CO_2 from the atmosphere. The carbon-negative systems are typically associated with such conventional carbon sink options as injecting CO_2 in the Earth's subsurface, or converting it into insoluble carbonate salts (mineral sequestration), that would sequester carbon for a considerable time duration (preferably, thousands to millions of years).

It should be noted that CCS applied to fossil fuels cannot generate negative carbon emissions, but co-firing fossil fuels with biomass at power plants equipped with CCS unit could lead either to reduced, zero or, even, negative emissions, depending on the biomass/fossil fuel ratio and the CCS system efficiency [3]. The Bio-CCS approach can be applied not only to biomass combustion systems but also to industrial processes where biomass is used as a feedstock and energy source resulting in flue gases rich with CO_2, e.g., pulp and paper plants, bioethanol production, biogas upgrading processes, among others. The important advantage of the Bio-CCS approach is that it enables mitigating carbon emissions that have already occurred (as opposed to capturing it from different point sources).

Currently, Bio-CCS based on sustainable biomass resources is considered the only near-term large-scale technology that could accomplish net negative carbon emissions in addition to any emission reductions achieved by replacing fossil fuels with that biomass [5]. IEA regards the Bio-CCS technology as a very important carbon abatement solution. In its 2011 GHG report, IEA underscored the potential climate mitigation impact of Bio-CCS systems stating that if widely deployed Bio-CCS technology could result in negative emissions of up to 10Gt of CO_2 per year [6]. It has been emphasized that achieving atmospheric CO_2 concentration stabilization goal of 450 ppm would depend significantly on the worldwide deployment of the Bio-CCS systems.

10.2.1 Bio-CCS Resources and Feedstocks

10.2.1.1 Resources

Biomass resources for the global deployment of Bio-CCS technology are immense. Biomass usage has grown steadily over the last 40 years, and by 2009, biomass accounted for about 10 % (or 50 EJ) of the annual global total primary energy supply (TPES) [7]. Global production of bioenergy comes from the following sources (in EJ) [5]:

Solid biomass	46.9
MSW used for heat and CHP	0.58
Biogas (secondary energy) for electricity, heat, and CHP	0.74
Biofuels (secondary energy): ethanol, biodiesel, and others	1.9

The IEA projects that the share of primary bioenergy in TPES will further grow and reach ~160 EJ by 2050, when it will provide about 24 % of TPES. Of the 160EJ of bioenergy, approximately 60 EJ will be used to produce transportation fuels, with remaining 100EJ (which corresponds to 5–7 billion dry tons of biomass) directed to generation of heat and electricity for residential, industrial, and other sectors [7].

10.2.1.2 Feedstocks

A wide range of biomass feedstocks is available for production of bioenergy (through combustion), and biofuels and biochemicals (through biological or chemical conversion), and for use in biorefinery concepts, e.g., through so-called *4F-Concept* (food, feed, fiber, and fuels) [5]:

- Agricultural residues (straw, prunings, corn stover, animal manure, etc.)
- Forest biomass (branches, foliage, roots, etc.)
- Energy crops, which include annual or perennial crops specifically cultivated to produce biomass with specific traits (biowaste streams, MSW, kitchen and garden waste, paper and cardboard, demolition and household waste wood, sewage sludge, packaging waste, market waste, food processing waste, etc.)
- Algae/aquaculture (microalgae, macroalgae, seaweeds)

More information on biomass as an energy source could be found in Sect. 6.2.3.

10.2.2 Bio-CCS Technological Routes

There are three main technological routes to the conversion of biomass into energy and fuels in combination with CCS: biochemical, thermochemical, and thermal (combustion); they are summarized in Fig. 10.2.

10.2.2.1 Biochemical Routes

Biochemical Bio-CCS routes involve fermentation or anaerobic digestion processes in which biomass or any biodegradable organic material is stepwise broken down into smaller molecules by microorganisms. Depending on the type of fermentation or digestion process, this approach could produce biogas (biomethane) or alcohols.

Biogas (Biomethane)

Although biogas could be formed via anaerobic digestion of a great variety of biomass sources, from a practical viewpoint, the following feedstocks present most interest: animal, human, food and organic waste streams, MSW, algae (see Sect. 9.3.2), and

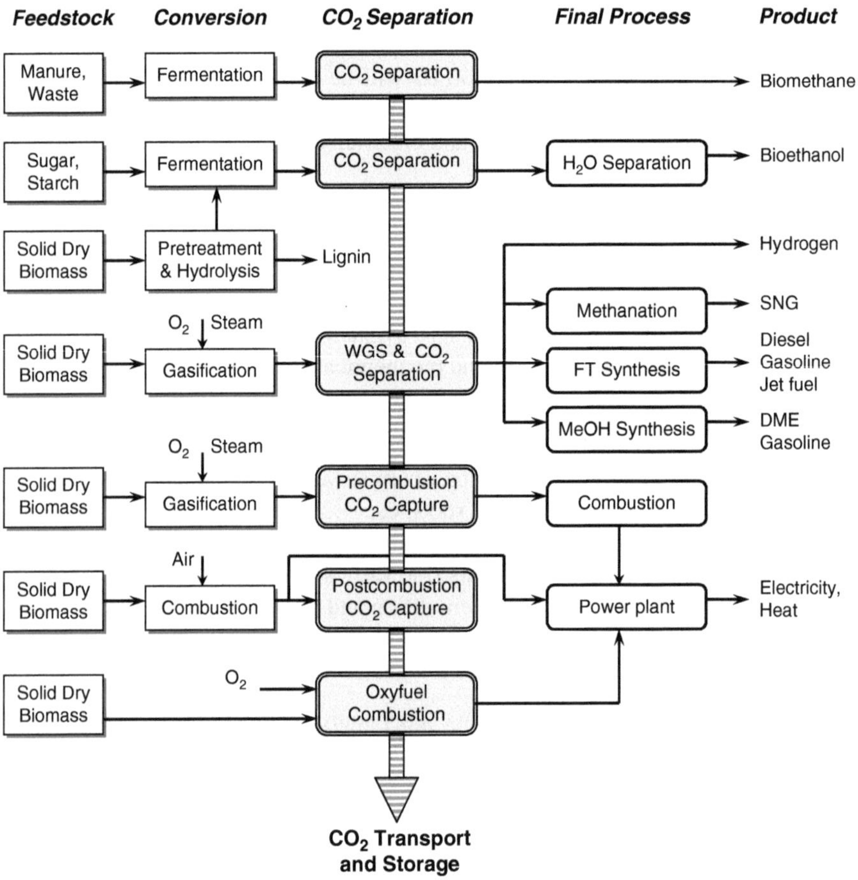

Fig. 10.2 Technological routes for the conversion of biomass into energy and fuels in combination with CCS. *Source* [5]

some green crops (but not wood). The process produces biogas: mixture of methane (45–70 vol.%), CO_2 (25–45 vol.%), and trace amounts of sulfurous compounds (H_2S, mercaptans), N_2, ammonia, silicon-organic compounds (if MSW is digested), and a solid residue (digestate). Methane is separated from biogas (after a preliminary removal of sulfurous and other harmful ingredients) by off-the-shelf technologies (adsorption, cryogenic distillation, membranes), and it can be injected into NG grid. Due to relatively small stream of CO_2 from biomethane operation, the economic feasibility of this Bio-CCS route will depend on the proximity of other major CO_2 sources.

Bioethanol

Conventional bioethanol fuel production process involves fermentation of starch-based feedstocks (e.g., corn, sugar cane) (see Chap. 6). A typical ethanol plant in the USA annually produces about 200 million liters of ethanol, with associated emissions of 140,000 t of near-pure CO_2 stream, that can be transported and sequestered [5]. The production of lignocellulosic ethanol requires a pretreatment step where cellulose is separated from lignin, and, after hydrolysis, simple sugars are further fermented to ethanol via conventional fermentation process. The main fraction (about 62 %) of carbon present in the lignocellulosic feedstock (e.g., straw) ends up in the lignin by-product, with about 25 % going to the ethanol product and approximately 13 % to the near-pure CO_2 stream [5]. Thus, both production processes produce sequestration-ready CO_2, which simplifies their Bio-CCS application. As in case of biogas, the economic feasibility of this Bio-CCS route will largely depend on the proximity of other CO_2 sources that could share a carbon storage site.

10.2.2.2 Thermochemical Routes

Most thermochemical Bio-CCS routes involve gasification of lignocellulosic or other nonfood-related biomass feedstocks using oxygen (or O_2-enriched air) and/or steam. The produced synthesis gas (sometimes also called biosyngas, which is mainly mixture of H_2, CO, and CO_2) is used in commercially available processes to form a variety of gaseous and liquid fuels such as:

- Hydrogen that can be used as fuel or a reagent
- Synthetic (or substitute) NG is produced via commercial methanation process
- Liquid hydrocarbon fuels are produced via FT synthesis (BTL process)
- Methanol can be commercially produced via methanol synthesis process

The H_2/CO ratio in the biosyngas can be adjusted depending on the targeted final product; this ratio for FT and methanol synthesis route is close to 2:1 (molar), whereas for SNG route it should be close to 3:1 (molar). This ratio is typically adjusted by WGS reaction with subsequent removal of CO_2. As a result, the process produces a relatively pure CO_2 stream that can be captured using the same precombustion CO_2 capture technology used in IGCC plants (i.e., physical solvents, such as methanol, glycols). Due to the use of large quantities of oxidants (e.g., O_2, steam) in the gasification processes, significant fraction of the feedstock carbon ends up in CO_2. Figure 10.3 shows the carbon balance in a typical BTL plant for production of diesel fuel using O_2-blown circulating fluidized bed (CFB) gasification technology.

Figure 10.3 indicates that only 37 % of carbon ends up in the diesel fuel product, with more than half of it converted into CO_2 in the form relatively pure stream that could be easily captured and sequestered via Bio-CCS (relatively small amounts of biochar recovered from the gasifier and CO_2-vent from the power island make up the difference) [5].

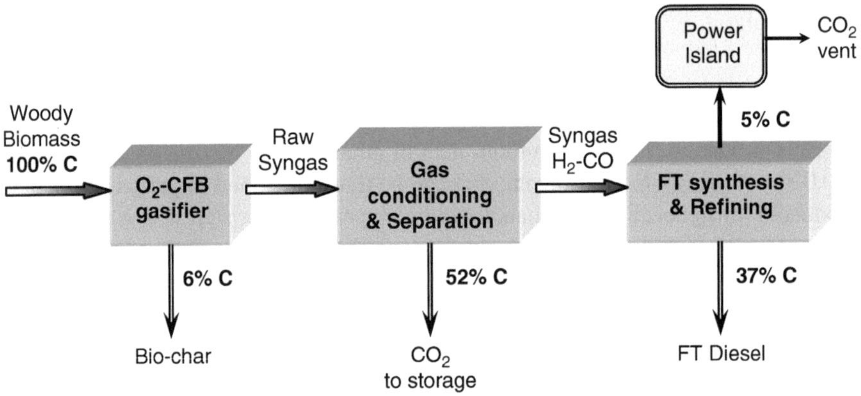

Fig. 10.3 The distribution of carbon in a typical biomass-to-liquid (BTL) plant for producing FT-diesel through biomass gasification technology. *CFB* circulating fluidized bed reactor. *Source* [5]

Biomass Combustion and Co-firing

Biomass could be used as fuel for electricity and/heat generation in pure form or as a mixture with fossil fuels, e.g., coal and peat (see also Sect. 6.2.3). When used in pure form, biomass combustion occurs in special modified pulverized feed boilers and medium-sized CHP plants. CFB boilers capacities typically range from 50 to 500 MW_{th}, and they are usually located close to urban areas and industrial facilities in order to supply heat [5]. The adaptation of oxy-fuel firing technology for biomass would greatly facilitate Bio-CCS application. Biomass could also be co-fired with coal; the ground biomass feedstock can be blended with pulverized coal and fed to the burners. The preferred biomass to coal ratio would depend on the biomass feed characteristics and the layout of the power plant. The practice of co-firing showed that achieving high biomass/coal co-firing ratios has proved difficult due to nonhomogeneous and fibrous nature of biomass. Thermal pretreatment (at 250–550 °C) of biomass allows mitigating this problem, as it increases the homogeneity and energy density of biomass [5].

Biomethane and Bio-SNG Combustion

Both fermentation-based biomethane (recovered from biogas after CO_2 separation) and gasification-based bio-SNG could be used as fuels in conventional gas-fired NGCC power plants or CHP plants without any restrictions or limitations. Conventional postcombustion CO_2 capture technology could be conveniently adopted to these power plants with subsequent CO_2 transport and storage.

Biomass-Based IGCC

The application of biomass feedstocks to IGCC power plants has been recently proposed [5]. Biomass-based IGCC (Bio-IGCC) plants can take advantage of the experience and technological improvements achieved in precombustion CO_2 capture technology used in coal-based IGCC.

10.2.2.3 Technological Specifics of Bio-CCS

CO_2 capture from flue gases of biomass co-firing could be accomplished with traditional carbon capture systems, among which amine scrubbing is the most advanced process. Although being relatively efficient (up to 90 % of CO_2 can be captured from flue gases), the amine scrubbing technology is still rather expensive and has high energy penalty (about 30 % of the power plant output) [2] (see Chap. 7). The developers of an alternative cryogenic capture technology claim that it could potentially reduce costs and energy penalty of the carbon capture process by factor two relative to other capture methods (e.g., amine scrubbing, oxyfuel combustion, membranes) [8]. Once completely developed, the process would be able to capture 99 % of CO_2 from stationary sources and could be applicable to postcombustion carbon capture at dedicated biomass combustion plants (or biomass-coal co-firing power plants), as well as suitable for retrofitting existing power plants.

The large-scale deployment of Bio-CCS systems is expected to face the scalability and logistics challenges [2]. Typical biomass-fired facilities (without co-firing) emit about 1 Mt CO_2 per year and are about one-tenth the scale of fossil fuel (e.g., coal-fired) plants (although some very large biomass-fired power plants exist in some Scandinavian countries). Besides, the scale issues, there are some logistical issues related to availability, transport, delivery, and intermediate storage of biomass feedstocks. Biomass pretreatment may be required to reduce moisture content and maximize specific heat content for large-scale biomass supply chains. A dedicated biomass-converting facility could be needed to secure biomass feedstocks for 30–50 years [2]. Furthermore, the composition of flue gas from biomass-fired plant could vary with feedstock and may differ from that of coal-fired plant; this could require some adjustments in CO_2 capture options (which could also vary with feedstock). The proximity of a storage site (e.g., geological storage site), biomass sources, electricity, and heat end users are also important factors for the economic success of a Bio-CCS project.

10.2.3 Carbon-Negative Potential of Bio-CCS

It is recognized that the technical potential of the Bio-CCS technology is constrained by the availability of sustainable biomass sources, CO_2 storage capacity, and technological performance of biomass conversion and CO_2 capture processes

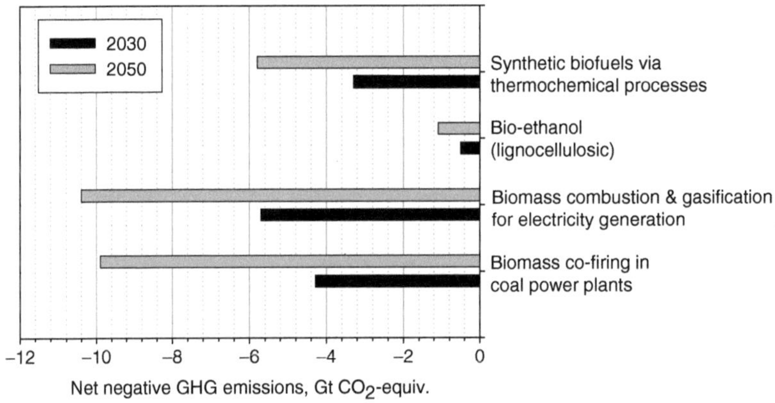

Fig. 10.4 Global technical potential in net negative GHG emissions for various Bio-CCS technologies in 2030 and 2050. The global supply of biomass feedstock is assumed 73 and 126 EJ/year in 2030 and 2050, respectively. Power plants are assumed to be equipped with Post-CCS, Pre-CCS, and OFC technologies. *Source* [5, 9]

(e.g., the net energy conversion efficiency, carbon removal efficiency) [5]. A study commissioned by the IEA GHG R&D Programme assessed technical potentials for Bio-CCS in the power generation and biofuels production sectors, including co-firing and co-gasification of biomass and coal [9]. Figure 10.4 compares the global carbon-negative potentials of various Bio-CCS technologies in 2030 and 2050.

The results of the IEA's study point to a large carbon-negative potential of Bio-CCS: taking only technical limitations into account, the maximum annual negative GHG emission potential reaches about 10 Gt in the power sector and about 6 Gt in the biofuel sector by 2050 (note that the technical potentials shown in Fig. 10.4 are calculated under an assumption that all available biomass feedstocks are allocated to one specific Bio-CCS route at a time; thus, the results cannot be totaled) [5, 9]. The realizable and economic potentials are estimated at 3.2 and 3.5 Gt of negative GHG emissions, respectively [9]. A relatively small technical potential for the biofuel routes can be explained by a relatively small fraction of CO_2 captured and, consequently, small storage capacity required.

There is a difference between *negative carbon emissions* from *carbon emission reductions*: the former provides an outlet of CO_2 from the Earth's atmosphere, whereas the latter decreases the inlet of CO_2 to the atmosphere; in that sense, the latter always depends on a reference scenario, while the former does not [5]. For example, if a large coal-fired power plant emitting 5 Mt per year CO_2 is replaced by a low-carbon technology that emits 1 Mt per year CO_2, then the emission reduction of 5 Mt − 1 Mt = 4 Mt per year is achieved. If the same plant is replaced by a Bio-CCS power plant that delivers 4 Mt per year of negative emissions, then the total emission reduction is 5 Mt + 4 Mt = 9 Mt per year, i.e., more than double compared to the first scenario. On the other hand, if Bio-CCS power plant replaces zero-emission technology (e.g., nuclear power plant), then negative carbon emission is equal to the carbon abatement.

The European Biofuels Technology Platform (EBTP) in its 2012 report identified abatement potential of Bio-CCS technologies for the European market in the near-to-medium term future [5]. The EBTP's 2030 target is 25 % biofuels in the transport sector, of which a quarter relate to bioethanol and three-quarters to BTL or FT-diesel. Based on the current technology and 90 % carbon capture rate, about 3.5 Mt CO_2 could be abated from the production of 1 Mtoe of FT-diesel and about 1.6 Mt of CO_2 abated from the production of 1 Mtoe of bioethanol via fermentation. The European Commission has already outlined a vision for an even higher share of aviation biofuels: 40 % of the aviation fuels market by 2050 [5].

10.2.4 Economics of Bio-CCS

Considering the substantial differences between various Bio-CCS technological routes, the economics of the large-scale Bio-CCS deployment has not been comprehensively analyzed and reported so far; nevertheless, some observations and preliminary estimates are available. According to existing projections, near-term Bio-CCS options include biofuel technologies that produce near-pure CO_2 streams, which allow avoiding substantial energy penalties and additional costs for CO_2 capture [5]. Such technologies as bioethanol and BTL production already include CO_2 separation as part of the production processes; once they reach a certain scale and can be clustered in terms of infrastructure, they could provide early Bio-CCS market penetration opportunities with very low additional costs [5]. IEA in its "Technology Roadmap for CCS in Industrial Applications" highlights biofuels production with CCS as one of the key "low-hanging fruits" for wide CCS deployment [10]. Preliminary estimates from the US-based industrial-scale ADM bioethanol production plant with CCS indicate that the cost of CO_2 captured, transported, and stored is lower than that for early movers in power generation with CCS (note that ADM project receives subsidies from US DOE to inject 2.5 Mt of CO_2 over 3 years) [11].

For electricity or CHP generation applications, Bio-CCS is considered more expensive than fossil-based CCS due to relatively high cost of biomass [5]. Besides, when 100 % biomass is combusted or it is co-fired with coal at high co-firing ratios the relatively lower volumetric energy density of biomass feedstock (compared to coal) could potentially lead to efficiency penalties and higher costs. Furthermore, the relatively high content of ash in biomass (especially, alkaline-rich ash) could lead to excessive corrosion when co-firing with coal in existing boilers, which will further drive the cost. On the other hand, co-firing biomass with coal or lignite at low-to-moderate ratios (less than 10 %) is not expected to adversely affect the equipment and require additional investment in CCS equipment compared to coal-CCS. Firing (or co-firing) biomethane or bio-SNG in NGCC power plants is also not expected to result in additional costs during deployment of NGCC-CCS [5].

It should be emphasized that the use of biomass residues for Bio-CCS applications (e.g., agricultural and forest product residues, and industrial and municipal waste) provides better lifecycle performance than dedicated energy crops, because most of the economic and environmental costs of the production of electricity or biofuels are borne by the food or end products [2].

The price of biomass feedstock will be one of the determining factors in the market penetration of Bio-CCS technologies. There are a number of factors and unknowns that would affect the future biomass prices, among them:

- Sustainable supply of biomass feedstocks
- Availability and scalability of novel feedstocks (e.g., microalgae, macroalgae, fast-growing aquatic plants)
- Global demand for biofuels, bioenergy, and biochemicals
- Demand in other biomass-based sectors (e.g., wood materials, pulp/paper)
- Environmental and biodiversity factors

In general, biomass prices are expected to rise in foreseeable future as the population and demand grows, unless novel biomass feedstock sources are sufficiently upscaled [5].

10.2.5 Current Status, Challenges, and Trends in Bio-CCS

Currently, more than a dozen of bioenergy projects aiming at deploying Bio-CCS technology are at the different stages of development. Among the projects in advanced stages of development and commercialization are [2]:

- A bioethanol plant with CCS in Illinois (USA), where about one million tons of CO_2 will be captured annually and transported by pipeline for storage in the Mount Simon Sandstone, an onshore deep saline formation at a depth about 2.1 km
- Bioethanol plants with beneficial reuse of CO_2 in Kansas and Texas, USA

Bio-CCS projects under construction and evaluation include the following:

- Bioethanol plant with geological CO_2 storage in Kansas, USA
- Gasification plant with geological CO_2 storage in North Dakota, USA
- Bioethanol plant with combination of geological storage and beneficial reuse in Rotterdam, the Netherlands
- A pulp and paper plant with geological CO_2 storage in Varo, Sweden
- Bioethanol plant with geological CO_2 storage in Sao Paulo, Brazil
- Bioethanol plant with geological CO_2 storage in Artenay, France

There are currently 73 dedicated biomass facilities in the USA; however, they run at only half the efficiency of coal-fired power plants [2]. This low efficiency rate results mainly from two factors: (1) poor heat integration (which could be potentially overcome), and (2) inefficiencies inherent in small-scale operations (mainly due to highly variable biomass feedstocks).

Reported studies indicate that co-firing biomass with fossil fuel (coal) is more efficient compared to operating solely on biomass (in the best cases, the efficiency could reach 95 % of a coal-fired power plant) [2]. However, there are significant technical issues to overcome when co-firing biomass with coal, such as pollutant formation, ash management, and balance-of-process issues (e.g., fuel preparation, ash utilization). Of particular concern is corrosion caused by some ash components (KCl is the most problematic corrosive agent) in boilers. Unfortunately, many fast-growing plants that are considered a sustainable source of biomass have high levels of potassium that results in KCl formation. Solving the corrosion problem will require more R&D efforts. Characteristics of different biomass feedstocks, e.g., particle size, and flame dynamics are important factors to consider in biomass co-firing plants. Although, from an engineering perspective, co-firing with up to 50 % biomass is possible, realistically, based on current biomass supply, this share is decreased to about 10 % (for the cases of small burners, or abundant biomass supply this number could increase) [2].

Bio-CCS is the only carbon-negative technology deployed at a full industrial scale, with the total capacity of about 0.5 Gt CO_2 per year [12]. Bio-CCS is projected to be widely deployed in industry by mid-century 2050 (59 % of total industrial CCS in 2050) [12, 13].

10.2.5.1 Challenges

Bio-CCS technology faces the following challenges:

- No adequate incentives for capturing CO_2 through Bio-CCS exist on a global basis because there is no recognition of "negative emissions" in the existing carbon reduction incentive schemes (e.g., the EU Emissions Trading System), and others [2].
- Fluctuating or nonexistent price on carbon in many countries.
- Availability and reliability of an affordable and sustainable biomass resource and its scale-up. Unsustainably produced biomass can have a negative impact in a number of different ways such as GHG emission from cultivation methods, water depletion, and biodiversity loss. As a result, the negative effects may outweigh the positive ones.
- Technological challenges specific to the biomass feedstock, e.g., ash management and equipment (boilers) corrosion.
- High cost and some technological challenges of CCS (reducing capture cost, proving up suitable storage sites, storage permanence issues).
- Competition with other low-carbon technologies.

The opponents of Bio-CCS consider this technology a dangerous hype. The authors of a recent study point out that large-scale bioenergy including biomass combustion and other processes generally could result in even more GHG emissions than the fossil fuels they are intended to replace (in part, due to the inefficiency of the process, e.g., 11–40 % more fuel would be needed to be burnt for the same energy output) [14]. The report underscores that the large-scale deployment of Bio-CCS would lead to "*a new form of underground land grab*" and to massively increased demand for biomass and attendant negative impacts on people and lands. Of particular concern is that Bio-CCS could be used as a rationale for construction of new "CCS-ready" coal and biomass burning facilities to stay within CO_2 allowances mandated by the EU while continuing to burn their polluting fuels. More detailed discussion of different aspects of Bio-CCS systems can be found in [12, 13].

10.3 Biochar as a Carbon-Negative Solution

Conversion of biomass to *biochar* with production of negative CO_2 emissions is considered by many as a promising approach in the context of near-term carbon mitigation policies. The sustainable technical potential of biochar, as well as issues related to biochar stability, environmental impacts, opportunities, and geographical locations for its optimal use, have been recently reviewed in a number of publications (e.g., [2, 15, 16]). The reported estimates indicate that GHG emission reductions can be 12–84 % greater if biochar is put back into the soil instead of being combusted to offset fossil fuel use [17]. Thus, biochar sequestration offers a realistic approach to utilize bioenergy while producing carbon-negative emissions.

Biochar (also known as *bio-carbon*, or, more commonly, charcoal) is a solid product of biomass pyrolysis that typically takes place at high temperatures and low-oxidant environment. Chemically, biochar primarily consists of carbon with small quantities of hydrogen, oxygen, and mineral matter. The relative yields of biochar and other products such as pyrolysis gases and liquids (biooil) vary with temperature and residence time. For example, *"fast" pyrolysis* process (short residence times and high temperatures of 700 °C and higher) favors production of liquid and gaseous products, whereas *"slow" pyrolysis* (temperatures of 400–500 °C and long residence times) is advantageous for biochar formation. As an example, pyrolysis of terrestrial biomass, such as plants, can be presented by the following generic equation [18]:

$$CH_{1.4}O_{0.59} \rightarrow 0.41C + 0.68H_2 + 0.59CO + 0.0005CO_2 + 0.001H_2O + 0.002CH_4 \quad (10.1)$$

where $CH_{1.4}O_{0.59}$ is a representative formula of terrestrial biomass; for simplicity, biochar is shown as C.

According to this equation, about 40 % of biomass carbon can be converted to biochar, with the remaining biomass carbon being converted to light gases (liquid product—biooil is not presented in the equation). It has been reported that biochar yields of about 50 % could be achieved during "slow" pyrolysis process [19]. Advantageously, biomass pyrolysis is a low-endothermicity process; in most cases, the energy content of gaseous and liquid pyrolysis products is severalfold greater than the amount of energy consumed by the pyrolysis process, which would greatly benefit the economics of biochar production.

Due to its chemical inertness (at ambient conditions), biochar can serve as a practically permanent sink for carbon captured from the atmosphere. Many experts view this approach as a beneficial intervention in the global carbon cycle that could potentially provide much greater storage permanence compared to storing CO_2 underground or under the ocean. Biochar can sequester carbon in soil for thousands of years [19]. Thus, as a carbon-negative solution, biomass-to-biochar technology would lead to a net CO_2 withdrawal from the atmosphere, while producing energy.

As an additional benefit, biochar can be used for the soil amendment applications to enhance or restore soil fertility, increase agricultural productivity, and provide protection against some soil-borne diseases (while providing a means of storing carbon for millennia). The stimulating effect of biochar on crop harvest (the technique known as "*Terra Preta*") has been known for centuries (pre-Columbian Amazonians used it to enhance soil productivity). Recent quantitative studies confirmed that adding biochar to soil can significantly increase seed germination, plant growth, and crop yields (e.g., crop yields could be increased by up to 200 %) [20]. It was shown that the application of biochar to soil enhances its water retention capacities, supports microbial communities, and activates the root activity; it also reduces nutrient leaching, soil acidity, and substantially diminishes irrigation and fertilizer requirements (the effects are mostly attributed to higher exchange capacities, surface area, changes in pH, and other factors).

It was reported that the presence of biochar in soil reduces or practically eliminates the emissions of such potent GHG as N_2O (up to 80 % reduction) and methane to the atmosphere [21]. Biochar can have a positive effect on the crop production in degraded and nutrient-poor soils, thus, effectively, increasing the amount of land suitable for agricultural production. These positive impacts of biochar on soil depend on many factors: physicochemical properties of biochar (e.g., nature of precursor, charring conditions), average surface temperature, humidity, precipitation, and regional soil conditions and types. [21].

Biochar can be designed with specific properties and qualities to match or target distinct characteristics of soil. For example, water-attracting and water-retaining properties of biochar can be enhanced by altering its porous structure and increasing surface area. As a result, the nutrients (P, N, K) and agrochemicals are retained for the plants benefit (preventing leaching of fertilizers into surface and groundwater), and the amount of applied fertilizers could be substantially reduced, thus providing an additional economic gain and further reducing CO_2 emissions into the atmosphere (originating from the fertilizer-manufacturing plants). On the other hand, if biochar properties are not well matched to the soil requirements, the biochar could exert a negative effect on soil fertility, e.g., through such mechanisms as nitrogen immobilization or aggravated pH constraints [15]. Experiments in the USA, Australia, Germany, and England are already showing some remarkable results—especially on otherwise poor soils, where biochar granules acted as a reservoir for water and fertilizers [22].

The above reported observations show that biomass-to-biochar approach could potentially have multiple favorable effects on the reduction of atmospheric GHG, namely, (1) through the withdrawal of CO_2 from the atmosphere and locking it in the form of biochar, (2) the enhancement of CO_2 absorption by plants via soil amendment (i.e., carbon from captured atmospheric CO_2 helps to remove even more CO_2 from atmosphere), and (3) reduction of GHG emissions from soil to the atmosphere. However, in order for biochar to be a net carbon sink, it should be produced from such biomass precursors as sustainably harvested crop-residues, waste biomass, biomass crops grown on abandoned land that has not reverted to forest, agroforestry timber, or animal manures where their overproduction could be a pollution problem. On the contrary, the biochar production systems that utilize long-lived carbon stocks (e.g., woodland), or those requiring land-use change with a large carbon debt, may cause a net increase in GHG emissions [2].

10.3.1 Storage Permanence of Biochar

The long-term stability of biochar in the soil environment is one of the major criteria determining its capacity as a net carbon sink: higher biochar stability results in longer carbon sequestration. Reported research shows that the stability of biochar depends on the chemical composition of its biomass precursor (or feedstock) from which it is produced and on the conditions of the biomass conversion process. In general, biochar consists of labile and recalcitrant components, and its overall stability depends on the half-life of both labile and recalcitrant fractions. The estimates of biochar half-life vary in a wide range from tens to millions of years, correlating with the oxygen-to-carbon (O/C) ratio: lower O/C ratio—higher half-life [2, 22]. The type of biomass feedstock and conversion technology used in the biochar production process also have an impact on its long-term stability. For example, woods were shown to produce more stable biochars compared to grasses or manure. It has also been reported that slow pyrolysis at temperatures higher than 500 °C produces biochars with estimated half-life of more than 1,000 years [23].

10.3.2 Biochar from Algae

Recently, there has been a growing interest in producing biochar from aquatic biomass. Many aquatic forms of biomass, such as macroalgae, *Lemna minor*, hyacinth, azolla, and microalgae, enjoy much faster growth rates compared to terrestrial biomass, thus, are able to faster remove CO_2 from the atmosphere. Muradov et al. reported on pyrolysis of *Lemna minor* (also known as duckweed) at the temperature range of 400–700 °C with biochar yield of 50.1 % (including a mineral component) [24, 25]. The results of pyrolysis of marine microalgae *Nannochloropsis* sp. have been reported by Porphy and Farid [26]. It was found that the biochar yield varied in the range of 30–68 wt% depending on the reaction temperature. Algae Systems (USA) is developing a process for production of carbon-negative fuels from algae fed by CO_2 captured directly from the air; the process produces biochar, diesel, and jet fuels. The developers of the process claim that the fuels produced (while being compatible with existing infrastructure) would have the net effect of reducing the CO_2 concentration of the atmosphere even after they were burned [4].

10.3.3 Economics

There are very few studies on the economics of biochar sequestration. Lehmann estimated that biochar sequestration in conjunction with bioenergy from biomass pyrolysis becomes economically attractive only when the value of avoided CO_2 emissions reaches $37 per ton [17]. Under this scenario, it would be feasible to produce both bioenergy and biochar in the near future. Currently, existing biochar sequestration projects are small-scale operations, making practically no impact on the global land–atmosphere carbon fluxes (although several companies in North America, Australia, and Europe produce and sell biochar in large quantities). Oil giant, Shell, shows a keen interest in biochar as carbon storage mechanism [22]. Some experts believe that with the right incentives, biochar could lock up as much carbon as the amount generated by aviation [22]. A number of African governments within the framework of Biochar Fund are implementing the concept to slow down deforestation, increase the food security of rural communities, and provide renewable energy; they are also attempting to get biochar accepted as a climate mitigation and adaptation technology.

10.3.4 Challenges

For all the apparent benefits, there are certain challenges facing widespread deployment of biochar as a carbon-negative driver:

- Disseminating the technology at affordable price could be an issue.
- Biochar would need clear global incentives, because current financial systems reward energy production from biomass and waste, but not carbon storage.
- More research and development efforts are needed to better understand processes occurring in biochars to optimize their effectiveness.

Summarizing, biochar is potentially an effective measure contributing to negative carbon emissions. Its advantage over other technologies is that it is a simple and passive method (i.e., once it was put in soil, there is no need for any further action), it could be distributed over large areas, and have multiple positive impacts. When comparing to other biomass-based approaches to carbon mitigation (e.g., Bio-CCS, biofuels), one must consider not only economics, energy, and GHG, but also wider issues, including soil conservation, hydrology, biodiversity, and nutrient cycling [2].

10.4 Chemical Carbon-Negative Systems

The advantages of chemical carbon-negative systems over biological systems are that CO_2 could be removed from the atmosphere much faster in industrial settings compared to biosystems that are limited by the slow rate of biomass growth (controlled by photosynthesis). On the other hand, chemical carbon-negative systems require expensive reagents, large air-scrubbing towers, and associated energy-intensive technological processes, whereas in the biosystems the sun provides all the necessary energy input. While some biological carbon-negative systems have already approached a commercial scale, chemical systems are still in an early stage of development. The main technological approaches to chemical carbon-negative systems include (1) direct capture of CO_2 from the atmosphere and its permanent storage, and (2) conversion of captured atmospheric CO_2 to stable carbonaceous products.

10.4.1 Capture of Atmospheric CO_2

Capture of atmospheric CO_2 (CAC) (sometimes also called "*air capture*," or "*air extraction*") is one of the main carbon-negative technological solutions designed to rapidly remove CO_2 from the atmosphere. Commonly, CAC is considered one of the strategies within the *geoengineering* concept (presented in Chap. 11); however, many proponents of this approach object to its inclusion in the family of geoengineering projects, arguing that, in contrast to many of them, CAC is not associated with potentially hazardous planetary-scale interventions. There is a notion among the CAC supporters that "*direct air capture would treat the disease, not merely the symptoms*" [27].

The main argument in favor of the CAC approach is that even if humankind starts immediately cutting CO_2 emissions at unprecedented rates (which is very unlikely), the net increase in the atmospheric CO_2 levels will still continue for the foreseeable future because fossil fuel reserves have been effectively increased (thanks to the development of revolutionary fracking technology), and such carbon mitigation technologies as CCS and renewables are moving at very slow pace and are not yet producing the desired effect. Thus, given the risks of further increasing CO_2 concentration in the atmosphere, there is a need to not only reduce present and future anthropogenic carbon emissions, but also tackle CO_2 which is already there.

The CAC is often viewed as an ultimate carbon mitigation technology due to the following advantages:

- It would alleviate the burden of equipping fossil power plants with expensive CCS systems.
- The cost of CAC would provide an absolute cap on the cost of carbon mitigation (i.e., CAC would limit the cost of a worst-case climate scenario) [28].

- It would remove any restrictions on the amount of CO_2 emitted by transportation and other small CO_2 sources.
- It could potentially reduce atmospheric CO_2 concentrations faster than the natural carbon cycle.

In some respect, CAC is similar to CCS in that it also involves capture, transportation, and permanent storage of CO_2 (note that CO_2 storage options in CAC are essentially the same as in CCS). However, CAC has important structural advantages over CCS systems as follows:

- CAC removes CO_2 directly from the atmosphere; thus, it manipulates the CO_2 atmospheric concentration rather than stack gases of fossil-based power plants.
- CAC significantly extends the concept of carbon capture from large centralized industrial CO_2 sources (that could be easily equipped with CO_2 capture units) to a myriad of small dispersed and mobile CO_2 sources such as cars, airplanes, homes, and businesses (that would be practically impossible to equip with CO_2 capture systems) [28, 29].
- CAC system is decoupled from a CO_2 emission source, which would allow setting the CO_2-capturing equipment to a certain fixed CO_2 concentration (e.g., 400 ppmv); since this value is globally constant, similar CO_2 capture devices could be installed anywhere in the world [30]. This factor is extremely important from practical and economical viewpoints, because CAC plants could be located close to CO_2 sequestration sites, thus eliminating CO_2 transport cost.

Given low concentration of CO_2 in the atmosphere (now, 400 ppm), its capture from air is more challenging than extracting it from diluted flue gases (5–10 vol.%) coming from industrial sources. Keith et al. assessed the ultimate physical limits on the amount of energy and land required for the practical implementation of CAC systems [28]. The thermodynamics of a gas separation process dictates the minimum amount of energy required to extract CO_2 from a mixture of gases, according to the equation:

$$H_{mix} = kT \ln \frac{p_f}{p_o} \tag{10.2}$$

where H_{mix} is enthalpy of mixing, k is Boltzmann constant, T is working temperature, p_o and p_f are partial pressure of CO_2 in the mixture (i.e., in air) and final pressure of CO_2, respectively.

At $p_o = 4 \times 10^{-4}$ atm (which corresponds to current atmospheric CO_2 concentration of 400 ppm) and $p_f = 1$ atm, the minimum energy needed to capture CO_2 from air would be about $\Delta H_{mix} = 1.6$ G /t carbon. If we add the energy consumption due to compression of CO_2 to 100 atm (required for the geological storage of CO_2), the overall minimum energy requirement for CO_2 capture from air and its geological sequestration would reach about 4 GJ/t carbon [28]. On the positive side: (1) because the enthalpy of mixing is a logarithmic function of partial pressure, the difference in

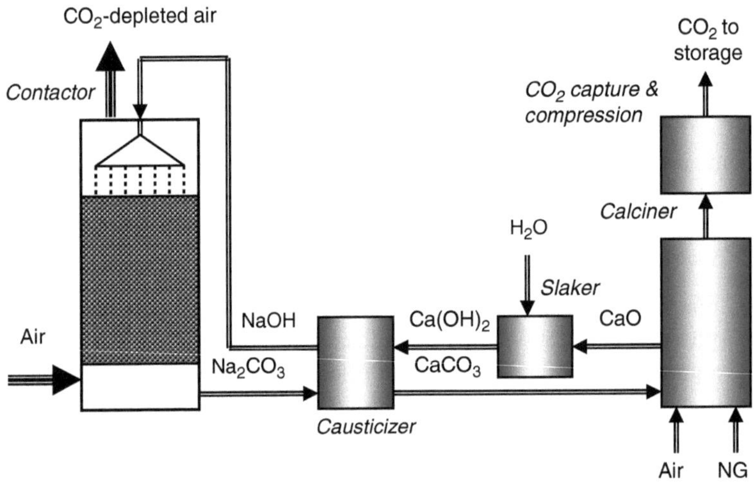

Fig. 10.5 Simplified process diagram of an atmospheric CO_2 capture system (packed bed tower option). *Source* [28]

energy penalties for CO_2 capture from air and from flue gases at atmospheric pressure would not be very significant (only by factor of about 2–3) [28], and (2) CAC systems would not be required to be 100 % efficient in removing CO_2 from the streams.

CAC is an area of very intensive worldwide research efforts. There are many technical approaches to the practical realization of the atmospheric CO_2 capture; generally, they fall into four main categories:

- Alkaline scrubbing solutions, e.g., NaOH, KOH, $Ca(OH)_2$
- Alkali metal carbonate scrubbing solutions, e.g., Na_2CO_3, K_2CO_3
- Chemical sorbents, e.g., MgO, CaO
- CO_2-selective sorbents, e.g., activated carbons, zeolites, nano-structured composites, metal-organic frameworks, amine-tethered sorbents, resin-based materials, and others

Most of the above options have been tested on laboratory scale units, with some of them reaching a field demonstration stage. The approaches based on CO_2 scrubbing using alkaline solutions (e.g., Na-Ca hydroxide-carbonate system reported by Zeman and Lackner [29]) are the most technologically advanced. Figure 10.5 provides a simplified schematic diagram of the atmospheric CO_2 capture plant based on Na-Ca hydroxide-carbonate system.

In this process, CO_2 is scrubbed from air by NaOH solution in a tower-type contactor apparatus, according to the reaction:

$$CO_2(g) + 2Na^+ + 2OH^- \rightarrow CO_3^{2-} + 2Na^+ + H_2O \qquad -110 kJ/mol\text{-}C \qquad (10.3)$$

The Na_2CO_3 solution is directed to a causticizer apparatus, where it reacts with calcium hydroxide at 70 °C regenerating NaOH and producing solid product $CaCO_3$:

$$CO_3^{2-} + Ca^{2+} \rightarrow CaCO_3(\mathrm{s}) \qquad 12kJ/mol\text{-C} \tag{10.4}$$

This process is similar to a kraft recovery process used in the pulp/paper industry. The $CaCO_3$ product is calcined at about 900 °C in a calciner to regenerate CaO and release CO_2:

$$CaCO_3 \rightarrow CaO(\mathrm{s}) + CO_2(\mathrm{g}) \qquad 179kJ/mol\text{-C} \tag{10.5}$$

A similar calcination process has long been practiced at an industrial scale for the production of lime and cement and in the pulp and paper industry. CaO reacts with liquid water in a slacker apparatus to produce $Ca(OH)_2$, thus closing the loop:

$$CaO(\mathrm{s}) + \mathrm{H_2O}(\mathrm{l}) \rightarrow Ca^{2+} + 2OH^{-} \qquad -82kJ/mol\text{-C} \tag{10.6}$$

In (10.3) through (10.6), (g), (l), and (s) correspond to gaseous, liquid, and solid phase, respectively. All the chemicals involved in the above process are inexpensive, abundant, and environmentally benign materials, and all the processes present current industrial practices and are well understood.

Physical dimensions and key technological parameters of different CO_2-scrubbing towers could be found in the literature [28]. For example, for the CAC system with CO_2 capture rate of 76,000 t C per year, the scrubbing tower dimensions are as follows: diameter of 110 m and height of 120 m [28]. Carnegie Mellon University researchers who studied an alkaline scrubber system reported dimensions for a full-size commercial plant as follows: a tower about 120 m high and about 100 m in diameter [31]. Baciocchi et al. reported a detailed analysis of the Na-Ca hydroxide-carbonate air capture system; based on their calculations, a unit handling 817,115 m^3/h of airflow would have cross section of about 20,000 m^2 (the area equivalent of four soccer fields) [32].

Alternative designs for CO_2 capture device could also be found in the literature; e.g., there has been a proposal to construct giant filters that would act like flypaper, trapping CO_2 molecules by alkaline solutions pumped through porous filters [33]. According to the reported estimates, a single commercial size wind scrubber with the height of 70 m and width of 55 m would capture about 90,000 t of CO_2 a year. Some experts, however, are skeptical of the idea, arguing that in order to capture all anthropogenic CO_2 an area at least the size of Arizona would need to be covered with the scrubber towers [33].

A range of estimates on the energy consumption rates by the air capture plants based on the Na-Ca hydroxide-carbonate system have been reported by different groups as follows (in kJ/mol CO_2): Zeman—442 [30], Keith—679 [28], Baciocchi—516 [32]. The thermodynamic efficiency of the air capture system was estimated at about 6.0 %, which is rather low, but would be comparable to other "end of the pipe" CO_2 capture technologies (e.g., monoethanolamine scrubbing) if adjusted to comparable CO_2 concentrations. The excessively high energy cost of the Na-Ca hydroxide-carbonate process was noted by Herzog and other authors [34]. Due to the high energy intensity of this process, the use of fossil electricity (especially if generated by coal-fired power plants) would dramatically reduce the net amount of captured CO_2 by the CAC system.

Different variations of the above system have been reported by a number of research groups. For example, Socolow described the process that captures CO_2 in lime water followed by the thermal treatment of the resulting $CaCO_3$ in a calciner producing high pressure CO_2 and regenerated CaO (similar to cement manufacturing technology) [35]. A group of Swiss researchers studied a variation of Na- and Ca-based cyclic CAC process, where air stream containing 500 ppmv of CO_2 was passed over Na_2CO_3 at 50 °C to form $NaHCO_3$ [36]:

$$Na_2CO_3(\mathrm{s}) + CO_2(\mathrm{g}) + \mathrm{H_2O}(\mathrm{g}) \leftrightarrow 2NaHCO_3(\mathrm{s}) \qquad H^\circ = -135.5\,kJ \qquad (10.7)$$

$NaHCO_3$ was thermally decomposed to Na_2CO_3, CO_2 and H_2O at 200 °C, followed by Na_2CO_3 recycling to the first reactor. The total thermal energy requirement for the Na-based cycles was estimated at 390 kJ/mol CO_2, while for $Ca(OH)_2$–$CaCO_3$–CaO thermochemical cycle, the required thermal energy input was much higher: 2485 kJ/mol CO_2 captured [37].

The ETH (Zurich, Switzerland) researchers are pursuing a different approach to capturing CO_2 by using chemical sorbents and concentrated solar power to provide a heat input for the process [38]. In the proposed process, CO_2 reacts with calcium oxide (CaO) pellets at 400 °C in the presence of small amounts of steam forming calcium carbonate ($CaCO_3$). At temperature of 800 °C (provided by the solar concentrator), $CaCO_3$ is converted back into calcium oxide and the stream of pure sequestration-ready CO_2 is released. The developers of the technology believe the device could be scaled up to take significant amounts of CO_2 out of the atmosphere, though, they don't know at what cost.

Recently, there have been important developments toward regenerable air capture systems using zeolites, amine-grafted sorbents and humidity-swing sorbents. It has been reported on testing of novel high microporosity materials based on zeolites Li-LSX and K-LSX, and amine-grafted silica for CO_2 capture from air (LSX stands for low-silica type zeolite X) [39]. The LSX-based sorbent showed the most promise for applications in dry conditions, reaching space velocities as high as 63,000 h^{-1}. Amine-grafted mesoporous silica (SBA-15) adsorbent demonstrated an adequate performance in wet conditions, albeit at a lower space velocity of 1,500 h^{-1}, due to slower uptake rates. The authors of the study claim that by using a combined temperature and vacuum swing cycle, they were able for the first time to obtain the CO_2 concentration in the desorption product above 90 % from a single cycle.

The moisture swing CO_2 capture technique reported recently by Lackner's group (USA) is a promising new approach to regenerating CO_2 sorbents; it trades heat input in a thermal swing operation (or mechanical energy input in a pressure/vacuum-based swing) against the consumption of water, whose evaporation provides the free energy that drives the cycle with associated potential cost savings [40, 41].

The technique is based on the use of specific resins with the surface groups comprising quaternary ammonium ions. The authors of the study emphasize that compared to water consumption in CO_2 capture via biomass production, the moisture swing method consumes two orders of magnitude less water. A similar approach to the air capture system involving a humidity swing cycle but using different sorbents was proposed by He et al. [42]. The scheme of reversible CO_2 capture using the humidity swing cycle is shown below:

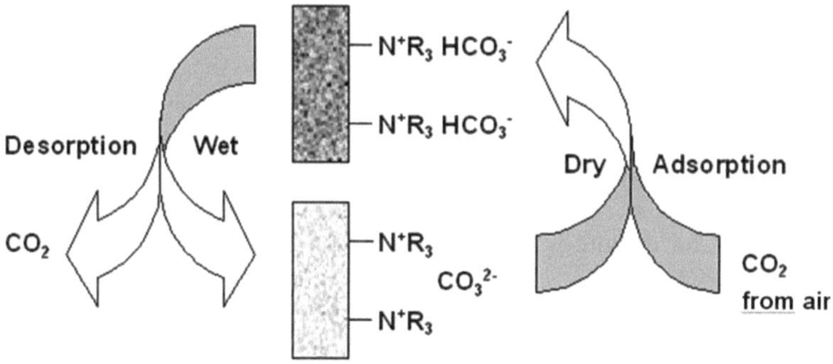

(10.8)

where [N^+R_3] are quaternary ammonium groups.

When exposed to dry air, the carbonate anion groups on the resin surface can capture CO_2 and produce bicarbonates, which exist in an equilibrium with carbonate ions. Adding moisture will shift the bicarbonate–carbonate equilibrium with the release of CO_2. Lackner proposed to use these resins in "*artificial trees*" that would pull CO_2 from the atmosphere. Despite the attractiveness of the concept, calculations show that its practical implementation might be challenging even in the foreseeable future: ten million of these "trees" would be required to drop atmospheric CO_2 concentration by 0.5 ppm per year, and each machine would require roughly 1.1 MJ of electricity for pumping and compressing air per kg of CO_2 captured [41].

10.4.1.1 Economics of CAC

Reported estimates on the economics of CAC systems vary in a wide range. According to Keith et al., for the alkaline-scrubbing CAC system with CO_2 capture rate of 76,000 t C per year, the capital and operation/maintenance costs would be US$12 million and US$400,000 per year, respectively [28]. The overall cost of the air capture system was estimated at US$500 per ton of C (or US$130 per ton CO_2) with a potential to drop to US$200 per ton of carbon [28]. For a similar alkaline scrubber system, Carnegie Mellon University researchers reported the cost of US$240 per ton of carbon removed from the atmosphere [31]. American Physical Society estimates the cost of air capture at roughly US$600 per metric ton of CO_2 [43]. According to Lackner, with today's technology, the cost of CO_2 capture could be below $100–US$200/ton CO_2, and with mass production it could further drop to as low as US$30/t CO_2 (or ~ US$100/t C) [43, 44]. For comparison, the cost of CO_2

removal via Bio-CCS systems is estimated at €50–100 per ton of carbon assuming large-scale deployment of Bio-CCS systems [6, 13].

Pielke conducted the analysis of the cost of global CO_2 stabilization through the deployment of air capture systems [45]. The analysis concluded that considering that 1 ppm CO_2 is an equivalent to 2.13 GtC and assuming the cost of CO_2 capture of $500/t C (according to Keith [28]), the cost of reducing atmospheric CO_2 by 1 ppm through CAC would amount to about US$1 trillion or about 2.5 % of global GDP (in 2007 values) [45]. Stabilizing atmospheric CO_2 at 450 ppm at the air capture costs of $500, $360, and $100 per ton C would require the cumulative costs of air capture over 2008–2100 period equivalent to 2.7, 2.0, and 0.5 % of global GDP (assuming 2.5 % global GDP growth). Thus, according to this idealized assessment, the cost of CO_2 stabilization at 450 ppm via air capture would be comparable to the costs of stabilization presented in IPCC-2007 report through other mitigation options [46]. If the cost of air capture systems will drop to $100/tC, then over twenty-first century, air capture would cost much less than alternative approaches considered by the IPCC.

In an analytical paper by Eisenberger et al., the authors discussed the prospects of air capture as carbon-negative technology and its future role in achieving lower-risk atmosphere [47]. The study concluded that the extensive deployment of air capture technology coupled with secure sequestration offers the advantages of centralization and control without direct intervention in the biosphere or a major collateral environmental impact, while maintaining anticipated economic growth. The authors pointed to two key features of the technology that would facilitate its deployment: (1) the utilization of widely available low-grade process heat to power the energy-intensive stages of the process, thereby greatly reducing the energy costs of CO_2 capture, and (2) the efficient integration with the industrial processes, enabling large air capture capacities at centralized facilities [47]. The large-scale deployment of CAC would compliment an orderly transition to nearly carbon emission-free energy by the end of century. The authors emphasized the need for substantially increasing the support for R&D in this area in order to reduce the cost of the air capture technology.

In the context of the uncertainty of future technological developments and location-specific prices for equipment and energy, the economics of CAC and CCS technologies appear to be of broadly similar magnitude [48]. However, regardless of the similarities and differences of CAC and CCS, both approaches represent technologies at the most expensive end of the carbon abatement cost curve, and, hence, they will determine the "ceiling" carbon price [49].

10.4.1.2 Commercialization Status of CAC

Although the methods and processes for capturing atmospheric CO_2 appear to be technically feasible from an engineering perspective, relatively high cost of the existing CO_2-capturing techniques limits the technological development of the CAC systems to mostly pilot-scale units and prototypes. Among most advanced projects is the one carried out by Carbon Engineering Company located in Calgary (Canada). Formed in 2009 with US$3.5 million support from Bill Gates and others, the company is testing prototype components of its air capture system; it is working on quite detailed designs and engineering aspects of the process and plans to build a complete pilot plant by the end of 2014 [50]. The technology is based on the use of caustic soda solutions to remove CO_2 from air (see the description of the method above). According to David Keith, the president of the company, the cost of CO_2 capture from air could be reduced to US$100 per ton of carbon.

10.4.2 Conversion of CO_2 to Elemental Carbon

As an alternative to permanently storing the captured atmospheric CO_2 (e.g., in geological formations), there have been proposals to convert it to stable products that would chemically lock carbon for thousands-to-millions of years. Some approaches to binding CO_2 into chemically stable products (e.g., through mineral carbonation, concrete curing, and others) have been discussed in Chap. 9.

An alternative approach to locking captured atmospheric CO_2 in the form of extremely stable and value-added product—solid carbon has been proposed by Muradov and Veziroglu [51]. The conversion of atmospheric CO_2 to elemental carbon provides means of practically permanent isolation of carbon from the atmosphere due to an exceptional chemical inertness of carbon at ambient conditions (coal has been stored underground for millions of years with practically no chemical changes). In contrast to conventional CO_2 storage options, the proposed approach would permanently withdraw carbon from the Earth's carbon cycle; it will not "leak" back to the atmosphere, whereas CO_2 stored underground or under the ocean will eventually do.

In principle, carbon can be extracted from CO_2 via a variety of chemical pathways. Of course, direct decomposition of CO_2 to carbon and oxygen ($CO_2 \rightarrow C + O_2$) would be the most attractive and desirable route (since it could replenish O_2 in the atmosphere), but the reaction thermodynamics is so challenging ($\Delta H^o = 395$ kJ/mol) that its practical realization is beyond existing and even emerging technological capabilities. Alternatively, the process could be conducted through cyclic two-step CO_2 decomposition:

$$CO_2 \rightarrow CO + 0.5O_2 \tag{10.9}$$

$$CO \leftrightarrow 0.5CO_2 + 0.5C \tag{10.10}$$

Despite the fact that the enthalpy of CO_2 dissociation (10.9) reaction ($\Delta H^o = 283$ kJ/mol) is substantially greater than that of water (242 kJ/mol), CO_2 molecule exhibits two very important attributes that make this cyclic process attractive and potentially feasible, especially, for a solar energy input. First, the energy requirements for CO_2 splitting are more favorable than for water splitting: it has a larger thermal ($T\Delta S$) component of the total energy ($\Delta H = \Delta G + T\Delta S$) compared to water. This means that more thermal energy as opposed to Gibbs free energy (e.g., electrical energy) can be used for CO_2 dissociation. Second, a certain amount of Gibbs free energy can be added to the system in the form of photon energy, since CO_2 becomes susceptible to near-UV light at high temperatures (>1,300 °C). The combination of these two effects could substantially decrease the dissociation temperature of CO_2 and increase the efficiency of the solar-to-chemical energy conversion process. The technical feasibility of CO_2 photo-thermal dissociation to CO and O_2 at elevated temperatures (2,000–2,500 °C) using a concentrated solar radiation source has been demonstrated [52], but the process still faces many technical challenges. The second step of the cyclic process (reaction 10.10) is based on an industrially important *Boudouard reaction*.

From the practical viewpoint, it would be more feasible to carry out CO_2 conversion to carbon by using reducing agents (e.g., non-fossil hydrogen). Hydrogenation of CO_2 into carbon with release of water is a well-known *Bosch reaction*:

$$2H_2 + CO_2 \rightarrow 2H_2O + C \tag{10.11}$$

This reaction has not yet been practiced on a large scale; recently it attracted the attention of researchers as part of a cyclic process aiming at oxygen recovery from CO_2 in a life support system for the space exploration applications. The Bosch reaction is strongly exothermic and is thermodynamically favored at relatively low temperatures and high pressure. The practical realization of the reaction faces several challenges, e.g., relatively low yield and slow kinetics due to a new phase formation and catalyst deactivation (due to carbon deposition).

Alternatively, the reaction could be carried out in two steps: CO_2 hydrogenation to methane, followed by its decomposition to carbon and hydrogen, which is recycled to the first stage:

$$4H_2 + CO_2 \underset{-2H_2O}{\rightarrow} CH_4 \rightarrow 2H_2 + C \tag{10.12}$$

The first step (the *Sabatier reaction*) is a well-developed process, but the second step: methane decomposition to hydrogen and carbon still requires more developmental efforts (see Sect. 8.2.6). Figure 10.6 depicts the conceptual scheme of a carbon-negative system based on conversion of captured CO_2 to solid carbon according to the hydrogenation route.

Atmospheric CO_2 is captured by a chemical scrubber, and the resulting concentrated CO_2 stream is directed to a CO_2-processing plant. At this plant, CO_2 can be converted to solid carbon via the reactions (10.11) or (10.12). It is important that hydrogen used in these processes is produced from water and noncarbon energy sources, e.g., noncarbogenic renewables (solar, wind) or a nuclear source. Carbon produced in the process could be either sequestered as solid carbon or transformed into structural long-lived materials (e.g., building and construction materials). Storing carbon is more environmentally safe and technically reliable compared to

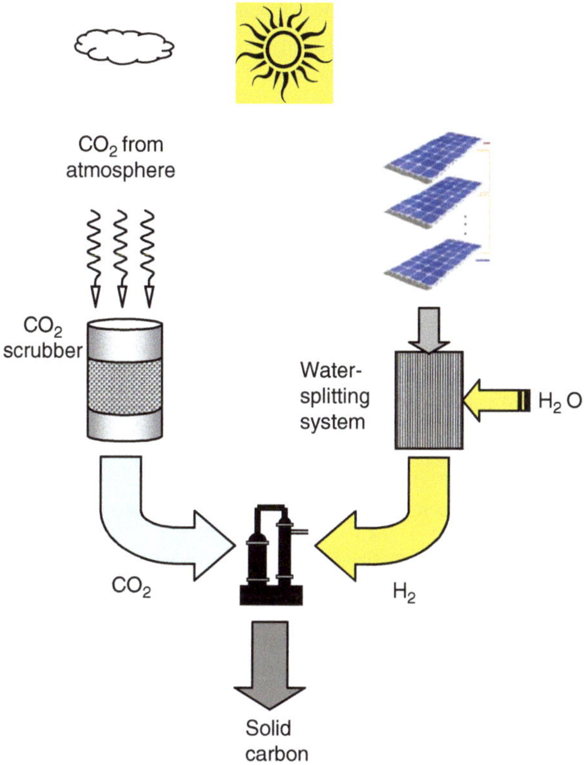

Fig. 10.6 Conceptual schematic diagram of a carbon-negative system involving conversion of atmospheric CO_2 into carbon. *Source* [51]

storing supercritical CO_2 (no leaks, no acidification, etc). In future (assuming less carbon-constrained world), the stored chemical energy of carbon can be recovered, e.g., by means of energy-efficient direct carbon fuel cells [51].

10.4.2.1 Concluding Remarks

Summarizing this chapter, given the slow progress of current climate change mitigation policies in stopping the rise in man-made CO_2 emissions, there is a growing recognition that the low-risk level of atmospheric CO_2 can only be achieved with the introduction of a significant carbon-negative component to the portfolio of carbon abatement options. Among the proposed carbon-negative solutions, Bio-CCS is the most advanced technology approaching a full-scale commercial deployment. Removal of CO_2 from the atmosphere by chemical means (air capture) is still in the early stage of development and requires a governmental and industrial support of R&D efforts and demonstration-scale projects in order to reduce its cost.

As with many other carbon mitigation technologies, air capture and other carbon-negative solutions will result in an additional cost to society, so there will be a need in appropriate policies and mechanisms to facilitate their adaptation and widespread deployment.

References

1. Moss R, Edmonds J, Hibbard K et al (2010) The next generation of scenarios for climate change research and assessment. Nature 463:747–756
2. Milne J, Field C (2012) Assessment report from the GCEP workshop on energy supply with negative carbon emissions. Stanford University. Global Climate and Energy Project. http://gcep.stanford.edu/pdfs/rfpp/Report%20from%20GCEP%20Workshop%20on%20Energy%20Supply%20with%20Negative%20Emissions.pdf. Accessed 15 Feb 2013
3. Global CCS Institute (2012) Bio-energy with CCS. http://cdn.globalccsinstitute.com/sites/default/files/publications/25921/fact-sheet-3-bioccs-v4.pdf. Accessed 26 Feb 2013
4. Algae Systems (2013) Carbon-negative technology. http://algaesystems.com/technology/carbon_negative/. Accessed 26 Feb 2013
5. European Biofuels Technology Platform (2012) Biomass with CO_2 capture and storage (Bio-CCS). June 20, 2012. http://www.biofuelstp.eu/bio-ccs-jtf.html. Accessed 12 Dec 2012
6. International Energy Agency (2011) Greenhouse gas R&D programme. Potential for biomass and carbon dioxide capture and storage. Cheltenham, UK. http://www.eenews.net/assets/2011/08/04/document_cw_01.pdf. Accessed 6 Jul 2011
7. International Energy Agency (2012) Bioenergy. http://www.iea.org/publications/freepublications/publication/bioenergy.pdf. Accessed 12 Feb 2013
8. Department of Energy (2012) Cryogenic carbon capture. National Energy Technology Laboratory. ARPA-E project DE-AR0000101. http://www.netl.doe.gov/publications/proceedings/12/co2capture/presentations/5-Posters/ D%20Frankman-SES-Cryogenic%20Carbon%20Capture.pdf. Accessed 10 Jan 2013
9. Koornneef J (2011) Ecofys. Global potential of biomass combined with carbon capture and storage. www.ecofys.com/en/project/biomass-combined-with-ccs-bio-ccs. Accessed 15 Dec 2012
10. International Energy Agency (2011) Technology roadmap. Carbon capture and storage in industrial applications. www.iea.org/roadmaps/ccs_industrial_applications.asp. Accessed 15 Mar 2012
11. Marshall C (2011) Ethanol carbon sequestration plant holds lessons for coal. New York Times. 12 Sep 2011. www.nytymes.com/cwire/2011/09/12/12climatewire-ethanol-carbon-sequestration-plants-hold-les-18588.html?pagewanted=2. Accessed 12 Dec 2011
12. Carbo M (2011) Global technology roadmap for CCS in industry. Biomass-based industrial CO_2 sources: biofuels production with CCS. ECN-E-11-012. Netherlands. http://www.ecn.nl/docs/library/report/2011/211012.pdf. Accessed 20 Sep 2013
13. Global CCS Institute (2010) Global status of BECCS projects 2010. Biorecro AB. http://cdn.globalccsinstitute.com/sites/default/files/publications/13516/gccsi-biorecro-global-status-beccs-110302-report.pdf. Accessed 12 May 2012
14. Schlossberg J (2013) Report: Bioenergy with carbon capture and storage a mistake. Climate Connections. http://climate-connections.org/2013/01/14/report-bioenergy-with-carbon-capture-and-storage-a-mistake/. Accessed 12 Apr 2013
15. Woolf D (2012) The role of biochar in a negative emissions portfolio. GCEP workshop on energy supply with negative CO_2 emissions. http://gcep.stanford.edu/events/workshops_negemissions2012.html. Accessed 10 Aug 2013

16. Woolf D, Amonette J, Street-Perrott F et al (2010) Sustainable biochar to mitigate global climate change. Nat Commun 1:56
17. Lehmann J (2007) A handful of carbon. Nature 447:7141
18. Reed T (1981) Biomass gasification. Principles and technology. Noyes Data Corp, New Jersey
19. Winsley P (2007) Biochar and bioenergy production for climate change mitigation. N Z Sci Rev 64:5
20. Glaser B, Lehmann J, Zech W (2002) Ameliorating physical and chemical properties of highly weathered soils in the tropics with charcoal—a review. Biol Fert Soils 35:219
21. Lehmann J, Gaunt J, Rondon M (2006) Biochar sequestration in terrestrial ecosystems—a review. Mitig Adapt Strat Glob Chang 11:395–427
22. Harrabin R (2009) Biochar: is the hype justified? BBC News. 16 Mar 2009. http://news.bbc.co.uk/2/hi/sci/tech/7924373.stm. Accessed 12 Apr 2011
23. Spokas K (2010) Review of the stability of biochar in soils: predictability of O:C molar ratios. Carbon Manag 1:289–303
24. Muradov N, Fidalgo B, Gujar A et al (2010) Pyrolysis of fast-growing aquatic biomass – *Lemna minor* (duckweed): characterization of pyrolysis products. Biores Technol 101: 8424–8428
25. Muradov N, Fidalgo B, Gujar A et al (2012) Production and characterization of duckweed bio-char and its catalytic application for biogas reforming. Biomass Bioenergy 42:123–131
26. Porphy S, Farid M (2012) Feasibility study for production of biofuel and chemicals from marine microalgae *Nannochloropsis* sp. Based on mass and energy analysis. ISRN Renew Energy. doi:10.5402/2012/156824
27. Krauss L (2013) We can't fight climate change without investing in this technology. Slate.com. 13 May 2013. http://www.slate.com/articles/technology/future_tense/2013/05/direct_air_carbon_capture_technology_must _be_developed_to_help_fight_climate.html. Accessed 14 Aug 2013
28. Keith D, Ha-Duong M, Stolaroff J (2006) Climate strategy with CO_2 capture from air. Clim Change 74:17–45
29. Zeman F, Lackner K (2004) Capturing carbon dioxide directly from atmosphere. World Resour Rev 16:157–172
30. Zeman F (2007) Energy and material balance of CO_2 capture from air. Environ Sci Technol 41:7558–7563
31. Carnegie Mellon University (2008) Snatching carbon dioxide from atmosphere. Climate Decision Making Center, Carnegie Mellon University. Accessed 12 Dec 2008
32. Baciocchi R, Storti G, Mazzootti M (2006) Process design and energy requirements for the capture of carbon dioxide from air. Chem Eng Proc 45:1047–1058
33. Behar M (2005) Now you CO_2, now you don't. Popular Sci August:53–58
34. Herzog H (2003) Assessing the feasibility of capturing CO_2 from air. Technical report LFEE 2003-002-WP. Massachusetts Institute of Technology, Cambridge, MA
35. Socolow R (2002) CO_2 capture: the long term view. NGO Focus Group meeting, Washington, DC
36. Nikulshina V, Ayesa N, Galvez M, Steinfeld A (2008) Feasibility of Na-based thermochemical cycles for the capture of CO_2 from air—thermodynamic and thermogravimetric analyses. Chem Eng J 140:62–70
37. Nikulshina V, Hirsh D, Mazzotti M, Steinfeld A (2006) CO_2 capture from air and co-production of H_2 via the $Ca(OH)_2$-$CaCO_3$ cycle using concentrated solar power—thermodynamic analysis. Energy 31:1379–1389
38. Kunzig R, Broecker W (2009) Can technology clear the air? New Scientist. 12 Jan 2009. http://www.newscientist.com/article/mg20126901.200-can-technology-clear-the-air.html. Accessed 15 Oct 2010
39. Stuckert N, Yang R (2011) CO_2 capture from the atmosphere and simultaneous concentration using zeolites and amine-grafted SBA-15. Environ Sci Technol 45:10257–10264
40. Wang T, Lackner K, Wright A (2011) Moisture swing sorbent for carbon dioxide capture from ambient air. Environ Sci Technol 45:6670–6675

41. Bielo D (2013) 400 PPM: Can artificial trees help pull CO_2 from the air? Scient Am. 16 May. http://www.scientificamerican.com/article.cfm?id=prospects-for-direct-air-capture-of-carbon-dioxide&print=true. Accessed 12 Aug 2013
42. He H, Li W, Zhong M et al (2012) Reversible CO_2 capture with porous polymers using the humidity swing. Energy Environ Sci 6:488–493. doi:10.1039/c2ee24139k
43. The American Physical Society (2011) Direct air capture of CO_2 with chemicals. June 2011. Accessed 22 Jan 2012
44. Lackner K (2009) Capture of carbon dioxide from ambient air. Eur Phys J. Special Topics:176
45. Pielke R (2009) An idealized assessment of the economics of air capture of carbon dioxide in mitigation policy. Environ Sci Policy 12:216–225
46. Intergovernmental Panel on Climate Change (2001) IPCC. Working Group III. Mitigation. http://www.mnp.nl/ipcc/pages_media/at4-chapters.html. Accessed 12 May 2008
47. Eisenberger P, Cohen R, Chichilinsky G et al (2009) Global warming and carbon-negative technology: prospects for a lower-cost route to a lower-risk atmosphere. Energy Environ 20:973–984
48. Capturing CO_2 from air (2011) Carbon Capture J. 9 Oct 2011. http://www.carboncapturejournal.com/displaynews.php?NewsID=851&PHPSESSID=m83hgv8gdtka3oahje30bt060. Accessed 18 Oct 2011
49. McKinsey & Co (2009) Pathways to a low carbon economy; Version 2 of the global greenhouse gas abatement cost curve. McKinsey and Company, Brussels
50. Eisenberg A (2013) Pulling carbon dioxide out of thin air. The New York Times. 5 Jan 2013. http://www.nytimes.com/2013/01/06/business/pilot-plant-in-the-works-for-carbon-dioxide-cleansing.html?_r=0&pagewanted=print. Accessed 15 Aug 2013
51. Muradov N, Veziroglu N (2008) "Green" path from fossil-based to hydrogen economy: an overview of carbon-neutral technologies. Int J Hydrogen Energy 33:6804–6839
52. Traynor A, Jensen R (2002) Direct solar reduction of CO_2 to fuel: first prototype results. Ind Eng Chem Res 41:1935–1939

Chapter 11
Emergency Carbon Management: Geoengineering

Abstract Geoengineering is a set of radical contingency actions to deliberately modify the Earth's energy balance in order to counteract the adverse impact of human activities on the global ecosystem and climate. The Greenhouse Gas Management (GGM) is one of the main geoengineering strategies aiming at reducing the atmospheric CO_2 levels within a reasonable time frame by enhancing an uptake and storage of carbon through a variety of biological (e.g., ocean fertilization) and chemical (e.g., enhanced weathering) engineered systems. Today, geoengineering remains a highly controversial issue; among major concerns are possible unintended planetary-scale adverse ecological impacts of the projects, and the disputable benefits of the geoengineering approach compared to other carbon mitigation strategies. Geoengineering poses acute and novel challenges that would require international cooperation, transparency, and the proactive and effective managing of research. The current status of major GGM geoengineering projects, their technical feasibility and economics, the challenges and risks associated with their global deployment are analyzed in this chapter.

11.1 Geoengineering: A Last Resort Option?

The term "geoengineering" (sometimes, also called "climate engineering") is generally defined as a set of measures to deliberately modify or manipulate the Earth's energy balance in order to counteract the adverse impact of human activities on the global ecosystem and climate. Originally, the notion of geoengineering was introduced by Cesare Marchetti in the mid-1970s as a means of controlling CO_2 levels in the atmosphere by collecting CO_2 from major emitters (e.g., power plants) and injecting it into deep seas (e.g., the Mediterranean undercurrent entering the Atlantic Ocean at the strait of Gibraltar) [1].

Today, geoengineering is a highly controversial concept enjoying a support from a number of distinguished scientists, but also facing fierce criticism and resistance

N. Muradov, *Liberating Energy from Carbon: Introduction to Decarbonization*, Lecture Notes in Energy 22, DOI 10.1007/978-1-4939-0545-4_11,

from a large number of environmental groups, governing bodies, and scientists reacting with alarm to any human interference with nature. The proponents of the geoengineering approach (among them, Nobel Laureate Dr. Paul Crutzen [2]) argue that since the traditional approaches to carbon mitigation, such as capturing CO_2 from stack gases of power plants, or sequestering CO_2 by terrestrial ecosystems, have proved too slow to make any difference in the rapidly deteriorating environmental situation, a radical contingency plan in the form of geoengineering will be needed. This emergency response action can supposedly "fix" the problem within a reasonable time frame by reversing the trend and, most importantly, preventing further slide to a dangerous *tipping point* (i.e., an uncontrollable increase in atmospheric CO_2 concentration through positive feedback mechanisms) (the detailed discussion of this subject can be found in Sect. 2.5.3).

The opponents of the geoengineering concept consider it a dangerous and untested "quick-fix" approach and are concerned that an excessive reliance on such a radical measure may hinder the resolve and worldwide efforts to deal with the original cause of the problem, i.e., human-induced GHG emissions [3]. Critics make a case that geoengineering not only could alter weather, ecosystems, geochemical, and biochemical cycles but also could potentially cause international conflicts if deployed on a large scale by separate nations or the groups of individuals. Currently, no international legally binding laws or regulations on geoengineering exist, although there are two nonbinding international agreements that discourage practicing geoengineering: the London Convention and Protocol (which regulates ocean dumping), and the United Nations Convention on Biological Diversity [4].

Since the Earth's climate is controlled by two fundamental forces: the amount of solar radiation hitting the Earth and the amount of solar radiation absorbed by its atmosphere, practically all proposed geoengineering projects fall into two main categories:

- Solar radiation management
- Greenhouse gas management

The proposed solutions in each category are diverse in terms of technological features, scale, and potential consequences to ecosystems. Figure 11.1 provides a sketch of some geoengineering projects from both categories.

11.1.1 Solar Radiation Management

The solar radiation management (*SRM*) approach aims at reducing the net incoming short-wave (UV and visible) solar radiation striking the Earth surface by deflecting the portion of incident sunlight back to the space, or by increasing the reflectivity (albedo) of the Earth surface. Among the two main geoengineering approaches, SRM has attracted most controversy, because many experts consider SRM a high-leverage intervention offering the dual prospect of either large benefits or great harms [5]. On one hand, SRM offers quicker fixes for cooling the Earth compared to other

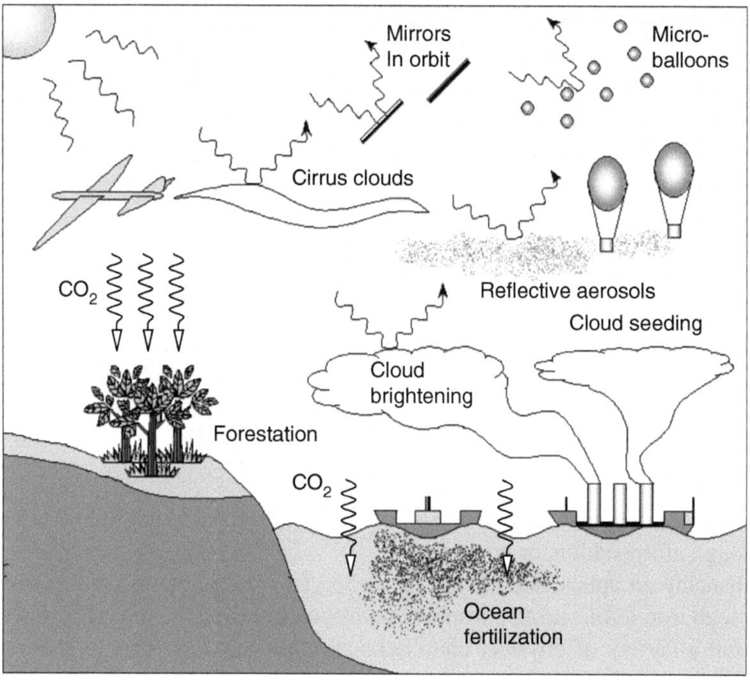

Fig. 11.1 Schematic representation of proposed geoengineering projects including solar radiation and greenhouse gas management options

geoengineering options, but on the other, it is potentially more invasive that could cause an enormous environmental harm, or even worsen the ecological situation, and, as a result, it might undermine other climate change mitigation efforts [4, 5].

The following is the list of major SRM geoengineering approaches and projects:

- Using reflective aerosols to back-scatter solar radiation.
- Brightening clouds by spraying them with sea salt to increase their reflectivity.
- "Seeding" cirrus clouds using airplanes.
- Using unmanned satellite-guided "cloud ships" that suck up seawater and spray microdroplets through giant funnels to form white clouds.
- Putting aluminum mesh structures into the stratosphere.
- Releasing myriad of highly reflective microballoons into the stratosphere.
- Placing giant mirror arrays into the space orbit.
- Whitening roofs, pavements, roads, and other surfaces.
- Reducing Arctic sea ice loss (called, "Arctic geoengineering").

The cooling effect due to the reflection of sunlight by aerosols is a well-known phenomenon, since it accompanies any major volcano eruption that emits immense quantities of airborne particles. It has been reported that after the eruption of Mount

Pinatubo volcano in Philippines in 1991 (when about ten million tons of SO_2 was injected into the stratosphere causing increased reflection of solar radiation by sulfate particles), in the following year, the Earth's surface cooled in average about half a degree Celsius [6]. The detailed discussion of SRM projects is out of scope of this book; interested readers can find comprehensive information on this topic in [7].

11.1.2 Greenhouse Gas Management

The main objective of the *greenhouse gas management* (*GGM*) (also called, *carbon dioxide removal*) geoengineering strategy is to reduce the levels of CO_2 and other GHG in the atmosphere either directly or indirectly (e.g., by manipulating natural processes), thus allowing more long-wave (infrared) radiation to escape into space. The main GGM options discussed in the literature include [7]:

- Enhancing an uptake and storage of carbon by terrestrial biological systems, e.g., through aforrestation, or land use.
- Enhancing an uptake and storage of carbon by oceanic biological systems, e.g., through iron fertilization, phosphorus/nitrogen fertilization, enhanced upwelling.
- Using a variety of physical and chemical engineered systems to remove CO_2 from air, e.g., through enhanced weathering, ocean alkalinity enhancement, capture of CO_2 from the atmosphere (this topic is discussed in Chap. 10).

Although most of the GGM geoengineering projects deal with CO_2, a number of proposed projects involve the removal of other potent GHG such as methane and halocarbons from the atmosphere. For example, the authors of one geoengineering project have proposed to remove atmospheric chlorofluorocarbons (CFC) using powerful lasers that can break large CFC molecules into smaller more benign molecules [8]. Methane removal from the atmosphere could be enhanced by its oxidative conversion to CO_2 via chain radical reactions initiated by reactive hydroxyl radicals generated by photochemical decomposition of ozone.

11.2 Ocean Fertilization

The *ocean fertilization* concept is based on the effect of enhancing the rate of atmospheric CO_2 uptake by the ocean through adding iron (Fe), nitrogen (N), phosphorus (P), and other nutrients stimulating algae growth. It is assumed that since the growth of algae species in the surface layers of the ocean is limited by the supply of these nutrients, this technique would drastically accelerate the action of the biological pump by shifting biological equilibrium and enhancing CO_2 absorption by the ocean.

The technical and economic aspects of the ocean fertilization concept sound quite reasonable. The quantity of nutrients needed for the necessary effect is determined by the ratio of nutrient elements to carbon in algal tissues (e.g., C:N:P:Fe $\approx$

106:16:1:0.001 [9]). Calculations and recent marine trials proved that 1 kg of iron can fix about 83,000 kg of CO_2 and generate over 100,000 kg of plankton biomass [10]. The reported estimates of the required amount of iron to achieve the large-scale ocean iron fertilization (e.g., 3 Gt of CO_2 per year) vary from 200,000 to 4 million tons per year, with the price tag reaching US$27 billion [10]. Assuming that the annual value of the global carbon credit is projected to exceed US$1 trillion (at the credit rate of $135 per ton CO_2-equiv.), the ocean fertilization approach appears to be a quite affordable strategy to offset about half of all anthropogenic CO_2 emissions [10]. For this reason, several private organizations have been making plans to conduct large-scale ocean fertilization projects to generate carbon offsets [11].

However, not all the experts are agreeable with the math and efficacy of the ocean fertilization approach. There are indications that only a small fraction of iron is utilized in the fertilization scheme, and most of it is rapidly remineralized (i.e., is returned to its inorganic mineral form) [7]. Moreover, there could be decrease in biomass production "downstream" of the fertilized region due to the imbalance of nutrients [12]. The skeptics also point to the generic limitations of the ocean fertilization strategy: given the immense amounts of man-made CO_2 released to the atmosphere every year (more than 9 GtC per year) and relatively low capacity of the technique (less than 1 GtC per year), fertilization can play at best only a modest role in the global carbon sequestration scheme [7].

Several ocean fertilization experiments have been conducted in the Pacific and Atlantic Oceans: IRONEX I (1993), IRONEX II (1995), SOIREE (1999), EisenEx (2000), SEEDS (2001), SOFeX (2002), SERIES (2002), SEEDS-II (2004), EIFEX (2004), CROZEX (2005), and LOHAFEX (2009). For example, in the IRONEX II experiment (conducted in the Pacific Ocean), 500 kg of an iron compound was spread over the ocean surface of about 72 km^2. The results were dramatic: the photosynthesis yield rapidly increased, phytoplankton biomass increased 30-fold within a week, and the partial pressure of CO_2 in the middle of the fertilized zone decreased [13]. The fertilizing impact of this rather short (18 days) transient experiment dissipated shortly after the cease of iron injection. LOHAFEX experiment was conducted in the southern Atlantic Ocean (in a 30 by 30 km ocean patch) despite a widespread international opposition [14]. Although this experiment triggered an intensive phytoplankton bloom, it did not succeed in terms of enhanced carbon sequestration because the bloom was not accompanied by diatoms growth (which is essential for sustaining the complex bioecosystem responsible for sinking of organic carbon).

Like many other geoengineering projects, ocean fertilization remains a highly controversial issue and it faces a stiff resistance from a majority of scientific and environmental communities. Many experts in the field stress that not only we poorly understand the unintended ecological impacts of ocean fertilization, but the efficacy of this concept as a carbon abatement solution is still in question [11]. In response to an increasingly strong opposition and concerns about the possible long-term ecological consequences of large-scale iron fertilization, in May 2008, UN Convention on Biological Diversity issued a decree obligating member states to limit the ocean fertilization activities to small-scale scientific studies [15].

The debate over iron fertilization technology sharply intensified in October 2012, when the details of an experiment involving the dumping of about 100 metrictons of iron-laden dust in the Northeast Pacific (off the coast of British Columbia, Canada), came to light [4]. The iron dumping was conducted by Haida Salmon Restoration Corp. near the Canadian islands of Haida Gwaii, using the amount of iron one order of magnitude larger than that used by any previous scientific ocean fertilization experiment. The main objective of this $2.5 million project was to stimulate a phytoplankton bloom to boost the region's ailing salmon fishery. The iron addition purportedly generated a phytoplankton bloom over the sea surface of 10,000 km^2 that attracted fish, seabirds, and marine mammals. Some in scientific community supported the project, saying that the ecological and legal consequences of the project deployment are small compared to its impact on future geoengineering policy. However, a worldwide controversy followed, with a number of environmental groups, governing bodies, and scientists voicing a serious concern over this world's largest geoengineering deployment. For example, ETC Group, a Canadian environmental watchdog organization, called the experiment "a blatant violation of moratoria" established by the London and UN Conventions. The participants of the London Convention meeting in November 2012 also expressed concern about the experiment [4]. There was an attempt in the UN Convention on Biodiversity to strengthen its 2010 decision opposing all geoengineering research [5].

Since testing the efficacy of the ocean fertilization approach on a sufficiently large scale would require the alteration of the ocean, and the long-term monitoring of its impact might not be feasible, researchers use different dynamic models that can simulate the ocean ecosystem responses. Many modeling studies conducted so far, however, implied that iron fertilization of the ocean as a carbon mitigation option may not hold to its promise. For example, in a modeling study reported by Zahariev et al., the authors concluded that even if the entire Southern Ocean were fertilized with iron, less than 1 GtC per year would be sequestered and only for a few years [16]. Other authors also reported that the ocean iron fertilization in the areas of the high nutrients content would unlikely sequester more than several hundred million tons of carbon per year [11]. Based on these considerations and some limited field-test studies there are persistent suggestions from oceanic and atmospheric science communities to discontinue all activities on ocean fertilization [15].

11.3 Enhanced Weathering

In nature, CO_2 is slowly removed from the atmosphere through different weathering processes involving the reactions of Ca- or Mg-based silicate rocks with CO_2 on the timescale of thousands to millions of years:

$$CaSiO_3\left(\text{or}\, MgSiO_3\right) + CO_2 \rightarrow CaCO_3\left(\text{or}\, MgCO_3\right) + SiO_2 \tag{11.1}$$

Due to the slow rate of this reaction, it removes about 0.1 GtC per year from the atmosphere, which is about two orders of magnitude less that the rate of anthropogenic CO_2 emissions [17].

A number of the geoengineering proposals aiming at artificially accelerating the natural weathering process are under consideration [18]. All the proposed methods are based on the principle of making soil or the ocean waters somewhat more alkaline. In one proposal, it is suggested to add abundant silicate minerals such as olivine to soil used for agriculture [18]. Calculations show, however, that this approach would require the amount of silicate minerals equal to the volume of 7 km^3 per year (about twice the current coal mining amount) to remove CO_2 at the rate it is currently emitted [7].

There are also proposals to enhance the weathering reactions through ocean-based variations of the reaction (11.1), e.g., involving the reaction of silicate compounds with CO_2 and water as follows:

$$CaSiO_3 + 2CO_2 + H_2O \rightarrow Ca^{2+} + 2HCO_3^- + SiO_2 \quad (11.2)$$

Compared to the solid phase process (reaction 11.1), this approach advantageously uptakes two CO_2 molecules per each silicate molecule. Similarly, carbonate rocks (instead of silicate) could be used to capture CO_2 with the resulting materials also placed in the ocean [7]:

$$CaCO_3 + CO_2 + H_2O \rightarrow Ca^{2+} + 2HCO_3^- \quad (11.3)$$

For example, Rau and Caldeira proposed to process carbonate rocks with CO_2 with a subsequent release of the resulting bicarbonate solutions into the ocean [19]. In an alternative approach, it was suggested to directly release carbonate minerals to the sea [20].

All enhanced weathering techniques utilize abundant naturally occurring minerals, and after reaction with CO_2 they produce benign products. Currently, no large-scale enhanced weathering projects are being practiced anywhere in the world. The main challenge is that the technology requires handling enormous amounts of materials, which would involve major mining and processing operations that could be comparable or even more expensive than conventional CCS systems.

11.4 Challenges and Risks of Geoengineering

11.4.1 Economics of Geoengineering

The economic assessments of the proposed geoengineering projects will play a crucial role in making policy decisions about the carbon mitigation potential of these projects. However, there are concerns that these assessments (although valuable from a scientific viewpoint) have a limited practical or policy-relevant value, mostly, due to the present lack of knowledge about real-world costs of the geoengineering projects and associated risks; this, essentially, means that uncertain input assumptions would result in uncertain outputs of any cost modeling [7]. Another reason for the uncertainties in the economic estimates has to do with systematic biases toward underestimating the costs of geoengineering projects and technologies [7, 21].

Some SRM-related geoengineering techniques, e.g., whitening of roofs, pavements, roads, and some open areas can be accomplished at very little cost (in fact, the projects like roof whitening may even offer some financial payback). Other projects, however, could be extremely costly. For example, the cost estimate for the full-scale implementation of the sulfate-aerosol geoengineering project reported by Crutzen indicated that it would cost annually about US$50 billion to put two million tons of sulfate aerosols per year in the stratosphere (or about US$50 per world capita, which does not seem to be very high price compared to the cost of alternative options) [2]. The Copenhagen Consensus Center and The Royal Society consider the "cloud ship" project to be more cost-effective than other geoengineering projects: the project would cost about US$9 billion to test and launch within 25 years, which is much less than the cost of releasing aerosols into the atmosphere (US$230 billion) and placing mirrors in space (US$395 trillion) [22].

According to Stern's estimates, the cost of conventional carbon mitigation is likely to be in the order of 1–2 % of global GDP, or about $1 trillion per year to avoid current carbon emissions [23]. This corresponds to the carbon price of about US$100 per ton of carbon (or ~$27 per ton CO_2). Thus, to be affordable, the cost of the geoengineering methods (for both SRM and GGM approaches) would need to be comparable to the mitigation costs of US$100 per ton of carbon, or in the order of US$1 trillion per year (to offset doubling of CO_2) [7].

The economic attractiveness of the GGM geoengineering strategy would clearly improve if there will be a well-established international valuation and trading system for carbon [7]. An international agreement to establish a stable minimum carbon price (carbon tax) would also substantially boost the economic attractiveness of GGM methods. Most experts, however, agree that it is unlikely that cost would play a crucial role in determining whether to deploy geoengineering systems. Most likely, the issues related to risk, politics, environmental ethics, and public support would be more decisive factors in the decision-making process [24].

11.4.2 Risk and Uncertainty Factors

Today, there is no general consensus on the efficacy of geoengineering: too many technical, environmental, political, and legal issues are involved. Even if a particular geoengineering technique is developed, successfully field-tested, and appears to be affordable, its large-scale deployment would be extremely challenging due to a global nature of the problem: no one country would be directly responsible for it. Thus, it is likely that, for time being, the geoengineering concept will remain at the stage of research and development of different technological options (so they will be available when they are needed). On the other hand, without meaningful field demonstrations it would be practically impossible to assess the economic viability of the technology and scaling-up issues: the classical "Chicken-Egg" dilemma.

11.4.2.1 Risk Factors

The risk factor associated with the wide-scale deployment of geoengineering projects is one of the prevailing arguments in the arsenal of opponents of this climate change mitigation solution. The major points are:

- Possible side effects of geoengineering projects are unknown; the full extent of hazards and damages (e.g., aerosols could destroy ozone layer, or have negative impact on agriculture, etc.) may be revealed at later time when the harm is already done.
- Its effectiveness could be less than expected (e.g., in the ocean fertilization experiments, the estimated amount of CO_2 removed from the atmosphere was less than expected).
- There may be reliability issues and difficulties with controlling the geoengineering systems (due to unpredictable impacts of certain external events, e.g., solar flares, volcanic eruptions, etc.)
- Possibility of weaponization by irresponsible groups and regimes (e.g., the use of cloud machines, creation of droughts, etc.).
- Uncertain (at best, but, likely, an adverse) impact on terrestrial ecosystems (e.g., plants, wild life, etc.).

One of the arguments against the geoengineering is that climate change itself is an unintended effect of the use of technologies once regarded safe and benign. Thus, responding to that with the large-scale deployment of untested technologies could simply exacerbate the problem (the cure could be worse than disease!) [7].

Among CO_2 removal geoengineering projects, ocean fertilization is the most controversial. Potential ecological consequences of large-scale ocean fertilization on the biosphere and biogeochemical cycling are mostly unknown, and could range from the changes in species diversity to the induction of anoxia and significant adverse effects on biocommunity structure function [13]. The fundamental question that should be addressed is whether any changes in the ocean ecosystem are justified relative to the benefits to society. The ocean plays an essential role in sustaining the Earth's biosphere; thus, any alterations in the ocean ecosystem must be taken with an extreme caution.

11.4.2.2 Knowledge Gaps

The knowledge gaps with regard to ocean fertilization as a carbon sequestration approach can be summarized as follows [13]:

- The long-term impact of ocean fertilization on the structure, balance, and function of marine ecosystems and transfer of carbon to the deep ocean is unknown. There is a possibility that fertilization with iron and phosphorus could lead to growth of toxin-producing cyanobacteria over other types of phytoplankton.
- The impact of sustained fertilization on the natural biogeochemical cycles of carbon, nitrogen, phosphorus, silicon, and sulfur is completely unknown; the perturbation of one elemental cycle can adversely affect other elemental cycles to the extent that cannot be predicted.

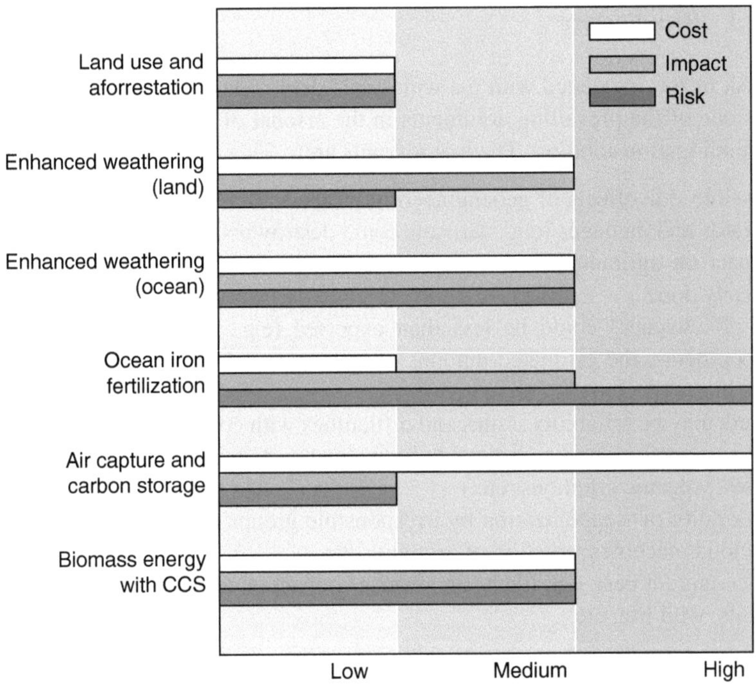

Fig. 11.2 Comparison of maximum effectiveness of different GGM geoengineering techniques deployed to remove 1 GtC per year. *White bars* show the cost of GGM projects; *light* and *dark gray bars* show impact and risk of anticipated environmental effects, respectively. *Source*: The Royal Society [7]

- The ocean fertilization could potentially lead to eutrophication causing oxygen depletion, which could kill many species requiring oxygen and can potentially result in enhanced production of methane (more potent GHG than CO_2).
- Iron infusions may unpredictably alter surface ecosystems disturbing food chains and adversely impacting fisheries and whale populations.

The concepts of "encapsulated" and "reversible" technologies are often used in the context of the risk factors of geoengineering projects [7]. Encapsulated technologies are those that do not release "foreign" material into the environment (e.g., air capture technology), as opposed to technologies that release materials into the wider environment (e.g., ocean fertilization, sulfur aerosols). The reversibility of technology refers to the ability of the geoengineering technique to cease its operation and terminate its effects in a short time (e.g., air capture is a reversible system). Examples of irreversible systems are sulfate aerosols and ocean fertilization, where there could be a time lag (or even, long-term impacts) after ceasing the operation.

Figure 11.2 compares different geoengineering techniques from both SRM and GGM categories, in terms of cost, anticipated environmental effect, and risk factors.

These factors were evaluated in the scale of low, medium, and high. For comparison, air capture and Bio-CCS systems (discussed in Chap. 10) are also included in the diagram (note that many literature sources treat air capture and Bio-CCS as geoengineering methods, e.g. [7]).

Figure 11.2 indicates that ocean fertilization method has low cost, but high enviromnetal risk, whereas air capture, in contrary, has high cost and low risk (since it is encapsulated and reversible type technology). Biological methods (land use, aforrestation, Bio-CCS) show advantages in terms of relatively low cost and risk.

Considering conflicting views and ongoing debates over the geoengineering projects, many experts call for global transparency and collaboration, as well as establishing international norms and regulations on geoengineering research [25]. The regulations would set quantitative limits on the scale of experiments and lessen the politically and ecologically disruptive potential of the states or individuals deploying the geoengineering technologies. These regulations would also encourage the scientific assessment of the risks and benefits of these geoengineering practices in case they are needed as a "least bad option" to mitigate (impending) climate change [4].

11.5 Concluding Remarks

The geoengineering approach poses acute and novel challenges that require the proactive and effective management of research [5]. The opponents of such research have to recognize the risks of completely suppressing the development of technologies that could potentially offer solutions in times of crisis. On the other hand, the supporters of such research must accept legitimate societal concerns over possible environmental perturbations and planetary-scale manipulations. Parson and Keith proposed to break this deadlock through a set of near-term policies [5]:

- Accept government authority over geoengineering research (scientific self-regulation is insufficient to manage the risks).
- Declare moratorium on large-scale engineering (e.g., for SRM) projects with the scale threshold of the project being defined by the area, time, and size of the radiative forcing perturbation (a possible threshold—the annual average $\Delta RF \sim 10^{-2}$ W/m^2).
- Set the small-scale threshold below which research may proceed (a possible threshold—the annual average $\Delta RF \sim 10^{-6}$ W/m^2).

These initial steps can help build the international norms of cooperation and transparency in geoengineering. If states would fail to build this cooperation and transparency now, when the stakes are relatively low, it could become even more difficult and thorny (as is the case with arms control) in the event of severe climate change.

References

1. Marchetti C (1976) On geoengineering and CO_2 problem, Report RM-76-17, International Institute for Applied Systems Analysis, Mar 1976
2. Crutzen P (2006) Albedo enhancement by stratospheric sulfur injection: a contribution to resolve a policy dilemma? Clim Chang 77:211–219
3. Cicerone R (2006) Geoengineering: encouraging research and overseeing implementation. Clim Chang 77:221–226
4. Lockwood D (2012) Geoengineering test fuels debate. Chem Eng News 90:26–27
5. Parson E, Keith D (2013) End the deadlock on governance of geoengineering research. Science 339:1278–1279
6. Lacis A, Mishchenko M (1995) Climate forcing, climate sensitivity and climate response: a radiative modeling perspective on atmospheric aerosols. In: Charlson R, Heinztenberg J (eds) Aerosol forcing of climate. Wiley, Chichester, pp 11–42
7. The Royal Society (2009) Geoengineering the climate: science, governance and uncertainty. ISBN: 978-0-85403-773-5, London
8. Stix T (1993) Removal chlorofluorocarbons from troposphere. 1993 IEEE international conference on plasma science, IEEE, Vancouver, BC, p 135
9. Sarmiento J, Gruber N (2006) Ocean biochemical dynamics. Princeton University Press, Princeton. ISBN 0 691 01707 7
10. Iron Fertilization (2010) http://en.wikipedia.org/wiki/iron_fertilization. Accessed 15 Dec 2010
11. Buesseler K, Doney S, Karl D et al (2008) Ocean iron fertilization – moving forward in a sea of uncertainty. Science 319:162
12. Watson A, Boyd P, Turner S et al (2008) Designing the next generation of ocean iron fertilization experiments. Mar Ecol Progr 364:303–309
13. US Department of Energy (1999) Carbon sequestration. State of the science. US DOE, Office of Science, Office of Fossil Fuels, Working paper on carbon sequestration science and technology. Washington, DC
14. Eureka Alert (2009) American Association for the Advancement of Science. http://www.eurekaalert.org/pub_releases/2009-03/haog-lpn032409.php. Accessed 12 Dec 2009
15. Strong A, Chisholm S, Miller C et al (2009) Ocean fertilization: time to move on. Nature 461:347–348
16. Zahariev K, Christian J, Denman K (2008) Preindustrial, historical and fertilization simulations using a global ocean carbon model with new parameterization of iron limitation, calcification and N_2 fixation. Prog Oceanogr 77:56–82
17. Intergovernmental Panel of Climate Change (2005) IPCC Special report on carbon dioxide capture and storage. In: Metz B, Davidson D, de Coninck H et al (eds) Working Group III of the intergovernmental panel on climate change. Cambridge University Press, Cambridge
18. Schuiling R, Krijgsman P (2006) Enhanced weathering: an effective and cheap tool to sequester CO_2. Climate Change 74:349–354
19. Rau G, Caldeira K (1999) Enhanced carbonate dissolution: a means of sequestration waste CO_2 as ocean bicarbonate. Energ Convers Manag 40:1803–1813
20. Harvey L (2008) Mitigating the atmospheric CO_2 increase and ocean acidification by adding limestone powder to upwelling regions. J Geophys Res Oceans 113, C04028
21. Flyvjerg B, Bruzelius N, Rothengatter W (2003) Megaprojects and risk: an anatomy of ambition. Cambridge University Press, Cambridge
22. Telegraph, UK (2009) "Cloud ship" scheme to deflect the sun's rays is favorite to cut global warming. www.telegraph.co.uk. Accessed 7 Aug 2009
23. Stern N (2007) The economics of climate change: the Stern review. Cambridge University Press, Cambridge
24. Intergovernmental Panel of Climate Change (2001) Climate change 2001: IPCC Working Group III: Mitigation. http://www.grida.no/publications/other/ipcc_tar/?src=/climate/ipcc_tar/wg3/176.htm. Accessed 14 Apr 2010
25. Blackstock J, Long J (2010) The politics of geoengineering. Science 327:527

Index

N. Muradov, *Liberating Energy from Carbon: Introduction to Decarbonization*, Lecture Notes in Energy 22, DOI 10.1007/978-1-4939-0545-4,

MIX
Papier aus verantwortungsvollen Quellen
Paper from responsible sources
FSC® C105338

If you have any concerns about our products,
you can contact us on
ProductSafety@springernature.com

In case Publisher is established outside the EU,
the EU authorized representative is:
Springer Nature Customer Service Center GmbH
Europaplatz 3, 69115 Heidelberg, Germany

Printed by Libri Plureos GmbH
in Hamburg, Germany